Ancient Egyptian Science

Volume Two

Ancient Egyptian Science

A Source Book
by Marshall Clagett

Volume Two

Calendars, Clocks, and Astronomy

American Philosophical Society
Independence Square • Philadelphia
1995

Memoirs
of the
American Philosophical Society
Held at Philadelphia
for Promoting Useful Knowledge
Volume 214

Cover illustration: Arrangement of the northern constellations on the astronomical ceiling of corridor B on the tomb of Ramesses VI. From: Neugebauer and Parker, *Egyptian Astronomical Texts*, Vol. 3, Fig. 30.

Library of Congress Cataloging in Publication Data

Clagett, Marshall
Ancient Egyptian Science, Volume 2
includes bibliography and indexes
illustrated

ISBN 0-87169-214-7

1. Egypt 2. Science, ancient, history of
3. calenders, ancient 4. clocks, ancient

ISSN: 0065-9738 89-84668

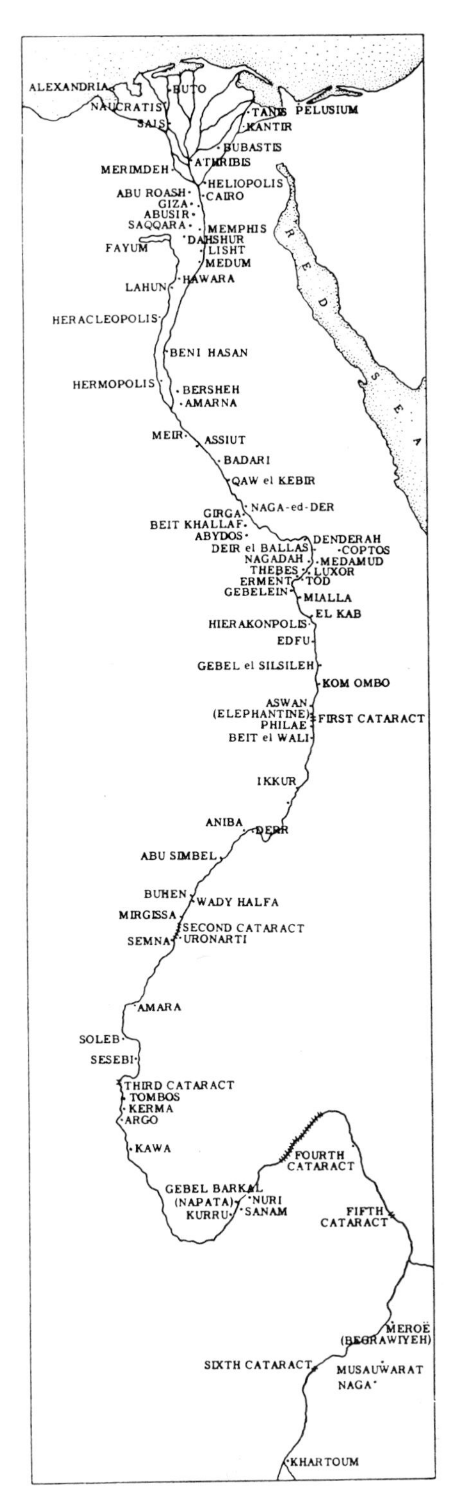

Map of Nile Valley

Map of the Nile Valley. Copied from W.S. Smith, *Ancient Egypt*, p. 192, with the permission of the Museum of Fine Arts, Boston.

Preface

This volume is the second of the three volumes on Ancient Egyptian Science which I hope to complete. I have not included everything which I projected for Volume Two in the Preface to Volume One. The chapter and documents regarding mathematics are missing. It would have greatly increased the length of the volume to have included them here and accordingly I decided to shift them to Volume Three. That shift makes no difference in the progression of subjects originally planned for the whole work, for I believe that the lack of theoretical discussions of mathematics by the ancient Egyptians in their rudimentary science made it imperative first to outline the principal uses of mathematics by the dwellers on the Nile before discussing its structure and content. One possible benefit of including mathematics in Volume Three is that it can be more closely related to my discussion of Egyptian techniques of representing nature and within that topic to appraisals of the ancient Egyptian lack of a direct angular or arcal measure to quantify stellar displacements and thus produce more accurate celestial diagrams, of the absence of an effective method of geometric projection, and finally of their lack of any extended use of perspective (which, however, surely did not hamper their considerable artistic skill).

The organization of the current volume is self-evident. Again I have given a lengthy introductory chapter which attempts to synthesize the three main subjects included in the volume: calendars, clocks, and astronomical monuments. It summarizes the principal conclusions which we can draw from the eighteen documents and the Postscript that constitute the bulk of the volume. The order of those documents follows that of the three subjects mentioned. There is, however, no hard and fast isolation of the topics one from another. Because of this there is much skipping around from date to date in the corpus of documents, but within each area of treatment there is fair chronology evident as befits a historical work covering three millennia of activity.

In the case of every individual document the effort is made to supply a meaningful date or dates. It is true that sometimes the carrier of the document, say a temple ceiling, has a date of construction or execution that is often much later than the document itself, as, for example, is the case of Document III.12 where the decanal transit tables (marked in my document by the letter "U") found in the ceiling of the Cenotaph of Seti I (ca. 1306-1290 B.C.) date from at least as early as the reign of Sesostris III in the 12th dynasty (ca. 19th century B.C.). Similarly, the earliest copies of the Ramesside Star Clock (Document III.14) are found in the Tomb of Ramesses VI (ca. 1151-43 B.C. in the 20th dynasty), but the carefully reasoned date implied by the document itself is some time between about 1500 and 1470 B.C. in the 18th dynasty.

I have given more than 150 pages of illustrations. For the most part they include the hieroglyphic (or rarely, hieratic or demotic) texts, some from a single legible copy, others from an edited text based on several copies. These illustrations will allow the reader who controls the Egyptian language to have ready access to the texts that lie behind my translation. But, as in the first volume, it is my hope that the translations themselves will give readers without detailed knowledge of the original language, i.e., most students of the history of science, a good sense of what the documents intend. In regard to the illustrations, I should note that occasionally a magnifying glass may be needed by the reader studying them. But even in the cases where considerable text appears on a single illustration, the reproductions are remarkably clear as a result of the careful photocopies prepared by my secretary, Ann Tobias, who often improved the contrast and clarity of the originals from which the illustrations were made.

I have given very full notes to illustrate the historical steps taken by earlier scholars to advance our knowledge of the subjects treated in this and the succeeding volume of my work. This was done not only to give the reader a good sense of the development of scholarship over the last two centuries, but also to give honor and credit where they are due. Since the appearance of Volume One, two towering figures in the study of Ancient Egyptian Astronomy have died: Otto Neugebauer, whose help and friendship I have acknowledged in the Preface to Volume One, and Richard Parker, a premier student of the Egyptian calendars and Neugebauer's coauthor of the penetrating and informative *Egyptian Astronomical Texts* in three volumes. Their respective talents complemented each other exceedingly well: Neugebauer's superb analytical powers and Parker's philological skill and extensive knowledge of the texts. The reader will be well aware of my debt to them. Among earlier authors, Renouf, Lepsius, Brugsch, Meyer, Sethe, and Borchardt stand out, but the reader will also find mentioned the works of many other later scholars (e.g., Hornung and Barta) and younger ones (e.g., Krauss) who have clarified and solved many of the puzzling problems concerning the topics of this volume. Unfortunately, it was only after I completed this volume that I obtained a copy of Christian Leitz's *Studien zur ägyptischen Astronomie* (Wiesbaden, 1989), and so I was unable in this volume to give it the careful study which it deserves (but see Chapter Three, note 49). I must also note with gratitude that James ("Jay") O. Mills of the Nekhen Excavations team has allowed me to publish as a Postscript most of his unpublished paper on a petroglyph with possible astronomical significance. This petroglyph was discovered by Mr. Mills and Ahmed Irawy Radwan during a survey operation in 1986. As in the case of Volume One, special thanks are tendered to Dr. Robert Bianchi for his helpful reading of this volume in its first version.

Closer to home, I must again thank my wife, Sue, to whom the whole work is dedicated, for her expert editorial help and, above all, for her constant encouragement. I have already thanked my secretary, Ann Tobias, for the magic she has worked with the illustrations. But she also undertook the formidable task of reading and rereading the manuscript numerous times, much to its improvement. I also repeat here my thanks to my home institution, The Institute for Advanced Study, for all of its intellectual and material assistance; and especially am I in debt to the Library staff, who uncovered copies of even the rarest articles and books. And finally I want to express my appreciation to the American Philosophical Society for continuing to publish this large work. In that organization thanks must especially go to Herman Goldstine, its Executive Director, to Carole Le Faivre, its Associate Editor, and to Susan Babbitt, who skillfully copy-edited this volume. All have contributed to its appearance, its accuracy, and its publication.

Once again I have provided the Society with camera-ready copy, using Printrix to print the copy and Fontrix to create special fonts (see the Preface to Volume One). But since the publication of the first volume, I have prepared additional, smaller fonts to represent the consonantal, phonetic transcription of hieroglyphs and to indicate a large collection of accented letters and letters with various diacritical marks in order to print accurately the notes and bibliography. My hieroglyphic fonts are those of the first volume designed by me and greatly improved and extended by Ann Tobias, though many more glyphs have been fashioned for this volume.

Marshall Clagett
Professor Emeritus
The Institute for Advanced Study
Princeton, New Jersey

Table of Contents

Part One

Chapter Three:

Calendars, Clocks, and Astronomy

Chapter Three

Calendars, Clocks, and Astronomy

Introduction to Egyptian Calendars

As we have seen in Volume One, Chapter Two, the Egyptians were wont to call the creator god, whoever he was, Lord of the Years or Lord of Eternity and Everlastingness. Though these titles primarily designated the creator god's creation of time as an aspect of or limit to world order or maat, they may also have oblique reference to the supposed role played by the creator god (or at least by the king with whom he was identified) in the establishment of a systematic calendar. However it originated, the civil calendar of 365 days was securely in place by the time of the Old Kingdom. An investigation of the origins of this calendar and its relationship to one or more lunar calendars is fraught with uncertainty and difficulty, but such an investigation will be most useful for anyone attempting to understand the steps taken in Pharaonic times to organize society and express its religious culture in a satisfying, efficient and productive way. Hence we must pursue this investigation as far as the scanty evidence permits.

In the last generation or so, the most influential work on the subject of Egyptian calendars has been that of Richard Parker.[1] Though Parker's views have wide acceptance, they are anything but certain, as we shall see presently. Parker believes, as did L. Borchardt in a tentative way earlier, that the Egyptians used the civil calendar of 365 days and two different lunar calendars. He describes the first lunar calendar and the civil calendar in a succinct manner:[2]

> Like all ancient peoples, the protodynastic Egyptians used a lunar calendar, but unlike their neighbors they began their lunar month, not with the first appearance of the new crescent in the west at sunset but rather with the morning when the old crescent of the waning moon could no longer be seen just before sunrise in the east. Their lunar year divided naturally, following their seasons, into some four months of inundation, when the Nile overflowed and covered the valley, some 4 months of planting and growth, and some 4 months of harvest and low water. At 2- or 3-year intervals, because 12 lunar months are on the average 11 days short of the natural year, a 13th or intercalary month was introduced so as to keep the seasons in place. Eventually the heliacal rising of the star Sirius, its first appearance just before sunrise in the eastern horizon after a period of invisibility, was used to regulate the intercalary month. Sirius, to the Egyptians the goddess Sopdet or Sothis, rose heliacally just at the time when the Nile itself normally began to rise, and the reappearance of the goddess heralded the inundation for the Egyptians. The 12th lunar month, that is the 4th month of the 3rd season, was named [*Wp-rnpt*] from the rising of Sothis and a simple rule was adopted to keep this event

within its month. Whenever it fell in the last 11 days of its month an intercalary month [*Thoth*] was added to the year, lest in the following year Sothis rise out of its month....

This luni-stellar year was used for centuries in early Egypt and indeed lasted until the end of Pagan Egypt as a liturgical year determining seasonal festivals. Early in the third millennium B.C. however, probably for administrative and fiscal purposes, a new calendar year was invented. Either by averaging a succession of lunar years or by counting the days from one heliacal rising of Sothis to the next, it was determined that the year should have 365 days, and these were divided into three seasons of four 30-day months each, with 5 additional 'days upon the year' or 'epagomenal days'. This secular year which is conventionally termed the 'civil' year remained in use without alteration to the time of Augustus when a 6th epagomenal day every 4 years was introduced. That the natural year was longer than the civil year by a quarter of a day was of course known to the Egyptians fairly soon after the civil year was inaugurated but nothing was ever done about it. Still it is a great achievement of theirs to have invented a calendar year divorced from lunar movement and to have been the first to discover the length of the natural year, which eventually led to the Julian and Gregorian calendars....

The first year in Egypt, the lunar one, had divided the month into 4 'weeks' based on

> 'first quarter', 'full moon', and 'third quarter.' The new [civil] month of 30 days was not divisible into 4 even parts but conveniently divided into 3 'weeks' of 10 days each, from later texts called 'first', 'middle' and 'last'. Thus in the entire year there were 36 weeks or decades, plus the 5 days upon the year.

Parker here maintains that the first of the Egyptian lunar calendars came into existence prior to the invention of the civil calendar. But he also suggests that a second lunar calendar developed as the result of the invention of the civil calendar to keep in step with it. In this second lunar calendar a thirteenth month was intercalated when the New Year's Day of that lunar year would have fallen earlier than the New Year's Day of the civil year without such intercalation.

Before considering Parker's views on the structure of the two lunar calendars, let us briefly review the Egyptian civil calendar. The details of this calendar are firmly known and not in dispute, except for the time and manner of its invention, which will be discussed later in the chapter.

As noted in the quotation from Parker above, the civil year consisted of twelve months of thirty days each plus 5 epagomenal days, totaling 365 days. The months were grouped in three seasons of four months each, perhaps taken over from the seasons of an earlier lunar year as Parker suggests,[3] but more likely drawn from an early 365-day calendar that was seasonally oriented, such as one based on the rising of the Nile. The first season was named Akhet (, transliterated as *ꜣḫt*, and usually translated "inundation" since originally, when the year coincided with the seasons, this season and indeed the year began with the sudden rising of the river and for much of that season the land lay under water), the second was Peret (, *prt*, meaning "emergence," no doubt for the emerging land and plant life), and the third was Shemu (, *šmw*, meaning "low water" and the time of "harvest"). After indicating the regnal year (see Vol. 1, Document I.1), dates usually specified the day, month, and season in the form of month-number, season, and day-number. For example, "IV *šmw* 20" signified the fourth month of the season of Shemu, day 20. Note that each month was divided into three "weeks" of ten days each (i.e., decades), both the month of 30 days and the 360 days of the twelve months of the year being neatly divisible by that number 10. That 10-day week dictated the use of 36 decans (stars or groups of stars) in the star clocks which we shall discuss later.

Month names instead of numbers were sometimes used in the New Kingdom and later.[4] The late forms of these names, based on their Greek spelling, are: *First Season*: 1. Thoth, 2. Phaophi, 3. Athyr, 4. Choiak; *Second Season*: 5. Tybi, 6. Mechir, 7. Phamenoth, 8. Pharmuthi; *Third Season*: 9. Pachons, 10. Payni, 11. Epiphi, and 12. Mesore.

Now the heliacal rising of Sirius became, at an unknown time, the event marking the beginning of the civil year, and because the year between successive risings of Sirius was about 365 1/4 days in Pharaonic times,[5] it is evident that the civil year of 365 days was about 1/4 day short of the Sothic year each year, so that the former's New Year's Day was about one day earlier after 4 years (hence 1460 Sothic years = 1461 civil years). Or, to put it in another way, the stellar event of the heliacal rising of Sirius occurred in the civil year about one day later after being on the same day for four years and thus in the course of about 1461 civil years it fell successively after every four years on each day of the civil year until finally once again the rising of Sirius took place on New Year's day of the civil year. This long-term procession is called a Sothic cycle. Now if a sixth epagomenal day were to be added to the civil year at the end of every fourth year in order to halt this march almost entirely, the civil year so modified would then be known as a Sothic fixed year, and the disparity of Sothic and civil years would almost disappear (again see footnote 5). Despite the belief of some earlier Egyptologists that the so-called Ebers Calendar (Doc. III.2) is evidence of such a fixed Sothic year, there is no sure evidence that the Egyptians widely or regularly used a separate calendar of the fixed year to which a sixth epagomenal day was formally added every fourth year, at least not before the time of Augustus. (We may except from this statement the abortive fixed-year calendar of Ptolemy III promulgated by means of the Decree of Canopus in 238 B.C., see Doc. III.10.) Still the Egyptians obviously knew that the New Year's Day of the civil year, initially set or better reset as I Akhet 1 by the rising of Sothis, was, quadrennium by quadrennium, steadily receding from the day of the rising of Sothis, appearing almost always one day earlier than that rising after each quadrennium, for at times they recorded (most surely as the result of observation during most of the Pharaonic period but perhaps occasionally by calculation by the time of the Ptolemaic period) the rising of Sothis on different days of the different civil years, as we shall see later when discussing Document III.10.[6] Furthermore, it could well be that, as was probably the case of the Ebers calendar, other specific calendars or tables correlating a specific Sothic year with some given civil year were used for finding dates in the civil year of seasonal prescriptions or for the construction or use of water clocks where hour scales varied from season to season. These uses we shall discuss later.

Parker's Account of the Old Lunar Calendar

Now let us examine the evidence that Parker presents for his view of the old lunar calendar.[7] Certainly we may agree that prior to the invention of the civil calendar a lunar month was in use, since the term for "month" *(ʾbd)* employed in the civil calendar was not only written with a crescent moon surmounting a star ( ) but appeared in civil dates in abbreviated form as a crescent moon ( ) with the seasonal number of the month stroked below it (see Fig. III.2, second register from the top where both hieroglyphic forms of the months of the seasons are given, but without the phonetic complement "d" [ ] of the first form). Hence the lunar form of the term was used even though the 30-day month of the civil year was not tied to the waxing and waning of the moon, as were the lunar months of 29 and 30 days.

The reader should be first reminded that Parker[8] (and before him Brugsch, Mahler, and Sethe)[9] concluded that an Egyptian lunar month began with the first day of invisibility of the waning crescent, i.e., in modern terms the day of mean conjunction of the moon and sun with the earth. This conclusion was arrived at by an analysis of the names of the days of the lunar month first collected by Brugsch (and I shall comment on this analysis in Document III.6 below) and confirmed, Parker believed, by his analysis of a late 25-year lunar cycle (see Document III.9). But granting the origin of the term "month" in a lunar calendar does not tell us anything about the nature of the old lunar year, i.e., whether it was a lunar year of twelve lunar months without intercalation (as for example was the case with the Moslem calendar later) or whether a system of intercalation was in use and if so whether it was the system, suggested first by Borchardt and improved on by Parker, in which intercalation of a thirteenth month was prompted by the heliacal rising of Sirius in the last eleven days of the twelfth month. This system would have produced a normal year of 12 months and an expanded year of 13 every three, or occasionally two, years.

Parker's conclusion that the old lunar year had an intercalary month was drawn, in the first instance, by analogy from the lunar year implicit in the 25-year lunar cycle given in Document III.9, which was written down in 144 A.D. or later but which Parker calculates was composed in about the middle of the fourth century B.C. In that 25-year table there were to be 16 "small years" of 12 months and 9 "great years" of 13 months. We shall talk about the details of that document later. Suffice to say now, it surely reveals the existence of an intercalary scheme at this later period of Egyptian history, but of course tells us nothing about the old lunar calendar, since the purpose of this later lunar cycle was to determine the position of lunar months within the civil year and hence the lunar calendar implicit in this table postdated the invention of the civil year.

Parker hoped to strengthen his analogical case for

the earlier use of intercalation by noting references to festivals of the "great year" and the "small year" in the calendar of Beni Hasan, a document of the Middle Kingdom (see Document III.1) that may confirm that at that time there was a lunar year of 12 months (the small year) and as well an intercalary one of 13 months (the great year).[10] But even this interpretation is not entirely certain, for the "small year" referred to may be the lunar year of 354 days and the "great year" the civil year of 365 days. Hence there is no completely sure way to tell whether the lunar calendar of the Middle Kingdom including these festivals was intercalated in the same way as the later lunar calendar, or what its relationship was to the old lunar calendar.

However, even if we assume that the analogical case for some kind of intercalation in the old lunar calendar is sufficiently strong, we still need to look at the evidence for the actual system of intercalation accepted by Parker, namely one depending on the heliacal rising of Sirius. The first type of evidence presented by Parker is the oft-expressed relationship between Sothis and the year or the opening of the year *wp rnpt* ([hieroglyph]).[11] In the *Pyramid Texts* (Volume One, Doc. II.1, Section 965) we read the following statement addressed to Osiris: "It is Sothis (Sirius), your beloved daughter, who prepares the yearly sustenance for you (or, has made your year-offerings or year-renewals) in this her name of '*Year*' (*rnpt*)." Then at the other end of Egyptian history, in the Temple of Hathor at Dendera, a reference to Isis-Sothis, says that "years are reckoned from her shining-forth."

Parker, in his *Calendars*, follows Sethe and Borchardt and points to an earlier possible reference to Sothis and *wp rnpt* on a first-dynasty tablet from the reign of Horus Djer (see Fig. III.3a).[12] He and his predecessors believed that on the right of the tablet the recumbent cow represented Sothis and that she has between her horns two vertical signs, the first one being an old form of the hieroglyph for "year" (*rnpt*) and the other a simple stroke. If this is correct, then the horns combined with the *rnpt* glyph suggest the meaning "opening or opener of the year," rendering the combined signs . Furthermore, if it is assumed that the cow represents Sothis, then the whole figure ought to be translated as "Sothis, opener of the Year." Below the cow are other signs that, they thought, ought to be read as the glyphs for *ꜣḫt*, the first season, i.e., Inundation. And still below them these early scholars thought they saw in the fragmentary copy of the tablet (see Fig. III.3b) traces of a crescent with at least two strokes below it, suggesting the meaning "month 2." However this is doubtful and no such traces can be seen on the main tablet itself. To the left of the cow is first a vertical spear-like bar that may be a separator or it may be a glyph representing the word "first" *(tpy)* or the number "1." Farther to the left are glyphs that relate to King Djer and a town that was earlier interpreted by Petrie as Dep (□) but is more likely to be read as Ap (□; see note 14 below), a town that is not otherwise known. Following this interpretation of the tablet, the whole expression, in a revised version, would then mean, if we discount the so-called month traces at the bottom right of the fragmentary copy of the tablet which do not show up on the tablet: "[The Year of the] first [celebration of] Sothis [as] Opener of the Year [in] Inundation [by] the Horus Djer [at] Ap."[13] But let me hasten to say that this interpretation can probably no longer be accepted (even in the revised version), and indeed the earlier version was later abandoned by Parker himself (see note 12 above), mainly as the result of the reexamination of the purport of the tablet by Gérard Godron.[14]

Godron's principal conclusions may be summarized: the inscription is not a date at all and has nothing to do with the opening of the year; the recumbent cow is Sekhet-Hor instead of Sothis; the uprights between the horns are not an earlier form of the "year" sign ſ plus a simple vertical stroke, but rather, as Petrie and Griffith thought, a single "feather" sign ꞵ since the two vertical signs are in fact connected at the top; the month signs on the lower right in the fragmentary copy of the tablet cannot be seen on the main tablet and so simply do not exist on it. Lastly, the sign interpreted on the one hand as the glyph for the season of Inundation or on the other for "marshes" could be either, and thus cannot be surely identified as the one or the other.

But such references connecting Sothis with the regulation or the beginning of the year, even including the doubtful reading on the tablet from Djer's reign, do not necessarily apply to the role of Sothis in determining the intercalation of a thirteenth lunar month, since the coincidence of the rising of Sirius and the inundation that signalled the beginning of the civil year was probably known from an early date even if that rising was not responsible for the determination and invention of the civil year of 365 days, as used to be thought. Hence all of these passages relating Sothis and the year could simply constitute recognition of that coincidence.

One further line of reasoning followed by Parker

involves the fact that on occasion the twelfth month (at least some time from the New Kingdom on) had the name *Wp rnpt* instead of Mesore, which former name Parker believed to be the name of that month in the old lunar calendar.[15] He reasoned that it was so named in the lunar calendar because this was the month in which the reappearance of Sirius would herald the succeeding month as the first month of the later lunar year if it happened prior to the last eleven days of the month, or if it occurred in those last eleven days it signalled the intercalation of a thirteenth month in order that during the next year the rising of Sirius would remain in the twelfth month rather than shifting to the first month and thus after the New Year's Day. Hence in this argument *wp rnpt* simply meant the actual appearance of Sirius and was equivalent to the later expression *prt Spdt*, "the going forth of Sothis."[16] So, according to Parker, it was only after the establishment of the civil calendar (and in fact perhaps not until the Middle Kingdom) that *wp rnpt* was used in its literal meaning of the New Year's Day of the civil calendar, and when this happened both meanings of *wp rnpt* were in use. He finds support for this double use of *wp rnpt* by slightly misinterpreting the opening reference to I Akhet 1 in an incomplete Ramesside calendar edited by Bakir, saying that on that day the second festival of *wp rnpt (wp rnpt sn-nw)* was celebrated.[17] The inference that Parker draws from this information is that another (that is, a "first") celebration of *wp rnpt (wp rnpt tpy)* took place to celebrate the actual appearance of Sirius in the twelfth month of the preceding lunar year. But this evidence of a "second" festival of *wp rnpt* does not ensure that what Parker calls the first celebration refers to the use of the appearance of Sirius as an intercalary device to keep the lunar calendar in step with the seasons or that it refers at all to a lunar calendar. Bakir's comment on the passage in question and on Parker's interpretation of it is of interest:[18]

> *Wpt rnpt pw ḥb snnw*: In the first place, it seems that either *wpt rnpt* or *wp rnpt* is a possible reading....It is clear from the reading of the whole passage that I differ from Parker's point of view....who regards the addition of *sn-nw* to *wp-rnpt* as a designation for [the "second" *wp rnpt*, that is] the "first" day of the "civil" year. He reads *Wp-rnpt snnw* [omitting *pw* and *ḥb* after *wp-rnpt*] and [so] translates "the second *wp-rnpt*." *Wpt rnpt* coincides, in this BOOK [of the Papyrus], with the "first" day of the year. Or as my reading rightly claims, there were two celebrations [on the same day]: one for the "first" [i.e.,] *tpy*-day, and the other for the "opening" [i.e.,] *wpt* of the year. It is also my contention that *pw* exists here [in the text and should not be ignored as Parker apparently does] and is to be regarded as copula since *wpt rnpt* is never written with an additional *p* and *w* [before *ḥb* whether *ḥb* is its possessor or its determinative]. Thus my interpretation of this passage runs as suggested: "the second feast is the opening of the year." Furthermore, if we go back to the inscriptions of the O. Kingdom mastabas, we find offerings presented on two separate festivals on the New Year's Day —the *wpt rnpt* and the *tpy rnpt*. To quote H. Winlock...: "the first of these festivals, in the XIIth Dyn.

> calendars, is also the 'coming forth of Sothis,' the second is, in all likelihood, the new year invented for the calendar when it became definitely and obviously separated from nature."

A quite different explanation of the significance of *Wp rnpt* as a name for the twelfth month has more evidence supporting it than that presented by Parker. Parker had viewed the Ebers Calendar (Doc. III.2) as being an effort to correlate a schematized old lunar calendar with the civil calendar in the ninth year of Amenhotep I's reign. He believed that the first entry which identified *wp rnpt* with *prt Spdt* is evidence that we have a lunar calendar in which the rising of Sothis is an indicator of the lunar New Year's day to follow in the next month. On the other hand, Kurt Sethe (see Document III.2 below) and, following him closely, Raymond Weill,[19] presented the view that the month names must be distinguished from the eponymous feast days from which they took their names, and that the eponymous feast days were not in their homonymous months but were in fact the culminations of those months and hence took place at the beginnings of the succeeding months. In this connection Weill (depending on Gardiner, Meyer [see especially Fig. III.6a], and Sethe) presents a list, developed from a number of calendars and calendaric statements dating from a period extending from the Middle Kingdom through the Ptolemaic-Roman epoch, which shows that almost invariably the feast days were celebrated in the months following those to which the feast days gave their names (also see Fig. III.6b). This would explain why the Feast of *Wp rnpt* begins the Ebers Calendar even though the month of *Wp rnpt* is later the name of the twelfth month. Hence the Ebers Calendar loses its significance as evidence for Parker's reconstruction of the old lunar calendar since in this year of the Ebers Calendar the Feast of *Wp rnpt* is correctly the first entry of the year occupying the first column. It is not, as Parker believed, an event of the twelfth month placed at the beginning of the calendar to inform us that the next entry (the Feast of Tekhi) is the beginning of a schematized lunar year. I shall elaborate my view of the Ebers Calendar later. Now I simply say that I find Sethe's presentation convincing and hence, in discussing Document III.2, I shall accept the view that the Ebers Calendar does not present a correlation of the old lunar calendar (even in its schematized form) with the civil calendar, but rather is an effort to correlate, for civil year 9 of the reign of Amenhotep I, an ad hoc fixed Sothic year with the civil calendar, a correlation needed when seasonal dates have to be converted to civil dates (such as is the case with seasonal medical prescriptions and in the construction and proper use of water clocks, a view I shall elaborate later).

Be that as it may, Parker turns to the listing of feast days in Old-Kingdom tombs to support his view that the first use of *wp rnpt* as a reference to the appearance of Sirius in the twelfth month was as a determiner of intercalation in the old lunar calendar:[20]

> So far we have dealt in the main with generalities. We have established the reasonableness of a primitive lunar calendar based on the rising of Sothis and also the fact that the rising was termed *wp rnpt.* In the following pages I shall present the evidence which has led me to the conclusion that the year which was opened by Sothis' rising

cannot have been the civil year or the fixed Sothic year [i.e., one with a sixth epagomenal day every fourth year] but must have been the natural lunar year....

Beginning in the 4th dynasty [and continuing through the 5th dynasty] the mastabas [i.e., tombs] of the Old Kingdom frequently exhibit a *ḥtp-dỉ-nsw* [or offering] formula involving Anubis (at times Osiris) and requesting..."invocation-offerings" on certain festivals [listed therein]. If a number of these lists [see Doc. III.1] are examined, it will be found that the feasts [in 21 lists] tend to follow a definite order [6 lists including all of the feasts in the same order and the others containing anywhere from 5 to 10 feasts preserving this same order] ...:

Order of feasts

(1) *wp rnpt*
(2) *Ḏḥwtyt*
(3) *tpy rnpt*
(4) *wꜣg*
(5) *ḥb Skr*
(6) *ḥb wr*
(7) *rkḥ*
(8) *prt Mn*
(9) (*ꜣbd*) (*n*) *śꜣḏ*
(10) (*tp*) *ꜣbd*
(11) *tp śmdt*
(12) *ḥb nb rꜥ nb* or variant....

The exceptions to be found are certainly insufficient to weaken the overwhelming evidence of a strict order to the calendar of feasts in which the dead expected to take part. This order, it is easily demonstrated, can be nothing other than chronological. Whatever may be the exact meaning of *wp rnpt* and *tpy rnpt*, they clearly belong at the head of a list. If we now check feasts 4 to 8 against the later temple calendars (Medinet Habu and the Greco-Roman temples), we find the following dates in the civil year on which they were celebrated:

(4) *wꜣg* I *ꜣḫt* 18

(5) *ḥb Skr* IIII *ꜣḫt* 26

(6) *ḥb wr* II *prt* 4

(7) *rkḥ (wr)* II *prt* 9 (Edfu) III *prt* 1 (Illahun)

(8) *prt Mn* I *šmw* 11 (Med. Habu)

Nos. 9-11 were monthly feasts, celebrated at least twelve times a year....

There seems full justification for considering the Old Kingdom list to be arranged chronologically. Moreover, there are reasons ...for thinking that all the feasts, with, of course, the exception of *wp rnpt*, were lunar at that time. The *wꜣg*-feast can be the movable lunar feast for which there is evidence from the Middle Kingdom...; *rkḥ* as the name of a lunar month cannot be other than lunar...; and the lunar character of *prt Mn* is brought out in the Medinet Habu calendar and later....The monthly feasts of *ꜣbd* and *šmdt*, and so probably of *sꜣḏ*, were lunar. Furthermore, there is no other plausible explanation for the sequence of *wp rnpt*, *Ḏḥwtyt*, and *tpy rnpt* than the assumption

> that the latter two also were lunar....
>
> The proposed original lunar calendar fits the chronological order perfectly, and I know of no other explanation. *Wp rnpt* was the rising of Sothis, the event which opened the new year but which, in itself, did not form part of it. *Tpy rnpt* was the first day of the new year, the first day of the month *tḫy* in which fell the feast of *w'g*; and the remaining feasts followed in chronological order. As for the feast of *Ḏḥwty*, between *wp rnpt* and *tpy rnpt*, this can be nothing other than the feast of the intercalary month which would occur at three-(at times two-)year intervals. As a special month it was fittingly dedicated to Thoth, the moon-god.

Now what are we to make of Parker's explanation of how these popular Old Kingdom feast days fit chronologically into and so help to support the specific intercalary lunar calendar described by Parker? It is certainly true that if the first three feasts were celebrating separate, distinct events on three distinct days which were items in a chronological order, then Parker has indeed a good case in view of the fact that the remaining annual feasts (Nos. 4-8) are quite probably in chronological order. For if the festival of the rising of Sothis had to fall before the festival of Thoth and that of the first of the year, then one could not simply answer that it was the rising of Sirius that had strayed from its position at the first of the civil year that was being celebrated *before* the first day of the year because if it were that civil-related rising of Sirius that was being celebrated even after moving away from the civil New Year's Day one would expect it to *follow after* the civil New Year's Day, as it inched through the civil year. But what about the festival of Thoth that appears regularly between the rising of Sothis and the New Year's Day? Was not Thoth the name of the First Month of the Year and should it not accordingly have followed the festival of the New Year's Day? No, Parker says, at the time of the old lunar calendar Thoth was the name of the intercalary month of the lunar year rather than that of the first month. Unfortunately Parker has had to assume what he should have been proving. That is to say, he first assumes that the first three feasts are chronological and relate to the old lunar year and then in order to support this he has to assume that Thoth was the name of the intercalary month rather than of the first month as it was later and for which there is a great deal of evidence. Thus in order to justify the chronological assumption he has to make a further assumption for which there is no other hard evidence, namely that the festival of Thoth must indeed precede rather than follow the First Day of the year. The kind of reasoning that Parker applied to the first three old feasts may leave the reader uncomfortable. But, in defense of Parker, it must be said that the evidence presented by Parker for the rest of the list's being chronological is substantial (though based on considering the order of these feasts celebrated in the civil year), and thus he believes that the first three feasts must also be chronologically ordered. Furthermore, since many of the feasts on the list appear to be lunar in origin, the first three festivals must refer to feasts in the old lunar calendar. If that is so, then the position of the festival of Thoth between the festival of the rising of Sothis and New Year's Day necessitates its reinterpretation in a lunar calendar, and the only sound explanation for the order of the three feasts in a lunar calendar would result if it assumed that the rising of Sirius signalled the coming New Year's Day at the beginning of the next lunar month and that the festival of Thoth was a festival of the intercalary month. This festival of Thoth (for which there is no other compelling evidence)[21] would be a quite different feast from the ordinary feast of Thoth celebrated in the civil calendar on I Akhet 19 (see Document III.5 below).[22] The objection implied by Gardiner that unlike the other festivals that of Thoth in the lunar calendar would not be an annual festival,[23] is simply set aside by Parker. Convinced that all of these twelve feasts were chronologically present in the old lunar calendar, he of course rejects Gardiner's quite different suggestion that the names of festivals (1)-(3) might reflect "three separate aspects of the beginning of the [civil] year, *Wpt rnpt* viewing it as the birthday of the sun-god Re^c^ (Mesore), *Ḏḥwtỉt* as under the sway of Thoth, the initiator of the year of 360 + 5 days, and *Tpỉ rnpt* as the most appropriate date for the accession of the earthly king."[24] Even if the aspects suggested by Gardiner are themselves wrong, the idea that more than one festival representing the beginning of the year, each with a different name, was celebrated on the same day (especially at different places) is certainly reasonable considering the calendars edited by Bakir and the evidence assembled by Schott.[25]

On considering further Parker's treatment of the first three festivals mentioned in the old mastabas, I remind the reader of an even more telling refutation of his explanation of the character and order of these festivals. Our examination of the Ebers Calendar shows [below and in Document III.2] that just the reverse of Parker's assumption appears to be true, namely, (1) the Feast of *Wp-rnpt* was not merely an indication of the rising of Sirius to hail the next month's beginning of the year or the necessity of intercalating a thirteenth month (Parker's view), but rather itself marked the beginning of the year, and (2) the Feast of Thoth mentioned as the second festival on the mastabas was not at all a celebration of an intercalary thirteenth month dedicated to Thoth which followed the rising of Sothis (within 11 days of the end of the twelfth month) but rather was the culmination of the annual celebrations of Thoth beginning before the end of the first month and culminating on the first of the second month of the year. The Feast of Thoth thus did indeed follow the Feast of the New Year but not as a festival of intercalation, and certainly not before the beginning of the year.

Without detailing the other lines of reasoning that convinced Parker of the correctness of his description of the old lunar calendar but which do not produce certain conviction, I believe we can reasonably conclude that he has given us an account that is only barely possible and is quite speculative in detail and not convincing in its over-all argument. His often used rhetorical expressions like such-and-such "cannot be other than," "can be nothing other than," "there is no other plausible explanation," "What can be more natural than," "It is a natural assumption," etc. when in most cases there could be alternative explanations make this reader uneasy.

In brief, it appears to me that Parker's opinion that the old lunar calendar was intercalary may be correct (though not certainly so) but that (1) the use of the

Sothic heliacal rising as the mechanism of intercalation, that (2) the intercalary month (if it existed) was named "Thoth," and that (3) the lunar calendar in schematized form is that given in the Ebers Calendar and in the astronomical ceilings of Senmut's tomb and the Ramesseum – are all unproved and indeed untenable.

Also, strictly speaking, the conclusion that the lunar month in the old lunar calendar commences on the first day of crescent invisibility is not certainly proved, although that seems to be the case for the later lunar calendar and appears to be in accord with a religious statement about the conception and birth of the moon god Khons (see Document III.6). I say "not certainly proved" because the three main pieces of evidence on which the conclusion is based, namely the names of the days of the lunar month, the 25-year lunar cycle of Papyrus Carlsberg 9, and a series of double dates (the lunar month days in the civil calendar) all date from a period more than a millennium or two later than the date when the old lunar calendar flourished. Indeed though some ten of the protective gods of the feasts of lunar days are found in astronomical ceilings of the 18th and 19th dynasties and the names of eight of the feasts of lunar days appear in the calendar of Medina Habu of Dynasty 20 (and perhaps earlier in the Ramesseum in Dynasty 19)–see Documents III.3, III.4 and III.5, in fact it is only in the Greco-Roman period that the full list of 30 lunar day-feasts appears, as is also true of the 25-year lunar cycle and the double lunar-civil dates. On the other hand, it does seem almost certain that the Egyptian day began at dawn the hour before sunrise rather than at sunset during ancient Egypt's recorded history,[26] and this fits better with a month whose first day is the first day of invisibility of the waning crescent before sunrise. We shall return to this question presently when we discuss the later lunar calendar.

One final word concerning Parker's reconstruction of the old lunar calendar. By means of it he has attempted to substitute a luni-stellar seasonal calendar for a fixed Sothic seasonal calendar. That is to say, Parker has replaced with his description of a schematized "old lunar calendar" the view of early Egyptologists that in order to keep track of their seasonal and other feasts the priests in the temples used some kind of fixed Sothic calendar whose New Year's Day was the heliacal rising of Sirius and to which a sixth epagomenal day was added every four years. Both reconstructions found the "heliacal rising of Sirius" as crucial and both employed the evidence of temple calendars (see Document III.5), astronomical ceilings, and the Ebers Calendar as useful. Needless to say, to question Parker's construction is not to affirm the Sothic fixed year, at least not in the complete and exaggerated form which held that a sixth epagomenal day was added formally to the year every four years.

Now two further subjects remain to complete the discussion of ancient Egyptian calendars: the later lunar calendar and the origin of the civil calendar and its occasional correlations with the rising of Sothis.

The Later Lunar Calendar

The essential features of the later lunar calendar are much more firmly established than those of the old. Key to the understanding of this calendar was the discovery and editing of Papyrus Carlsberg 9 by Neugebauer and Volten in 1938 (see Document III.9). As Parker has noted, this is "the only truly mathematical astronomical Egyptian text yet published."[27] He continues with the following characterization of the document:

> This papyrus was written in or after A.D. 144, and it furnishes a simple scheme, based upon the civil calendar, for determining the beginning of certain lunar months over a 25-year cycle. Underlying this cycle are the facts that 25 Egyptian civil years have 9,125 days and that 309 lunar months (divided into 16 years of 12 months and 9 years of 13 months) have 9,124.95231 days. The earliest cycle mentioned in the text began in the sixth year of Tiberius, A.D. 19, and the latest in year 7 of Antoninus, A.D. 144. Actually, the calendar does not indicate when every month began but lists only six dates for every year, those falling in the second and fourth months of each season. It also indicates the years of 13 months ("great" years) according to the following scheme: 1st, 3d, 6th, 9th, 12th, 14th, 17th, 20th, and 23d year of each cycle.
>
> Table 3 [equivalent to my Fig. III.8a taken from Neugebauer and Volten] gives the complete cycle as stated in the papyrus. Anyone using this calendar would begin a lunar month, without regard for actual observation, on II *ꜣḫt* 1 (which for A.D. 19 would be September 18; for A.D. 44, September 11; for A.D. 69, September 5; and so on up to A.D. 144, August 17), on IIII *ꜣḫt* 30 (for A.D. 19, December 16; for A.D. 144, November 14), on II *prt* 29 (for A.D. 20, February 13; for A.D. 145, January 12), and so forth. What is

significant about the cycle for our purpose is that it gives a continually recurring series of lunar dates, easily and accurately translatable into the Julian calendar for this period.

The first use that Parker makes of this lunar cycle is to test his view (propounded first by Brugsch and followed by Mahler and Sethe) that the lunar month began with the first invisibility of the waning crescent moon in the morning. The original editors of the text of the cycle, after analyzing the values for the year 144 A.D., arrived at the conclusion that for this table the lunar month began with new crescent visibility in the evening.[28] Though this seems to go against Parker's view, derived originally from an analysis of the names of the days of the lunar month, he agrees that the cycle when applied to the year 144 A.D. does indeed seem to show that the month is starting with the first visibility of the new crescent.[29] But he goes on to observe that the cycle produces an error of one day in about 500 years. Hence if it had been in use for about 500 years, the accumulated error would produce values that coincided with the day of first crescent visibility rather than with first invisibility of the waning crescent. Thus it becomes crucial to know when the cycle was invented and thereafter continued in use. Since the document shows no evidence of contemporary Hellenistic astronomy, it seems a safe bet that its fundamental structure and calculating techniques are of more ancient Egyptian origin. Analysis of a series of double dates (i.e., lunar dates specified in the civil calendar) pushes back the use of the cycle into Ptolemaic times.[30] Furthermore, if we apply the table to the first two years of the cycle beginning in 856 B.C., we find (Parker, *Calendars*, p. 17) that "the month starts in every case exactly one day before the morning of invisibility and two or three days before the evening of crescent visibility." Hence it is clear that the table would be of no use that early and thus had not yet been invented. But if we apply the table to the first two years of the cycle beginning 500 years later in 357 B.C., (*ibid.*) "ten out of twelve months start on the morning of invisibility, and only one on the evening of visibility." If one applies the table to the first two years of the seventh cycle in the period between 357 B.C. and A.D. 144, i.e., of the cycle beginning in 207 B.C., (*ibid.*) "seven months out of twelve begin with the morning of invisibility, one begins with crescent visibility and the other four are in between. It is clear that, the closer in time the cycle approaches A.D. 144, the less frequently will the months start with invisibility and the more frequently with visibility." Thereupon, considering also his examination of the double dates already mentioned, which push the use of the cycle back into Ptolemaic times, Parker reasonably concludes (*ibid.*, p. 23) that the time most appropriate for the cycle's inauguration was around 357 B.C., "at least in the form in which we have it, with invisibility of the old crescent as its foundation."[31]

Satisfied with the yield of Document III.9 to the problem of when the lunar month began, Parker makes other contributions to a description of the lunar calendar that underlay this table. In the first place, he uses the techniques apparent in the cycle's determination of the beginnings of the six even lunar months of each year of the 25-year cycle and specific lunar dates to produce values for the odd months, thus giving us the completed 25-year cycle (see Fig. III.9).[32] Having already stated that the document with the

25-year cycle tells us which years have twelve months and which contain thirteen, Parker then deduces with little trouble the principle of intercalation followed in the lunar calendar underlying the 25-year cycle: "Whenever the first day of lunar month Thoth would fall before the first day of civil Thoth, the month is intercalary."[33] A few final observations by Parker may be quoted:[34]

> The outstanding fact of the cyclic calendar is that it begins on I *ꜣḫt* 1. We have come to the decision that in this form it originated in the fourth century B.C....Despite the fact that in A.D. 144 the cyclic calendar was clearly no longer in agreement with lunar phenomena, *it was not corrected.* This is certain, since lunar date 1 falls in A.D. 190 and still fits exactly into the scheme. Obviously, then, there was no provision in the calendar itself which required that periodically it be adjusted. Furthermore, if the present cycle were the result of a correction in the fourth century B.C. of an already existing cycle, instituted, let us say 500 years earlier, that earlier cycle would have had to begin with I *ꜣḫt* 2, so that its correction would result in I *ꜣḫt* 1. This is exceedingly unlikely....
>
> The question may still be legitimately asked, however, whether any sort of schematic lunar calendar was in earlier use....[35] We have decided that the first schematic calendar was introduced in the fourth century B.C. and that prior to that time there is no evidence that any other method than observation was used to begin the month. We

have discovered the rule that regulated intercalation in the 25-year cycle and have seen that it is one which could easily have been operative before the cycle was installed. The essential point to observe is that the lunar calendar was governed by the civil calendar, since whenever the first day of lunar Thoth would fall before the first day of civil Thoth, the month was intercalary. Since the civil calendar moved forward through the natural year, the lunar calendar attached to it must likewise have moved with it. Moreover, *this lunar calendar cannot have existed before the civil calendar was introduced.*

The Origin of the Civil Calendar

I have already outlined the main characteristics of the remarkable civil calendar of 365 days (12 schematized months of 30 days each, arranged according to four months in each of three seasons, plus 5 epagomenal days) invented by the ancient Egyptians. The most difficult problem regarding this calendar is the time and circumstance of its inauguration. We should first examine the earliest historical evidence of its existence. Setting aside the Djer tablet of the first dynasty, which we earlier characterized as probably not containing a reference to Sothis and the opening of the year, we can point to two indications in the early *Annals* that seem to be plausible evidence that the civil calendar of 365 days was in use at least by the time of the beginning of Shepsekaf's reign in the fourth dynasty and irrefutable evidence for that of Neferirkare's in the fifth (see Volume One, Document I.1, note 13), since the year shared between each of these pharaohs and his predecessor (probably in the first case and certainly in the second) adds up to twelve months and five days. Another probable reference to the civil calendar occurs in the *Annals* for the reign of Neferirkare where it is noted that in the first year of his reign ceremonies took place for the "Birth of the Gods," which in all likelihood were ceremonies devoted to the gods of the epagomenal days (*ibid.*, n. 112). The full civil year, perhaps headed by the epagomenal days ( ) and including the three seasons with their months, may be found in an annual schedule of priests performing services for Hathor in the fifth-dynasty tomb of Nekankh at Tehne (see Document III.1, line 11 of the extract from that tomb; and see the contrary opinion of E. Winter in note 2 to the Introduction to Document III.1). There is as well a reference to the epagomenal days in the *Pyramid Texts* (Sect. 1961b-c, Pepi II's pyramid at the end of the sixth dynasty): "He sees preparation for the festival of the burning (*lit.*, making) of the braziers, the birth of the gods before you in the five epagomenal days ( )."[36]

Accepting these references as firm evidence that the civil calendar of 365 days was in use in the Old Kingdom, we are brought to the question of when it was first used. The exposition of the details of the calendar by Eduard Meyer in 1904[37] had wide acceptance for a generation or longer. His fundamental conclusion was that the civil year with its structure of twelve months of 30 days each, ordered into three seasons, plus the five epagomenal days remained essentially unchanged from the time that it was adopted and recognized to have a New Year's Day at the heliacal rising of Sirius throughout the whole Pharaonic period and indeed until 139 A.D. when, we are told by Censorinus 100 years later (see Document III.10, Section 6, caps. 18 and 21), the rising took place on the first day of the civil year, i.e., an Apokatastasis ("re-establishment or return [of Sirius]") took place on the first day of the month Thoth, and hence this was the first year of the Great Year (=period) called (cap. 21) "Solar Year, Year of the Dog-star [i.e., Canicular or Sothic Year] or Year of god."[38] To this was added the assumption that the rising of Sirius was determined at one place (most likely Memphis or Heliopolis) and then promulgated throughout Egypt.

A second assumption was that the Egyptians concerned with the calendar recognized after some time that the rising of Sirius was delayed one day after every four years in the civil calendar (which is very close to the actual delay during the Pharaonic period) and consequently they learned how *to calculate* simply the day when the heliacal rising of Sirius would occur in any given civil year, and hence *observation* of the rising was no longer the basis for determining and disseminating the day of rising in the civil calendar, that determination being calculated on the basis of an assumed delay of one day in the rising after every four years.

If these assumptions were correct, and given the fact that a Sothic year calculated on the basis of a one-day delay after every four years of successive risings of Sirius is nearly identical in length with the Julian year of 365 1/4 days, we could determine, from the Egyptian historical records of the risings of Sirius in the civil year, the Julian dates of these recordings so long as we knew which Sothic period the record fell in. If we work backward from 139 A.D. we would find the possible quadrennial dates of the beginnings of the three Sothic periods preceding the one beginning in that year: B.C. 1321-1318, 2781-2778, and 4241-4238.[39] Meyer believed that the period beginning from 2781-2778 B.C. was too recent for the establishment of the calendar and so he accepted the earlier period. He thought that he had confirmed this when he deduced from rather inconclusive considerations of the Gregorian dates for recent risings of the Nile that it was in the Sothic period beginning from 4241-4238 that the date of Sirius' appearance best conformed with the date of the Nile's rising.[40] However, with the development and acceptance of the "short chronology" of Egyptian history which can be coordinated with the short chronology elsewhere in the Near East, the general opinion has switched to the third millennium for the establishment of the civil calendar since, at the beginning of the earlier period, Egyptian society was at an undeveloped level of sophistication and was not a unified state. This was given a significant boost by two influential articles of Otto Neugebauer.[41] In the second of these articles he summarized the main conclusion of the first:[42]

> ...the simplicity of the Egyptian calendar is a sign of its primitivity; it is the remainder of prehistoric crudeness, preserved without change by the Egyptians, who are considered to be the most conservative race known in human history.
>
> ....there is no astronomical phenomenon which possibly could impress on the mind of a primitive observer that a lunar month lasts 30 days and a solar year contains 365 days. Observation during one year is sufficient to

> convince anybody that in about six cases out of twelve the moon repeats all of its phases in only 29 days and never in more than 30; and forty years' observation of the sun (e.g., of the dates of the equinoxes) must make it obvious that the years fell short by 10 days! The inevitable consequence of these facts is, it seems to me, that *every theory of the origin of the Egyptian calendar which assumes an astronomical foundation is doomed to failure.*
>
> Four years ago I tried to develop the consequences of this conviction as far as the Egyptian *years* are concerned. I showed that a simple recording of the extremely variable dates of the inundations leads necessarily to an average interval of 365 days. Only after two or three centuries could this 'Nile calendar' no longer be considered as correct, and consequently one was forced to adopt a new criterion for the flood, which happened to be the reappearance of the star Sothis.

He went on to point out that the seasons of the Egyptian year are agricultural and not astronomical. The adoption of a uniform month of 30 days is not to be laid to astronomical observation but to a deliberate schematization of the real lunar months for administrative purposes in a highly centralized society. He further noted that a schematic calendar of twelve months of 30 days is also found in Babylon, though the practical needs in Babylon, organized in city-states, did not lead to a single centralized civil calendar.

Following Neugebauer's rejection of the astronomical origin of the 365-day year in favor of its origin in the seasonal rising of the Nile and adopting the increasingly short chronology for the establishment of the unified state, we could arrive at the conclusion that sometime during the first dynasty a 365-day calendar was adopted, say about 3000 B.C. (see the chronology in the Appendix to Volume One), and that it was maintained for about 220 years until it was decided to use the heliacal rising of Sothis at the beginning of Inundation as the New Year's Day. A strong point in support of giving a role to the Nile's rising is the fact that from at least the reign of Djer in the first dynasty, a yearly record of the Nile height was kept (and later recorded in the *Annals* which we have given as Document I.1 in Volume One; see that document, notes 13 and 14). The first of the year and the rising of the Nile are identified in a later reference (see note 25, quotation from a Cairo Calendar of Lucky and Unlucky Days). To be certain about these conclusions would be foolhardy, since we have piled conjecture upon conjecture, starting with Meyer's assumptions about the Sothic cycles, the initial observations of the heliacal rising of Sirius at a given, unchanged location, and the adoption by Egyptians of the scheme of calculating the risings of Sirius upon the rule that it was delayed one day after every four years.[43]

Before leaving the problem of the origin of the civil year, I must recount in more detail the views of Parker on its origin,[44] views which were alluded to in our long quotation from Parker at the beginning of the chapter. He remarks first that prior to Neugebauer's articles, the students of the Egyptian civil calendar had believed it to have been astronomically determined in that it was inaugurated with the heliacal rising of Sirius as its New Year's Day, and that despite the fact that New Year's Day fell progressively in advance of Sirius' rising, no attempt was made to adjust or correct the calendar. He also noted that Sethe believed that the lunar year and the civil year were parallel in their development. As for Neugebauer's views, Parker finds them unsatisfactory, primarily because Neugebauer had not known the structure of the old lunar calendar based on the rising of Sirius, that is, the old lunar calendar discerned by Parker (and which we have characterized above as wanting in surety). Here then is the explanation of the origin of the civil calendar in Parker's own words:[45]

> During the protodynastic period the only calendar in use was the lunar calendar already described. By this time Egypt had become a well organized kingdom, and the economic disadvantages of a lunar year of now twelve months, now thirteen, all of which began by observation, must have pressed themselves upon the government. In an effort to alleviate the situation and to provide a simple and easily workable instrument for the measurement of time, they hit upon the idea of a *schematic lunar year*, or as it might be termed an *averaged lunar year*. There are two ways by which the length of this schematic lunar year might have been determined as 365 days. On the one hand, since the current calendar was based on Sothis, one might simply have counted, for one or two years, the number of days between successive heliacal risings of the star. On the other hand, the lunar year might have been averaged. It would have been little trouble to refer to the various records of one kind or another from

which data on the number of days in recent calendar years might be derived.

To show how this might have happened, Parker gives a hypothetical table of the lengths of 25 successive lunar years by which he suggests that by accumulative averaging from year to year it will soon be clear that 365 days is the proper length or that by averaging after each year in which the intercalary month was added the same figure will be evident even after only 11 years. Parker continues:

> Thus easily, by either method, could the Egyptians of the protodynastic period have arrived at 365 days as the proper length for a schematic year which could be adopted for administrative and economic purposes. After the analogy of the ordinary lunar year the schematic year would be divided into twelve months and three seasons, each month having 30 days for simplicity and regularity. The extra five days of the schematic year were regarded as an abbreviated intercalary month and were placed before the year, just as the intercalary month headed the lunar year whenever it occurred.
>
> Now as the new year was after all only a schematic lunar year, it must have been planned to have it run concurrently with and as far as possible in concord with the old lunar calendar, which was by no means to be abandoned. Just what were the actual circumstances of its introduction we shall probably never know. It is possible, at least, that it was introduced in a year with an intercalary month and that the five

> epagomenal days were concurrent with the last five days of that month, so that the first day of both the old lunar year and the new schematic year fell on *tpy rnpt,* the first of *tḫy*; or it may have been handled without regard for the epagomenal days, with the new year, at its installation, simply having its first day coincide with *tpy rnpt.*
>
> If we cannot be certain of the circumstances of the introduction, we can at least set a range in time within which it took place. According to the rule for intercalation in the lunar calendar, when the first month after *wp rnpt* began within eleven days of that event it was intercalary. *Tpy rnpt* then might be as close to *wp rnpt* as the twelfth day following or, with a full intercalary month beginning on the eleventh day following, as far away as forty-one days. It is almost a certainty that the first day of the first month of the civil year, as we shall henceforth term the schematic lunar year, also fell within these limits. As this first day had come to be the date of the rising of Sothis in ca. 2773 B.C., the civil calendar must have been introduced between ca. 2937 and ca. 2821 B.C., with the probability that it was in the direction of the former rather than the latter date.[46]

This is a remarkably consistent account. It has the advantage of dovetailing nicely with what we know about the historical circumstances and chronological limits of early Pharaonic Egypt from other sources. The only major difficulty with it is that it is entirely dependent on Parker's use of the heliacal rising of Sirius to govern the old lunar calendar, and, as I have already remarked, this is anything but certain.

One last piece of evidence may bear on the origin of Egyptian knowledge of the length of the solar year as 365 days. This is a petroglyph, most probably of Predynastic date, which was discovered at Nekhen by Ahmed Irawy Radwan and James O. Mills. It may indicate a record of annual solar risings from solstice to solstice (see the Postscript at the end of this volume). The length of the solar year would have been much more quickly established in this way than by averaging the wild swings of Nile-risings. Needless to say, even if the solar year was determined in this manner, we still cannot be sure when the official civil calendar with its beginning assigned to the heliacal rising of Sirius was established.

Sothic Dates and the Ebers Calendar

We have gone as far as the evidence allows in considering the origins of the civil calendar. It now behooves us to note briefly the three extant double dates that refer to the heliacal rising of Sirius in the civil calendar (see Document III.10). In our discussion we no longer stick exclusively to the original assumptions noted above in describing Meyer's chronological procedures, but apply to the formulations and tables of P.V. Neugebauer the various possible values for the latitude of the observer (φ) and the so-called arc of vision (β), i.e., the minimum angular distance between the sun and Sirius after which Sirius is first visible on the horizon in the dawn.[47] The latter is somewhat misnamed, for at the minimum angle the star is not visible and the subsequent visibility in fact depends on variable conditions of the observation, as we shall see.

The first of these Sothic dates (proceeding from earliest to latest), is based on two entries from a temple register from Illahun (see Doc. III.10 and Figs. III.92a and III.92b): "Year 7 [of the reign of Sesostris III], [Month] III [of Season] Peret, Day 25....You should know that the Going forth (i.e., heliacal rising) of Sothis takes place on [Month] IV [of Season] Peret, Day 16" and "Year 7 [of the same reign], [Month] IV [of Season] Peret, Day 17.... Receipts [from the] Festival Offerings of the Going Forth of Sothis .... Loaves, assorted, 200; beer, jars, 60 ...." It has been widely accepted that the monarch in question in both references is Sesostris III. Thus we have two entries for the seventh year of his reign, the first of which predicts the rising on the sixteenth day of the fourth month of Peret and the second of which notes on the seventeenth day of that month the offering receipts for the festival of that rising [which presumably did take place the day before, i.e., on the sixteenth day]. We do not know with certainty that the prediction was made as the result of the technique of calculating by means of assuming a one-day delay of rising after every four civil years, or whether it was made as the result of a continuously kept record of observed annual risings on which accurate guesses were made beforehand. In the last case the remarks of Edgerton are pertinent for the determination of the Julian date to which it corresponds:[48]

> The view that a heliacal rising of Sothis was predicted for the sixteenth day of the eighth month in the seventh year of Sesostris III seems, therefore, to be historically well

> attested. Judging by what is known of the history of Egypt during and before the Twelfth Dynasty, it is almost certain that the underlying observations must have been made somewhere between latitude (φ) 29.2° (el-Lahun) and 30.1° (Heliopolis). I pointed out five years ago that the modern experimental determination of the *arcus visionis* (β) for Sothis in Egypt rests on inadequate data; we must continue to use the resulting figures (β may range at least from 8.6° to 9.4°) but we must bear in mind that a wider range is not excluded....
>
> The earliest possible equation for the Sothic date, if φ=30.1° and β=9.5°, is 1876 B.C., July 18, Julian; while if φ=30.1° and β=9.4°, the earliest possible equation is 1875 B.C., July 18, Julian. The latest possible equation, if φ=29.2° and β=8.6° is 1864 B.C., July 15 Julian. The sixteenth day of the eighth calendar month in the seventh year of Sesostris III may have fallen in any July from 1875 (less probably 1876) to 1864 B.C. inclusive. The result may conveniently be expressed as 1870 B.C. ± ca. 6 years.

As a matter of fact, Parker, by considering lunar data, arrived at what he thought was a more certain date, namely 1872, with the first year of Sesostris III thus being 1878, a date rather generally accepted for the next generation.[49]

The second Sothic date is that of the ninth year of the reign of Amenhotep I, III Shemu 9 (see Document III.2, the Ebers Calendar below). Again Edgerton, following the considerations of P.V. Neugebauer's *Astronomische Chronologie* and the article the latter co-authored with Borchardt (see note 47), has given us the highlights of determining the limits of that date in the Julian calendar:[50]

> The most important unknown quantities [for determining that Julian date] appear to be the *arcus visionis*...and the terrestrial latitude of the ancient observations....The experimental determination of the *arcus visionis*, made in 1926 on Borchardt's initiative, marked a great advance on previous knowledge but still falls far short of certainty. We are given only five individual observations, made by five different persons at five different places. Since the *arcus visionis* varies with shifting atmospheric conditions and also with the eyesight of the observer, it is evident that a very much larger number of observations must be made before any historical conclusions resting on this basis can really deserve the high respect which we habitually record them. But for lack of anything better, we shall proceed to use what we have.
>
> Borchardt and [P.V.] Neugebauer calculated that the actual *arcus visionis* for their five observers in five different parts of Egypt in July and August, 1926, ranged between 8.6° and 9.4°, with an average of 9.0°, while they believed it possible that under more favorable weather conditions an *arcus visionis* as low as 8.5° or even 8.3° might have been possible....
>
> I shall assume with Borchardt that the

> ancient observations were made at Heliopolis, latitude 30.1° [and use his range of values for β].... The conclusion, for historical purposes, is that 9. 11. W. [i.e., III Shemu 9] of year 9 of Amenhotep I may have fallen in any July from -1543 to -1536 [Julian] inclusive, if the ancient observations were made at Heliopolis. It is possible that the actual *arcus visionis* of the ancient observations may have been either greater than 9.4° or less than 8.6°, in which case the actual year may have been earlier than -1543 or later than -1536.
>
> If the ancient observations were made at Thebes, the year might lie anywhere between -1525 (earliest year if β=9.4°) and -1518 (latest year if β=8.6°). Other localities in Egypt would yield yet other results. Sais as an observation point would yield earlier years than Heliopolis, and Assuan would yield later years than Thebes. No one believes that the observations were made at Sais or at Assuan [however, now see Document III.10, n. 4, for Krauss's views], but this disbelief rests on purely historical considerations, which have nothing to do with astronomy.

For this same Sothic date, Hornung gives the same limits 1544-1537 B.C. (i.e., expressed in historical rather than astronomical form) if the observations were made at Memphis or Heliopolis, and the limits 1525-1517 B.C. if they were made at Thebes.[51] The chronology I have given in the Appendix of Volume One assumes Thebes as the center of observations and the lower limit of 1517 B.C., thus suggesting 1525 B.C. as the first year of the reign of Amenhotep I. Needless to say, if the observations were made at Aswan, as Krauss proposed, then the rising in question could be as late as 1506 (Document III.10, n. 4), which date rests on the doubtful assumption that the Ebers Calendar correlated a lunar and a civil year.

Whichever exact Julian date is appropriate for this second Sothic date, the document on which it appeared, namely the Ebers Calendar (Document III.2), was the object of much speculation and interpretation. Before the time of Borchardt most Egyptologists believed that the Ebers Calendar had the purpose of specifying in civil year 9 of Amenhotep I first days of the twelve schematic months (each having 30 days) of a fixed Sothic year (see the Introduction to Document III.2 below, where the views of the earlier Egyptologists are described). In fact, even before the discovery of the Ebers Calendar, Lepsius and Brugsch seem convinced on the basis of very scanty evidence that the early Egyptians kept a Sothic calendar of 365 1/4 days along side of their civil calendar of 365 days.[52]

The Ebers Calendar on the verso of the first column of the Ebers medical papyrus was discovered (or at least its significance was discovered) in 1870 (see Figs. III.10-12 and the discussion in the Introduction to Document III.2 below). After the rubric specifying the regnal year of Amenhotep I, twelve lines follow. The first column has a series of twelve monthly festivals beginning with that of *wp rnpt*, the second had the corresponding civil dates (each falling on the "ninth" day of successive months), and the third is occupied on the first line by the expression the "Going Forth of Sothis" (=the heliacal rising of Sirius) and on each succeeding line by a dot. The usual interpretation of this document before Borchardt was that it gave evidence of the Sothic fixed year regularly coexisting with the civil year, since it began with the festival of the rising of Sothis and proceeded by monthly festivals successively thirty days apart. One obvious difficulty was that the day of each successive festival fell on the same day of each month of the civil year (the "9th") without taking account of the epagomenal days inserted in the civil year. One would have thought that after falling on the 9th day of the fourth month of Shemu, it ought next to have fallen on the 4th day of each of the remaining civil months because of the insertion of the five epagomenal days at the end of IV Shemu. There is no completely satisfactory explanation of this inaccuracy, but Edgerton, following the early views of Lepsius, gives us a possible one:[53]

> But the inaccuracy is one of the given facts in our document, and instead of trying to interpret it out of existence by philologically improbable translations, it is our duty to face the fact and to try to account for it. The explanation may well start from the purpose which led the ancient scribe to write out this calendar in a prominent and readily accessible place (the first page) on the back of a medical treatise. The purpose, as Lepsius pointed out, was to give the physician an easy means of knowing at what seasons in the year certain prescriptions were to be used. No prescription in the entire document is restricted to a shorter period than two months. In calculating the calendric equivalent of a season of the year which was two months long, an inaccuracy of five days would probably not seem very serious to the

> Egyptian medical practitioner. I still do not understand why the scribe was so thoughtless as to write "9. 1. W.[=Wandering, i.e., Civil Year]," etc., where "5. 1. W.," [! "4. 1. W." etc.] would have been just as easy, and obviously more precise—but I think I have seen more deplorable examples of thoughtlessness in modern works whose reputation for accuracy stands deservedly higher than that of the Papyrus Ebers.

Up to the time of Borchardt's *Mittel* the date for each civil month was read as "9" (*pśḏ*). But Borchardt read the hieratic sign used in these dates as standing for "New Moon Day" *(pśḏntyw)* or sometimes perhaps *pśḏ-nḫb* (where the sign for *-nḫb*, i.e., ⊂⊃, could be a determinative) rather than for "9" and accordingly he believed the festivals of the first column to be those of twelve successive lunar months beginning with the festival of *wp rnpt* (see Fig. III.12). But since the hieratic sign, without an accompanying sign for ⊂⊃, which was interpreted by Borchardt as "New Moon Day," is precisely the same as the unadorned sign used for "9" in the regnal date given in the rubric, and its use for "New Moon Day" (when unaccompanied by the sign for *-nḫb*) cannot be attested in any hieratic documents, Borchardt's reading must certainly be rejected.[54] While recognizing the cogency of this rejection, Parker retained Borchardt's belief that the first column gave festivals in the lunar year, as we have already noted.[55] In explicating his view, he properly felt the need to explain why the festival of the month *wp rnpt*, identical later with that of Mesore, the last month, was in the lead position, while *tḫy*, the usual first month, was shifted to the second position, to be followed by a consequential shift of one position of all of the months. His solution to this problem also, he thought, disposed of the problem of the missing epagomenal days. Here is what he says (*Calendars*, p. 42):

> The event which regulated the original lunar calendar was the rising of Sothis, called *wp rnpt*. The date of this event would, then, correctly go at the head of a calendar governed by it. But this event also gave its name to the last month of the year. In the first column of the Ebers calendar, therefore, the last month of the year appears at the head of the months merely because its eponymous feast determined the following year. The correct interpretation of the second line of the calendar seems to me to be that the date III *šmw* 9 is common both to the going-forth of Sothis and to the beginning of the lunar month wp rnpt. From this date as a starting-point was projected a schematic lunar calendar of full months of 30 days. The failure of the scribe to reckon the epagomenal days is accordingly not deserving of blame, as it would be were we concerned with a fixed year. The schematic lunar calendar was regarded as exactly that; it was to be merely a guide to the proper identification of the lunar month which included the 9th day of any civil month in its first few days.

First consider Parker's last argument against the omission of the epagomenal days. The problem is not that of the omission of the epagomenal days from the first calendar, the schematized, lunar calendar according to Parker, but rather is why the epagomenal days were omitted from the civil calendar with which it was being correlated! The main part of his explanation, which serves to explain the appearance of the last month first, is also lacking in conviction, for if we look at the first "month" listed, namely *wp rnpt,* we see that it is not the month (𓆳) that is given but rather the "[eponymous] festival of *wp rnpt* (𓆳𓊽)," as Parker well knew and as all the early discussants of the Ebers Calendar realized. Indeed, on the basis of the evidence analyzed by Sethe and Weill (see Document III.2), it seems to me certain (as it did to them) that not just the first but all of the twelve entries in the first column of the Ebers Calendar ought be considered as the eponymous feast days rather than the months themselves. If this is so, then the problem of why the months are out of order does not exist. For as Sethe has already shown, the eponymous feast day ordinarily appears in the month following as the culmination of the preceding month (see Fig. III.6b).[56] Furthermore, the fact (1) that the Ebers Calendar (in the third column) identifies the heliacal rising of Sothis with the eponymous feast day of *wp rnpt,* which indeed was the first day of the civil year when the latter was organized or perhaps reorganized in the third millennium B.C., and the fact (2) that that feast day could not have been the first day of the old lunar calendar outlined by Parker (which, unlike Parker's suggested schematized lunar calendar presented here, was itself not schematic but had months of variable length), seem to dispose of any view that the calendar which was being correlated with the civil calendar was the old lunar calendar with its intercalation scheme. In fact, all that remains of

Parker's old lunar year in the Ebers Calendar are the names of the eleven monthly feast days succeeding the Feast of *Wp-rnpt* and indeed there is nothing to distinguish the Ebers Calendar understood by Parker as a schematized lunar calendar from an ideal, ad hoc Sothic calendar. And hence what appears to be the case is that we have in the first column a truncated Sothic year beginning with the rising of Sothis and containing 11 more monthly feast days 30 days apart but lacking any epagomenal days.

As the result of this discussion we have so far concluded that the year being correlated with civil year 9 of the reign of Amenhotep I was an ad hoc fixed year embracing twelve monthly feast days 30 days apart, which year began with the day of the feast of the Opening of the year, namely the day on which Sirius rose heliacally. I say *an ad hoc fixed year* rather than *the Sothic fixed year based on regularly adding a sixth epagomenal day every fourth year* because there is no evidence whatsoever of the use of a sixth epagomenal day in Egypt until the abortive calendar stipulated by the Decree of Canopus of 238 B.C. This calendar was apparently ignored but perhaps served as a model for the establishment of the Alexandrine year (see Document III.10, Decree of Canopus).

The effort by Weill,[57] who firmly believed that the Egyptians maintained a fixed Sothic Year of religious feasts along side the civil year (see the Introduction to Document III.5, note 5) is not convincing. Still, as the Sothic dates being here analyzed reveal, the date in the civil year of the rising of Sothis was recorded on a number of occasions, and thus was probably kept track of on a regular basis. This was most probably by observation but perhaps also occasionally by calculation, even though no trace of a formal calendar with a sixth epagomenal day added every fourth year or with any more complicated scheme of correction based on yearly observation can be found.

In concluding my account of the Ebers calendar, I wish to mention what I think may be a correlative correct reason why the Ebers calendar was composed, namely to assist in the construction or use of a water clock, an idea first suggested to me on reading Siegfried Schott's *Altägyptische Festdaten*, but which I have developed in a somewhat different direction. I shall describe how the Ebers calendar would have been used for this purpose later in the chapter under the rubric "Outflow Water Clocks," and at that time it should be clear that the major arguments against accepting the Ebers calendar as a correlation between an ad hoc fixed Sothic year and a given civil year can be answered.

Leaving the Ebers calendar, we may complete this treatment of Sothic dates by merely referring to the last such date from Pharaonic Egypt, namely that found in a calendar in Aswan from an unknown year of the reign of Tuthmosis III: III *šmw* 28 (ca. 1464 B.C.).[58] Further details concerning heliacal risings of Sirius in ancient Egypt and Sothic periods are presented below in Document III.10.

The Night Hours

So far we have concerned ourselves with the grand units of time represented in calendars with their years, seasons, and months determined in some fashion by the motions of the moon, the sun, and Sirius. Now it behooves us to consider the day and the night and their division into hours. For the early Egyptians nighttime and daylight were each divided into twelve hours, thus producing hours that varied in length according to the seasons (I shall have more to say about this when I discuss water and shadow clocks and the lengths of days and nights for which such might be useful). But we do not know the precise time when the number of divisions became 12. We saw in Volume One, Chapter Two and its documents, that by the time of the New Kingdom a series of works had been composed that described the passage of the sun through twelve divisions of the Duat or Netherworld, each division occupying a single "hour." But long before these works the Egyptians had divided the night into hours. Two passages in the *Pyramid Texts* taken from the wall of the pyramid of Wenis, the last king of the fifth dynasty, confirm this.[59] The first (Utterance 251, Sect. 269a) reads: "Oh you who are over the hours, who are before Re, prepare a way for Wenis." The second relates (Utterance 320, sect. 515a) the following: "Wenis has cleared the night; Wenis has dispatched the hours." The hours mentioned in these passages are no doubt nighttime hours, since the contexts seem to confirm this. In both passages the word for hours (*wnwt*) is determined by three stars, suggesting to us that the most primitive meaning of "hours" was "nighttime hours."[60] Later, daytime hours were sometimes determined by the sun (☉), as in the hour names listed in Fig. III.42, or determined by a star and complemented by the sun when "hour of the day" is meant (e.g., see Fig. III.38, line 1).

It seems certain that the division of night into twelve hours arose from the 12-month year and the 10-day week and thus was a consequence of the installation and acceptance of the civil calendar. There were of course 36 decades (i.e., 10-day "weeks") in the course of the 12 months embracing 360 days. At some time during the third millennium B.C. star watchers divided the night into twelve hours, determined by the heliacal risings of stars or groups of stars called decans.[61] Parker describes the invention and development of the decanal star-clock system succinctly:[62]

> We do not know exactly when the stars were first used to tell time at night but it was certainly by the 24th century B.C. [evinced by the quotations from the *Pyramid Texts* above] and quite possibly soon after the introduction of the civil calendar....
>
> It was not until roughly 2150 B.C. that we know for certain that these night hours totalled 12. This we learn from diagrams of stars on the inside of coffin lids, which were earlier termed 'diagonal calendars' but which are better called 'star clocks'. Figure 1 [=my Fig. III.13] is a schematic version of such a clock, followed in some degree by all our examples. It is to be read from right to left. A horizontal upper line T is the date line, running from the 1st column, [i.e.,] the 1st decade of the 1st month of the 1st season, to the thirty-sixth column, [i.e.,] the last decade of the 4th month of the 3rd season. With decade 26 there begins in the 12th hour a triangle of alternate decans to tell the hours through the epagomenal days....
>
> The mechanism of such a clock was the risings of certain selected stars or groups of stars that conventionally are termed 'decans', at 12 intervals during the night, and at 10-day

intervals through the year. If we now examine the star clock of Idy...[my Fig. III.14] even though it lists only 18 decades we can see very graphically how the name of a decanal star in any hour is always in the next higher space in a succeeding column [i.e., in the next decade], so that a star in the 12th hour rises over 120 days to the first hour and then drops out of the clock....

We have already seen in connexion (!) with the star Sirius-Sothis that it eventually disappears because it gets too close to the sun, and then after some days it reappears on the eastern horizon just before sunrise, its heliacal rising....

Of course, a star which has just risen heliacally does not remain on the horizon but every day, because of the Earth's travel about the Sun, rises a little earlier and is thus a little higher in the sky by sunrise. Eventually another star, rising heliacally is likely to be called the Morning Star. In the early third millennium B.C., we may conjecture, the combination of 10-day weeks in the civil calendar and the pattern of successive morning stars led some genius in Egypt to devise a method of breaking up the night into parts, or 'hours'. He observed a sequence of stars each rising heliacally on the first day of a decade or week. From a text in the cenotaph of Seti I at Abydos [see Document III.12]... we learn that stars were chosen which approximated the behavior of Sothis in being invisible for 70 days. The star which had risen heliacally

> on the 1st day of the first decade, thus marking the end of the night, would 10 days later be rising well before the end of night. The interval between its rising and the new star's heliacal rising would be an 'hour'. Inevitably this pattern of heliacal risings at 10-day intervals would lead to a total of 12 hours for the night.

Parker goes on to observe that all of the decans marking the hours would fall in a band south of and parallel to the ecliptic. This is illustrated by Fig. III.15, where Orion and Sothis are marked on the band. The hours so marked off by the successive risings of these decans would obviously not be of equal length, for as the length of the night increased the length of the hours would increase and as it decreased the hour lengths would decrease. As Parker notes, it is the combination of the lengthening and shortening of the night with the periods of morning and evening twilight and the oscillation of the star clock "that explains why the night was divided into 12 hours though there were always 18 decanal stars from horizon to horizon in the night sky."[63] The time from sunset to the appearance of very bright stars (the so-called civil twilight) averaged approximately 1/2 hour, while the time from sunset to that when all the stars were visible (the astronomical twilight) may be over an hour and the decans determined the time of total darkness exclusively. The first hour of the night was of indeterminate length, beginning with total darkness and ending with the appearance of a decanal star in the eastern horizon. That hour became progressively shorter each night of the decade, until it was replaced by the rising of another decan as the determiner of the first hour during the next decade. The end of the twelfth hour, at the beginning of a decade, was determined by the heliacal rising of its decan and followed soon thereafter by morning twilight. Obviously, as the decade advanced the twelfth hour was followed by a progressively longer period of darkness, which was apparently incorporated into the twelfth hour, its length being considered as lasting until light.

Decanal Clocks

Now we are prepared to examine in somewhat more detail the early diagonal clocks based on the risings of decans. Our generalized decanal clock given in Figure III.13 is based on the detailed analysis of twelve rather incomplete and defective clocks on coffin lids that date from about the time of Dynasty 9 or 10 to that of Dynasty 12. The twelve clocks can be placed into five groups, each of which has been cleverly dated by Neugebauer and Parker.[64] The least corrupt of the texts on the various lids are those of Coffins 1 and 2 (see particularly the coffin of Meshet given in Document III.11 and also Fig. III.16 for Coffin 2), which have 34 decans instead of the full 36 decans. Indeed in Document III.11 we are told in column 40, lines 11-12, that 36 decans ought to be there ("Total of those who are ⌜in their places⌝, the gods of the sky, 36").[65] The first line of the diagonal clocks, labelled T on the model clock of Fig. III.13 and reading from right to left, contains the headings for the thirty-six decades or 10-day periods we conventionally call "weeks" of each month (labelled "first," "middle," or "last"). The months are grouped into the three seasons of four months which we have described above when discussing the civil calendar. Seven of the twelve coffins contain the date line (e.g., see Document III.11 and also Coffin 2 in Fig. III.16).

Then under each decade-column we find twelve decans. If we look carefully at the clock of Meshet as depicted in Fig. III.86, which is described in detail in Document III.11, and also at another sample decanal clock, that of Idy, shown in Fig. III.14, we note that the decan in the twelfth row of the first column, whose heliacal rising marks the end of the twelfth hour of the nights of the first decade and whose hieroglyph is composed of two pairs of facing men ("crew?"), we see that the decan shifts up to the eleventh row in the next decade, while a new decan, namely *knm*, occupies the twelfth row of that decade. In the twelfth decade or column the decan of the two pairs of men has reached the first row of the twelfth decade and thus marks the end of the first hour during that decade. The decan then passes from the eastern sky. Skipping over to the bottom of the 25th column in our ideal clock schematized in Fig. III.13, the 36th decan (*pḥ̥wy ḫ'w*, shown in the bottom of the 23rd decade in Fig. III.16) would have arisen heliacally to mark the end of the twelfth hour during the 25th decade. That decan will ultimately mark the end of the first hour during the last or 36th decade of the normal year of 360 days. The twelve decans appearing in the twelfth row successively from the 26th decade onward (i.e., the 11 decans from the 26th through the 36th decades and the 12th decan originally no doubt in the 37th column but later on in the 40th column) serve to determine the hours of the five epagomenal days. Hence that last column (whether the 37th or 40th) strictly speaking is a pentad (i.e., a period of 5 days) rather than a decade.

These twelve additional decans form a kind of triangle on the diagonal chart as they progressively rise through the hours from decade to decade, and the hypotenuse of that triangle is marked in Fig. III.13 by a bold step-like broken line. The result is that the twelve hours of the night during the last 11 decades of the regular year of 360 days and the pentad of the epagomenal days are marked by some combination of the regular 36 decans and 12 supplementary decans (starting with a sum of 11 of the former and 1 of the latter in the 26th decade and progressing to the sum of 1 of the former and 11 of the latter in the 36th decade) and finally in the last column (the pentad of the epagomenal days) by the twelve supplementary decans alone. Incidentally, as I have said, it seems likely that the last column was initially merely the 37th column. But apparently this was expanded into 4 columns, the first three occupied by a list of the regular 36 decans and the 40th by the 12 supplementary decans.

The identification of the various decans is extremely difficult (see Fig. III.18 for a table of the decans based on the star clocks). Not only are all of the extant star clocks corrupt textually to some degree, but even when we are certain of the decanal names and their translations, their proper astronomical identity is for the most part unknown.[66] Decans numbered 26-29 are related in some fashion to Orion (as are the epagomenal decans C and D) and 30-31 to Sothis (as are decans F and G).[67] One interesting detail of the star clocks is the strip (V) of four sky-related figures depicted between columns 18 and 19. As shown on Coffins 2 and 3 (see Figs. III.14 and III.16-III.17), they are the goddess Nut, holding up the sky; the Foreleg of an Ox (=the Big Dipper); Orion, looking back toward the

Foreleg and holding an *ankh-(ᶜnḫ-)*sign in his right hand and a *was-(wꜣs-)*scepter in his left; and Sothis, the goddess of the star Sirius, facing Orion and holding a *was*-scepter in her right hand and an *ankh*-sign in her left. Finally we should note that between rows 6 and 7, each coffin lid contains a strip of non-astronomical offering prayers for the deceased.

The fact that the diagonal clocks were organized to tell night hours during the civil year of 365 days instead of the Sothic year of approximately 365 1/4 days guaranteed that they would quickly lose even the moderate usefulness they might have had when a particular clock was composed. Decans would have to be shifted in place and some new ones used. Evidence for a revision toward the end of the twelfth dynasty exists, but by that time a new system based on the transiting of the meridian by decans was developed, as we shall see below. Be that as it may, it must be remembered that the extant diagonal clocks on the coffin lids were funerary in purpose and hence largely symbolic, such a clock being merely one more necessary convenience for the deceased to have at hand for eternity just as he also had at hand the offering prayers to sustain his supposed physical wants during that same eternity. Hence the accuracy of the coffin clocks was probably of little moment.

Transit Decanal Clocks

We learn of the new system of using the transiting of the meridian by decans, i.e., their culminations, to mark the nighttime hours from later texts which may be collectively called the Book of Nut or the Cosmology of Seti I and Ramesses IV (see Document III.12). These texts accompanied a large vignette of the sky-goddess

Nut in a vaulting position, with her feet in the east and her extended arms in the west, both ends touching the earth, and her body supported in the center by the air-god Shu. Examining Document III.12 will give us a good picture of the transit clock. They tell us that the decan stars, on the model of Sirius, disappear in the Duat, i.e., the Otherworld, for 70 days, after which time they rise again. The account is of a simplified scheme of 360 days, an old star dying and a new star being born every 10 days. After spending 70 days invisible in the Duat, a decan rises and spends 80 days in the eastern heaven before it culminates or transits the meridian, after which it "works" for 120 days, its work being to mark a given hour by transiting for ten days starting with the 12th hour during the first decade and ending with the first hour during the 12th decade. Then, its "work" being over, it spends the remaining 90 days in the western sky before dying once more. The change from rising to transit resulted in changes in the decanal list, as Parker observes when describing and evaluating the transit clocks.[68]

> Since a star spends 80 days in the east before working, it is clear that it is transiting when it marks an hour. This change to transiting required a wholesale readjustment of decanal stars. When rising, two decanal stars could mark an hour between them but if they were at opposite sides [i.e., boundaries] of the decanal belt, thus some distance apart on the horizon, they could, when transiting, either pass the meridian together or so far after one another as to result in an extremely long hour. Of the 36 decans of the star clocks nearest in time to the transit scheme only 23

> remain exactly in place in the transit list. The other 13 have each a different place or drop out and are replaced by new decans.

The simplification of the clock on the basis of 360 days reveals to Parker[69]

> the strongly schematic character of such a star clock and poses the question of its real utility. In actual practice one may speculate that all an observer would need to do would be to memorize a list of 36 decanal stars, preferably rising ones as being the easier to observe, watch to see which one was in or near the horizon when darkness fell and then use the risings of the next 12 to mark off the night hours. Whether this was done in fact cannot be said with certainty, but it is true that while we have lists of decans on various astronomical ceilings or other monuments to the end of Egyptian history, we have nothing at all approaching a star clock in form after the time of Merneptah (1223-1211 B.C.), and that that was a purely funerary relic is indicated by the fact that its arrangement of stars dates it 600 years earlier into the Twelfth Dynasty.

However long the practice of telling time by the observation of decans lasted, the title of "hour-watcher" (*ỉmy wnwt* or *wnwty*) persisted until the Ptolemaic period, but perhaps with the general meaning of astronomer once water clocks, shadow clocks, and sundials came into common use.[70] This is borne out by an inscription from a sighting instrument made for an observer named Hor, of the sixth century B.C., which has on its bottom the following statement: "I knew the movements of the two disks (i.e., the sun and the moon) and of every star to its abode; for the ka of the hour-watcher *(ỉmy wnwt)* Hor, son of Hor-wedja."[71] Still, even if the title had a more general meaning, one would suppose that the astronomer might have assisted in overseeing the remains of the old star clocks that were added to temple monuments.

The Ramesside Star Clock

Before examining water clocks (in most of the ancient period used exclusively for telling time at night) and shadow clocks and sundials which allowed for different systems of the measurement of hours in the day, we can note the passing by the middle of the second millennium B.C. of the decanal clocks based on risings or transits for anything but decorative purposes. Still we should briefly examine a new form of star clock usually designated as the Ramesside Star clock because it appears on the ceilings of a number of royal tombs of the Ramesside period. Two sets of the tables appear in the tomb of Ramesses VI, and one each in those of Ramesses VII and Ramesses IX.[72] Though all of the sets are corrupt in some fashion, a prototype clock can be constructed from them (consult Document III.14). Recall that the ancient decanal clock in its ideal form had 36 columns of 12 hours whose ends were marked by the rising of decans and that in its modified form the hours were marked by successive transiting of the meridian by decans. But the new Ramesside clocks had instead 24 tables (two per month) of 13 stars, with the first star marking the beginning of the night. Unlike the earlier transit clocks it did not confine itself to the transiting of the meridian but also included the transiting of lines before and after (but presumably near) the meridian.

Each of the 24 tables was accompanied by the figure of a seated man (the target figure) with a grid or chart consisting of nine vertical lines and thirteen (sometimes erroneously only twelve) inner horizontal lines (see Figs. III.19a and III.19b). The inner seven vertical lines are lines of transit, the fifth (of the nine lines) representing the meridian circle, and the other six inner lines representing lines of transit before and after the meridian; these lines are apparently at equal increments of distance from the meridian. The inner horizontal lines represent first the beginning of the night, for which the first star is given, and then, successively, the ends of the twelve hours of the night. The twelve hours are accompanied by the names of stars, each one marking the end of an hour when it crosses one of the vertical lines that are related to the target figure positioned within or below the grid.

Apparently a water clock was used to set the original hour ends, the north-south meridian having been established. At the end of an hour, as determined (without great accuracy) by the water clock, a marking star was looked for, preferably one at the meridian but if there was no significant star there on the meridian then one transiting one of the other six inner vertical lines that were mentioned above as being before or after the meridian. Preceding the list of stars, each table bore a date in the civil calendar, the first table having the date "I Akhet [1]," which signified that it was valid for the period from the first through the 15th of that month, the second "I Akhet 16, [2nd] half-month," signifying its validity from the 16th day of the month through the 15th day of that second half of the month (i.e., the 30th day of the month; see Document III.14, note 7). Though they are not separately mentioned, it seems probable that the epagomenal days (along with the last fifteen days of the fourth month of Shemu) were to be covered by Table 24.

From (1) these dates as given in the four extant copies and (2) the notation that Sothis culminated at the beginning of the night in Table 12, dated II Peret 16, and (3) from our knowledge of the relevant Sothic cycle, we can deduce a date of about 1470 B.C. for the star clock, i.e., more than three centuries earlier than the Ramesside monuments on which the clock appears.[73]

After the date in the tables a phrase "Beginning of the night" is given, followed by the name of a star transiting at that time, and this refers to the first interior horizontal line. Then the names of the twelve hour stars are listed over the succeeding horizontal lines. The name of each star is followed by a brief phrase descriptive of its position relative to the seated figure, i.e., the target figure: "opposite the heart" *(r ꜥḳꜣ ỉb)* if on the center line or meridian, "on the right eye" *(ḫr ỉrt wnmy)* if on the first line left of the center line, "on the left eye" *(ḫr ỉrt ỉꜣby)* if on the first line right of the center line, "on the right ear" *(ḫr msḏr wnmy)* if on the second line left of the center line, "on the left ear" *(ḫr msḏr ỉꜣby)* if on the second line right of the center line, "on the right shoulder" *(ḫr ḳꜥḫ wnmy)* if on the third line left of the center line, and "on the left shoulder" *(ḫr ḳꜥḫ ỉꜣby)* if on the third line right of the center line. These verbally indicated positions are in addition marked on the accompanying grid by star marks (see Figs. III.19a, III.19b, III.99b, and III.99c).

Parker presents his view of how these transit clocks were used:[74]

> .... On a suitable viewing platform, probably a temple roof, two men would sit facing one another on a north-south line. The northernmost would hold a sighting instrument like a plumb bob (called by Egyptians a *mrḫt* [see Fig. III.20a and Fig. III.20b]) before him and would call out the hour when a star had reached either the meridian or one of the lines before or after as sighted against the target figure. The effort for such precision points to the use of the water clock as an independent means of marking when an hour had ended, and emphasizes as well the reluctance of the Egyptians to abandon telling time at night by the stars. Indeed, a water clock of the Ptolemaic period [Borchardt's Auslaufuhr 3] has an inscription on its rim that its purpose is to tell the hours of the night only when the decanal stars cannot be seen (Borchardt [*Zeitmessung*], 1920, p. 8 [cf. also p. 9 for Auslaufuhr 10]).

If the lists of stars in the 24 tables are examined (see Document III.14), it will be evident that only three of the stars used in the Ramesside clock are the same as or near to those used in the decanal clocks: "Star of Sothis" *(sbꜣ n spdt)*, i.e., Sirius, "Star of Thousands" *(sbꜣ n ḫꜣw)*, and "Star of Orion" *(sbꜣ n sꜣḥ)*, the last two not identifiable with any specific star. Incidentally, it will be seen from the list of the stars found in the 24 tables that many of them are stars from or related to other well-known Egyptian constellations or stars, such as the "Giant" (16 stars), the "Bird" (4 stars), "Orion" (2 stars, one known as the "Predecessor of [the Star of] Orion"), and "Sothis" (2 stars, one of which is called "The One

Coming After [the Star of] Sothis"). The conclusion of an inquiry into the star list seems to be that for the most part the stars of the Ramesside tables lay roughly parallel to but (except for slight overlapping) outside and south of the decanal belt (see Petrie's effort, here reproduced as Figs. III.84a and III.84b, to reconstruct the constellations and stars of the Ramesside Clock along with the northern constellations and some of the decanal stars).[75]

So far we have seen that the decanal and Ramesside clocks differ (1) in all but three of the stars they used, (2) in the periods during which each list of twelve hour stars was effective (36 ten-day periods in the decanal clock and 24 half-month periods in the Ramesside clock), and (3) in the fact that the former (in its transit version) used only transits of the meridian to mark the hours, while the latter used transits of the meridian and of lines before and after the meridian. Notice further that the decanal star marking successively the twelfth hour, eleventh, tenth, etc. through the first hour in successive ten-day periods in the decanal clock does not change, but that in the Ramesside clock the hour stars do not always move regularly up the hour-scale from one half-month period to the next. (4) In addition, the hour lengths themselves (however much they vary from each other), once established, remain essentially fixed in the decanal clock but often varied in length in its successor as the result of star substitutions "and shifting hour boundaries on, before or after the meridian."[76] Incidentally, as in the case of the earlier star clocks, the twelve hours of the night marked in the tables of the Ramesside clock, like those marked in the decanal clocks, were not equinoctial hours but were variable, seasonal hours that divided into twelve the period of total darkness (without twilight) as that period varied from half-month to half-month.[77] Neugebauer and Parker are careful to point out the inaccuracies inherent in the procedures of transiting in the Ramesside clock:[78]

> When we describe the Ramesside star tables as lists of transits the reader should not be misled into associating with this term any astronomical accuracy we are accustomed to assume for transit observations. Actually the sources of inaccuracy in the Egyptian procedure are only too apparent. That two persons, sitting opposite each other, cannot resume exactly the same position night after night is clear. To fix accurately the moment of transit, when even very small motions of the eye of the observer will displace the apparent position of a star, is impossible. Even if we assume optimistically that the target figure was replaced by a fixed statue the least shift of the observer's location would influence badly the accuracy of timing. Schack-Schackenberg assumed that a frame with vertical strings was used to give precision to such vague terms as "on the shoulder," "on the ear," etc. Though we have no evidence to support such an assumption the results would still be crude, if only from the lack of definition of the position of the observer's eye with respect to distance, altitude above the horizontal level and deviation from the exact meridian line. And whatever results might have been obtained in one place could not be repeated with any

> accuracy by another observer and target somewhere else. To this has to be added the well-known inaccuracy of time measurement by water clocks, which must have been the instrument on which the star tables were built.
>
> In short: Just as in the case of the decans the crudeness of the underlying procedures is so great that only under severely restrictive assumptions could numerical conclusions be abstracted from the given lists. If we add the fact of obvious errors and carelessness in details in the execution of the texts as we have them one would do best to avoid all hypothetical structures designed to identify Egyptian constellations from the analysis of the Ramesside star clocks.

A more recent, quite different (and I believe less satisfactory) analysis and evaluation of both the data of the Ramesside star clock and the procedures used in establishing and using it has been given by E.M. Bruins.[79] As will have been evident to the reader, the account I have given here and more fully in Document III.14 rests primarily on the pioneering treatment of Peter Le Page Renouf in 1874 and the thoroughgoing editorial and analytical efforts of Neugebauer and Parker.[80]

Outflow Water Clocks

In the description above of the Ramesside star clocks, it was suggested that water clocks may have been the instruments with which the hours of star tables were established or at least they may have been used as substitutions for star clocks. Two general types of water clocks were known and used in Ancient Egypt:

Outflow Clocks and Inflow Clocks, with remains of the former outnumbering those of the latter.[81] The earliest and most complete specimen of an outflow clock was one found in broken pieces at Karnak in 1904 by G. Legrain and restored for the Cairo Museum (see Fig. III.21a).[82] About 14 inches high, it dates from the reign of Amenhotep III (ca. 1391-1353 B.C.), is made of alabaster, and has the form of an inverted truncated cone. Thus it is shaped rather like a flower pot. The vessel was filled with water which flowed out gradually from a small aperture or orifice near the bottom of the vessel. The Egyptians used the truncated cone for the form of this and other outflow clepsydrae to compensate for decreasing water head as the water level descended and so to obtain a more even flow. By noticing the dropping water level against a scale of 12 hours (in fact 12 scales, one for each month) on the inside of the vessel the hour was obtained. However, the conical form only approximated the paraboloidal interior surface that would produce steady outflow and thus a better estimate of the time. In fact a vessel with a height of 14" and a conical shape like that of Karnak would have produced a better approximation to an even flow (i.e., an equal descent of the water level in equal times) if the angle of slope of the vessel had been slightly less (103° instead of about 110°).[83] We shall discuss the form of the clock in more detail later.

Let us examine the outside of the vessel in the four views comprising Fig. III.21a. Astronomical and other decorations of colored stone and fayence inlays are set out in three rows or registers (see Fig. III.22) that comprise the so-called celestial diagram, which we shall examine in more detail when we examine Egyptian astronomy later. The first two rows are interrupted by a vignette of King Amenhotep III making an offering to Reharakhti with Thoth behind the king. The uppermost row (starting on the right) represents the decans. Just past the midpoint we see Isis as Sothis (Sirius) in a bark. Then continuing to the left we see the planets and the so-called triangle decans. The middle row has the more prominent northern constellations in the center plus some deities on both sides. The bottom row includes six frames or fields that display the king with the twelve gods of the months, two gods per field. Between the first (just left of center) and the sixth (to the right) was located the outflow aperture. Examination of later specimens of water clocks and of the astronomical ceiling of the Ramesseum of Ramesses II (Fig. III.2), which like other such ceilings mimics to a considerable extent the exterior decorations of water clocks, makes it probable that a small statue of Thoth as the cynocephalic baboon and thus as the patron of time was once mounted over the aperture (see also Fig. III.23).[84]

Turning to the inside of the Karnak clock (Figs. III.24a-b), we see the aforementioned twelve scales of varying total length, each divided by depressions more or less equally spaced that divide the night in the specified month into 12 hours. The month for which each scale was originally intended at the time of its construction (or better, at the time of the construction of the water clock from which it was copied) is indicated on the top rim of the clock (and thus with the hieroglyphs inverted) in the conventional form of months of the civil year: Month 1, Season Akhet; Month 2, Season Akhet;...Month 4, Season Shemu. No effort was made to give separate scales for the epagomenal days. Presumably either the scale for the twelfth month or that for the first month was used to tell the hour during the epagomenal days. Under each month the hour scale consists, as I have said, of a linearly distributed series of small circular depressions each about 1/5" in diameter.[85] When the water level is at one of the depressions it indicates the hour at that time of the year to which the scale belongs. While the depressions of each scale are by no means precisely placed, the total lengths of the scales were quite accurately measured, as we shall see later. The longest scale is about 14 fingerbreadths, which is applicable to the month containing the winter solstice and thus the longest night of the year. The shortest scale is about 12 fingerbreadths and is used for the month containing the summer solstice and thus the shortest night. Each fingerbreadth is equal to 18.75mm (i.e., about 3/4"). The 14:12 ratio is not a very good approximation of the length of longest night to shortest night for Egypt. The ratio 14:10 would have been a better one. I shall have more to say about the 14:12 ratio later. But we should observe now that, since the civil year was about 1/4 day shorter than the fixed seasonal year, such a clock progressively becomes out of synchronism with the solar seasons (i.e., a shift of one month in about 120 years). Hence as time went on the scales were necessarily being applied to months other than those to which they were assigned in the clock. In fact, since the shortest scale of 12 fingerbreadths was applied in the Karnak clock to the civil month II Shemu, i.e., the tenth month of the civil year, it is evident that when the clock was built in Amenhotep III's time it was already out of date, since the assignment of its shortest scale to the tenth month was not valid for the period of Amenhotep III but rather for that of Amenhotep I more than one hundred years earlier (see Fig. III.37) when a certain official named Amenemhet describes having constructed such a clock and dedicated it to Amenhotep I. So it is evident that the maker of the Karnak clock simply copied the monthly assignments of the twelve scales from an earlier clock, and thus the scales were about one month off. We shall speak more of Amenemhet's "invention" shortly. Incidentally, it would have been easy to keep a clock like that of Karnak approximately up to date merely by shifting the month names on the rim of the clock after about 120 years. There is, however, no evidence that any effort was made to cover over the month names on the rim of the Karnak clock itself and replace them with new month names.

Sloley sums up the use of the Karnak clock as follows:[86]

> In use, the vessel was filled to the full-line, not visible in the Karnak example, where it had probably been painted on. Traces of the line are to be seen in the other fragments. At the end of the first "hour" as measured by the clock, the water level would have sunk to the first mark, and so on successively through the various "hours" to the last mark. Except in the case of one month, the last mark is not shown as its position is covered by the row of alternate signs [i.e., glyphs signifying] Life and Stability which decorate the interior [as a bottom register]. The flow of water from the orifice averaged 10 drops a second.

In fact when the vessel was filled and then started to empty when the aperture was opened, the water at first flowed in a fairly constant stream; but, towards the bottom, it slowed and dribbled out. Hence true equal hours were not produced, as we shall discuss further on. As Sloley further notes:[87]

> Although the hours were nearly correct at the middle of the scale, the earlier were too long and the later too short (Fig. 16 [cf. my Fig. III.31]). The Egyptians did not know this. They had no [sure] means of determining whether their hours were equal or not. To divide a space of time equally is a very difficult problem without [a] regularly moving mechanism or some precise means of observing the movements of the stars.

There are two pieces of evidence of written material that deserve examination in our effort to understand the development of water clocks. The first document bearing on water clocks is an inscription from a ruined tomb near the top of the hill of Sheikh ᶜAbd el-Gurna in Western Thebes. This was the tomb of the official mentioned above, Amenemhet, who lived under the reigns of the first three kings of the 18th dynasty: Ahmose, Amenhotep I, and Tuthmosis I (see Document III.15 below and Fig. III.25). In the much worn inscription the deceased tells us (lines 7-9) that "while reading in all of the books of the divine word" (i.e., the whole of Egyptian literature) he found "[the (longest) night of wintertime to be] 14 [hours long] when the [shortest] night of the summertime is 12 hours [long]" and that the hour lengths from month to month increase and then decrease. This fits in precisely with the Karnak water clock as we have described it: the monthly scales of the lengths of the nighttime hours to be used in a water clock vary from 12 fingerbreadths on the summer solstice to 14 fingerbreadths on the winter solstice with the intervening monthly lengths first increasing uniformly and then decreasing uniformly (with the fingerbreadth = 1/4 palm = 1/28 cubit = ca. 3/4"). Compare this description with the measurements of the monthly hour scales in the Karnak clock which have been detailed by Borchardt,[88] and reported succinctly by Sloley:[89]

> The fundamental scale (12*f*) is the shortest of the Karnak clock.... The aperture of the Karnak clock is a distance of 4*f* below the lower end of the 12*f* scale.
>
> The variations in the scales from month to month must now be considered.
>
> In the Karnak example, the scale lengths in terms of fingerbreadths "*f*," are approximately as follows:—
>
> 10th month....12 *f*
> 11th and 9th months....12 1/3 *f*
> 12th and 8th months....12 2/3 *f*
> 1st and 7th months....13 *f*
> 2nd and 6th months....13 1/3 *f*
> 3rd and 5th months....13 2/3 *f*
> 4th month....14 *f*
>
> The errors are almost negligible except in the case of the figure 13 *f*, which is .25 *f* too great.
>
> These figures show that a uniform change of 1/3 *f* per month was intended.

The second piece of literary evidence that might bear on the development of the water clock dates from about the same period as the first, i.e., the reign of Amenhotep I, is less certainly connected with a water clock of the Karnak type or in fact any water clock. However, a good case for its pertinence to water clocks can be made. This is the Ebers calendar, which has been mentioned earlier and translated below as Document III.2. This document has a first column consisting of the eponymous feast days of 12 months of a truncated Sothic year beginning with the rising of Sirius (and, like water clocks, without epagomenal days) and a second column listing the ninth day of each of the corresponding months of the civil year, this year being the ninth year of the reign of Amenhotep I. I believe that this calendar or equivalency table was composed to aid in constructing or using a water clock like the Karnak clock. As in the case of the Karnak clock the prospective clock maker could have placed the 12 months of the civil year about the rim of the vessel. Then he could have found out from the Ebers or similar Sothic-civil double calendar which civil month in the ninth year of Amenhotep's reign is equivalent to the Sothic month that includes the shortest night, i.e., the month that includes the summer solstice. No doubt he would know that this would be in the 11th Sothic month, the first month of that year being the time of the appearance of Sirius. Then he would simply put the shortest monthly scale (say the 12*f* scale) under the proper month name on the water clock he is constructing, that is, under the current civil month II Shemu, as the Ebers table would show him. He could then do the same thing for the longest hour scale (say, the 14*f* scale), and for all the intervening scales, assuming, as the maker of the Karnak clock did, that the hour scales uniformly varied in length by 1/3 *f* each month. This in fact may have been what Amenemhet did, i.e., construct such a clock of the Karnak variety after studying the literature and devising the monthly scales. Perhaps its originality lay in his placing all twelve scales in one clock with a single aperture.

Or it is possible that the hour watchers at this time had at their disposal a kind of fixed water clock where the scales had been inserted under the rubrics of the months of the fixed Sothic year. Then the watcher could simply look in the second column of the Ebers or similar calendar for the civil day and month he was using the clock. Opposite it he could then find the equivalent Sothic month and thus was able to know which hour scale on the water clock to use. The only difficulty in this last reconstruction is that so far we have not found any early water clock of the fixed season variety. Furthermore, if such a clock with monthly scales of hours already existed, we must wonder why Amenemhet would have bothered to single out his construction of a water clock as an achievement to record in his tomb.

A few remarks regarding other outflow clocks are in order. Borchardt has shown that the fundamental scale of 12 fingerbreadths that was used in the Karnak clock for the shortest night appears in other clocks to have been used as the scale for the months embracing the equinoxes, so that the scale for the shortest night was 11 fingerbreadths, for the equinoxes 12, and for the longest night 13.[90] Although in the Karnak clock the monthly scales vary in length by the constant amount of 1/3 of a fingerbreadth per month, these variations do not represent the actual changes of the lengths of the night from month to month, regardless of whether one defines the night as lasting from first darkness to last darkness or sunset to sunrise. The clocks that have the 12 fingerbreadths scale as the length of the hourly scale at the time of the equinoxes—all from the Alexandrine period (Borchardt's clocks nos. 2-4)—attempt to correct the constant changes of the scale lengths to take into account the fact that the length of night changes most slowly during the months immediately preceding and following the shortest or the longest night, i.e., the nights of the summer and winter solstices, and most rapidly in the months immediately preceding and following the months containing the equinoxes. Instead of the constant relationship 1:1:1 they have what appears to be an incremental relationship of [1]:2:3,[91] i.e., where the increments or decrements of the scale lengths of the months preceding or following the lengths of the scales of the solstitial months are 1/3 of a fingerbreadth, followed by 2/3 of a fingerbreadth, then by 1 fingerbreadth, or of course the reverse relationships where the increments or decrements are related as 3:2:[1] preceding or following the equinoctial months.

Another fragment of an outflow clock (Borchardt's no. 9; see Fig. III.27) yields a curious set of scale lengths: ?, 11, 11 1/2, 12, 13, 13 2/3, 14, where 12 fingerbreadths give the length of the equinoctial scales. The second half of the set of scales shows the 3:2:1 relationship noted in clocks 2-4. But the top part of the scale deviates from it, and I see no ready explanation for this deviation, for one would expect it to show the same 3:2:1 relationship as the bottom half. There is always the possibility that the clock was carelessly constructed or that it was simply made to serve as part of the tomb furniture for some worthy deceased and hence was not actually meant to be used in this world.

In discussing the monthly hour scales I have paid little attention to the distance between the hourly marks, except to note that certainly in the early outflow clocks and probably in all Egyptian water clocks the assumption was that they were to be equally spaced, that is, that the water level sank equal distances in equal times. However this was not a fact, the segment of the vessel bearing the scales being a truncated cone that could only produce an approximated even flow instead of being the segment of a paraboloid which would yield steady flow. Presumably the users of such a clock must have realized that the flow was not uniform. Indeed there is evidence that points in that direction from a papyrus of the third century A.D. which sought to determine the volumetric discharge of water between the various hourly marks.[92] The text is corrupt and includes some wrong numbers; it takes π as 3 and the basic formula it uses for the volume of a frustum of a cone is incorrect. The pertinent part of this determination begins in Grenfell and Hunt's text in line 31:

> They give the number[s] for the construction of horologes as follows, making the [diameter of the] upper part of the frustum 24 fingerbreadths, [that of the base] 12 fingerbreadths, and the height 18 fingerbreadths (see Fig. III.28).

The volume of each segment of the frustum having a height of one fingerbreadth is determined, according to the papyrus, by finding the mean sectional area (i.e., $2 \times r \times 1/4 \times 2\pi r = \pi r^2$, with the value of π taken as 3) and multiplying this by the height (i.e., 1 fingerbreadth). This involves the incorrect assumption that that volume of the frustum is equal to a cylindrical segment whose radius is equal to that of the mean area of the segment of the frustum. Since the over-all height of the frustum is 18 fingerbreadths, the volume would be the sum of all the segments of 1 frustum height. It is not at all clear how this incorrect volumetric analysis was to be used, but possibly the clock makers thought they could adjust the hour markers to produce what they thought were equal quantities being discharged in equal times, though of course that would not be completely accurate so long as they were using the wrong formula for the volume of a frustum of a cone.

But whatever the explanation of this curious fragment, it certainly seems that in Pharaonic Egypt the hour-watchers thought the hour markings ought to be equidistant and that the shape of their pot produced this uniform flow. As we have said earlier, precisely equal flow (other things being equal) could only have been produced by a vessel with a paraboloidal surface (see the parabolic section of such a paraboloid in Fig. III.29). However, as further shown in Fig. III.29, a vessel with a slope of 2:9 and thus an angle of inclination of about 77° measured between the pot wall and the base extended (see the left side of the figure) would produce a very good approximation. Less satisfactory is a pot with a slope of 1:3 (like that given in the Oxyrhynchus papyrus; see the right side of Fig. III.29). And even less satisfactory is the angle of inclination of the Karnak clock, namely 70° 25'. The angles determinable for the other clocks whose fragments we possess are of still lesser inclination to the base extended than that of the Karnak clock.[93]

Borchardt completes his discussion of hourly scales by showing quantitatively that neither the Karnak clock nor the clock discussed in the Oxyrhynchus papyrus will produce equal hours. This is clear from the table I have reproduced here as Fig. III.30 and the linear scales of Fig. III.31. The question of comparative hour lengths and their applicability to the possible day and/or night periods leads to his principal conclusion that the Karnak clock measures the hours of what he calls the "astronomical night," i.e., the period from the beginning of darkness to its end, with the morning and evening twilights excluded.[94]

I call the reader's attention to one last point concerning the various outflow clocks discussed by Borchardt, namely their dating, where possible, by the position of the shortest and longest monthly scales in the civil calendar (for a useful table showing where the seasons fall in the civil year at different Julian dates see Fig. III.37).[95] There is no particular need to review that discussion here, since I have given the specific dates as we have mentioned the various outflow clocks and have more than once indicated how they were found, most particularly when I described the probable use of the Ebers calendar for the construction and use of water clocks.

Inflow Water Clocks

The only Egyptian water clock of the inflow type is that found by G. Maspero at Edfu in 1901 (see Fig. III.23), though a number of small models, which were probably used as votive offerings, are extant (Fig. III.32).[96]

The Edfu clock is a cylindrical vessel of limestone unadorned on the exterior except for a small representation in relief of Thoth as the cynocephalic baboon placed immediately above an aperture for draining the water in the clock.[97] On the interior surface of the vessel is a poorly drawn grid embracing vertical monthly hour scales intersected by horizontal sloping lines at the hour points (see Fig. III.33). As an inflow vessel, the Edfu clock performs its time-telling duty by having water flow into the vessel from a reservoir and reading its level against one of the monthly hour scales. The great advantage that such a clock has over the outflow type is that if the reservoir is kept full constantly the head of inflowing water will remain the same and hence in theory the water flowing into the clock will rise at a constant rate. Before starting the clock at the beginning of the night, the water is drained from the vessel to the lowest mark on the hourly scale appropriate for the month that it is being used. It would, of course, be very difficult to read the water level against the scales on the interior of the cylindrical vessel. Hence the hour-watcher would surely benefit from some kind of movable float with an extended pointer which could be set at the proper month. This pointer could then indicate the hour on the exterior surface of a column that possessed a duplicate hourly grid and was attached to the top of the vessel.[98] However no such attachments have been found.

Turning to the grid with its vertical monthly scales and intersecting sloping lines, we must first note that it preserves still the Old Egyptian system developed by Amenemhet and appearing in the Karnak clock of having 12-fingerbreadth and 14-fingerbreadth scale lengths for the months containing the summer and winter solstices, and 13-fingerbreadth scales for the months with the equinoxes. But a significant change in the distribution of the vertical scales, that is, in the horizontal distances between the scales, is signalled by the interior diagram on the Edfu clock, where the distances vary erroneously as 3:2:1; 1:2:3; 3:2:1; 1:2:3, beginning from a 14-fingerbreadth scale, though according to the phenomenon of slower increments and decrements before and after the solstitial months the distances ought to have been 1:2:3; 3:2:1; 1:2:3; 3:2:1. It was Alexander Pogo who first explained the significance of this feature of distance displacement, suggested its possible origin in some earlier prismatic inflow clock, and revealed the error made by the Edfu clock maker in applying the 3:2:1 relationship to the distances from the solstitial scales:[99]

> [Following the discovery of the 12- and 14-fingerbreadth scales and the concurrent assumption that the monthly scales varied uniformly in length from month to month by 1/3 of a fingerbreadth:] The next step in the development of water clocks —made, apparently, by a designer of an inflow clock— was the correct observation that the length of the nights changes slowly during the month immediately preceding and during the month immediately following the shortest or the longest night; the advanced design of sun dials made such an observation possible at a relatively early date.
>
> Here is a simple method for incorporating this observation into the scales of a prismatic inflow clock with a square base; see Figure 1 [my Fig. III.34]. The corner scales of the prism ADGJ are 14, 13, 12, and 13 fingerbreadths, respectively; each corner scale is divided into 12 equal parts, and the corresponding points [of division] are joined by inclined straight lines which form "stripes" on all four walls; additional vertical lines are then drawn through the centers, C and K, and E and I, of the sides of the prism, and form

the 13 1/2 and the 12 1/2 scales, respectively; finally, the distances AC and AK, and GE and GI, are divided into thirds, and the vertical lines B and L, and F and H, form the 13 5/6 and 12 1/6 scales respectively.

Here is another method for drawing the diagram of Figure 1 [my Fig. III.34] on the inside of a square inflow clock. The corner scales of the prism CEIK are 13 1/2, 12 1/2, 12 1/2 and 13 1/2 fingerbreadths, respectively; the equinoxial [!] scales, D and J, occupy the middle of the side walls of the prism, while the solstitial scales, A and G, with their satellites B and L, and F and H, are drawn on the inside of the front and back walls, respectively; the inclined lines form "chevrons" on the front and back walls, and ordinary "stripes" on the side walls....

When prismatic inflow clocks were replaced by cylindrical ones, the diagram remained unchanged, and the distances between the scales, 1:2:3; 3:2:1; 1:2:3; 3:2:1, preserved on the cylindrical diagram of the Edfu clock, point, it seems to me, clearly to the prismatic prototypes of the cylindrical inflow clocks. The Edfu diagram, Figure 2 [see my Figs. III.33 and III.35], is a somewhat careless and not too intelligent copy of a diagram of the type reproduced in Figure 1 [my Fig. III.34]; it was copied either directly from an obsolete prismatic inflow clock, or, more likely, from a cylindrical inflow clock which was "out-of-date." The fact that the Edfu diagram begins the 1:2:3 count from the

> autumnal-equinox scale placed over the cynocephalos orifice seems to indicate that the maker of the clock was trying to copy a diagram which originated on the inside of a prismatic clock of the "chevron" type, CEIK, where the cynocephalos orifice was placed, for reasons of symmetry, below a solstitial scale; the engraver of the Edfu diagram did not realize that the 1:2:3 count of distances between the scales must begin at a solstitial scale, and that the position of the cynocephalos orifice is irrelevant; he saw that the diagram he was trying to copy started the 1:2:3 count from the orifice, and he faithfully reproduced this irrelevant detail. The lengths of the Edfu scales increase and decrease by the traditional amounts, 1/12, 1/6, and 1/4 of the difference between the longest and the shortest night, but the wrong starting point of the 1:2:3 count leads to slow changes of the lengths of the scales near the equinoxes, and to rapid changes near the solstices—an absurdity which failed to shock the maker of the Edfu clock...[who] did not understand the meaning of the 1:2:3 diagram.

While there is no archaeological evidence to support Pogo's ingenious reconstruction other than the existence of models of prismatic inflow clocks (which of course had no interior grids), it seems to me inherently probable. However, we might well temper somewhat his judgment that the inflow clock with its interior grid was a truly original discovery and thus an outstanding event in the history of science (see note 103 below), for he does not mention the fact that the makers of some extant outflow clocks (earlier than the Edfu clock) had already altered the lengths of their monthly scales so that the increments followed a 1:2:3 relationship, thus making a start toward recognizing the observation that the lengths of the nights changed most slowly before and after the solstices and more rapidly before and after the equinoxes.

Furthermore, Pogo is less than successful in trying to push the invention of the prismatic inflow clock and interior diagram back to Dynasty 18. While recognizing that the models of the inflow clocks are more than a millennium later than the Karnak outflow clock, he argues the possible earlier existence of a prismatic inflow clock of the "chevrons" type from a relief in the Luxor temple dated to the time of Amenhotep III (see Fig. III.36).[100] In this relief, Amenhotep III is shown presenting a votive offering ( ) to the goddess Mut which Pogo says "might be described as a prismatic *ḥn* receptacle, with a squatting cynocephalos attached to its front wall, presented on a *nb* [sign, 𓎟]." Pogo finds it "difficult to escape the impression that the object on the *nb* sign represents a prismatic outflow clock." Looking closely at the *ḥn* prism (or so he calls it), , he sees the V-shaped notch on top of it as a possible "allusion to the 'chevrons' characterizing a CEIK [prismatic] diagram;" and if this is so, then we have further confirmation of his belief that the offering is a prismatic clock of the "chevrons" type, the chevrons being on the side walls rather than on the front and back walls. His interpretation of the notch at the top of the *ḥn* sign in this manner seems to me entirely fanciful since the V-notch is strictly a part of the sign from its earliest use in the Old Kingdom,[101] which certainly predates any known use or record of inflow clocks. The whole votive offering in the Luxor relief is referred to as a *šbt* in the column of glyphs under the offering. This column is partly worn away but what remains may be rendered: "...the giving of the *šbt* upon the land so that he (Thoth?) may cause the giving of life forever." Hence we cannot with any assurance describe the clock (if that is what it be)[102] as a *prismatic* influx clock, at least not of the "chevrons" variety. All we really know from this offering scene is that it involves "the giving of life forever." Since the *ḥn* sign was used often from Dynasty 18 onward to mean an extended or infinite time period,[103] we could perhaps say that the three signs comprising the votive offering, [glyph], meant "Thoth, Lord of extended time." This rendering would underline its significance in the presentation scene as concerned with "the giving of life forever."

Because of the clear link between the monthly scales of the outflow and inflow clocks, it seems probable that in Egypt, at least during the Pharaonic period we are concerned with, the inflow clocks were primarily used to tell the time of Borchardt's "astronomical night," i.e., the period of total darkness without the morning and evening twilights. However when the inflow clocks spread from Egypt [and no doubt especially from Alexandria] through the classical world,[104] changes and improvements were made so that this was no longer true. Sloley has described briefly these later water clocks and mentions their use in public places.[105]

Shadow Clocks

The determination of hours in the daytime during

the Pharaonic period was based on the sun's motion and its projection of shadows. Hence the basis of marking daytime hours is quite distinct from that of marking the nighttime days, which, we have seen, was rooted in the risings or meridian transits of stars and in the outflow or inflow of water from water clocks. The only assumed similarity is that the length of daylight (initially including morning and evening twilights) was assumed to be divided into 12 hours just as was the length of nighttime.

There are two main types of sun-clocks. The first is an instrument to measure the length of shadows projected upon it and is usually simply called a shadow clock; the second measures the hours by the changing direction of a shadow projected during the daily course of the sun's motion; they are of course known as sundials. Let us first examine shadow clocks.

Telling time by shadows appears to be alluded to in a well-known passage from *The Prophecies of Neferti* quoted above in Volume One (page 402) to the effect that during the time of unhappy conditions preceding the Middle Kingdom "Re separates himself from men; he shines that the hour [of dawn] may be told, but no one knows when noon occurs, for no one can discern [*or* measure?] his shadow."[106] The use of "measure" or "number" instead of "discern" by some authors may result from their desire to push the history of the shadow clock at least to the Middle Kingdom.

The first (and indeed the only) Egyptian technical description of an ancient Egyptian shadow clock is found in an inscription in the cenotaph of Seti I (ca. 1306-1290). This is given in translation below (Document III.16), with the text transcribed in Fig. III.38. The shadow clock *(sṯ't)* so described (see Fig.

III.39) consists of a horizontal board (*mrtwt*) 5 palms (= 20 fingerbreadths = ca. 15") in length with a vertical head (*tp*) affixed at one end. Attached to the vertical head is a crossbar (*mrḫt*)[107] 2 fingerbreadths in height whose vertical plane is perpendicular to the horizontal board. Marked ("branded") on the board are signs that measure the hours.[108] The first is 30 units from the vertical head, the second 12 units nearer, the third 9 nearer, the fourth 6 nearer, and the distance from the fourth sign to the vertical head is 3 units, the signs being placed to measure four "hours." The horizontal board was first aligned to the east with the head pointing in that direction. Accordingly, the crossbar then pointed toward the north and the horizontal board was probably leveled by a plumb line, as is evident from the shadow clocks that have come down to us, though no such procedure is mentioned in the text. When the shadow falls on the first mark it begins to measure the first of four hours before noon. When it reaches the second mark the first hour is completed, and so on until the shadow length on the horizontal board comes to an end and noon has been reached. Then, as the text tells us, the shadow clock is reversed so that the head points to the west, and the lengthening shadow falling on the marks indicates the passage of the next four hours. Now it should be noticed that only eight hours have been marked off for measurement by the clock. But the text goes on to tell us that "2 hours have passed in the morning before the sun shines [on the shadow clock] and two hours [will] pass before the sun enters [the Duat or realm of darkness] to fix the place of the hours of night [i.e., before darkness arrives and the hours of the night begin]." Hence it is clear that the author of this piece assumed that the daylight (with the twilights) was divided into 12 hours. The selection of twelve hours seems to have been made in imitation of the division of the night into 12 hours necessitated by the civil calendar with its division into 36 decades. According to Neugebauer and Parker,[109] the "most plausible assumption [concerning the four extra hours not measured by the clock] is that sunrise and sunset, which are observable phenomena, mark the dividing points between the two extra hours before and after the central 8." This leads to their belief that this account describes a clock that is much earlier than the time of the cenotaph of Seti I, for the first extant shadow clock, preserved from the time of Tuthmosis III (ca. 1479-1425 B.C.), though very much like the shadow clock here described, has 5 instead of 4 hour marks, thus supposing a 10 hour clock with 2 extra hours and thus a day of 1+10+1 hours, the 12-hour day perhaps beginning with sunrise and ending with sunset. Or it could be that the 5 marks simply indicated a decimal division of the day with two additional intervals to connect it to the 12-hour nighttime.

Now it is evident that the marks on the Seti clock (those accompanied by Roman numerals on straight line E in Fig. III.40) indicate the central hours at the time of the equinoxes, while the marks with Arabic numerals mark the central hours in the 5-mark clock from the time of Tuthmosis III. But even if these marks correctly measured equal hours at the equinoxes (which they did not), they would not have accurately marked the lengths of those hours at other times of the year in view of the changing declination of the sun throughout the year. This is evident again from Fig. III.40, where the marks of the hours fall on points on the hyperbolic curves W and S representing in each case the shadow end's movement at the winter solstice (W) and the summer solstice (S), and the shadow end passing through a point on the curve marking a given hour at either of those times of the year would not fall on the horizontal board, i.e., on line E, in such a way that it would pass through the corresponding hour mark on E. Hence the original hour markings would not serve to mark the hours correctly at the summer and winter solstices. However, since the Egyptians did not appear to be exacting about the equality of the hour lengths, as we have already shown to be the case for the hours measured by water clocks, they probably used the early shadow clocks during the whole year without concern. But see Document III.16, note 2, for Bruins' clever suggestion that in Seti's clock a properly thin strip was added to or removed from the crossbar to change its height and thus to accommodate the clock to the changing declination of the sun at the times of the winter solstice, the equinoxes, and the summer solstice. Still, it should be remarked that none of the extant shadow clocks included a crossbar or thin strips to be added to that bar but only an end block with a groove for hanging a plummet line on it. In fact it is not clear from the text in the cenotaph of Seti I that the shadow clock included a cross bar mounted on the vertical end piece as is usually depicted (see Fig. III.39). For the name *mrḫt* translated in Document III.16, line 10, as "crossbar" could just as well refer to the vertical block alone, especially if the block contained a groove for the plummet line. Furthermore, the illustration of the shadow clock which accompanies the text (see Fig. III.38) does not reveal any crossbar distinct from the vertical head; it merely shows the whole instrument in profile as an L with a short vertical upright. Of course, if the vertical upright and an attached crossbar were the same thickness, the crossbar might well not show in a profile view. But the principal reason for assuming that a crossbar (much longer than the vertical block) was attached to the vertical block was that the shadow clock was apparently used throughout the year. And so in all probability a longer crossbar was needed to ensure that part of the line terminating the shadow always fell on the rather narrow horizontal board during the four measured hours at all times of the year.

Now I am prepared to examine actual shadow clocks that have been preserved from ancient Egypt.[110] I have already briefly mentioned the earliest one dating from the reign of Tuthmosis III (see Fig. III.41, West Berlin Mus. Nr. 19744, with separate endpiece above it from the reign of Nebmaatre, i.e., Amenhotep III, No. 14573). A clock from Sais (Fig. III.41, West Berlin Mus. Nr. 19743) dates from 500 or more years later; it is quite similar to the Tuthmosis clock but larger and more intact. Following Borchardt's treatment,[111] I shall concentrate on the latter.

The base is a ruler of about 30 cm. in length. At one end is a small vertical, rectangular block mounted perpendicularly on the base ruler. The vertical block has been pierced through its length by a narrow boring and a groove has been scratched out on one of the pierced surfaces, the boring and the groove being used to mount and guide a plummet line for the purpose of ensuring that the shadow clock lies on a level surface. Borchardt surmises that the plummet would not be needed very often since a known level surface such as the roof of a temple would have been readily available. On the upper surface of the base ruler we find five circular markings that indicate the passage of the hours (no circle is found at the end of the sixth hour since the vertical upright serves instead), the intervals between the hours decreasing in linear fashion from the mark of the first hour toward the end of the rule up to the upright which marks the end of the sixth hour (i.e., the intervals are diminished by decreasing multiples of the distance from the vertical upright to the mark ending the fifth hour). Accompanying these markings are the names of the first six hours (that is, the names of their protective goddesses), the first hour lasting until before the sun projects a shadow on the rule which reaches the first marking. The names follow (see Fig. III.42 for their hieroglyphic signs): 1st Hour—The Rising One, 2nd Hour—The Introductory (or Guiding) One, 3rd Hour—The Protector of Her Lord, 4th Hour—The Secret One, 5th Hour—The Flaming One, 6th Hour—The Standing One.[112]

It is obvious by the description I have just given that this shadow clock is used in a manner closely similar to that of the Seti clock we have already discussed. The base ruler lies in the east-west direction and the upright block with an assumed linear crossbar mounted on it accordingly lies in the north-south direction. Borchardt assumes such a crossbar mounted on the vertical block, for, as I have said before, the vertical block by itself is so narrow that the shadow would at all hours fall on the hour marks only at the periods of the equinoxes. At the periods of the solstices it would fall widely north and south of the narrow baseboard and therefore not on the hour marks. Therefore, Borchardt suggests that perhaps a cross bar ought to have been mounted on the vertical board so that in all seasons some piece of the shadow of the crossbar will fall on the graduated baseboard (see

Borchardt's drawing given below as Fig. III.43).

Borchardt further points out that the graduation of hours on the baseboard does not produce equal hours even at the times of the equinoxes. This is immediately evident from the fact that the increments of the intervals between the hour marks decrease by linearly decreasing multiples of a constant distance. This is an approximation much too simple to represent the complicated changes in shadow length (and their intercepts on the baseboard) produced by the angular changes in the sun's height in the course of the sun's apparent daily motion. This is not surprising in view of the complete absence of simple trigonometric calculations among the ancient Egyptians, for in order to determine shadow lengths and above all their linear intercepts on the ruler with the hourly graduations at the times of the equinoxes we would need to know the trigonometric functions of 15°, 30°, 45°, etc. Their values could not be exactly expressed using a simple arithmetical series of numbers. We might, however, know more about the ancient theory lying behind the graduation of the later shadow clocks if we had an intact version rather than charred remains of a papyrus found in Tanis and dating from about the first century A.D. (see Fig. III.44). They depict a shadow clock, with the numbers of the hours marked above the rule and under them are names of the protective gods of the hours which are inscribed on the rule. The vertical end piece which casts the shadow is not shown but the rays proceeding from it to the hour marks are shown, the heavy lines picture those portions of the rays present on the fragments.

Returning to the Berlin shadow clock analyzed by Borchardt, we show in Fig. III.45 the results on his investigation of the approximate deviation of its hour graduations from equal hours. Moreover, in Fig. III.46 is shown a system for using a 5-edged beveled crossbar (Borchardt's "sacred cubit") in three different positions to produce shadows which mark equal hours. At sunrise the crossbar would be in position 1. The first hour would be completed when the shadow's end fell on the first hour mark. The bar would then be changed to position 2, and the second hour would be completed when the shadow fell on the second hour mark. At this time the crossbar would be shifted to position 3. For the fourth hour the bar would be shifted back to position 2, for the fifth back to position 1, and for the sixth hour it is left in position 1 or shifted to position 3. The diagram of Fig. III.46 indicates how these positional changes produce the appropriate shadows. One cannot emphasize too strongly, however, that there is no actual evidence that such a beveled crossbar was used, and I suspect that the apparent indifference of the ancient Egyptians to the exact divisions into equal hours of any of their clocks makes their use of this device unlikely. Indeed, Borchardt himself does not insist that such a crossbar was used, though he believes that its use was a possibility.[113]

Before leaving our treatment of shadow clocks with their base rules oriented in an east-west direction, mention must be made of a model of a sun clock prepared in soft white limestone and now in the Cairo Museum (No. 33401).[114] This is later than the Berlin shadow clocks we have already described. It represents three different kinds of shadow clocks (see Figs. III.47 and III.48), the first with the scale on a flat, level surface like the Berlin shadow clocks, the second in which the shadow falls on stepped surfaces, and the third in which the shadow falls on an inclined plane. The clocks with a stepped or inclined surface for the reception of the shadows makes possible the reduction in the length of the shadow-receiving surface that was necessary when using a fat, level baseboard to measure the early and late hours. Notice that in representing the clock with the level surface, the vertical shadow producing block has no crossbar attached. But presumably no such bar was needed (at least from the end of the second through the sixth hours) because the graduated shadow-receiving surface is not narrow as in the case of the Berlin shadow clocks but is instead wide, so that some portion of the shadow's end-line would probably fall on the surface regardless of the season of the year if we had a real clock rather than merely an illustrative model. In view of the fact that we have only this model and we do not have any idea as to the latitude for which each of its three clocks was designed for and that we are not sure how faithfully the model reflects the proportions of the real clocks it represents, any extended analysis of real or potential hourly graduations like that given by Borchardt can hardly be more than educated guesswork illustrating the way in which the clocks functioned. However, after noting that the hourly graduations on the level-surface and stepped-surface clocks, even if they were correct for the equinoxes, would be incorrect for any other days of the year, Borchardt suggests that the inclined-plane clock (without graduations of any kind on the model) could have had scales that were to be mounted and usable for other days of the year as well (see Fig. III.49).

It should be realized that the particular inclined-plane clock of the Cairo model like the other two clocks represented on it was also to be fixed in an east-west orientation so that like them the hourly graduations represented the east-west intercepts of the shadow produced by the sun falling on the vertical edges. However there are evidences of another type of inclined-plane clock that was meant to be movable and was not maintained in an east-west orientation (see Fig. III.50 for several examples).[115] At whatever time the clock was to be read, its vertical block was pointed directly at the sun so that the sun's rays were always perpendicular to the shadow producing edge of the vertical block with the consequence that the shadow of the block fell completely and exclusively on the inclined plane (see Fig. III.51), and indeed the lack of any side shadow would be an indication that the sun-pointing clock was properly oriented. In view of the orientation of the clock directly toward the sun, the hourly scales on the inclined surface measured the direct length of the shadows rather than the intercepts of those shadows on an east-west oriented rule as was the case in the earlier shadow clocks we have been describing.

The first of these sun-pointing shadow clocks (the one from Qantara, Fig. III.50, 1) is datable to about 320 B.C. (plus or minus 60 years). It is shaped like the halves of the inclined-plane clock represented on the Cairo model. On the left vertical block from which the inclined plane begins to rise there is a groove for a plumb line used to level the clock. On the inclined surface are seven scales of varying length, each of which has six hour points. Beyond the incline is a right hand block on the top surface of which are the names of the Egyptian months given in Greek so that the longest scale (on which the midday shadow would be the shortest) is beneath the month P(!Ph)armu[thi], the second scale beneath the months P(!Ph)ame[noth] and Pach[on], the third beneath the months Mechi[r] and Pay[ni], the fourth beneath the months Tybi and Epe[iph], the fifth beneath the months Choia[k] and Me[sore], the sixth beneath the months Hathyr and Thoyt, and the last scale, i.e., the shortest one (which has the longest midday shadow), beneath the month Phaophi. The fact that the longest scale (which would be the scale to be used at the summer solstice) is assigned to the month Pharmuthi of the 365-day civil year indicates that the clock scales were designed for the year of 320 B.C. (with a leeway of 60 years before and after that date), as I have said.

The only other of these sun-pointing clocks complete enough to date even roughly is that of Paris clock shown in Fig. III.50, no. 2. The Egyptian month names on this clock are also written in Greek. The longest scale there is for Payni and the shortest for Choiak. Since the summer solstice fell at the end of the 10th month (Payni) in the fixed Alexandrine year, it is evident that the Alexandrine calendar underlies the scales of the Paris clock. Hence all we can say about the date of that clock is that its scales must date to the Roman period sometime after Augustus' introduction of the Alexandrine calendar into Egypt in 25 B.C.

One last point can be made concerning these sun-pointing shadow clocks. It evident that the Egyptians constructed the various scales so that straight lines could connect the corresponding hour points, e.g., a straight line would connect the points designating the first hour in all seven scales, as seems to be confirmed in a fragment in the Petrie Museum at University College, London (see Fig. III.50, no. 3 and

Fig. III.52). In actuality this ought not be the case, as is indicated in the dotted lines drawn in Figs. III.53a and III.53b through the theoretically correct points on the scales of the Qantara and Paris sun-pointing clocks (the first taking scales for the Qantara clock assumed to be constructed for latitude 31° north and the second scales for those of the Paris clock with an assumed latitude of 29°).[116]

We note finally that hieroglyphics representing some of the shadow clocks we have discussed were used in the Ptolemaic period either as determinatives for the word "hour" or by themselves as ideograms for that word (see Fig. III.54).

Egyptian Sundials

The sun clocks described in the preceding section measured shadow lengths projected by the sun off the edge of a vertical upright and they depended on the altitude of the sun. The other major kind of sun clock found in ancient Egypt was the vertical-plane sundial, which, of course, measured the hours by the changing angular direction of the shadow of a gnomon produced by the sun during its daily motion. I shall consider here two examples which are clearly of Egyptian origin. The fragment of a third dial found at Dendera, which includes only the radial lines marking the six hours after noon (Fig. III.55a), seems to be completely similar to the first two examples and so I shall not examine it.

The first of the Egyptian sundials is a small ivory disk found at Gezer in Palestine[117] (see Fig. III.55b). It dates from the reign of Merenptah (ca. 1224-1214 B.C.), whose cartouche it bears on the back of the disk and who is also pictured there making an offering to Thoth. Above the scene are the cartouches of the king. On the front of the disk we see the remains of 10 radiating lines (from a set that almost certainly originally contained 13 such lines, with the angles between two successive lines each being about 15°). The diameter of the dial is about 2 1/4 inches, with its center being evident from a small circle on the back. One can see from the edgewise view in Fig. III.55b that the disk was bored through its entire length so that the dial could be hung up by a chord passing through the bored channel. The inscriptions of figures and radial lines were originally filled with green enamel, some of which remains. Presumably, when intact, a style (i.e., a gnomon) must have been set into a hole in the center of the diameter of the semicircular dial face for the purpose of casting the angularly changing shadow.

The second dial (i.e., its face) is just like the first one but more complete (Fig. III.56). It is made of a green fayence that suggested to Borchardt that it was probably fashioned in the Greco-Roman period.[118] The dial was acquired at Luxor and is a part of the West Berlin Museum collection (No. 20322). Since the dial is intact (except for the gnomon) it contains 13 lines radiating from the center hole. The first and last line together make a straight line, to which the 7th line (marking midday) is almost perpendicular. The dial is without any decoration. As one can see in the drawing of Fig. III.57, left, the dial was hung vertically by a cord supported from two eyes in the rear rather than from a cord passing through a channel bored lengthwise through the top of the clock as in the case of Merenptah's sundial. A now missing gnomon for projecting the movable shadow was inserted in the center hole and so stood perpendicular to the dial face. There remains above the center hole a rectangular recess. That recess apparently allowed for hanging a plumb line necessary for positioning the dial correctly. It is clear from the way the sundials were made that for their use the straight line consisting of the first and last radial lines had to be level and that the dial surface had to hang in an east-west plane. The plumb line accomplished the first of these conditions when that line was parallel to the 7th radial line. The second condition was met by using it in a place where the east-west line had been previously established.

This clock like all the other clocks we have examined did not measure equal hours at any time of the year. At the time of the equinoxes, the 1st and 12th hours were each too long, as were the 2nd and the 11th; the 3rd and the 10th hours, as well as the 4th and 9th, were each approximately one daylight hour long; and the 5th and the 8th hours, as well as the 6th and the 7th, were each too short. This is clear from Fig. III.57, right, which indicates the results of Borchardt's investigation of the Berlin clock for an assumed latitude of 25.5°. In that figure the bold radial hour lines are those inscribed on the clock and thus end at the circumference of the disk. Near the end line of each hour (except the 6th and the 12th) is a set of three extended lines: two broken lines and one that is continuous. The broken line at the greatest angle indicates the proper marker for the hour at the winter solstice, the middle continuous line represents the desired marker for that hour at the time of the equinoxes, and the broken line at the least angle is the suggested marker for that hour at the time of the summer solstice. This figure then not only shows us that the radial lines did not mark equally long hours at the time of the equinoxes, but, as expected, they also failed to do so at the solstitial times.

It is evident from my treatment of the various kinds of Egyptian clocks (star clocks, water clocks, shadow clocks, and sundials) that there is great diversity in their accuracy and their methods of telling time. Indeed these various kinds of clocks share virtually nothing in their theoretical underpinnings, except of course the division into day and night hours and the seasonal variability of hour lengths (which is ignored in all but the water clocks and the inclined-plane, sun-pointing clock). It seems apparent that from at least the New Kingdom onward all of the types of clocks we have described were in use despite their diversity. This is one more instance of the fact that emerged in the first volume in my treatment of cosmogony and cosmology: the Egyptians often accepted essentially contradictory and mutually inconsistent practices, doctrines and modes of thought without great concern.

Traces of a 24-hour day with Equal Hours

As I have more than once stressed, the division of hours generally accepted by the ancient Egyptians and reflected in the development of their clocks consisted of seasonally variable hours, with separate systems of 12 night hours and 12 day hours, the division of daylight into 12 hours perhaps arising by analogy from the 12 hour division of the nighttime required by the structure of the civil calendar. But I have also mentioned, in describing Amenemhet's invention of his variable scale water clock in the reign of Amenhotep I, Document III.15, note 4, that the fact that the length of the longest hour of a winter night was compared to that of the shortest summer night as 14:12 perhaps hinted at a later theoretical discussion of the possible division, month by month, of the whole period of day and night into 24 equal hours. It is this later division that I now mention. Two documents (and the fragment of a third) refer to such divisions. The first (my Document III.7 but only given here in the body of Chapter Three and not in the section of documents) is a table added on folio XIV verso of Cairo Papyrus No. 86637, a Ramesside papyrus of about the twelfth century B.C.[119] The date of the hour table given in the text is earlier, falling between about 1400 B.C. and 1280 B.C.[120] Its hieratic text is shown in Fig. III.58a and its hieroglyphic transcription is given in Fig. III.58b. The table may be rendered as follows:[121]

1. I *ꜣḫt*: hours (*wnwt*) of daylight (*hrw*), 16; hours of nighttime (*grḥ*), 8.
2. II *ꜣḫt*: hours of daylight, 14; hours of nighttime, 10. Phaophi.
3. III *ꜣḫt*: hours of daylight, 12; hours of nighttime, 12. Athyr.
4. IV *ꜣḫt*: hours of daylight, 10; hours of nighttime, 14. Khoiak.
5. I *prt*: hours of daylight, 8; [hours of] nighttime, 16. Tybi.
6. II *prt*: hours of daylight, 6; hours of nighttime, 1(8). Mekhir.
7. (III) *prt*: hours of daylight, 8; hours of nighttime, 1(6). Phamenoth.
8. (IV) (*p*)*rt*: hours of daylight, (10); (hours of nighttime, 14). Pharmuth.
9. (I *šm*)*w*: (hours) of daylight, 12; (hours of nighttime, 12). Pakhons.
10. II *šmw*: hours of daylight, 12 (*sic*; but should be 14); hours of nighttime, (*blank*; but

should be 10). Payni.

11. III *šmw*: hours of daylight, 16; hours of nighttime, 8. Epeiph.

12. IV *šmw*: [hours of] daylight, 18; hours of nighttime, 6. Wep-renpet.

Aside from implying the division of the whole day into 24 hours, this calendar also appears to assume that equal hours are being used. We have seen that the Egyptians were unable in their clock to produce truly equal hours. But they perhaps produced the inaccurate approximations (indeed their preposterous data) "by the consistent use of a water clock during night and daytime after the division into 12 of both parts had been already established."[122] Most puzzling is the fact that the table reports values that seem to yield 3:1 as the ratio of the longest daylight to the shortest, a ratio that Clère declared to be a fantasy. As Neugebauer and Parker note,[123] this ratio corresponds to no locality in Egypt "if 'day' would mean the interval from sunrise to sunset." However they suggest that if the longest daylight was merely the complement of the shortest "night" of 12 decanal hours and the latter is worked out in accordance with their analysis of the elapsed times of decanal hours, this would produce a shortest night of about 6 hours in agreement with the table we have been examining. However it still does not explain the extreme ratio of 3:1, "particularly in view of the scales in the ratio of 7 to 6 of the Karnak water clock which makes the difference between winter and summer night smaller than it should be." Further note should be taken of the table's technically incorrect assumption that the length of daylight from month to month increases and decreases in a linear fashion, which, as we have seen, was also assumed in the scales of water clocks. The constant monthly increment or decrement in our table is 2 hours. At any rate the fact that in none of their clocks did the ancient Egyptians produce series of truly equal hours, makes any modern assumption that they used accurate equinoctial hours from month to month over the whole year in this and the succeeding table exceedingly unlikely.

A second table of the lengths of daylight and nighttime at 15-day intervals (my Document III.8, presented only here in this chapter and not in the section comprising documents) appears on a stone plaque discovered in Tanis in 1947.[124] Though the date of the plaque is not certain, it probably dates from the time of Necho II in Dynasty 26.[125] The incomplete plaque and another fragment discovered in 1948 contain four Texts. Text No. II is the table under consideration. A drawing of the table and its restoration by Clère is given in Fig. III.59. Two additional columns, numbers 11 and 12 (not shown in Fig. III.59), contain a small part of a similar 24-hour table, of which not enough remains for any useful analysis.[126] In translation, the table of Text No. II runs as follows:[127]

Knowledge *(rḫ)* of the quantity *(ꜥꜣ(w))* [i.e., the length] of the Daylight (*mtr(t)*) in relationship to the Night *(grḥ)*.

I [i.e., 1st Month of] Akhet, Day 1: daylight, 10 1/4 hours (*wnwt*); nighttime, (13 3/4 hours)

(I Akhet, Day 15: daylight), 11; nighttime, 13

II Akhet, Day 1: daylight, 11 1/2; nighttime, (12 1/2)

(II Akhet, [Day] 15: daylight, 12);

nighttime, 12

III Akhet, Day 1: daylight, 12 1/4; nighttime, 11 3/4

(III Akhet, [Day] 15: daylight, ...; nighttime, ...)

(IV) Akhet, Day 1: daylight, 13 3/4; nighttime, 10 1/4

IV Akhet, (Day 15: daylight, ...; nighttime, ...)

(I Per)et, Day 1: daylight, 14; nighttime, 10

I Peret, Day 1(5): (daylight, ...; nighttime, ...)

(II Peret,) Day 1: daylight, 14; nighttime, 10

II Peret, Day 15: da(ylight, ...; nighttime, ...)

(III Peret, Day 1: daylight, 13;) nighttime, 11

III Peret, Day 15: daylight, 12 1/4 1/6; nighttime, (11 1/3 1/4)

(IV Peret, Day 1: daylight, ...; nighttime, ...)

(IV Peret, Day) 15: daylight, 12; nighttime, 12

I Shemu, Day 1: (daylight, ...; nighttime,...)

(I Shemu, Day 15: daylight, 11 1/3 1/4; nig)httime 12 1/4 1/6

II Shemu, Day 1: daylight, 22 *(sic)*; nighttime, (...)

(II Shemu, Day 15: daylight, ...; nighttime, ...)

(III Shemu, Day 1: daylight, ...; nighttime, ...)

(III Shemu, Day 15: daylight, 9 1/3; nighttime 1)4 1/2 1/6

IV Shemu, Day 1: daylight, (...; nighttime,

...)

(IV Shemu, Day 15: daylight, ...; nighttime, ...)

Clère suggests that unlike the Ramesside table, which is a schematized table where the values seem to have been arbitrarily chosen, with their succeeding increments and decrements being a fixed two hours each month, the Tanis table despite its gross incompleteness and inaccuracies has no trace of schematization of the values. In elaboration he continues:[128]

> One does not find either regular progression or parallelism in the sequences. The few observable concordances (for example, the same duration of 12 h. 25 m. for the daylight one month before, and for the night one month after the autumnal equinox, 15,IV.*prt*) are certainly fortuitous—except, doubtless, for the duration of 12 hours, which, placed at a 6-month interval and a mid-distance between the solstitial days, ought to have been intentionally regularized to bear witness of the understanding that the Egyptians had of the phenomenon of the equinoxes....The duration-values of the text of Tanis certainly rest upon observations and, if they differ from the actual values, it could only be that some of the errors were owing to the imperfection of the measuring instruments employed to determine them. There could, of course, also have been copyist errors.

It was also observed by Clère that the use of fractional hour measurements in this table demonstrated that the Egyptians did not simply use successive divisions by two or three in their tables as some people have thought, but that their hour divisions were duodecimally based, doubtless in imitation of their division of the year into twelve months and the day and night into twelve hours.[129] The obvious difficulty of measuring accurately time units as small as twelfths of an hour when the Egyptians could not even accurately measure the length of an hour itself strongly contradicts the suggested observational base of the table and points rather to some more arbitrary mathematical schematization of the table that Clère initially rejected. Finally we can note that Clère produced a table of his own comparing the values for 13 days that survive in the Tanis table with actual values of the duration of daylight on those days estimated for several cities in Egypt (see Fig. III.61).

We can complete our consideration of the Tanis table by examining the analysis of it made by Neugebauer and Parker. It stands on its own and is worth quoting in full, especially since in the course of their analysis they reject Clère's suggestion that the table was established by means of observation. Here then are the remarks of Neugebauer and Parker:[130]

> This list concerns the length of daylight and night, expressed in equinoctial hours. Unfortunately only 13 of the original 24 data for the length of daylight are preserved. They are indicated in our Figure 10 [see my Fig. III.60], by black dots when both daylight and night are given, and their total is 24 [hours]. In all other cases only one of the two values is preserved.
>
> Four values of the thirteen conform to a pattern one would expect for lower Egypt.

The longest daylight, at the beginning of I and II *prt*, is 14^h, corresponding to the standard Hellenistic norm which assumes a ratio $14^h:10^h$ for the extrema. The equinoxes are located in II *ꜣḫt* 15 and IV *prt* 15 in conformity to the solstice data.

It would be natural to expect from this arrangement a linear scheme for the remaining values, following a well-known pattern of Hellenistic astronomy. Such a scheme is indicated in our figure by the dotted line. It would require a constant difference of ± 24 minutes for each 15-day interval. None of the preserved data, however, agrees with such a scheme. A constant increase occurs only once (I *ꜣḫt* 15 to II *ꜣḫt* 15) but with 30 minutes instead of 24; a constant decrease of 25 minutes per month (instead of 48) is found between III *prt* 15 and I *šmw* 15. The remaining values follow no recognizable pattern, not even as far as the necessary symmetries are concerned. The minimum is 9 $1/3^h$ instead of the expected 10^h; finally 22^h in II *šmw* 1 is an obvious scribal error (for 11?).

The absence of any strict regularity caused Clère to assume that the data of the text were based on observations. It is, of course, impossible to disprove such an explanation. On the other hand a month by month observation of lengths of daylight and night with relatively high accuracy (down to twelfths of hours, i.e., 5 minutes) falls completely outside the experience we have elsewhere with the handling of the problem of variable daylight. The norm is always a simple ratio of longest to shortest day and for the intervening data either linear variation or linear variation of the rising times. The occurrence of small fractions also speaks more in favor of some arithmetical process, perhaps incorrectly applied, than of direct measurements with very finely calibrated water clocks.

That the lengths of daylight and night were not observed independently is obvious since such observations could not result in a constant total of exactly 24 hours. Hence either daylight or night would need to be observed and the other part found by subtraction from 24. Since the water clock seems to be the only instrument available for this purpose one could perhaps consider the length of night as the basic interval. But whatever the measured interval may be it seems extremely unlikely that observations would result in equinoxes spaced exactly six months apart or in a neat longest daylight of 14^h from I *prt* 1 to II *prt* 1. An original poorly computed and badly copied text seems to be a historically more likely hypothesis.

Astronomical Ceilings and Other Monuments

The last subject of this chapter centers on the development and existence of a series of astronomical registers and related insertions. These appear primarily as reliefs or paintings on various surfaces, including coffin boards, tomb and temple ceilings, water clocks, and other objects. Neugebauer and Parker catalogue 81 relevant astronomical monuments and objects.[131] The subjects represented on the various registers include the decans in a belt south of the ecliptic (i.e., some varying number of the standard decanal stars and constellations used on the diagonal clocks we discussed earlier), planets, certain northern constellations with attending and flanking deities. Various of these objects are also found on zodiacs which are combinations of Hellenistic concepts with traditional Egyptian elements; the zodiacs all date from the Greco-Roman period. Needless to say I shall only describe and analyze a few representative monuments, emphasizing what the monuments tell us about the detail and extent of ancient Egyptian knowledge of the heavens.

Before expanding my discussion of the standard arrangement of astronomical elements in celestial diagrams, a few pertinent words on the orientation of monuments from Neugebauer and Parker are in order:[132]

> The term "orientation" should not be understood in an exact astronomical sense. Temples were usually directed with their main axis perpendicular to the Nile, thus generally east-west or west-east. But the course of the Nile can deviate considerably from its northerly direction and thus may bring the "northern" side of a ceiling decoration actually to the west, as for example in the temple of Hathor at Dendera which lies south of the Nile on its great bend from Qena to Nag^c Hammadi and so looks almost due north.
>
> ....rather strict rules governed the position of the various elements of an astronomical composition with respect to the cardinal points. The principal areas are north and

> south of an east-west axis. When the areas of a ceiling or other monument are east and west of a north-south axis, then the eastern area has the same decorative elements as would the southern area, and similarly for west and north. To state it in another fashion, north and west oppose south and east....
>
> On astronomical monuments which are fixed in place, then, we consistently find the northern constellations on the north or west or northwest, while the decans, since they lie in a belt south of the ecliptic..., are consistently to the south or east or south-east.

The arrangement of the various astronomical elements developed into an almost standard form that we can with some looseness call the Ancient Egyptian Celestial Diagram. In fact there are about six families of the standard form, but not much will be gained by describing all of the divergencies. Hence we shall confine our preliminary discussion to the standard arrangement, with only general and occasional remarks to family differences.

Initially we should recall, from our treatment earlier in the chapter, that the principal astronomical diagrams before the development of the standard celestial diagram were those of the diagonal clocks of the coffins. We remember that, in the case of the clocks dependent on risings (that go back to the ninth or tenth dynasty), the ideal diagonal clock (never realized but inferred from those coffins that remain) in its columnar structure included the names of 36 decans applicable to the 36 decades of the year plus 12 decans for the epagomenal days. Between the 18th and 19th columns a strip (designated as the "V strip") included depictions of Nut, the Foreleg of an Ox (the Big Dipper), Orion, and Sothis (the goddess of the star Sirius). Some different decans were selected for use in the clocks dependent on their culminations (meridian transits).

In the ordinary celestial diagram which we are now considering all 36 decans of the diagonal clocks (or more or less than the 36) were listed but not in the diagonal form used on the coffins (see Figs, III.2, III.4, III.21, III.22, III.65b and the detailed description of the ceilings of the tombs of Senmut and Seti below). Usually one (or occasionally more than one) star accompanies the decanal name as a determinative. If more than one, the stars indicated an attempt to represent a decan consisting of more than one star. Furthermore as our detailed presentation in Documents III.3 and III.4 show, groups of stars appear in the columns below the decans. However, such star groupings or clusters are of little assistance in identifying the constellations they represent because the number and arrangements of stars for each group varies widely from monument to monument. Exceptions are those below the decans related to Orion and Sirius, which can be identified, at least roughly. The names of the deities associated with the decans are usually included, but again not with great uniformity. Sometimes depictions (and not just the stars from which depictions were conceived) of the gods or possibly larger constellations associated with a number of the decans were also included, e.g., a bark, a sheep, an egg, Orion in a bark, and Isis-Sothis in a bark.

In addition to the main 36 decans, the artists included the names of some (and perhaps in the original celestial diagram, if it were extant, all) of the 12 triangle decans used in the old star clocks for the hours during the epagomenal days; these triangle decans were also accompanied by stars, their associated gods, and less frequently other material (such as a depiction of the constellation of the Two Tortoises under the decan *šṯwy*). After the last regular decan, that of Isis-Sothis (i.e., Sirius), and before the triangle deities the names of the three outer planets (Jupiter, Saturn, and Mars, or sometimes only Jupiter and Saturn, Mars being omitted) were given and following the triangle decans the two inferior planets Mercury and Venus were added. As depictions of the exterior planets we find two or three figures of Horus in a bark. For Venus the deity names *bꜣḥ̥* and *wsỉr* (Osiris) are given and a heron is depicted. So much then for the celestial bodies south of the plane of the ecliptic (those of the decanal belt) and the bodies which in their motions intersect the plane of the ecliptic (the planets). All of the decanal and planetary material ordinarily occupies one grand southern panel.

In addition, many of the monuments include in a northern panel a group of constellations oriented about the Big Dipper. These are usually called the "Northern Constellations," though the changing arrangements of these constellations and their accompanying deities from monument to monument makes their identification (except for the Big Dipper) extremely difficult if not impossible. Hence without a sure identification of these constellations, we cannot assert that all of them are necessarily circumpolar, as is usually said, though I suspect that they are since the Egyptians conventionally distinguish two groups of stars, namely the Imperishable stars, which are the circumpolar stars that do not set, and the Unwearying Stars, which are the decans and planets (see Volume One, Index of

Proper Names, under "stars") and this is just the distinction that seems to be made in the northern and southern panels of the celestial diagram.

On each side of the northern constellations is a register of panels which contains protective deities of the days of the month with disks on their heads, beginning on the right side with Isis (sometimes), the four sons of Horus, and four more protective day-deities, and on the left side as many as ten or eleven other protective day-deities. We owe to Brugsch the discovery of the origins of the assignment of these day-gods.[133]

Below the northern constellations we occasionally find a depiction of Thoth as the cynocephalic baboon, and we have already discussed its significance on the Ramesseum astronomical ceiling and on water clocks (see notes 21 and 84). Flanking Thoth the standard diagram (beginning on the left side of Thoth) includes panels for the twelve months showing the deity of each month facing the king.

So much then for a generalized account of the so-called celestial diagram. Now we should look briefly at two specific monuments, which are described in greater detail in Documents III.3, and III.4.

The Ceiling of the Secret Tomb of Senmut

The oldest monument[134] that includes enough of the celestial diagram to allow us to get a good sense of the whole is contained on the ceiling of the tomb of Senmut (Theban tomb no. 353).[135] This so-called "secret," unfinished and empty tomb of Senmut was discovered near the entrance of the temple of Hatshepsut during the excavations of the Metropolitan Museum in western Thebes in 1925-1927.[136] It is, of course, not to be confused with his completed tomb (Theban tomb no. 71) on the hill of Sheikh cAbd el-Qurna. Presumably it remained unfinished because Senmut, who supervised the building of Hatshepsut's temple and apparently had long and close service "on the queen's right hand," fell out of favor with his queen.

One of the most interesting features of the tomb is a sketch of Senmut on the wall of the stairway leading to the first chamber (see Fig. III.63). Next to the sketch we read "Overseer of the Estate of Amun, Senmut." But it is the astronomical ceiling of the chamber at the bottom of the third flight of stairs that principally concerns us (see Fig. III.4 and Document III.3). It consists of a southern panel (Fig. III.4, top) and a northern panel (Fig. III.4, bottom) separated by five lines of prayers for Senmut, the middle of which includes the titulary of Queen Hatshepsut. The southern panel is devoted to the decans (listed from right to left in Fig. III.4), the northern panel to the so-called Northern Constellations which occupy the center of the panel and which are flanked at the top by well drawn circles representing the 12 months of the year (with their names superscribed) and at the bottom by the day deities which I mentioned above.

So far as the decans of the southern panel are concerned, the relatively early date of the Senmut ceiling decorations (ca. 1473 B.C.) in comparison to the other monuments containing the celestial diagram seems to suggest that the Senmut decorations represent the prototype for a decanal family, which comprises 18 lists that can be grouped into one main group and two subgroups.[137] The earliest extant list is that on the Senmut ceiling and the latest list is that of Harendotes dating from about 246-221 B.C. A careful analysis of the decans of the Senmut ceiling by Neugebauer and Parker resulted in their reconstruction of the original thirty-six decans (one of which is missing in the Senmut ceiling), and most importantly in the straightening out of the confusion in the decans associated with Orion (mentioned in more detail in Document III.3).[138] The reader will notice in Fig. III.4, top, that the decan names have stars as their determinatives. For example, in the first column (starting from the right) the name of the first decan *(tpy-ᶜ knmt)* has one star as a determinative. This is followed by the names of the gods associated with the decan: Hapy and Imseti which are directly followed by two stars. Then in the second column at top appears the third decan *(ẖry ḫpd knmt)* with the second decan *(knmt)* below it, each decan followed by a star. Below them the name Isis appears and below it, covering both columns in a descending line, appear five stars plus a sixth one below in column 2, and considerably below them near the bottom of the two columns is a descending line of five more stars. These star groups apparently indicate some general constellation embracing all three of the decans connected with *knmt*. I leave the details of the remaining columns of regular decans to the considerations found in Document III.3, except to note that at the bottom of several columns we see what are apparently grand constellations with pictorial representations and accompanying lines of stars. Thus under columns 7-12 is depicted the Ship, 14-16 the Sheep, 21-23 the Egg (but considerably above the line of the other grand depictions), 24-28 Orion in a boat, and 29 Isis in a boat, with a feathered crown surmounted by a disk. She represents Sothis, whose name here, as often in the decan lists, is written at the top of the column as Isis-Sothis *(ꜣst-spdt)*. The ultimate origin of the decan list of this ceiling certainly lies in the decans arranged in some diagonal star clock. This is shown by the fact that, beginning with column 7, a horizontal line is drawn all the way up to the 29th column and each of the columns so underlined contains a successive 12th hour decan from a diagonal star clock. This line then represents the bottom line of a star clock.

The next column (col. 30) in the southern panel following those containing the main decans is devoted to the planet Jupiter, and we read in the hieroglyphics "*Horus who bounds the two lands* is his name" followed by a disordered phrase which probably should be rendered "southern star of the sky." Below is the falcon-headed Horus in a boat and bearing a star on his head. The column next to Jupiter is that of Saturn, where we read "*Horus bull of the sky* is his name," followed by a disordered epithet, which should probably read "the eastern star which crosses the sky." Once more the falcon-headed Horus with a star on his head is depicted in a boat at the bottom of the column. A column devoted to Mars often follows in other monuments (see Fig. III.65b and my discussion of the ceiling of the tomb of Seti I below). In the Senmut tomb it is omitted.

Following the exterior planets are six of the 12 triangle decans in five columns. I mention only the first decan *(šṯwy)*, which has at the bottom of the column two tortoises, and the second decan *(nsrw)*, which has below its name the comment: "It is a cluster *(ẖt pw)*." This is no doubt an indication that this decan is a group of stars. Finally the southern panel terminates with columns devoted to the interior planets Mercury *(sbg)* and Venus *(ḏꜣ)*. At the bottom of the columns of Venus a heron with a star on its head is pictured.

Turning to the northern panel of the Senmut ceiling (Fig. III.4 bottom) we should first concentrate on the northern constellations in the middle of the panel (cf. Fig. III.66), those usually designated as the circumpolar constellations. At the top center we find the Big Dipper or Great Bear represented here as a bull with a recognizable bull's head joined to a curious ovoidal body, with two short protruding legs and a tail of three stars, the last being a red encircled dot from which two diverging vertical lines extend to the bottom of the panel (see Fig. III.66 [O]). Presumably the use of this form of the bull as well as a fully formed striding bull (in some copies of the celestial diagram) and a hybrid where the bull's head is joined to a bull's foreleg (in other copies of the diagram) merely reflect the earlier Egyptian representation of the Great Bear or Big Dipper as a bull's foreleg *(msḫtyw)*, as it is always depicted on the Middle Kingdom coffins (see Figs. III.14, III.16 and III.17, where the four stars of the dipper and three of its handle represent the basic points of the foreleg). Pogo, by referring to a sketch of the northern constellations with the horizon at Thebes, dated about 2000 B.C. (see Fig. III.64a; and cf. Fig. III.64b for the northern constellations at about 3500 B.C. not long before the beginning of the dynastic period when the "hour watchers" began to delineate the main features of the celestial diagram), comments on the representation of the Great Bear as follows:[139]

> The three stars attached to the bull of the Senmut ceiling correspond to the position[s] of Delta, Epsilon, and Zeta Ursae Majoris. Around 3000 B.C., Zeta was the only 2nd

> magnitude star within ten degrees of the pole; the upper culmination of Eta Ursae Majoris coincided, for the latitude of Thebes, with the setting of Sirius.
>
> Observations of [the] culminations of circumpolar stars for the determination of the meridian were certainly made by Egyptians when Eta and then Zeta Ursae Majoris were the nearest of the bright circumpolar stars. The transit staff in Berlin [see my Figs. III.20a and III.20b] is of relatively recent date, but the plumb-line holder which apparently belongs to it looks like a distant descendant of the cross-shadow ruler [i.e., shadow clock] which bears the name of Thutmose [Tuthmosis] III [and is described above in my account of shadow clocks];...

The cord-stretching procedures accompanying the foundation of temples no doubt used such determinations of the meridian.[140] Cord-stretching goes back to the first dynasty and is mentioned in the Palermo Stone (see Volume One, p. 50 and the Index of Proper Names: "Stretching the Cord").

Continuing our perusal of the constellations that accompany Meskhetyu at the center of the northern panel of the Senmut ceiling (Figs. III.4, bottom, and III.66, [O]), note that there are seven figures that surely represent constellations of stars: [1] Serket (the scorpion goddess above Meskhetyu and looking down on Meskhetyu and the other figures), [2] Anu (a falcon-headed god below and facing Meskhetyu, with arms extended and holding a cord or spear directed at and touching Meskhetyu), [3] Isis-Djamet (the hippopotamus goddess—with a crocodile on her back–standing at the bottom of the group and to the right, with front paws holding on to a mooring post and a small, vertical crocodile), [4] Hotep-redwy (whose name seems to be an epithet of the Crocodile god Sobek–"restful of feet") is the crocodile which is toward the left and opposite the bottom of the hippopotamus and which is lunging at [5] Man (with upraised arms as if to spear the crocodile; to the left behind the crocodile is a small figure of a man that seems not to be a part of the whole group), [6] Divine Lion (with a crocodile-like tail, resting on his haunches above the figures of the crocodile and Man), [7a] Haqu–the Plunderer (a crocodile with a curved tail, located above the Divine Lion); the figure follows the name perhaps as a determinative; since [7a] appears only on the Senmut ceiling, it perhaps reflects or has some relationship with the next figure. The last constellation, [7] Saq (another crocodile with a curved tail, located at the top of the panel, to the left of Serqet). As the result of the size of the two flanking rows of the monthly circles, there is considerable space between the group at the top comprising Meskhetyu and figures (1), (2), and (7) and the group at the bottom including (3), (4), (5), (6), and (7a). One supposes that the totality of figures representing the northern constellations in its original form was a much more compact group, and indeed Neugebauer and Parker have attempted to reconstruct what they believed to have been the original, more tightly knit grouping (see Fig. III.66).

As one examines the various extant versions of the celestial diagram given by Neugebauer and Parker in the third volume of their *Egyptian Astronomical Texts*, Chapter IV, and my remarks in Documents III.3 and III.4 below, it is evident that there was considerable variation in the arrangement of the northern constellations in their depictions on the sundry astronomical monuments (e.g., compare Fig. III.66 with Figs. III.67-69).[141] Hence, as in the case of decans given on the southern panel, it is exceedingly difficult (if not impossible) to identify the Egyptian constellations exactly with modern depictions, except, of course, for the certain identification of Meskhetyu with the Big Dipper. Still, the identifications proposed by Pogo are reasonable, if not certain.[142]

> The star surrounded by a circle on the Senmut ceiling corresponds, obviously, to the early recognizable bright star, Zeta Ursae Majoris—the one with the conspicuous companion. The scorpion goddess Selqet (*Serqet*) stands behind the bull Meskheti in such a way that it seems as if she were trying to grasp the two cords stretched from the culminating star Zeta—over the invisible pole—down to the northern horizon....
>
> By 1500 B.C., the celestial pole was closer to Ursa Minor than to Ursa Major; culminations of Beta Ursae Minoris could be —and possibly were— used for the determination of the meridian. It is therefore not surprising that craftsmen of the XIXth dynasty began to consider the meridian cords as mere reins attached to the tail of the bull;...
>
> The hippopotamus with the crocodile on her back probably corresponds to the stars in Draco indicated in our sketch (Fig. 4 [=my Fig. III.64a; cf. Figs. III.64b and III.70a for the locations of Draco in 3000 B.C. and the present day]). One of the hands of the

> hippopotamus rests on the "dipper" of Ursa Minor, which is just east of the meridian with our polaris near the horizon of Thebes. The hand of the hippopotamus always rests, in later representations, on an object which is wider at the top than at the bottom [i.e., a mooring post]; the reins always lead to that support of the hippopotamus in more or less fancy curves—a tradition blindly followed by craftsmen who could not be expected to realize that in the dim past a vertical line joined Zeta or even Eta Ursae Majoris, when they culminated, with Ursa Minor near the horizon.
>
> The identification of the circumpolar constellations...which appear west of the meridian, near the horizon, is rather difficult. The stars Omicron and 23 Ursae Majoris may have something to do with the crocodile, and the stars Lambda-Kappa-Alpha Draconis, with the man facing the crocodile (his left hand, his head, and his right hand respectively).

The identification of the falcon-headed god Anu with Cygnus argued by Wainwright[143] seems unlikely because of the considerable distance of Cygnus from the Big Dipper (cf. Fig. III.64a and Fig. III.70a).[144] I suspect that the same reasoning puts into doubt some of the identifications of Herbert Chatley, which he calls "probable": the Hippopotamus with Boötes (the Guardsman), Hercules, Lyra, as well as Draco; the crocodile and accompanying god or man with Cassiopeia and Perseus; the Lion with Auriga (to which he appends a "?"); and the Goddess Serqet with Coma Berenice (again with a "?").[145] The methods used by Rekka

Aleida Biegel to establish the stars that make up these and other constellations in the Egyptian celestial diagram[146] have been shown by Pogo, in a severe review of her dissertation, to be quite fallacious and inadequate.[147]

Not much more need be said in this brief account of the celestial diagram as represented on the Senmut ceiling. I have mentioned the twelve circles surmounted by month names that appear on each side of the northern constellations. They proceed in the first row from right to left and then in the second row from left to right. Hence the four circles at the top right refer to the four months of the season Inundation, the two at the top left and the two below them on the left refer to the season of Peret, and the four circles on the right to the season of Shemu. Each circle is divided into 24 unidentified sectors. Since the time of the first descriptions of the tomb the sectors have been thought to mark in some fashion the 24 hours of the day and night periods, but there have not been any satisfactory explanations as to how they marked the hours.[148] Perhaps the sectors were meant to carry two sets of 12-hour meridian or other transits of bright stars, as in the Ramesside transit clocks, which we have seen probably originated at least as early as the time of Senmut in the 18th dynasty. If so, then each circle was meant to contain the star-transits marking the 12 nighttime hours of the first day of the month in one half the circle and the star-transits marking those same hours of the 16th day of the month in the other half of the circle.

There is no need to discuss here the nature and number of deities that flank the lower part of the central northern constellations since they seem to have no astronomical significance. The reader may compare the Senmut ceiling with the illustrations of two other members of the Senmut family of depictions given in Figures III.2 and III.22.

The Vaulted Ceiling of Hall K in Seti I's Tomb.

We have taken some pains to describe the Senmut ceiling because it represents an early and influential family of the Egyptian celestial diagram. A few words can be added concerning another influential depiction of that diagram found in the ceiling of hall K in the tomb of Seti I (1306-1290 B.C.) in the Valley of Kings in Western Thebes (see Document III.4 for a more detailed account). The sepulchral hall with the whole astronomical ceiling is shown in Fig. III.65a. It is evident that the northern and southern panels (often designated as the Eastern and Western Parts; see Document III.4 below) are separated by an empty band in the center. Let us look first at the decan list of the southern panel (see Fig. III.65b). Neugebauer and Parker call it "the prototype for the third family of rising decans."[149] The list includes 39 decans, but three of them appear to be triangle decans that have strayed out of position. Sometimes the columns are widened to include more than one decan (e.g., columns 11, 12, 20, and 22), no doubt because the decans included are part of a larger constellation (the prime example being the various decans in column 22 that are a part of the constellation of Sah, i.e., Orion). As in the Senmut list mistakenly omitted decans are occasionally inserted below the top decan (or decans) in a column (see columns 11, 14, 20, and 22). In most cases the omitted decan is placed beneath the decan it should precede. But not always. For example, in column 11, the added decan *(ṯms [n] ḫntt)* was properly squeezed in below the two decans *(ḫntt h̥rt* and *ḫntt ẖrt)* given at the top of the column.

The orientation of the southern panel in the Seti ceiling is more astronomically correct than in the Senmut ceiling, as Pogo has remarked:[150]

> As stated before, the orientation of the southern panel [on the Senmut ceiling] is such that the person in the tomb looking at it has to lift his head and face north, not south. The list of the decans preceding the Sah-Sepdet [Orion-Sirius] group occupies, therefore, the eastern part of the southern panel, whereas the planets and constellations which follow Sepdet in the traditional arrangement are listed in the western part. The southern strip of the Ramesseum, like the southern panel of Senmut, must be read, in the temple [see Fig. III.2], by a person facing north. On the ceiling of Set I, on the other hand, the orientation of the southern panel is astronomically correct, so that Orion precedes Sirius in the westward motion of the southern sky.
>
> The irrational orientation of the southern panel has caused some confusion in the representation of Sah [Orion] on the ceilings of Senmut and of the Ramesseum, both of which obviously follow the same tradition. On the ceiling of Seti I — which reflects another tradition — Osiris-Sah, participating in the nightly westward motion of the sky, is running away from Isis-Sepdet; he turns his head eastward to look at the pursuing goddess; the bow of her boat is almost ramming the

> stern of his. With the reversed orientation of the southern panel [in the Senmut tradition], Orion, the most conspicuous constellation of the southern sky, appeared to be moving eastward, i.e., in the wrong direction; to save the situation, Orion was turned around—within or with his boat. The Senmut and the Ramesseum ceilings represent Orion in this "reversed" position, adapted to their orientation of the southern panel; the Senmut draftsman made the bows and the sterns look alike; in the Ramesseum, the bow of Orion's boat approaches the bow of the boat of the Sirius-goddess; the element of pursuit is lost in both cases; instead of looking back at the bright star behind him, the god is turning his head away from the goddess. Mythologically, both the Senmut-Ramesseum and the Seti traditions may be equally valuable; astronomically, the Seti representation is far more satisfactory.

The first 25 columns (starting from the right) are divided into three registers and the last 10 into 4 registers. In the decanal columns, the names of the decans are included in the first register; stars indicating (but not always accurately) the number of decans in a column are in the second register; and the names and depictions of most of the patron deities of the decans (having anthropomorphic bodies and a mixture of human and animal heads) are in the third register. There is considerable crowding of the figures of the gods and goddesses so that some of the decans are not represented and some of the figures do not represent accurately the decan given above it. The large figures of constellations mentioned in our description of the southern panel of the Senmut ceiling are mostly present on the Seti ceiling, except for the figure of the sheep (under columns 14-16 in the Senmut list) and the figure of the two tortoises under the triangle decan *šṯwy* (in column 32 of the Senmut list). One other difference in the two lists is the inclusion in column 26 of the Seti list of the planet Mars ("eastern star of the sky, his name is Horus of the Horizon; he travels backwards [i.e., is in retrograde motion]")[151], with the depiction of the falcon-headed god in a bark, while this planet is missing from the Senmut panel. Hence Seti's list includes all five of the easily visible planets: Jupiter, Saturn, and Mars before the six triangle decans and Mercury and Venus after them.[152] The reconstructed list of triangle decans on the Seti ceiling is the same as that of Senmut.

Finally, in regard to the southern panel, there is considerable divergency in the number and arrangement of stars that accompany the main 36 decans. As in the case of the Senmut ceiling, it is extremely difficult to identify with any certainty the decans listed on the Seti ceiling, other than those connected with Orion and Sirius at the end of the list, and even those surely connected with Orion are not identifiable with specific stars in the Orion constellation.[153]

A few words remain to be added concerning the northern panel on the Seti ceiling (see Fig. III.65b, bottom, and especially Fig. III.69). Among the differences in the northern constellations of the Seti ceiling from those on the Senmut ceiling, are the following: (1) Meskhetyu, the constellation of the Great Bear, is now represented by a complete bull striding on a platform; (2) An (written as An rather than as Anu), the falcon-headed god, appears to be supporting the bull's platform; (3) behind the bull a Man (or god) with a disk on his head has been added; he holds reins that connect the bull with a mooring post resting on the bottom of the figure; (4) in front of the bull a falcon has been added; (5) both paws of the Hippopotamus appear to be on the mooring post; there are seven circles (stars) running down the Hippo's extended headdress; (6) the crocodile on the back of the Hippo here is much larger than it is on the Senmut ceiling; (7) the other crocodile on the left is more horizontal; as before, the crocodile lunges at Man (7a) in whose outline are 10 circles (stars); the Lion (8) above the crocodile is almost completely outlined by stars; (9) there is only one depiction of the crocodile with a curved tail and it is located in front of the Lion and acts as a determinative for its name; (10) the scorpion goddess Serqet is at the top left rather than above the bull; and all of the figures except the crocodile on the back of the Hippo, the crocodile on the bottom left, and the Falcon, have circles within them representing the stars on the basis of which the constellations were imagined; those in the figures with anthropomorphic bodies being more numerous than those in the figures with animal bodies (see the use of the circles as stellar connecting points for the construction of the constellations by Biegel mentioned in notes 146 and 147).

I ignore once more the deities flanking the northern constellations on the northern panel of the Seti ceiling since they do not seem to have any astronomical significance. Finally we can note that there is no register on the Seti ceiling that includes the months of the year and their deities.

Egyptian Zodiacs

One last kind of astronomical decoration on monuments (temples, coffins, and tombs) needs to be treated, Egyptian zodiacs, all of which originated in the Greco-Roman period. The earliest known Egyptian zodiac (the so-called Esna A zodiac) was a rectangular one from the temple of Khnum 2 1/2 miles northwest of Esna, that is, from the part of the temple originating in the reigns of Ptolemy III-V. Hence it dated from about 200 B.C. That temple was destroyed in 1843 to build a canal and the zodiac disappeared along with it. But the zodiac is still preserved in a plate from the *Description de l'Égypte* (see Document III.17 and Figs. III.75a and III.75b).

The pictorial representations of the zodiacal signs in the Egyptian zodiacs (though not their full range of Hellenistic additions) were surely of ultimate Babylonian origin. This is true of both the early rectangular form of the zodiacs and of the round zodiacs where the signs constitute a circular belt (as in the Dendera B zodiac presented in Document III.17 and depicted in Figs. III.76a and III.76b). Though it is tempting to think that the Egyptian round zodiacs represented the divisions of the zodiac in terms of degrees (as some of the earlier investigators seem to have assumed), it is evident from the analysis of the celestial diagrams I discussed above that there was no use of a system of degrees in ancient Egypt to measure celestial arcs. The Egyptian zodiacs not only included the signs of the zodiac but the old hour decans, the planets, and other Egyptian stars or constellations as well (like the Big Dipper and the Hippopotamus). By the time of the preparation of these zodiacs, the risings or transits of decans were probably no longer used to mark the hours of the night. In the earliest Egyptian zodiac, the rectangular zodiac Esna A mentioned above (Fig. III.75a) and also in the rectangular zodiac in the ceiling of the outer hypostyle hall at Dendera (designated as Dendera E), the decans for the most part seemed to be associated with or represented divisions of the areas assigned to the zodiacal figures, each sign being divided into three decans (except in a few cases of four decans), which, when absorbed in Greek and Roman zodiacs, were merely the names of the three 10-degree divisions of each of the zodiacal signs. Hence, in this new form the decans continued to play a part in the later astrology of Greece, Rome, India, Islam, and medieval and Renaissance Europe.[154]

A more detailed examination of the Egyptian elements in these zodiacs is given in Document III.17 below. We can simply note here that these elements, and particularly the decans, were drawn primarily from two families of the celestial diagram: those called by Neugebauer and Parker "the Seti I B" and the "Tanis" families and a few constellations from those we have discussed as northern constellations on astronomical ceilings. And indeed the earliest complete Greek version of decanal names, that of Hephaestion of Thebes (fourth century A.D.), appears to have been chosen exclusively from the Seti I B and Tanis families (see Fig. III.103).

Conclusion

So far I have discussed the principal Egyptian texts and monuments bearing on astronomy. The reader may have been conscious of the fact that almost all of the documents have been anonymous. A spectacular exception was the tract on the waterclock composed by one Amenemhet and dedicated to Amenhotep I (see Document III.15). Another interesting example is given below as Document III.18, where in an inscription on his statue (ca. 2nd century, B.C.) an astronomer and snake charmer named Harkhebi enumerated his astronomical, calendrical, and time-telling activities, including the observation of the stars and announcements of their risings and settings, his purification on the days when the decan Akh rose heliacally beside Venus, his observations of other heliacal risings, and particularly his foretelling of the heliacal rising of Sirius at the beginning of the [civil] year, and so on. These are essentially the activities we have described in this volume, and the documents that now follow and constitute the major part of this volume will throw further light on them.

It should be clear from my summary account that the ancient Egyptian documents do not employ any kinematic models, whether treated geometrically or arithmetically. However they did use tabulated lists of star risings and transits (as is revealed clearly in Documents III.11, III.12, and III.14), all tied to their efforts to measure time by means of the apparent motions of celestial bodies.

On more than one occasion in this chapter, I have remarked on the absence in early Egyptian astronomy of the use of degrees, minutes, and seconds to quantify angles or arcs, though slopes were copiously used in the construction of buildings, water clocks and shadow clocks; such slopes were measured by linear ratios.[155] And obviously, in the above mentioned tables for hour charts, right ascension was being approximated in terms of hours (though we are not completely sure how precisely). So far as I can determine, there were no standard, quantitative evaluations of declinations given in the tables. But of course in determining meridian transits the observers were keeping track of the maximum height of the transiting stars each night throughout the year.

Finally, the reader should realize that by the time of the Ptolemaic and Roman occupations of Egypt there were numerous other late Hellenistic and Roman astronomical and astrological texts which, though including some Egyptian elements, are not discussed in my volume. This is because they were primarily motivated by Greek and Roman astronomy or astrology (e.g., documents such as planetary tables consisting of the dates of entry of planets into zodiacal signs). I consider them outside the limits of my account of ancient Egyptian astronomy, but the reader can find expert treatment of such texts, or guidance to them, in the last chapter of Volume Three of Neugebauer's and Parker's *Egyptian Astronomical Texts*.[156]

(For notes to Chapter Three, see page 131.)

Notes to Chapter Three

1. See *The Calendars of Ancient Egypt* (Chicago, 1950); "Ancient Egyptian Astronomy," *Philosophical Transactions of the Royal Society of London*, A.276 (1974), pp. 51-65; and a somewhat more recent reiteration of his main conclusions, "Egyptian Astronomy, Astrology, and Calendrical Reckoning," C.C. Gillispie, ed., *Dictionary of Scientific Biography*, Vol. 15 (New York, 1978), pp. 706-10, full article, 706-27.

2. "Ancient Egyptian Astronomy," pp. 52-53. Parker in his *Calendars*, pp. 30-31, gives a historical prĕcis of the earlier theories of the original lunar calendar. Indeed he declares that "Borchardt (1935 [=*Mittel etc.*, pp. 5ff., 24; cf. also *OLZ*, XXVIII, 1925, 620, and *ZÄS*, LXX, 1934, 98-99]) was the last chronologer to discuss all forms of the Egyptian year, both civil and lunar. According to him [Borchardt], the first year was lunar, and it began with the next lunar month after the heliacal rising of Sothis. When, because of the yearly shift forward of the lunar calendar by about eleven days [its year averaging only 354 days in length], the first month of the year would fall before the rising of Sirius, the month was, instead, intercalary. The name of this intercalary month was *wp rnpt*. Proof of the existence of such a calendar was to be found in the Ebers calendar [see Doc. III.2] and in the names of certain of the lunar months. After this year had been in use for some time, the civil year was inaugurated, with its beginning marked by the heliacal rising of Sothis. Still later, the lunar year concurrent with the civil year was developed. By the end of the New Kingdom, or possibly later, the original lunar calendar fell entirely out of use and was superseded by the later lunar calendar, which was probably present in the Middle Kingdom and certainly present in the New Kingdom." As Parker notes later in *Calendars*, p. 74, n. 17, "In the very same context Borchardt qualified his clear statement that the original lunar year began with the first month after the rising of Sothis by placing its beginning 'around' the rising of Sothis, or 'around' the longest day of the year."

3. There is a hymn to Amun in the temple of Hibis, dating from the time of Darius I, in which the moon as Amun's left eye at night is said to be "ruler of the stars, who divides the 'two times'

(i.e., day and night), the months, and the years." See H. Brugsch, *Thesaurus inscriptionum aegyptiacarum,* 2. Abtheilung (Graz, 1968, unaltered repr. of Leipzig ed. of 1885-91), p. 511. Parker notes this passage as one indicating the relationship of the moon to the year (*Calendars*, 32). The only reason I quote it here is the possibility that perhaps the dual form of "time" or "period" (*drwi* or *trwi*) indicated here rather peculiarly by the addition of two solar signs surmounting the normal plural form of three strokes ( ) instead of the common dual form of two strokes might be an error for the form with one solar sign surmounting the three strokes. If so, then it would become a plural form for "seasons" and we could translate the end of the passage as "who divides the seasons, the months, and the years," and indeed that is how Brugsch translated it. But in any case, the hymn is too late to be of any significance as evidence for the old lunar calendar.

4. J. Černý, "The Origin of the Name of the Month Tybi," *ASAE*, Vol. 43 (1943), pp. 173-81. Citing A. Erman, "Monatsnamen aus dem neuen Reich," *ZÄS*, Vol. 39 (1901), pp. 128-30, he says (p. 173): "Only more recently has it been possible to trace these month-names back to actual names occurring in business texts of the Ramesside period." He also mentions the earlier monthly festival names of the Ebers calendar and those of the Calendar of Lucky and Unlucky Days that Abd el-Moshen Bakir was later to edit, *The Cairo Calendar No. 86637* (Cairo, 1966), p. 54, Verso XIV. The use of the month name instead of its number in a regnal date occurs only later. An inscription on a statue of Iti (an official of the reign of King Shabaka of the 25th dynasty) in the British Museum (No. 24429) may be the first recorded regnal date in which a month (Payni) is named. See T.G.H. James and W.V. Davies, *Egyptian Sculpture* (London, 1983), p. 53.

5. I remind the reader of the following more precise values: The Julian Year is 365.25000 days. Because of the astronomical conditions during the whole Egyptian period, the Sothic year is virtually identical in length to the Julian year (the total delay of the rising of Sothis between 4231 B.C. and 231 B.C. is about one day; after that time the progression of delay accelerates so that by 1926 A.D. an additional delay of about 2.64 days has occurred; see also the data mentioned in note 7 of the Introduction to Document III.10). The solar, natural (better, tropical), year, i.e., the interval between two consecutive solstices, is 365.24220 days. The difference between the Julian and tropical years is 0.00780 days

per year. It is this difference that the Gregorian reform of 1582 sought to account for by dropping ten days from October and by removing the extra leap year day when a century year is not divisible by 400 without fractional remainder. Incidentally the tropical year is shorter than the sidereal year of 365.25636 days, which latter year is the interval between consecutive conjunctions of the earth and the sun with a fixed star. The difference between the tropical and the sidereal years results from the procession of the equinoxes, which amounts to a displacement of the equinoxes in the sidereal year of about 1 day in 72.5 years.

6. There is one inscription recorded in the reign of Ptolemy IV at Aswan that seems to speak of the procession of Isis-Sothis through a total of 730+360+12+1/2 (i.e., 1102 1/2) years, which accords well with the assumption that the current Sothic cycle of which the Ptolemaic period was a part began in the quadrennium 1321-18 B.C. of the Julian calendar. This strengthens the view that the ancient Egyptians, at least by the time of the Ptolemaic period, kept an accurate record of the annual Sothic risings over long periods of time. See below, Document III.10, Section 5, and Fig. III.1 as well.

7. See Parker, *The Calendars of Ancient Egypt*, Chaps. I and III, and his reply to Gardiner, "The Problem of the Month-Names. A Reply," *Revue d'Égyptologie*, Vol. 11 (1957), pp. 85-107. See also Parker's concise account in "The Calendars and Chronology," in *The Legacy of Egypt*, 2nd ed., edited by J.R. Harris (London, 1971), pp. 13-26, where all three calendars are discussed: the old lunar calendar, the civil calendar, and the later lunar calendar. By its very conciseness, this account is the clearest of his several efforts to summarize calendaric developments in Ancient Egypt.

8. *Calendars*, Chap. I, where he gives a history of earlier treatments of the beginning of the month and a detailed discussion of the Egyptian names of the days of the lunar month. Cf. Document III.6 below.

9. *Ibid.*, pp. 9 and 70 (notes 7, 8. and 9), citing treatments by Brugsch and Mahler, and Sethe. See also E. Mahler, *Études sur le calendrier égyptien (Annales du Musée Guimet, Bibliothèque d'études*, Vol. 24), Paris, 1907, p. 7.

10. Parker, *Calendars*, p. 72, n. 49, and "The Problem of the Month-Names," p. 91. For the list of feasts in the Beni Hasan calendar, see K. Sethe, *Historisch-biographische Urkunden des Mittleren Reiches*, I (=*Urkunden* VII) (Leipzig, 1935), pp. VII 29-30,

and Document III.1, "The Middle Kingdom" (Section 1).

11. *Calendars*, pp. 32-34.

12. Parker, *Calendars*, p. 34, "The right half of a tablet of the 1st dynasty...bears the figure of a recumbent cow (Isis-Sothis) which has the sign for 'year' with a stroke between its horns. ... Underneath is a sign which is apparently *ꜣḫt*; and the whole can be plausibly read 'Sothis, the opener of the year; the inundation'." But later [in O. Neugebauer and R.A. Parker, *Egyptian Astronomical Texts*, Vol. 3 (Providence and London, 1969), p. 201] he accepts the conclusion of Godron that the cow is Sekhet-Hor (see footnote 14 below) rather than Sothis. The case for identifying the glyphs on the tablet with Sothis and the beginning of the year was succinctly made earlier by L. Borchardt, *Die Annalen und die zeitliche Festlegung des Alten Reiches der Ägyptischen Geschichte* (Berlin, 1917), p. 53, n. 1. He concludes in this note: "Mir scheint es, dass die Inschrift bedeuten soll, dass das Hundssternneujahr zum ersten Male (der Strich neben dem Jarhreszeichen zwischen den Hörnern der Kuh?) in den zweiten Monat des Wandeljahres fiel, also dasselbe, was wir oben aus dem 'ersten Male des Festes der Zeitordnung' abgelesen hatten."

13. In order to make the best case for the earlier interpretation of this tablet (though as I later suggest there is scarcely any support for the basic idea that the tablet represents a reference to Sirius and the opening of the year), I have left out any reference to the so-called month signs, and I have deliberately cast the translation in a form that suggests that the tablet was a marker that dated the particular object to which it was attached to the year of the king's reign which was named from a special activity of the king, namely from his first celebration at Ap of Sothis as the opener of the year. Presumably he would celebrate the cow goddess in other years, but that this was the "first" year of his doing so. Examples of such tablets are noted in Vol. I, Chap. 1. The name that I have suggested does not mean that Sirius actually rose in Akhet at the time of the celebration but merely that Sothis as the hallowed opener of the Year in Akhet was celebrated by Djer in Ap for the first time. The form is clearly like that recorded in the *Annals* translated as Doc. I.1.

14. G. Godron, "Études sur l'Époque Archaique," *BIFAO*, Vol. 57 (1958), pp. 143-49, full article, pp. 143-55. His evidence is persuasive and thus makes it quite doubtful that the tablet is a reference to the Sothis as the opener of the year. Still, if the older

interpretation is correct, then this tablet would tell us either that the civil calendar of 365 days with some tie to the rising of Sirius as its one-time New Year's Day was in existence early in the first dynasty or that the luni-stellar calendar proposed by Parker was being used at the time of Djer. See also W.M.F. Petrie, *The Royal Tombs of the Earliest Dynasties*, Part II (London, 1901), p. 22, who makes no mention of the possibility that this tablet is concerned with the opening of the year. Regarding its meaning he says: "They [the two fragments of the tablet] seem to name 'Hathor in the marshes of King Zer's city of Dep,' or Buto....The figure of Hathor with the feather between the horns is already known." So the upright strokes between the horns were for Petrie and also Griffith simply the sign for "feather" rather than a stripped palm branch (for "year") with an accompanying vertical stroke. Griffith in the Petrie volume, p. 49, reads the glyph for the city not as "Dep" but as "Ap".

15. *Calendars*, pp. 41, 45. The earliest source in which *Wp rnpt* is clearly the twelfth month is the astronomical ceiling of Senmut (see Fig. III.4), which Parker believed represented the old lunar calendar. The latest source is on the wall of the Temple at Edfu. See H. Brugsch, "Ein neues Sothis-Datum," *ZÄS*, Vol. 8 (1870), pp. 109-10. In between is a geographical fragment from Tanis in which the last month of the year (IV *šmw*) is called *Wp rnpt* (see Fig. III.5).

16. Parker, *Calendars*, pp. 33-34, gives a series of passages in which the meaning of *wp rnpt* and *prt Spdt* are equivalent. But granting that equivalence does not mean that the reference is to the lunar calendar since in every case the reference might be to the New Year's Day of the civil year when the rising of Sirius and New Year's Day were coincidental at the beginning of a Sothic cycle, a coincidence that continued to be celebrated on the New Year's Day of the civil year, long after the initial coincidence of the rising of Sothis and the first day of the year. Obviously at any other time the actual rising of Sothis would not be the New Year's Day but mention of it in the phrase "opening of the year" could refer to the initial coincidence at the beginning of a cycle, an event worth keeping track of for seasonal celebrations.

17. Parker, "The Problem of the Month-Names," p. 93. Cf. Bakir, *The Cairo Calendar*, Plate IA, line 2. The text actually says "*Wpt rnpt pw ḫb snnw*."

18. Bakir, *The Cairo Calendar*, p. 61.

19. *Bases, méthodes et résultats de la chronologie égyptienne* (Paris, 1926), pp. 112-26.

20. Parker, *Calendars*, pp. 34-36.

21. One piece of so-called evidence for the existence of the intercalary month cited by Parker (*Calendars*, p. 43) is the appearance of the Thoth symbol of a cynocephalic baboon seated on a djed column in the central dividing position of the penultimate lower register of the Ramesseum astronomical ceiling (Fig. III.2). Now there are thirteen divisions of the ceiling, twelve of which are assigned to months on the second upper register; the thirteenth division in the middle of that register contains no month designation but is empty. And since that division separates the twelve months of a year (a year which Parker by comparing it with the year given on the Senmut ceiling asserts is the old lunar year) Parker concludes that the blank division is that of the intercalary month whose name Thoth is confirmed by the Thoth symbol in the lower register. But, as I shall remark later, the month and season designations in the second register are simply those of the civil calendar and furthermore there is another explanation for the appearance of the Thoth symbol in its central position in the lower register: as in its appearance over the spouts of water clocks, it merely symbolizes Thoth's role as the inventor and patron of the determination of time.

22. For the ordinary Thoth-Feast on I Akhet 19, also see S. Schott, *Altägyptische Festdaten*, in *Akademie der Wissenschaften und der Literatur in Mainz, Abhandlungen der Geistes- und Sozialwissenschaftlichen Klasse*, Jahrgang 1950, Nr. 10, p. 962.

23. Gardiner, "The Problem of the Month-Names," p. 23.

24. *Ibid.*, pp. 27-28.

25. Schott, *Altägyptische Festdaten*, pp. 890-91, 959-60. See also Bakir, ed., *The Cairo Calendar*, p. 11, where for I Akhet 1 of the incomplete calendar we read: "Day 1: Second Feast Being the Opening of the Year...Feast of Osiris; Feast of Isis; Feast of every god; Feast of Sobek...." The author of this calendar may well be confusing the feasts of the epagomenal days (all five of these days being considered as a kind of extended New Year's Eve; see Schott, pp. 886-87) with the ordinary feast or feasts of New Year's Day. Note that in the second calendar of the Cairo document, we find on I Akhet 1 (*ibid.*, p. 13): "The Birth of Rē'-Ḥarakhte; ablution throughout the entire land in the water of the beginning of the High Nile which comes forth as fresh Nun." This latter statement

may well reflect the very ancient view that the year was to begin with the sudden rising of the Nile. In the many references given for I Akhet 1 by Schott, he cites Hatshepsut's remark (*Urkunden* IV 261) that her father "knew that a crowning on New Year's Day was good (i.e., favorable)," thus lending some support to Gardiner's idea that New Year's Day was considered an appropriate day for the accession of the king, and as such produced a feast distinguishable from other feasts on that day. For the connection of Thoth with the reckoning of time, see the collection of his epithets that I have assembled in Volume One, p. 304.

26. K. Sethe, "Die Zeitrechnung der alten Ägypter im Verhältnis zu der der andern Völker," *Nachrichten von der Königlichen Gesellschaft der Wissenschaften zu Göttingen, Philologisch-historische Klasse aus dem Jahre 1920* (Berlin, 1920), pp. 130-38.

27. *Calendars*, p. 13.

28. O. Neugebauer and A. Volten, "Untersuchungen zur antiken Astronomie IV," *Quellen und Studien zur Geschichte der Mathematik, Astronomie, und Physik,* Abt. B, Vol. 4 (1938), pp. 401-02.

29. Parker, *Calendars*, pp. 15-16.

30. *Ibid.*, pp. 17-22. He concludes (p. 22) by saying that "Of the entire seventeen calculations, nine [indicate] months [that] start on the morning of old crescent invisibility, five on the day of new crescent visibility, two on the day in between, and one on the day before the morning of invisibility. This is precisely the sort of irregularity that one would expect from a cycle scheme, rigidly adhered to, in contrast to a method of strict observation for starting the month....In my mind there is not the slightest doubt that the cyclic calendar was in use during the period covered by the lunar dates."

31. O. Neugebauer in his later work *A History of Ancient Mathematical Astronomy*, Part 2 (Berlin, Heidelberg, and New York, 1975), pp. 563-64, accepts both the view of the reckoning of lunar months from the time of invisibility and Parker's general treatment and dating of the 25-year cycle. He does, however, add the following statement: "As we have remarked before, the dates furnished by the cycle agree with the beginnings of the Egyptian lunar months in the fourth century B.C. Since dates provided by the cycle and actual lunar dates will only agree in the mean and since the differences between cycle dates and facts will vary only

very slowly, one cannot exclude a date of origin of the cycle, say, in the fifth century B.C. If this date were correct it would constitute a curious parallel to the contemporary development of mathematical astronomy in Mesopotamia. In neither case is there the slightest indication of Greek or any other foreign influence. Nevertheless it looks as if the creation of the Persian empire stimulated intellectual life everywhere in the ancient world of which the Hellenistic world was to become the heir." Regarding the beginning of the lunar month with crescent invisibility he says (p. 563, n. 3) that it "is in all probability caused by the Egyptian reckoning of the day from sunrise, a procedure which in itself is most natural and does not require any astronomical motivation."

32. *Calendars*, pp. 24-26.

33. *Ibid.*, p. 26.

34. *Ibid.*, pp. 27, 29.

35. Parker, *ibid.*, pp. 27-29, rejects the schemes presented by L. Borchardt and G. Wheeler. W. Barta, "Die ägyptischen Mondaten und der 25-Jahr-Zyklus des Papyrus Carlsberg 9," *ZÄS*, Vol. 106 (1979), p. 10 (full article, pp. 1-10) essentially agrees with Parker and doubts E. Hornung's opinion that the 25-year cycle might have already been known in the 18th dynasty.

36. Ed. of K. Sethe, *Die altägyptischen Pyramidentexte*, Vol. 2 (Leipzig, 1910), Utterance 669, Sect. 1961b-c, p. 472. See also references to the epagomenal days in the Introduction to Doc. III.1 nn. 2-3, and to the document itself, Mid. Kingd., sect. I, from the tomb of Khnumhoptep at Beni Hasan and that of Amenemhet at Thebes.

37. *Ägyptische Chronologie* (Berlin, 1904); *Nachträge zur Ägyptischen Chronologie* (Berlin, 1908).

38. *De die natali*, cc. 18, 21. See Document III.10, Sect, 6, for a translation of the Latin text of Censorinus' discussions of the Sothic year. There has been a great deal of discussion of these passages, and I refer the reader to R. Lepsius, *Die Chronologie der Aegypter. Einleitung der erster Theil. Kritik der Quellen* (Berlin, 1849), pp. 167-71; Meyer, *Ägyptische Chronologie*, pp. 23-29; L. Borchardt, *Quellen und Forschungen zur Zeitbestimmung der ägyptischen Geschichte*, Vol. 1, *Die Annalen und die zeitliche Festlegung des alten Reiches der ägyptischen Geschichte*, p. 55; and to Weill, *Chronologie égyptienne*, pp. 9-15, 59-60, 65-66.

39. I say "possible quadrennial dates" since the fixed Sothic year is 365 1/4 days and hence the rising of Sirius marking the

beginning of a Sothic period (and particularly of the first Sothic period when the civil calendar was made dependent on the rising) could have taken place on any one of the four years when it fell on the first day of the new civil year. The quadrennium is usually called "tetraeteris" from its Greek designation.

40. Meyer, *Ägyptische Chronologie*, pp. 42-44.

41. "Die Bedeutungslosigkeit der 'Sothisperiode' für die älteste ägyptische Chronologie," *Acta orientalia*, Vol. 17 (1938), pp. 169-95, and "The Origin of the Egyptian Calendar," *Journal of Near Eastern Studies*, Vol. 1 (1942), pp. 396-403. Both articles were reprinted in O. Neugebauer, *Astronomy and History, Selected Essays*, (New York/Berlin/Heidelberg/Tokyo, 1983).

42. "The Origin of the Egyptian Calendar," pp. 396-97.

43. Neugebauer's own views of the introduction of the civil calendar are epitomized as follows (*ibid.*, n. 3, pp. 397-98): "The old story about the 'creation' of the Egyptian calendar in 4231 B.C. can now be considered as definitely liquidated. An objection has been raised against my theory of a 'Nile-year' resulting from averaging the strongly fluctuating intervals between the inundations. This objection is that there is no proof of the existence of 'Nilometers' at so early a period....However no precise Nilometer is required for my theory. The sole requirement is that somebody recorded the date when the Nile was clearly rising. As a matter of fact, *every* phenomenon which occurs only once a year leads to the same average, no matter how inaccurately the date of the phenomenon might be defined. The averaging process of a few years will automatically eliminate all individual fluctuations and inaccuracies and result in a year of 365 days. Fractions, however, would be obtained only by much more extensive recording and by accurate calculation. The actual averaging must, however, be imagined as a very simple process based on the primitive counting methods as reflected in the Egyptian number signs: the elapse of one, two, or three days recorded by one, two, or three strokes. After ten strokes are accumulated, they are replaced by a ten-sign, thereafter ten ten-signs by a hundred symbol, etc. This is the well-known method of all Egyptian calculations. This method finally reduces the process of averaging to the equal distribution of the few marks which are beyond, say, three hundred-signs and five ten-signs; in other words, there is no 'calculation' at all involved in determining the average length of the Nile-years. Of course, we need not even assume the process of counting all the single days

every year; the averaging of the excess number of days over any interval of constant length (say twelve lunar months) gives the same result. This equal distribution of counting-marks finally makes it clear that no fractions will be the result of the process....(pp. 401-02, n. 17). When I reviewed the content of this paper at the meeting of the American Oriental Society in Boston, Professor H. Frankfort asked whether the institution of the schematic calendar could be assumed to belong to the reign of Djoser. I think that no serious objection can be raised against such an assumption, because the only condition for the creation of the schematic calendar is a sufficiently well-organized and developed economic life. On the other hand, means to determine such a date by *astronomical* considerations do not exist. The problem of the invention of the schematic months must not be confused with the problem of the period in which the 365-day year was introduced. The two institutions are absolutely independent — at least in principle. The 365-day year must have been created at a period when the inundation coincided roughly with the season called 'inundation'. Such a coincidence held for the centuries around 4200 and again for the centuries around 2800. The latter date (i.e., the time of Djoser) has been considered by Winlock...as the date of the definite establishment of the Egyptian year." At this point Neugebauer refers to the paper of A. Scharff, *Historische Zeitschrift*, Vol. 161 (1939), pp. 3-32, which argues against the earlier date.

44. *Calendars*, Chap. IV.

45. *Calendars*, p. 53.

46. These possible dates for the inauguration of the schematic year are based on the coincidence of the heliacal rising of Sirius and the New Year's Day in 2773 B.C. For it is evident that if the New Year's Day of the civil year advanced 41 days (i.e., 11 + 30 days) it will have begun that advancement 164 (i.e., 41 x 4) calendar years earlier, that is in 2937 B.C. But if it advanced 12 days, it began that advancement 48 (i.e., 12 x 4) years earlier, that is in 2821 B.C. But all of this rests on his suggested mechanism of intercalation of a thirteen month, i.e., such a month was added when Sirius rose within the last 11 days of the twelfth month. Parker finds support for his views by analyzing the 59 deities that occur first on the base of a relatively late statue of Mut of the 22nd dynasty (*ibid.*, 54-56). He suggests that the list reveals the dual character of the concept of "year." R.K. Krauss, whose

dissertation is mentioned below in note 49 and discussed in Document III.10 (Intro. n. 1 and Sect. I of the Document), on pp. 24-29 of that dissertation discusses Parker's theory of the origin of the 365-day calendar and, on p. 28, finds it "contradictory and not practicable."

47. P.V. Neugebauer, *Astronomische Chronologie*, Vol. 1 (Berlin, 1929), 149-55, 160-62; P.V. Neugebauer and L. Borchardt, "Beobachtungen des Frühaufgangs des Sirius in Ägypten im Jahre 1926," *Orientalistische Literaturzeitung*, Vol. 30 (1927), cc. 441-48. See also the brief, but neat summary in E. Hornung, *Untersuchungen zur Chronologie und Geschichte des Neuen Reiches* (Wiesbaden, 1964), pp. 17-21. Consult also Weill, *Bases, méthodes et résultats de la chronologie égyptienne*, pp. 189-204.

48. W.F. Edgerton, "Chronology of the Twelfth Dynasty," *JNES*, Vol. 1 (1942), pp. 308-09 (full article, pp. 307-14).

49. W.C. Hayes, "Chapter VI. Chronology," *The Cambridge Ancient History*, Vol. 1, Part 1 (Cambridge, 1970), p. 182 says "The date in question [i.e., the Sothic date] is preserved for us in a temple papyrus from El-Lāhūn and can with great probability be pinpointed to 1872 B.C. by reference to four lunar dates contained in documents from the same archive." As I have said, it was Parker, *Calendars*, Excursus C, p. 66, who used the lunar dates to fix the Sothic date as July 17, 1872 B.C. (Julian Year). Parker's conclusions have been disputed by R.K. Krauss, *Probleme des altägyptischen Kalenders und der Chonologie des mittleren und neuen Reiches in Ägypten* (Diss. Berlin, 1981), who used six lunar references and proposed 1836 as the year of the rising, having assumed that Elephantine was the site of the observation. For a summary of the arguments of both authors, see Document III.10, note 3. A method of deciding between Parker and Krauss, which rests on original proposals by L. Borchardt, and which (without my yet being able to examine the evidence) I cannot properly evaluate, has been proposed by R.A. Wells, "Some Astronomical Reflections on Parker's Contributions to Egyptian Chronology," in L.H. Lesko, ed., *Egyptological Studies in Honor of Richard A. Parker* (Hanover and London, 1986), p. 170 (full article, pp. 165-71): "Nevertheless there is a potentially very important astronomical means available which may not only assist scholars in choosing between Parker's or Krauss's dates but also provide a firmer network of dates for a number of other dynastic periods. I refer to the possibility of computing the foundation dates of certain temples, whose principal

axes may have been aligned toward specific rising or setting stars, by using the equations of precession which account for motion of the earth's axis about the North Celestial Pole." Wells notes the title of his preliminary investigation of the suggested method.

Very strong support for the use of star risings or transits with exactly measured temple alignments to determine foundation dates is found in the more recent *Studien zur ägyptischen Astronomie* (Wiesbaden, 1989) of Christian Leitz, who uses this technique and other astronomical considerations independent of the conventional employment of (1) the heliacal rising of Sirius on given days of the civil year in various reigns and (2) known lunar and civil-year double dates, that is, the main techniques I describe here in this volume. I note that Leitz believes that he has shown that the recent efforts of Krauss and others to locate the place of observation and thus the calendar reference point at Aswan are incorrect (p. VII): "Die vorliegende Arbeit bringt 8 voneinander unabhängige astronomische Argumente für Unterägypten, sodass in den Augen des Verfassers das Problem des Kalenderbezugspunktes als gelöst erscheint —nicht notwendigerweise aber das der absoluten Chronologie Altägyptens." My first reaction is that I cannot agree with all of his various assumptions and analyses of pertinent texts, but I should note that I had completed my volume when I finally obtained a copy of Leitz's book. Hence to be fair, I must leave my estimate of the soundness of his work to a future date. I do note, however, that my late colleague Otto Neugebauer vehemently rejected (for the most part) the value of temple alignment data for these discussions.

50. "On the Chronology of the Early Eighteenth Dynasty (Amenhotep I to Thutmose III)," *The American Journal of Semitic Languages and Literatures*, Vol. 53 (1936), pp. 192-93 (full article, pp. 188-97).

51. *Untersuchungen zur Chronologie*, p. 21.

52. R. Lepsius, *Die Chronologie der Aegypter* (Berlin, 1849), p. 151-56, 165-67; H Brugsch, *Matériaux pour servir à la construction du calendrier des anciens égyptiens* (Leipzig, 1964), pp. 28-29. Lepsius argues from Horapollo's reference to the Egyptians' "year of the four years," which he (Lepsius) takes to be the quadrennium of three years of 365 days and one of 366, and from the references in the Beni Hasan calendar to the festival of New Year's Day, to that of the "Beginning of the Sun Year" (his translation of *wp rnpt*), and to the festivals of the great and small years (see Doc.

III.1 below), that the Egyptians knew of the quadrennium of intercalary days and thus had side by side the civil year of 365 days and the Sothic year of 365 1/4 days.

53. Edgerton, "Chronology of the Early Eighteenth Dynasty," pp. 191-92. For the pertinent article by Lepsius, see the Introduction to Doc. III.2, note 4.

54. Edgerton, *ibid.*, pp. 190-91.

55. Parker, *Calendars*, pp. 38-42.

56. If we study Fig. III.10 we see under the glyph [glyph] in [glyph] enough space to suggest that that determinative [glyph] was to be understood in the next line and indeed in most lines that succeed it. Note that in the list of Fig. III.6a we see mention of the Feast of Re or Re-Harakhti, celebrated on the first day of the first month rather than as the first day of the twelfth month Mesore. This of course becomes equivalent to having the Festival of *wp rnpt* as the first day of the first month rather than as the first day of the twelfth month when that month was known as *wp rnpt*. It is the latter feast day which is present in the Ebers Calendar. See Weill, *Chronologie égyptienne*, pp. 114-15.

57. *Chronologie égyptienne*, pp. 8, 127-33, 145-58, 175.

58. Brugsch, *Thesaurus inscriptionum*, p. 363a (=my Fig. III.93); cf. R.A. Parker, "Sothic Dates and Calendar 'Adjustment'," *Revue d'Égyptologie*, Vol. 9 (1952), p. 103.

59. As in Volume One I use the text of K. Sethe, *Die altägyptischen Pyramidentexte*, Vol. 1 (Leipzig, 1908), pp. 145, 264.

60. See K. Sethe, *Übersetzung und Kommentar zu den altägyptischen Pyramidentexten* (Hamburg, 1962), Vol. 1, p. 276; Vol. 2, pp. 383-84.

61. While the names of the decans of these Egyptian star clocks provided the source for the decans found in Hellenistic zodiacs, the latter no longer refer to the same temporal decans but simply refer to divisions dividing the zodiac. I shall discuss the relation of the Egyptian decans with zodiacs below when discussing the later astronomical monuments.

62. Parker, "Ancient Egyptian Astronomy," pp. 53-54.

63. *Ibid.*, p. 55. For a detailed and lucid discussion of the astronomy that lies behind the diagonal star clocks, see Neugebauer and Parker, *Egyptian Astronomical Texts*, Vol. 1, Chap. 3, and the somewhat earlier account by Neugebauer, "The Egyptian 'Decans'," *Vistas in Astromomy*, Vol. 1 (1955), pp. 47-51.

64. Neugebauer and Parker, *Egyptian Astronomical Texts*, Vol. I, pp. 29-32. The authors treat each of the twelve coffins in detail, noting the name and titles of the deceased, the date, the provenance, present location, previous publication and discussion, schematic diagram of the arrangement of the decades, and critical apparatus. Plates are given at the end of the volume. In my account here I have ignored the decan list from the ceiling of the tomb of Senmut (which will be treated later in the chapter) and the textual fragment from the Cosmology of Seti I and Ramesses IV (*ibid.*, pp. 22-23 and 32-35). For extracts from the latter, see Document III.12 below.

65. *Ibid.*, p. 5.

66. Neugebauer and Parker, *Egyptian Astronomical Texts*, Vol. 1, pp. 2-3 and 23-26; S. Schott, "Die altägyptischen Dekane," in W. Gundel, *Dekane und Dekansternbilder: Ein Beitrag zur Geschichte der Sternbilder der Kulturvölker* (Glückstadt und Hamburg, 1936), pp. 1-21. Neugebauer, before his death, told me that he had no confidence in the elaborate identifications of the decans proposed by R. Böker in his posthumous publication: "Über Namen und Identifizierung der ägyptischen Dekane," *Centaurus*, Vol. 27 (1984), pp. 189-217. Böker attempts to locate the 36 decans by considering the positions of the full moon at 10-day intervals and noting that the limits between decans were in many cases exactly preserved in Greek astronomy. But as I have said at the end of the chapter, the Egyptian decan-names as used in Greek astronomy had little conformity with the star and constellation decans used by the ancient Egyptian stargazers to mark the hours of the night, since by the time of their use by the Greeks the names were merely used to mark the 10-degree divisions of the zodiacal signs, and the ties of the decan names with the old decan stars used were surely not determinable with any exactitude. The remarks I quote from Neugebauer and Parker in the Introduction to Document III.17, text to which note 4 pertains, suggest further why exact astronomical locations cannot be found from the evidence we possess.

67. Neugebauer and Parker, *Egyptian Astronomical Texts*, Vol. 1, pp. 24-26.

68. Parker, "Ancient Egyptian Astronomy," p. 56.

69. *Ibid.*

70. Neugebauer and Parker, *Egyptian Astronomical Texts*, Vol. 3, p. 213, gives references to star observers or hour-watchers

in Dynasty 26 and in the Ptolemaic period. See also Document III.18 below, and note the earlier references for the Middle Kingdom in my first volume, Chap. I, n. 25. Consult also W. Helck, *Urkunden des ägyptischen Altertums: IV. Abt. Urkunden der 18. Dynastie*, Heft 19 (Berlin, 1957), p. 1603, line 4; Heft 21 (1958), p. 1799, lines 17-18, and p. 1800, lines 10-11.

71. L. Borchardt, "Ein altägyptisches astronomisches Instrument," *ZÄS*, Vol. 37 (1899), p. 11 (whole article, pp. 10-17).

72. For a detailed treatment of these clocks, consult Neugebauer and Parker, *Egyptian Astronomical Texts*, Vol. 2 (Providence, R. I., and London, 1964).

73. Neugebauer and Parker, *Egyptian Astronomical Texts*, Vol. 2, p. 9.

74. Parker, "Ancient Egyptian Astronomy," p. 58.

75. Parker, "Egyptian Astronomy, Astrology, and Calendrical Reckoning," p. 715.

76. Neugebauer and Parker, *Egyptian Astronomical Texts*, Vol. 2, p. 4.

77. See the discussion of the astronomical considerations that bear on the lengths of these hours and other aspects of the Ramesside star clock *ibid.*, pp. 9-18.

78. *Ibid.*, p. x.

79. "Egyptian Astronomy," *Janus*, Vol. 52 (1965), pp. 161-80 (but particularly, pp. 173-80). Bruins assumes a more highly sophisticated knowledge on the part of Egyptian astronomers of the relationship between solar and sidereal motions as they bear on "the equation of time" than seems likely from the nature of the textual remains of the clocks and of the few surviving rather confused and simplistic passages displaying the technical, astronomical knowledge of the ancient Egyptians. In terms of procedures Bruins suggests the use of the so-called fist-decan, that is the use of the knuckles of the extended fist as a sighting "instrument," a usage never mentioned, so far as I know, in the ancient Egyptian literature. His argument leads to the conclusion that the sightings of transits were not made by two astronomers sitting face-to-face, as is usually believed, but that only one astronomer was involved (pp. 173-74): "Once a 'meridian' for the transits has been fixed the fist-decan allows [one] to measure time differences of 4 min. from midnight. Two remarks are then to be made: firstly that [use of] the fist-decan invites [i.e., encourages the observer] to [assume] a division into three domains left and right from the mid-knuckle and

thus provides us with an interpretation of the origin of the *seven* [internal vertical] lines around [and including] the meridian [as a central line] in the coordinate system of the star clocks; and secondly that if such observations are made their interpretation can never be given correctly neglecting the time equation as *true* transits are observed with a high precision, an error of some minutes! Considerably more accurate observations can start [i.e., be undertaken when] carried out by *one* astronomer. Here we should remark that we completely disagree with Neugebauer and Parker [in their two-person theory]....It seems obvious to us that the observations were carried out by one observer, who, having at his disposal the North-South direction has only to stretch his arm forward in that direction to be able to measure about 1° along the equator! In our opinion the 'target figure' of the star clocks *is not an assistant* of the observing astronomer, but the astronomer *himself!* The painter depicted the seated astronomer and *what he sees* is, independently, drawn 'behind him' in the charts...A strong indication of the independent representation of the seated astronomer and 'what he sees' is contained in the fact, that in the description of the star clocks, in the chart the 'left' and the 'right' are just opposite to the 'left' and 'right' of the 'target figure'. The painter pictured the astronomer 'en face', then turned around and pictured what he saw." (I have added a few bracketed phrases to smooth the flow of the English.)

Surely this is a perverse theory to explain why the parts of the body were used to mark the transit lines, since with extended fist the observer has no accurate sense of where the lines cross his eye, or ear, or shoulder. And one cannot really use the non-perspective aspects of Egyptian representation to explain the contrary senses of right and left on the grid and the human body, for the Egyptian painter had no difficulty in correctly orienting figures in the same or parallel planes. Starting from his unsupported and implausible interpretation of the sighting procedure, Bruins goes on to explain how completely he disagrees with Neugebauer and Parker: "Where these authors assume crude procedures and ill-conceived schemes, which attempted to combine contradictory elements, we, on the contrary, conclude from the mere existence of the charts that time intervals of about a twelfth of a decan—as the stars are in the charts put *'upon'* and *'between'* the coordinate lines—some sidereal minutes were considered and quite a different problem or observation from combining star transits and seasonal hour lies at the base of the star clocks. That small parts of a decan were considered stands to reason: if the differences covered by the charts were more than a decan...the astronomer should have observed the star of the next or former decan! The observations are therefore undoubtedly meant to have an accuracy of some minutes of sidereal time. As we have already indicated, an assumption of a constant duration of twilight is identical with removing from observed facts the change in duration of the day and night in the automatically constant decanal hours. Moreover, true star positions need *to imply automatically the time equation* and as Neugebauer and Parker neglected this time equation completely their method of interpretation is completely ruined by their first assumption and even if this were not the case, [it] is again ruined by their neglecting the time equation."

80. See the Introduction to Document III.14 and Neugebauer and Parker, *Egyptian Astronomical Texts*, Vol. 2, pp. ix-x.

81. For a detailed description of both types, see L. Borchardt, *Die altägyptische Zeitmessung* (Berlin and Leipzig, 1920), pp. 6-26. To Borchardt's list of fragments of outflow clocks (see next note below) may be added those cited by Didier Devauchelle in W. Helck and W. Westendorf, eds., *Lexikon der Ägyptologie*, Vol. 6, (Wiesbaden, 1986), cc. 1156-57 ("Wasseruhr"). Cf. Neugebauer and Parker, *Egyptian Astronomical Texts*, Vol. 3, pp. 12, 42, and 60. See also R.W. Sloley, "Ancient Clepsydrae," *Ancient Egypt*, Vol. ix (1924), pp. 43-50, and by the same author, "Primitive Methods of Measuring Time with Special Reference to Egypt," *The Journal of Egyptian Archaeology*, Vol. 17 (1931), pp. 174-76 (full article, pp. 166-78). Incidentally, in Fig. III.21b, I have given considerably better photographs of the exterior, interior, and edge of water clock No. 938 at the British Museum in London (Borchardt's Auslaufuhr 6) than Borchardt's single photograph of the exterior. Note that Borchardt gives a drawing of those graduations of the interior surface of Auslaufuhr 6 that remain (=my Fig. III.21c). I have also added, in Fig. III.21d, exterior and interior views of a fragment of a second water clock (No. 933) at the British Museum.

82. Borchardt, *Die altägyptische Zeitmessung*, pp. 6-7. On pp. 7-10 Borchardt describes briefly fragments of 12 other water clocks. See also the list given by Devauchelle in the article mentioned in the preceding note. See A. Wiedemann, "Bronze Circles and Purification Vessels in Egyptian Temples," *PSBA*, Vol. 23 (June, 1901), pp. 270-74 (full article, pp. 263-74), where

fragments of inflow clocks are interpreted as purification vessels. See the descriptions and depictions of four of the already known fragments: A. Roullet, *The Egyptian and Egyptianizing Monuments of Imperial Rome* (Leiden, 1972), pp. 145-46, item nos. 326-328, 330, and figs. 334-36, 337-38, 339-42, 344. Incidentally, Dr. Karl-Heinz Priese, the Director of the Ägyptische Museum (Staatliche Museen zu Berlin) tells me that he has no remaining evidence of the clock with the inventory no. 19556, mentioned by Borchardt as "Auslaufer 11 (Berlin)" and by Roullet as her item 326.

83. Sloley, "Primitive Methods," p. 175.

84. Borchardt, *Die altägyptische Zeitmessung*, p. 6. See also note 21, above. The representation of Thoth in the center of the third register of Fig. III.2 shows the Thoth baboon surmounting a djed column, which latter has the meaning of "stability" or "endurance" and is part of the decoration of the bottom of the interior of the Karnak clock where it appears below every other monthly scale without having anything to do with the clock's purpose of telling time, except in a very general way. This then explains its appearance below the baboon in the Ramesseum ceiling shown in Fig. III.2. In Fig. III.23 depicting the inflow clock found at Edfu by G. Maspero in 1901 we see the Thoth baboon still attached to the vessel. Cf. the Thoth baboon present on a fragment of a water clock dating from Dynasty 26 (Neugebauer and Parker, *Egyptian Astronomical Texts*, Vol. 3, p. 43 and plate 22B), where the baboon appears in the same position in the bottom register. Similarly the three models of inflow clocks presented by A. Pogo, "Egyptian Water Clocks," *Isis*, Vol. 25 (1936), p. 416, Fig. 4, and plate 4 opposite p. 416 (full article pp. 403-425) (cf. my Fig. III.32), all have the Thoth baboon, as does the Cairo model depicted by Sloley, "Primitive Methods," plate XXI,1. In the *Hieroglyphica* of Horapollo, ed. of C. Leemans (Amsterdam, 1835), Bk. I, cap. 16, we are told that "in their water clocks the Egyptians carve a seated dog-faced baboon and they make water flow from his member." G. Daressy, "Deux clepsydres antiques," *Bulletin de L'Institut Égyptien*, 5me Série, Vol. 9 (l'Année 1915), p. 7, n. 1 (full article, pp. 5-16), states that "according to a general belief in antiquity, the cynocephalic baboon urinates at regular intervals twelve times a day." One would suppose that this merely means that in expelling water from the clock beneath the seated baboon, the baboon is effectively "urinating" the twelve hours of the day.

Horapollo, *Hieroglyphica*, I,16, says that the baboon "alone of all animals, during the equinoxes, cries out each hour twelve times a day." But the sundry references to the role of the baboon in announcing the hours do not limit the baboon's custodial role of the hours to the equinoxes.

85. Sloley, "Ancient Clepsydrae," p. 45.

86. *Ibid.* Sloley claims in a note [no. 5] to this passage that Horapollo (I,16) "states that a metal tube of narrow bore was fitted to the aperture." But it is evident that Horapollo is describing a device like a spigot to control the flow of water into an inflow clock rather than a spout fitted to the aperture of an outflow clock, as the editor of the text, C. Leemans, suggested in his description of a model of an inflow clock in Leiden (see Pogo, "Egyptian Water Clocks," p. 416). Hence I believe that George Boas' translation of the passage has it right (*The Hieroglyphics of Horapollo* [New York, 1950], p. 69): "In order that the water may not be too copious...a contrivance exists [to regulate it]. Through this the water is let into the clock, not too fine a stream, for there is need of both [fine and broad]. The broader makes the water flow swiftly and does not measure off the hours properly. And the thinner stream gives a flow which is too small and slow. [Hence] they arrange to loosen the duct by a hair's breadth and they prepare an iron plug for this use, according to the thickness of the stream [required]."

87. Sloley, "Primitive Methods," pp. 175-76.

88. Borchardt, *Die altägyptische Zeitmessung*, p. 12. Borchardt's table of measurements is given in millimeters and fingerbreadths, with approximate errors in both measures.

89. Sloley, "Ancient Clepsydrae," p. 47.

90. Borchardt, *op. cit.* in note 88, p. 13. This is the case for Borchardt's outflow clock no. 2 and probably for no. 3 and no. 4. Incidentally in clock no. 2 the aperture for the outflow is 3 fingerbreadths below the bottom of the 12-fingerbreadth scale.

91. The number [1] is a proper deduction from the numbers available from the fragments of those clocks. See Fig. III.26.

92. This fragment was first published by B.P. Grenfell and A.S. Hunt, *The Oxyrhynchus Papyri*, Part III (London, 1903), No. 470, pp. 141-46. It was examined in great detail by Borchardt, *op. cit.*, pp. 10-12, with some changes in transcription. Cf. Sloley, "Ancient Clepsydrae," p. 46. Borchardt, p. 14, suggests that the so-called "sacred cubits (ruler-like boards)" preserved in various

temples, which contain a table that may indicate monthly variations of shadow lengths for shadow clocks, also have a table that may express monthly volumetric changes in the water content of outflow water clocks as they empty, although he has no explanation for the specific meaning of the sums of unit fractions that constitute the table.

93. See Borchardt, *Die altägyptische Zeitmessung*, p. 11, where he gives the angles for the various clocks. Sloley simply gives the direct angle of inclination of the wall with the base, which is of course merely 180° minus the angle given by Borchardt, so that the best angle of a conical clock would be about 103°, while the actual angles of the extant clocks range between 109° and 112° (See Slolely, *Ancient Clepsydrae*, pp. 45-47).

94. Borchardt, *Die altägyptische Zeitmessung*, pp. 16-17, reviews the various definitions of day and night hours, assuming the following possibilities: (1) the daytime between sunrise and sunset, resulting in the "civil day," or (2) the nighttime hours between sunset and sunrise, resulting in the "civil night," or (3) the duration of daylight that includes morning and evening twilight, in his terminology the "astronomical day" or finally (4) the time of darkness after the subtraction of morning and evening twilight. As Borchardt says, theoretically we could measure any one of these four durations by a water clock. But actually examination of the clock fragments and their scales shows us that some are clearly for nighttime in some fashion, namely clocks numbered 1, 4, 7, and 8, which display the star decans in some form in the top exterior register or explicitly say that the clock is to be used at night. In addition, we can further deduce for the Karnak clock (no. 1) that the false position of the end of the sixth hour in the Karnak scale would immediately be evident if used in the daytime (i.e., the midday observation of noon would be off; see Fig. III.31). Borchardt summarizes his view concerning the Karnak clock as follows (p. 19): "Als sicheres Ergebnis dieser Untersuchung ist aber hervorgetreten, dass die Auslaufuhr I (Kairo) nicht für bürgerliche, sondern für astronomische Nächte **oder** Tage zu gebrauchen war. Dass sie nicht für astronomische Nächte **und** Tage zu verwenden war, ergibt die weitere Erwägung, dass sie dann, wie die Zusammenstellung zeigt, Tag und Nacht zusammen zur Zeit der Wintersonnenwende zu kurz (bei -12° um $0{,}25^h$ = 15'), zur Zeit der Sommersonnenwende zu lang (bei -12° um $0{,}25^h$ = 15') angegeben hätte, was zu einer unhaltbaren Verschiebung des Tages- und

Nachtanfangs, also der Zeitpunkte, an denen die Wasseruhr frisch zu füllen war, hätte führen müssen. Die Messung und Teilung des astronomischen Tages hat nun wenig Wert, wohl aber ist die der astronomischen Nacht wichtig. Wir können also von den beiden Möglichkeiten ruhig die eine fallen lassen und behaupten, dass die Auslaufuhr I (Kairo) nur für astronomische Nacht bestimmt war."

95. Borchardt, *Die altägyptische Zeitmessung*, pp. 19-21. The helpful table I have given as Fig. III.37 was taken from A. Pogo, "Egyptian Water Clocks," *Isis*, Vol. 25 (1936), p. 410. (Note that the full article occupies pages 403-425.)

96. The most complete treatment of Egyptian inflow clocks is that of Pogo mentioned in the preceding note. As we shall see, its most interesting section treats of the grid diagram found on the interior surface of the Edfu clock. A less complete account of inflow clocks is found in Borchardt's *Die altägyptische Zeitmessung*, pp. 22-26. Although they have been superseded, the original articles first describing the Edfu clock by G. Daressy are worth mentioning: "Grand Vase en pierre avec graduations," *Annales du Service des Antiquités*, Vol. 3 (1902), pp. 236-39, and "Deux clepsydres antiques," *Bulletin de l'Institut Égyptien*, 5me série, Vol. 9 (1915, Cairo, 1916), pp. 5-16. The second of these articles supersedes the first.

97. See the description by Daressy, "Deux clepsydres antiques," pp. 6-7: "Le vase est entier...il se compose d'un fût en calcaire dur sensiblement cylindrique de 0 m. 285 mill. de diamètre à la base et d'une hauteur de 0 m. 30 cent. Au-dessus une moulure fait le tour de la partie supérieure, et au sommet (0 m. 38 cent.) le diamètre est porté à 0 m. 345 mill. Intérieurement on a d'abord un évidement circulaire de 0 m. 05 cent. de hauteur et 0 m. 225 mill. de diamètre, puis la largeur du creux se réduit à 0 m. 169 mm. sur 0 m. 275 mill. de hauteur. Sur le côté, au niveau du fond, un trou est percé, traversant la paroi et légèrement incliné vers le bas; il a 0 m. 005 mill. de diamètre à l'intérieur mais s'élargit jusqu'à avoir 0 m. 01 cent. extérieurement. L'accroissement de diamètre n'est pas régulier: il augmente brusquement près de la sortie, comme si l'on avait dû fixer là un tube ou un autre appareil. Cet orifice est à 0 m. 043 mill. au-dessus de la base, et immédiatement au-dessus est sculpté en relief un cynocéphale assis de 0 m. 10 cent. de hauteur."

98. See Borchardt, *Die altägyptische Zeitmessung*, Tafel 10.

99. Pogo, "Egyptian Water Clocks," pp. 407-09, 411-12.

100. *Ibid.*, pp. 420-22. For another example of the

presentation of this clepsydra-like object, see H.H. Nelson (and edited by W.J. Murnane), *The Great Hypostyle Hall at Karnak*, Vol. 1, Part 1: *The Wall Reliefs* (Chicago, 1981), plate 191 (*The University of Chicago Oriental Institute Publications Volume 106*). Here Seti I (ca. 1306-1290 B.C.) presents the object (as in the relief of Amenhotep III called a *šbt*) to the goddess of magic Weret-Hekau.

101. *Wb*, Vol. 3, p. 100, reference 12; K. Sethe, *Urkunden des Alten Reichs* (Leipzig, 1933), p. 152, Doc. 41, line 2.

102. Didier Devauchelle, in the article "Wasseruhr," *Lexikon der Ägyptologie*, Vol. 6, c. 1156, suggests that the usual identification of the *šbt* with a water clock perhaps should be rejected, and he cites literature that tends to support this view. See particularly R.A. Caminos, *The New-Kingdom Temples of Buhen*, Vol. 2 (London, 1974), p. 82, n. 4, and the papers of C. Sambin-Nivet given below in the bibliography.

103. *Wb*, Vol. 3, p. 106 *passim*. The combined signs representing the votive offering, , appear as a determinative for *šbt* in the New Kingdom according to *Wb*, Vol. 4, 438, item 8-n. In the Middle Kingdom the determinative shows only Thoth as the cynocephalic baboon in a naos (*ibid.*, 8-m).

104. Pogo, "Egyptian Water Clocks," pp. 412-14, indicates, as an example of the probable influence of Egyptian clocks and hour determinations on classical authors, the formation of the duodecimal rule given by Cleomedes, who lived sometime between the end of the first century B.C. and the end of the first century A.D.: "Our prismatic diagram reproduced in Figure 1 [my Fig. III.34] accounts not only for the cylindrical diagram of the Edfu clock—it throws new light on several passages of classical literature dealing with the seasonal rate of increase and decrease of the length of the day and of the night. Cleomedes [κυκλικὴ θεωρία μετεώρων, I, 6]... notes that the difference between longest and shortest day is thus distributed on the six months following winter solstice:

during the 1st month, the day increases by 1/12 of the difference;

during the 2nd month, the day increases by 1/6 of the difference;

during the 3rd month, the day increases by 1/4 of the difference;

during the 4th month, the day increases by 1/4 of the

difference;

during the 5th month, the day increases by 1/6 of the difference;

during the 6th month, the day increases by 1/12 of the difference.

The numerators of these fractions [all being unity] immediately suggest an Egyptian origin of Cleomedes' rule. A glance at our diagrams...shows that these six fractions are the direct result of the 1:2:3 process of distributing the twelve monthly scales on the four inside walls of a square inflow clock; it was the intention of the Egyptian inventor of the 1:2:3 diagram to produce scales showing slower monthly variations near the solstices and faster ones near the equinoxes; the monthly rates, 1/12, 1/6, and 1/4, happen to be an inevitable consequence, for the ordinates of the diagram, of a simple and symmetrical geometrical construction based on the favorite Egyptian fractions, 1/2 and 1/3, applied to the abscissae....

...; both the Egyptian inventor of the 1:2:3 diagram for prismatic inflow clocks, and the Egyptian astronomer who first formulated the duodecimal rule concerning the monthly rate, were so close to the true solution of the problem that their achievements deserve a place of honor in the history of Egyptian science."

105. "Ancient Clepsydrae," pp. 49-50, and "Primitive Methods of Measuring Time," pp. 176-77.

106. Concerning the word *ṯnỉ* here rendered as "discern [*or* measure ?]," Borchardt, *Die altägyptische Zeitmessung*, p. 27, and n. 1, uses "zählt." This usually translates *ṯnw* ("numbers"), whereas the translation "discern" more often renders *ṯnỉ*. Borchardt was obviously looking for a meaning more precisely connected with the measuring of shadow lengths, i.e., where the shadow length grows continually shorter up to the time of noon. Indeed Borchardt goes on to suggest that the meaning of the passage is that because of an eclipse of the sun it was not possible to fix the time by means of shadow length numbers. The passage may however simply mean that the sun-god at the time of social upheaval willfully projected too weak a light to produce a shadow, say by producing overcast days. In either case it does assume that time telling was associated with shadow length.

Borchardt (*ibid.*, pp. 27-32) discusses two sets of monthly tables seemingly related to shadow lengths for the telling of time.

The first is contained on the "sacred cubits" or rules I mentioned above in note 92, the oldest specimen being from the time of one of the Osorkons (10th-9th centuries B.C.). It contains (when complete) three shadow lengths for each month expressed in cubits and palms; but why only the shadow lengths for three hours per month are given and how the shadows are produced and measured are not clear. The second is a table from the temple at Taifa in Northern Nubia that dates from late Roman times which has the Egyptian months in Greek letters and the shadow lengths for the various hours are expressed in "feet." Again we do not know how the shadow lengths are produced and for what location they were originally prepared and indeed whether they have any relation to ancient Egyptian techniques of telling time by shadow lengths. Borchardt makes an heroic but admittedly failing effort to explain these tables.

107. As I have already noted the word *mrḫyt* or its equivalent *mrḫt* was not only employed in connection with the sighting and leveling instrument used with a shadow clock (as it is here) but also as a term for the water clock of Amenemhet, and thus perhaps as a general term for any clock (see Document III.15, line 14).

108. See Frankfort, *The Cenotaph of Seti I*, Vol. 1, pp. 78-79.

109. Neugebauer and Parker, *Egyptian Astronomical Texts*, Vol. 1, p. 118.

110. Cf. Borchardt, *Die altägyptische Zeitmessung*, p. 32; see also his article "Altägyptische Sonnenuhren," *ZÄS*, Vol. 48 (1910), pp. 9-17.

111. *Die altägyptische Zeitmessung*, pp. 32-36.

112. See also H. Brugsch, *Thesaurus inscriptionum aegyptiacarum*, I. Abtheilung (reprint, Graz, Austria, 1968), p. 31, for a comparative table of hour names and their eponymous goddesses from Dendera. Cf. J. Dümichen, "Namen und Eintheilung der Stunden bei den alten Ägyptern," *ZÄS*, Vol. 3 (1865), pp. 1-4.

113. *Die altägyptische Zeitmessung*, p. 36.

114. *Ibid.*, pp. 37-43. The clock was first published by C.C. Edgar, *Sculptors' Studies and Unfinished Works: Catalogue général des antiquités égyptiennes du Musée du Caire, Nos. 33301-33506* (Cairo, 1906), p. 51 and plate 21. It was correctly described physically by Edgar, but he was unable to identify it as a clock model.

115. *Die altägyptische Zeitmessung*, pp. 43-47.

116. *Die altägyptische Zeitmessung*, pp. 45-46. W.M.F. Petrie, *Ancient Weights and Measures*, p. 45, mentions the Paris clock as well as the London clock. Concerning the latter he says: "Below this copy [of the Paris scales on Plate xxvi] is a dial [i.e., a shadow clock] cut in black steatite, the full inscriptions on which are copied in pl. xxv. It was made for Sennu, who held many priesthoods, but the inscription does not relate to the dial itself. At the lower point a mass has been broken off which rose up, doubtless to carry the edge which was to cast the shadow on the slope....The graduation is not exact and the latitude cannot be deduced from the maximum readings. When the dial was moved about, it was provided with a plumb bob, hanging down the projection which is now lost: this enabled the dial to be set upright."

117. This dial was first described by R.A.S. Macalister, *The Excavation of Gezer 1902-1905 and 1907-1909*, Vol. 1 (London, 1912), p. 15; Vol. 2 (London, 1912), p. 331, but he merely called it an ivory pectoral and did not realize its purpose as a sundial. The best description of it is by Borchardt, *Die altägyptische Zeitmessung*, p. 48.

118. *Ibid.* His description and analysis of the Berlin sundial occupy pp. 48-50. See also his earlier description of the clock, "Eine Reisesonnenuhr aus Ägypten," *ZÄS*, Vol. 49 (1911), pp. 66-68, where he suggested that this type of sundial was of Greek origin, an opinion he abandoned in his *Zeitmessung* in consideration of the Cairo example discovered in Palestine, which bears the cartouche of Merenptah (see n. 117, and the text above it). It is worth noting that the construction of vertical-plane sundials of the Egyptian type (the so-called "protractor" type) continued among the Greeks and Romans (see S. L. Gibbs, *Greek and Roman Sundials* [New Haven and London, 1976], pp. 45-46, 348, 354, 356, 360, 366). There is no trace among the early Egyptians of the far more sophisticated Greek and Roman planar, spherical, and conical dials described and catalogued by Gibbs.

119. The table was first described by J. Černý, "The Origin of the Name of the Month Tybi," *ASAE*, Vol. 43 (1943), pp. 179-81; the full article covers pp. 173-81. The whole papyrus was edited by A.M. Bakir, *The Cairo Calendar No. 86637* (Cairo, 1966). For the table, see pp. 54 and Plates XLIV and XLIVA.

120. Neugebauer and Parker, *Egyptian Astronomical Texts*, Vol. 1, p. 119.

121. I have indicated the main errors and omissions shown in

Figs. 58a-b. As for the month-names at the end of each line (but the first), I have not attempted to specify what elements of the month-words and their determinatives are missing but have simply used the conventional spellings. In this translation I have in some places, for typographical reasons, substituted parentheses for square brackets. But it should be clear to the reader from the diagrams of the text what parts are missing and what parts are not readable.

122. Neugebauer and Parker, *Egyptian Astronomical Texts*, Vol. 1, p. 120.

123. *Ibid.* For the earlier remarks of J.J. Clère, see "Un texte astronomique de Tanis," *Kêmi*, Vol. 10 (1949), p. 10 (whole article pp. 3-27).

124. Clère, *op. cit.* in note 123, p. 3.

125. Neugebauer and Parker, *Egyptian Astronomical Texts*, Vol. 3, p. 44.

126. The figure *ibid.*, Plate 23, adds columns 11 and 12 and marks them as [Text] III. A brief treatment of these columns is given by Clère, *op. cit.* in note 123, pp. 20-21, and by Neugebauer and Parker, *Egyptian Astronomical Texts*, Vol. 3, p. 47.

127. Compare the translations of Clère, *op. cit.* in note 123, p. 9, and Neugebauer and Parker, *Egyptian Astronomical Texts*, Vol. 3, pp. 45-46. Again, as in the earlier table, I have used, for typographical reasons, parentheses instead of square brackets in my translations. The reader can easily see from the figures which readings are difficult to read and which are simply missing from the fragment. Clère's clear discussion (pp. 11-18) of the terminology used in the table is helpful. He shows the essential equivalence of *mtrt*, which I have rendered as "daylight" though it commonly means "noon," with *hrw*, the common word for "day" used in the Ramesside table above, which, incidentally, I have also translated as "daylight."

128. J.J. Clère, "Un texte astronomique de Tanis," pp. 10-11.

129. *Ibid.*, pp. 16-17.

130. *Egyptian Astronomical Texts*, Vol. 3, pp. 46-47.

131. *Ibid.*, Vol. 3, pp. 1-104.

132. *Ibid.*, pp. 4-5.

133. Brugsch, *Thesaurus inscriptionum aegyptiacarum*, 1. Abt., pp. 52-53, relates the names of the feasts of the protective gods of the days of lunar months and the pictorial representations of these day-gods at the time of Dynasty 19. In Brugsch's chart on page 53

of his volume, the Roman numerals show which days of the month the deities belong to. See also Document III.3 (following the section on the Northern Constellations) and notes to Document III.6 below for the earlier references to some of these protective gods on the ceiling of the tomb of Senmut and other astronomical ceilings.

134. I have omitted any discussion of the earlier loose boards from the coffin of Heny, which probably date from Dynasty 11 and thus suggest that the standard form of the astronomical diagram originated on coffins in the 11th dynasty, or before. For their discovery in 1922, see G.A. Wainwright, "A Subsidiary Burial in Ḥap-Zefi's Tomb at Assiut," *ASAE*, Vol. 26 (1926), pp. 160-66, and for the astronomical inscriptions see B. Gunn, "The Coffins of Ḥeny," *ibid.*, pp. 166-71. See also A. Pogo, "The Astronomical Inscriptions on the Coffins of Ḥeny (XIth dynasty?)," *Isis*, Vol. 18 (1932), pp. 7-13, and the brief treatment in Neugebauer and Parker, *Egyptian Astronomical Texts*, Vol. 3, pp. 8-10. The pertinent astronomical inscriptions are on board nos. 1 (inside, see Fig. III.62a) and 4 (inside, Fig. III.62b). Board no. 1 has the remains of the Northern Constellations arranged about the Big Dipper (with some of the flanking day-deities) and board No. 4, we are told by Gunn (p. 169, para. 4, and see Fig. III.62b), had "traces of three registers divided up by vertical lines: in the uppermost, names of stars or constellations...; in the second register, illegible signs in red; in the lowest, varying numbers of stars in blue." In the visible part of board no. 4, we can identify the signs [hieroglyphs] for Sepedet *(Spdt)* or Sothis (=Sirius). The signs before this name ought to have some relationship to the decans made up from Orion (but this cannot be confirmed), while those after Sirius appear to be parts of the name of the planet Jupiter. Unfortunately the boards appear to have been reburied after their inscriptions were transcribed by Gunn and hence are not now available.

135. Aside from the description in Neugebauer and Parker (*ibid.*, pp. 10-12, and passim throughout the volume), we should cite the most important of the earlier discussions, namely that of A. Pogo, "The Astronomical Ceiling-decoration in the Tomb of Senmut (XVIIIth Dynasty)," *Isis*, Vol. 14 (1930), pp. 301-25.

136. H.E. Winlock, "The Egyptian Expedition 1925-1927," *Section II of the Bulletin of the Metropolitan Museum of Arts* (February, 1928), pp. 32-44, full article pp. 3-58.

137. Neugebauer and Parker, *Egyptian Astronomical Texts*, Vol. 3, pp. 11, 105-06.

138. *Ibid.*, pp. 106-18.

139. Pogo, "The Astronomical Ceiling-decoration in the Tomb of Senmut (XVIIIth Dynasty)," pp. 309-10. Though there has long been no doubt about the identity of the bull, or primitively the bull's foreleg, with the Great Bear, the summary of the evidence for the identification is neatly presented by C.A. Wainwright, "A Pair of Constellations," *Studies Presented to F.LL. Griffith* (London / Oxford, 1932), pp. 373-75 (full article, 373-82). As Wainwright notes, the alternative, early Egyptian representation for the constellation is as an adze (which is named, like the constellation, *msḫtyw*), the instrument used for the ceremony of the Opening of the Mouths of the dead and of the statues of gods, and indeed the constellation looks somewhat like such an instrument ( ) when it is high up in the sky above the pole. In fact, when the constellation is mentioned in the *Pyramid Texts*, Sect. 458c, it is determined by an adze as well as a star. After a rather long period in which the Big Dipper was depicted as some form of a bull, in the Ptolemaic period it again began to be depicted as the Foreleg. Interestingly enough, not long before the Ptolemaic period, that is in the time of Nectanebo II (360-43 B.C.), a bull's sarcophagus from Abu Yasin has two parallel strips (i.e., two registers) that purport to show the positions of the dipper (depicted as a foreleg) for the beginning, middle and end of the night on the first night of each month of the civil year (see Fig. III.74). As Neugebauer and Parker point out (*Egyptian Astronomical Texts*, Vol. 3, p. 51), the strips are astronomically useless since they do not show the proper steady rotation of the dipper: "Obviously the positions in each triple should represent a rotation about the pole...of approximately 90° between the first and the second and the same between the second and the third. Instead we find identical positions, e.g., in the first [top] strip fields 2 and 3; 4 and 5; 10, 11 and 12; 14 and 15; 16 and 17; and in the second [lower] strip fields 1 and 2; 3 and 4; 6 and 7; 10, 11, and 12; 14 and 15. Similarly the evening positions in months I and II of *ꜣḫt* cannot be almost 180° different, etc. Since each field in the second strip should represent a situation six months later than in the field above it, one should find positions about 180° different from one another. In fact this is only three times the case (fields 7, 13 and 17). Hence the archetype of our text cannot be dated astronomically and it might antedate the fourth century B.C. by a considerable interval."

140. Borchardt, "Ein altägyptisches astronomisches

Instrument," pp. 12-13 mentions inscriptions from Ptolemaic temples that include references to observations of the stars (and particularly to the Great Bear) during the ceremonies of stretching the chord: "Spannen der Schnur im Tempel zwischen den beiden Fluchtstäben. Zu opfern eine Gans....Zu sprechen: Ich fasse den Fluchstab, packe das ende des Schlägels und ergreife die Schnur zusammen mit der Weisheitsgöttin [i.e., Seshat; see Volume One, Fig. I.31c and the present volume, Figs. III.73a-b]. Ich wende mein Gesicht nach dem Gange der Sterne. Ich richte meine Augen nach dem Grossen Bären*. Der...steht neben (?) seinen Zeiger (...[*mrḫt*]...). Ich lege die vier Ecken deines Tempels fest."

*Borchardt has "kleinen Bären" but the text has *msḫtyw*, i.e., the Great Bear. By the time of the Ptolemaic period it could well be that the Little Bear was used in observations but not in quotation of the traditional text (cf. Pogo, "The Astronomical Ceiling-decoration," p. 310, n. 25). Hence there can be little doubt that in the early dynasties the Great Bear was used in both observations and text. So the traditional earlier statement of the king ("Zu sprechen...." in the translation above) was continued on into the Ptolemaic period. This traditional text was treated earlier by Brugsch, *Thesaurus inscriptionum aegyptiacarum*, 1. Abth., pp. 84-85.

141. Though the treatment of the northern constellations by Neugebauer and Parker is by far the most complete, the reader should realize that Pogo's discussion in "The Astronomical Ceiling-decoration in the Tomb of Senmut (XVIIIth Dynasty)," pp. 308-312, was a fundamental step forward. In addition, attention ought to be called to the lengthy, early treatment by Brugsch, *Thesaurus inscriptionum aegyptiacarum*, 1. Abth., pp. 121-31, who published a number of drawings of the northern constellations from the different monuments and highlighted their differences.

142. Pogo, "The Astronomical Ceiling-decoration in the Tomb of Senmut (XVIIIth Dynasty)," p. 311.

143. Wainwright, "A Pair of Constellations," pp. 375-79. L.S. Bull, "An Ancient Egyptian Astronomical Ceiling-decoration," *Bulletin of the Metropolitan Museum of Art*, Vol. 18 (1923), p. 286 (full article, pp. 283-86), says, concerning the bull and the god or man he calls "the personage grasping the 'reins'" between the bull and the head of the Hippopotamus in the depiction of the northern constellations in Seti I's tomb (see Fig. III.69): "A glance at the 'stars' on the figure of the bull and of the personage grasping the

'reins' shows a strong resemblance between their relative positions and those of the familiar stars which form the Great Bear." He adds that he is indebted to George Ellery Hale, Director of the Mount Wilson Observatory, for this observation. Incidentally, the reins holder perhaps arose from the earlier depiction of Serqet in the family of diagrams represented in Senmut's tomb, though in the Seti I diagram the figure of Serqet also appears with her name on the other side of the bull, while the reins holder is an entirely distinct figure that has no name.

144. Neugebauer and Parker, *Egyptian Astronomical Texts*, Vol. 3, p. 183, n. 2.

145. H. Chatley, "Egyptian Astronomy," *The Journal of Egyptian Archaeology*, Vol. 26 (1940), p. 123 (full article, pp. 120-26).

146. R.A. Biegel, *Zur Astrognosie der alten Ägypter* (Zurich, 1921). She declares (pp. 15-16, referring to her figures 6a-b and 7 appended after p. 36; cf. my Figs. III.71a-b and III.72): "Sämtliche helleren Sterne sind berücksichtigt worden, dazu fast alle schwächeren Sterne im Bereich der Zeichnung. Wir sehen, dass das Siebengestirn des Grossen Bären zerlegt wird in das Viereck *αβγδ* und den Teil *εζη*. Die Figur 6 [= my Figs. III.71a-b] kommt auf der Sternkarte mehr nach rechts zu liegen. Die Gestalt der Stierhüterin [i.e., Hippopotamus] deckt sich umgefähr mit unserem Sternbild Bootes, das Krokodil, Figur 1 [=my Fig. III.69], wird gebildet aus einigen Sternen von Bootes und einigen von Corona Borealis und Serpens. Das Herz des Löwen fällt mit einem Stern [=27 Lyncis, Grösse 4.9] unseres Sternbildes der Lynx zusammen, die vordere Klaue identifizierte ich mit f Ursae majoris [Grösse 4.9]. Die Sternchen, welche auf der ägyptischen Darstellung den Körper umgeben, sind wohl alle Sterne der Lynxgruppe. Die alten Völker betrachteten diese Gruppe als unwichtig, sie wurde nicht benannt. ....Die grössten Sterne der Canes Venatici (α, β und den Stern 14) habe ich als Teile der Kette identifiziert. Das Messer (?), auf welches der Nilpferd sich stützt, liegt in der Gruppe Coma Berenicae."

147. A. Pogo, "Zum Problem der Identifikation der nördlichen Sternbilder der alten Ägypter," *Isis*, Vol. 16 (1931), pp. 102-14. He is highly critical of Biegel's belief that the Egyptians considered the sky as a heavenly globe with the observer at the center and also of her self-serving use of some bright stars while ignoring others and her use of very faint stars as stellar connecting points to construct

the network of lines constituting Egyptian constellations. She also uses the mirror image of a common representation of the northern constellations in the celestial diagram found in the tomb of Seti I. This is equivalent to the arrangement found in the vaulted ceiling of hall J in the tomb of Tausert (1198-96 B.C.), for which see Neugebauer and Parker, *Egyptian Astronomical Texts*, Vol. 3, Plate 9. Concerning the passage of Biegel's work identifying the northern constellations which I have quoted in the previous note, he says (p. 114): "Die Lage der Ursa Maior in den roten Linniennetzen 6b und 7 [=my Figs. III.71b and III.72] entspricht der unteren Kulmination des Siebengestirns. Der Himmelsabschnitt von 2h bis 14h und von +25° bis +80° (Epoche -1300), dessen verzerrte Polarprojektion in Zeichnung 7 (=Biegel Fig. 7 and my Fig. III.72) dargestellt ist, befand sich leider unter dem Horizont von Ägypten zur Zeit der unteren Kulmination von Ursa Maior. Das Biegelsche Nilpferd im Bootes lag auf dem Rücken oder auf dem Bauch—oder es stand gar auf dem Kopf,—wenn es den Ägyptern sichtbar sein konnte; ein aufrecht stehendes Nilpferd haben die Ägypter in die Bootes-Sterne überhaupt nicht hineinphantasieren können. Wenn ein sternkundiger Ägypter den Biegelschen Himmelsabschnitt zwischen Ursa Maior und dem Äquatorialgürtel sehen wollte—und zwar am gestirnten Himmel und nicht etwa auf einem Himmelsglobus der Zukunft,—musste er die Zeit der oberen Kulmination von Ursa Maior abwarten und sich auf den Rücken legen, den Kopf nach Norden, die Füsse nach Süden. Hatte er nun das Spiegelbild der bekannten Darstellung der Sternbilder in der Nähe des Nordpols im Gedächtnis und besass er ebensoviel Einbildungskraft wie R. A. Biegel, so konnte er zwischen Ursa Maior und dem Zenit des Nilpferd im Bootes erblicken, auf ein mit Coma-Sternchen geschmücktes Messer gestützt; den Horus An in Coma und Leo Minor—den halbwegs zum Äquator verschobenen Wender der Polgegend; und anderer Wunder mehr, auf deren Schilderung ich hier verzichte. Sollte nun dieser auf dem Rücken liegende Ägypter auf den Gedanken kommen, aufzustehen und sich die nie untergehenden Gestirne des Nordens anzusehen, so würde er das Siebengestirn (Schenkel, Stier-Ovoid, Stier) in oberer Kulmination nach links dahingleitend erblicken, darunter die Lanze des Wenders [Anu], die Gegend des Pols durchstechend, noch weiter nach unten, rechts vom Nordpunkt des Horizonts, eine auf der Spitze stehende Sterngruppe (Pflock, Messer, steifes Krokodilchen, Gefäss usw.) und weiter rechts vom Meridian das aufrecht einherschreitende Nilpferd...." We also note the attempt (wholly erroneous, I believe) of an unnamed reader of the *Chronique d'Égypte*, No. 32, Juillet 1941, pp. 251-52, to identify the northern constellations under discussion.

148. H.E. Winlock, "The Egyptian Expedition 1925-1927," p. 37: "Across the sky the twelve ancient monthly festivals are drawn, each as a circle with its round of twenty-four hours...." Cf. G. Roeder, "Eine neue Darstellung des gestirnten Himmels in Ägypten aus der Zeit um 1500 v. Chr.," *Das Weltall*, 28. Jahrgang (1928), Heft 1, pp. 3-4 (whole article, pp. 1-5), where he suggests (in a rather confused fashion) that the sectors of circles were to include data identifying, for the first day of each month, the daytime hours (in the upper half circles) drawn from sundials, and the nighttime hours (in the lower semicircle) drawn from astronomical tables or monthly star-charts specifying the stars that culminate every hour between sunset and sunrise on the first day of each month. But this is very loose and it is by no means clear how the 24 sectors would be used to specify both day and night hours. It would be more likely that each circle contained star transits for the first and sixteenth day of each month (i.e., 24 transits), as I suggest in the text. See also Pogo, "The Astronomical Ceiling-decoration in the Tomb of Senmut (XVIIIth Dynasty)," pp. 312-13: "Whether the unfinished monthly circular tables on the Senmut ceiling represent an ambitious attempt to introduce a new feature in the decoration of sepulchral halls or a hesitating attempt to continue a dying tradition, only future finds could show....Incidentally...[the circles of Fig. III.4] show that the draftsman inscribed his circles in squares, the diagonals coinciding—as a rule but not without exceptions—with the corresponding subdivisions; part of the blame should go to the slight curvature of the ceiling."

149. Neugebauer and Parker, *Egyptian Astronomical Texts*, Vol. 3, p. 15.

150. "The Astronomical Ceiling-decoration in the Tomb of Senmut (XVIIIth Dynasty)," pp. 315-16.

151. It was only from the Ptolemaic period that Mars was called by the Egyptians "Horus-the-red" *(Ḥr-dšr)*.

152. Pogo, "The Astronomical Ceiling-decoration in the Tomb of Senmut (XVIIIth Dynasty)," pp. 322-23, 325, comments on the appearance of the planets on the three ceilings (those of the tomb of Senmut, the Ramesseum, and of the tomb of Seti I) as follows: "Jupiter, the wandering star of the south, is the first planet after

Sirius on the three ceilings. The associated divinity is Horus, in a boat, with a star on his head. Saturn, the next planet, is a western wandering star on the Seti ceiling; this 'bull of the heaven' is wandering in the east on the ceilings of Senmut and of the Ramesseum. The associated divinity is [again] Horus, in a boat, with a star on his head. Mars is the next planet on the ceilings of Seti and of the Ramesseum. In both cases, he is in retrograde motion; he is an eastern star on the Seti ceiling, a western star in the Ramesseum. Mars is not mentioned on the Senmut ceiling; apparently, he was in conjunction with the Sun, when the decoration was going on. The associated divinity [once again] is Horus, in a boat, with a star on his head. It is probable that Mars was in conjunction with the Sun, and that Saturn was an eastern evening-star, when Senmut's draftsman was at work; that Saturn was in the west, and Mars in the east, at sunset, when the inscriptions were added to the figures on the Seti ceiling; and that Saturn was an eastern and Mars a western evening-star, when the Ramesseum craftsmen were at work on the planetary part of the ceiling. The laconism of the Jupiter rectangle of Seti I, and the erasure in the Ramesseum of the reference to the south might be due to conjunctions with the Sun or to the embarrassing appearance of Jupiter as an eastern or western evening-star, when the craftsmen of the two astrographic monuments of the XIXth dynasty were dating their masterpieces.... (p. 325) The planet Mercury comes next [after the triangle decans, called by Pogo 'meta-Sothic constellations'] on the three ceilings. The associated divinity, on the Senmut monument, is Set[h]. As could be expected, this devilish divinity is replaced in the Ramesseum by a Horus, although a reminder of Set[h] is apparently preserved in the scepter-head attached to the name of the planet. On the Seti ceiling, the name of Set[h] appears over the juxtaposed stars of the slant-eared divinity associated with the planet. The planet Venus occupies its traditional position at the end of the list; the associated divinity is always Osiris, symbolized by the phoenix Bennu."

153. For efforts to identify the decans, see note 66 above. As we said, Neugebauer and Parker were skeptical of establishing certain identifications of the decans because of the great divergencies apparent in the many monuments. Pogo, in his account of the differences in the selection, order, and grouping of the decans between the Senmut-Ramesseum and Seti I C traditions of the decans ["The Astronomical Ceiling-decoration in the Tomb

of Senmut (XVIIIth Dynasty)," p. 321], makes some interesting remarks: "...our 6-inch celestial globe [with several pairs of polar holes drilled to correspond to intervals of time showing appreciable processional changes on the celestial globe, so that the support of the polar axis could be modified to permit the variation and measurement of the altitude of the pole] has permitted us to identify, tentatively, several decans with certain conspicuous stars or groups of stars, which appeared in ten-day intervals on the horizon of Thebes about 3000 B.C. and about 1500 B.C. [e.g., the constellation of the Ship (decans 12-17) on the Senmut ceiling seems to overlap with our Scorpio; the Kenmu stars of decan 17 seem to belong to the Milky Way in Sagittarius, etc.]....A glance at Fig. 3 [=my Fig. III.65b; cf. Fig. III.65c, column 20] will show that the 'Egg' [constellation] has the shape of a loop of stars on the Seti ceiling. The V-shaped group of stars, between Orion and the [column below the beginning of the] 'Egg' [on the Senmut ceiling, columns 21-24], reappears in the shape of a V on the Seti ceiling [column 21]. Our 6-inch globe shows that about 3000 B.C., when the vernal equinox was in Taurus and the celestial equator passed through the Pleiades, the heliacal rising of the Pleiades and of the Hyades occurred almost exactly east of Thebes and was followed—two decads after Aldebaran or four decads after the Pleiades, and practically in the same part of the horizon—by the heliacal rising of Betelgeuze. With these explanations in mind, it will not be difficult to see on the ceilings of Senmut and Seti I [and especially on the latter] a fairly correct representation of the Pleiades and the Hyades, preceding Orion and Sirius. The general confusion which characterizes, in all decanologues, the Orion region—so rich in stars and star-lore—frustrates all attempts of a closer identification of the individual decans."

154. See W. Gundel, *Dekane und Dekansternbilder* (Glückstadt and Hamburg, 1936), 2. Abt., and the monograph of his son H.G. Gundel, *Weltbild und Astrologie in den griechischen Zauberpapyri* (Munich, 1968), *Münchener Beiträge zur Papyrusforschung und antiken Rechtsgeschichte*, 53. Heft (1968).

155. F. Petrie, *Wisdom of the Egyptians* (London, 1940), p. 21: "*Angular measure.* Measurements of angles were unknown in Egypt or elsewhere till perhaps 200 B.C., after which Eratosthenes used a divided circle .... It was probably some Babylonian who made the advance of a graduated circle, giving direct measurements of altitude, which has culminated in the modern transit circle and

equatorial. As Babylonian degrees left no trace in Egypt, it is only by hour divisions that the right acension is designated, and there is no record of declination."

156. For a brief account of Egyptian astrology, see Parker, "Egyptian Astronomy, Astrology, and Calendrical Reckoning," pp. 723-25. See also the works cited in the preceding note.

Part Two

Documents

Document III.1: Introduction

Feast Lists in The Old and Middle Kingdoms

My interest here in the lists of feast days that appear as part of numerous offering-formulas on the walls of the tombs of dignitaries of the Old and Middle Kingdoms is only with their significance for the understanding of Egyptian calendars. Hence, in introducing my first document, I shall not discuss the nature and purpose of the lists except to say that the offering-prayers of which they are integral parts have the objective of assuring the provision of invocation-offerings to the tomb occupants on numerous feast days throughout each year and thus maintaining and enriching their afterlife.

The Egyptians in the Old Kingdom (and indeed substantially in the whole Pharaonic period) did not employ month names to date documents but rather month numbers, four for each of the three seasons of the civil year, as has been said repeatedly in Chapter Three. Hence, since no revealing month names were mentioned in the Old Kingdom, we are forced to inquire what these formulaic offering-statements with their lists of feast days tell us, if anything, about the early calendars. One thing seems clear from the most common form of the lists in the offering-statements: the first eight feasts were annual ones, while feasts 9-11 were monthly feasts celebrated twelve times a year, and the last item is a reference to daily feasts every day.

In considering these feast lists, we should first recall Parker's two main conclusions concerning them from our earlier discussion:[1] (1) the twelve entries that appear to make up an almost canonical list of feasts in the Old Kingdom (see the numbered list appearing in tomb C. 9 and constituting section II in Document III.1; cf. the similar lists mentioned in note 2 to that document) were given in chronological order (or at least the first eight were in order during each year and the next three in order during each month), and (2) they were feast days that were originally celebrated in the old lunar calendar and so they can help us to understand the structure of that calendar.

The first opening day feast *(wp rnpt)* Parker interpreted as celebrating the heliacal rising of Sirius or Sothis in the twelfth month of the lunar year, which phenomenon, according to Parker, announced for the next month the beginning of the succeeding lunar year if the rising took place before the last eleven days of the month in which it appeared, or it announced the necessity of intercalating a thirteenth lunar month if the rising took place within the last eleven days of the twelfth month. I earlier reasoned against this view of the opening-day feast. I believe that the calendaric evidence shows that it never meant anything but the "Opening of the Year" in its various usages, i.e., (1) as the common opening of the Sothic and civil years when the heliacal rising of Sothis was first established as identical with the beginning of the civil year, or (2) as the opening day of any ad hoc Sothic year after the civil year of 365 days and the Sothic year of 365 1/4 days had diverged significantly from each other, or finally (3) later as a ceremony celebrated at the beginning of the civil year even when it had no temporal connection with the rising of Sothis.

I further believe that Parker's interpretation of the second feast day (that of Thoth) was equally wrong, namely that it celebrated an intercalated month of which Thoth was the protecting deity. As I argued, there is no hard evidence supporting this view and its corollary that this feast day was not the ordinary Feast of Thoth which appears in later calendars and was celebrated on the 19th day of the first month of the civil year.

The third feast, that of the Head or the First of the Year *(tp rnpt)*, was probably not celebrated exclusively on the first day of the old lunar year, as Parker assumed, but was more probably the feast celebrating the beginning of the civil year, which was distinct from the feast celebrating the Opening of the [Sothic] Year (*wp rnpt*) when the Sothic and civil years first diverged significantly from each other. But later it is apparent that the feasts of both *wp rnpt* and *tp rnpt* were sometimes celebrated on the same first day of the civil year (see Chapter Three, Bakir's quotation above note 18).

So it appears that a strict chronological order for the first three annual feasts in the lists from early mastabas is not proved, though of course Parker is no doubt right that the rest of the annual feast days (i.e., feast days 4-8) were chronologically ordered. Furthermore, certain of them may have originally derived from feasts celebrated when the old lunar year was exclusively followed. We can also grant that feasts 9-11 were monthly celebrations and were probably all of lunar origin. Even if they were lunar in origin, we must not conclude from that, or indeed from the probable chronological positions of feasts 4-8, that feasts 1-3 were celebrated on three separate days in the old lunar calendar, particularly in view of the contrary evidence we have adduced against considering them as separate ceremonies on three chronologically ordered days.

Finally, entry 12 in our list of feasts is one prescribing "every feast every day." From the various examples of this entry in the early inscriptions we can detect two somewhat different meanings. The first one is simply a generalized meaning, namely that the offering should be made at any and all feasts during the year. In that sense the phrase includes the 11 feasts already specified and others not specified. The second meaning amplifies the first and implies the third of the three categories of feasts, namely "day feasts" (those celebrated every day at the same time) as distinct from "annual feasts" (those celebrated once a year) and "monthly feasts" (those celebrated once a month). At any rate, entry 12 offers us nothing very specific toward understanding the early lunar calendar. So, all in all, the evidence for the character of the early lunar calendar from the list of these feasts is minuscule and adds little to support Parker's specific views.

As a third example under the rubric of the Old Kingdom I have included an inscription from a round alabaster offering table made for Hetepherakhti, a Judge for Nekhen (Hieraconpolis) and a Prophet of Maat in Dynasty 5. The two inscriptions on round offering tables which I have included (see Figs. III.79 and III.80) do not have complete lists of the feasts; nor are the feasts in the strict normal order. But Hetepherakhti's list does have entry no. 12 in its general form ("every feast every day"), which presumably made it as effective magic as the offering inscriptions that embrace all or most of the specific feasts.

Our last item from the Old Kingdom (beginning at section IV) is not a list of feasts but is a testament concerning two almost duplicate schedules of monthly priestly and mortuary assignments that are tied to the income-return of two parcels of land, each 60 arouras in size. They had originally been created as two distinct benefice-producing parcels of land by King Menkaure (Dynasty 4): one an endowment for priestly duties in the temple of Hathor of Royenet at Tehne and the other an endowment to ensure mortuary duties on behalf of the deceased in the tomb of Khenuka, a dignitary of Menkaure's time. Later the two benefices were both conveyed by King Weserkaf (Dynasty 5) to his palace steward, Nekankh. Then in the testamentary inscriptions included in the tomb of Nekankh we find the benefices broken up and assigned month by month to his children in the fashion indicated in the document (though only the assignments for their monthly duties as priests of Hathor are given below; those for the monthly assignments of mortuary duties at Khenuka's tomb are here omitted as being virtually identical to those for the priests of Hathor, but they are described in the bracketed material concerning lines 27-40).

The monthly assignments begin with one that may have embraced the epagomenal days as well as the first month of the civil year, I Akhet. I have assumed this conjunction against the contrary opinion of E. Winter[2] by my translation "epagomenal days" in line 11, even though I believe that Winter's arguments have considerable merit. Along with most interpreters of this passage I have assumed in my translation that this is a sound rendering of the literal phrase "those upon the year" *(ḥrw rnpt)*, which I have taken to be an abbreviatory expression for the common denomination of epagomenal days: "the five days upon the year" *(hrw 5 ḥrw rnpt)*. The literal expression "5 upon the year" (without the glyph for "days") which I have translated as "Five Epagomenal Days" in feast (13) of the tomb of Khnumhotep II at Beni Hasan (given below as the first extract under the rubric "The Middle Kingdom") seems to be using the abbreviatory expression found in the extract from Nekankh's tomb and I believe helps to confirm the commonly held view that the phrase used in the inscription from the tomb of Nekankh should indeed be translated as "epagomenal days."

These priestly assignments in the passage from the Nekankh tomb extend through the twelfth month, IV Shemu. The possible mention of epagomenal days embedded in a calendaric structure of numbered months and three seasons characteristic of the civil year seems to show that the priestly duties were organized according to a schedule fashioned for the civil year of 365 days. This is not surprising in view of the evidence that I have already given in Chapter Three, which indicates that the authors of the Palermo Stone were, by the time of the Old Kingdom, operating with the full civil year and that the epagomenal days of that year were clearly mentioned in the *Pyramid Texts* (text quoted over note 36).

The schedule of assignments in the Nekankh tomb, if we accept that the text refers to epagomenal days, appears to show (against Winter's denial quoted in note 2) that those added days tended to be tied rather to the first month of the first season (I Akhet) of the civil year than to the last month of the year (IV Shemu).

So, as we are told in the last paragraph of the section on the mortuary duties at the tomb of Khenuka (see line 41), invocation-offerings were to be given for Khenuka, his parents, and progeny at the Feast of Wag, the Feast of Thoth, and on every feast day. This shows, then, that the feast lists of the sort we have already included in sections I-III of our document, whatever their origins, were solidly a part of the civil calendar as early as the fourth dynasty since those celebrations were surely specified in the original foundation established by Menkaure in the fourth dynasty and just as they were here mentioned by Nekankh in his testament in the next dynasty.

Proceeding to the rubric "The Middle Kingdom," we first note that the list presented in the section numbered I, namely, that from the tomb of Khnumhotep II in Beni Hasan, while including a number of the feasts found in the Old Kingdom mastabas, has some noteworthy new entries: to wit, the celebrations of (3) the Great (or Long) Year (perhaps an intercalated lunar year of thirteen months or, perhaps even the civil year itself), of (4) the Small (or Short) Year (probably a lunar year of twelve months, as Lepsius suggested in 1849 in his *Die Chronologie der Aegypter*, p. 146), of (5) the end of the year (a pre-New Year festival?), and of (9) the 5 epagomenal days. The festivals of these added days were not so novel since we know that festivals were celebrated on the epagomenal days during the Old Kingdom. In addition to the evidence of the epagomenal days on the Palermo Stone, the probable evidence of their inclusion in the priestly assignments given in the tomb of Nekankh, and the certain evidence of them in the *Pyramid Texts* – all mentioned in the paragraphs above, we can point to their silent presence in the extract from a Middle Kingdom day-book from the Temple at Illahun, which I have included in Document III.1 (see note 10 to that document) and discuss in the next paragraph. I can also mention the later specific notice of festivals on four of the epagomenal days in the reign of Tuthmosis II.[3]

Evidence that supports the probability of Lepsius' interpretation of the above-noted Small or Short Year as a lunar year of twelve months is found in a papyrus-fragment of the above-mentioned temple day-book from Illahun, also of the Middle Kingdom.[4] The translation of this fragment ends our Document III.1 (for a hieroglyphic rendering of it see Fig. III.83a). It records six alternate service periods, in an over-all period extending from II Shemu 26 in civil year 30 of a twelfth-dynasty king (first identified by Borchardt as Sesostris III but later as Amenemhet III by Parker, though it now appears that Sesostris III might well have been the reigning pharaoh) through I Shemu 16 in year 31 of his reign. It was during these monthly service periods that a scribe of the temple of the pyramid of Amenemhet III, Heremsaf, received income, though the actual numbers specifying Heremsaf's income are missing. From the dates of the priestly overseers given it is evident that the periods of service of the phylum, a separate group or department of priests and its overseer who served together, were reckoned on the basis of lunar months, regardless of whether we interpret the document in either of two ways. The six months of service, according to one interpretation of the document (in which it is assumed that the first day of the lunar month recorded falls on the *day after the last day of the preceding month*) so that each such recorded month comprises 30 days and each of the six alternating unrecorded months of non-service comprises 29 days (except in one case of 30 days). The recorded and unrecorded months in this interpretation are joined in the table of Fig. III.83b, where, by adding the intervening months to the recorded months, we appear to have a lunar year totaling 355 days.[5] In a second interpretation of the document each recorded month is believed to begin on the same day that the preceding unrecorded month ends, each such recorded month being 29 days, while each intervening month would be 30 days (except for one which would have the impossible length of 31 days). The result of this juncture of recorded and unrecorded months is given in Fig. III.83c (again see note 4), and once more this adds up to a short 12-month lunar year. Still a third interpretation of this data by reference to dates calculated by Parker (closer to the second than the first interpretation) also produces a short lunar year (once again see note 4). Hence, no matter how it is interpreted, this document demonstrates the use of a short 12-month lunar year, which may well be the Small or Short Year referred to in the feast list from Beni Hasan. Since the latter follows the Feast of the Great or Long Year in the feast list, we are tempted, as was Parker and most other recent scholars, to conclude that the Feast of the Great Year was the Feast of a 13-month lunar year and that of the Small Year was the feast of a 12-month lunar calendar, since these terms "great" and "small" years (or very similar ones) were used for the long and short lunar years in the Egyptian 25-year lunar cycle at a much later time (see Document III.9 below). But of course, if intercalation was not used at the time of the Illahun document, then the Great or Long Year referred to there might have been the civil year of 365 days, as

I said in Chapter Three above.

But against this interpretation it could be said that, since the Egyptians ordinarily confined the term *rnpt* to the sum of the months without inclusion of epagomenal days (see note 2), the short "year" (the quotation marks indicating the Egyptian sense of *rnpt*) would have to be twelve months and the long "year" thirteen. But to this objection it can be reasonably answered that a short lunar "year" of twelve months would have 354 or 355 days, while the long "year" as part of what we call the civil year has twelve months totaling 360 days (the five epagomenal days being considered by the Egyptians as an extension to the twelve months of the "year"). Thus, with respect to the total number of days in the months alone, the longer "year" (i.e., *rnpt*) becomes the monthly parts of the civil year and not a lunar year of thirteen months.

The principal works used for the preparation of the various sections of this document were the following:

A.E. Mariette, *Les Mastabas de l'Ancien Empire. Fragment du dernier ouvrage de Auguste Edouard Mariette. Publié d'après le manuscrit de l'auteur par Gaston Maspero* (Hildesheim and New York, 1976). Abbreviated M in our document. We have used the tomb numbers from this work preceded either by C or by D according to Mariette's estimate of the age of the tombs.

M.A. Murray, *Saqqara Mastabas*, Parts 1-2 (London, 1905-37).

H. Schäfer et al., *Ägyptische Inschriften aus den Königlichen Museen zu Berlin*, Vol. 1 (Leipzig, 1933). Includes inscriptions of short texts ranging from the Old Kingdom to the end of the time of the Hyksos.

R.A. Parker, *The Calendars of Ancient Egypt*

(Chicago, 1950).

K. Sethe, *Urkunden des Alten Reichs*, Vol. 1, 2nd ed. (Leipzig, 1933) (=*Urkunden* I).

K. Sethe, *Historisch-biographische Urkunden des Mittleren Reiches*, Vol. 1 (Leipzig, 1935) (=*Urkunden* VII).

S. Schott, "Altägyptische Festdaten," *Akademie der Wissenschaften und der Literatur in Mainz. Abhandlungen der Geistes- und Sozialwissenschaftlichen Klasse*, Jahrgang 1950, Nr. 10. Published as a separate volume by the Messrs. Scheel (27 Oct. 1950).

The English translations are my own except where indicated. At the beginning of each section I have given specific references to the text being translated.

Notes to the Introduction of Document III.1

1. See note 20 to Chapter III above and the text which refers to it.

2. E. Winter, "Zur frühesten Nennung der Epagomenentage und deren Stellung am Anfang des Jahres," *Wiener Zeitschrift für die Kunde des Morgenlandes,* 56. Band (1960), pp. 262-266. His conclusion (p. 266) is not unreasonable though I think it is not decisive, as I have indicated in the text: "Es scheint mir demnach die Lesung von *ḥrw rnp.t* in der Tehne-Inschrift als 'Überschrift' wesentlich wahrscheinlicher als die bisherige Deutung als Epagomenentage. Nun ist aber unsere Stelle der einzige Beleg für die Annahme, die Epagomenentage hätten einstmals am Anfang des Jahres gestanden. Akzeptiert man also die Lesung von *ḥrw rnp.t* als 'Überschrift', so würde das zur Folge haben, dass die Epagomenentage schon ursprünglich dem Ende des Jahres angefügt waren und niemals an dessen Spitze standen." The principal difficulty for me is that there ought to be some assignment for the epagomenal days, as there was later at Illahun and at Medina Habu. Also, as I pointed out in the text to which this note refers, the

abbreviatory phrase "those upon the year" was used in the Middle Kingdom extract from Beni Hasan to stand for the epagomenal days (which was clear over 140 years ago in Lepsius' account of the list from Beni Hasan, *Die Chronologie der Aegypter,* Abt. 1 [Berlin, 1849], p. 146, and see his chart of the epagomenal days on p. 147). But Winter is quite right in reasoning that if epagomenal days are not meant by *ḫrw rnpt*, then a major piece of evidence for considering that ancient Egyptians thought of the epagomenal days as being inserted at the beginning of the year is eliminated (still, see the next note). In this connection I can point out that the term *rnpt* translated as *year* is usually thought of as applying to the 12 months only, so in fact it is of little moment whether Egyptians said that the epagomenal days were after the end of one year or before the beginning of another year. But it is clear that they always thought of the civil year in our sense of the term as consisting of 365 days and when the supplies for the daily feasts are listed (as at the Temple of Medina Habu) the amounts for the whole production of Upper and Lower Egyptian grain consist of those for the days of the 12 months and the 5 epagomenal days ("one year and 5 days"; see Document III.5, List 6, line 259, reported from Plate 146; the same expression "one year and 5 days" is continually repeated in that list). There is a reference to a work *The Five Epagomenal Days* in Document III.12, Text J, line 12. We find such a work extant in part in the hieratic papyrus Leiden I 346 (III recto, lines 4-12). It was probably written down in the early years of the 18th dynasty and published by B. H. Stricker, "Spreuken tot Beveiliging gedurende de Shrikkeldagen naar Pap. I 346," in *Oudheidkundige Mededelingen uit het Rijksmuseum van Oudheden te Leiden,* Vol. 29 (1948), p. 65 (see my Fig. III.81b), full article pp. 55-70 (with photographic plates of recto folios I-III). As far as I can tell, it has no particular calendaric or astronomical significance.

3. The feasts on the epagomenal days celebrating the births of the gods, or at least four of them (namely, those for the births of Osiris, Horus, Isis, and Nephthys, the Feast of Seth being unmentioned) were listed as preceding the day of the New Year in the tomb of one Amenemhet, a Theban scribe who lived in the reign of Tuthmosis III. In describing them, A. Gardiner, *The Tomb of Amenemhēt (No. 82)* (London, 1915), p. 97, calls them "special occasions." See also the account of Lepsius, *op. cit.* in note 2, pp. 146-47, and much later the discussion of R. Weill, *Bases, méthodes et résultats de la chronologie égyptienne* (Paris, 1926), pp. 63-65.

4. For the text and a discussion of this fragment, see L. Borchardt, "Der zweite Papyrusfund von Kahun und die zeitliche Festlegung des mittleren Reiches der ägyptischen Geschichte," *ZÄS*, Vol. 37 (1899), pp. 92-94 (full article, pp. 89-102). Also cf. Schott, *Altägyptische Festdaten*, pp. 923-25. Above all, see R.A. Parker, *The Calendars of Ancient Egypt* (Chicago, 1950), pp. 63-67, for an extended interpretation of the lunar months in this fragment.

5. Before discussing the interpretations of the lunar dates included in our document, Parker (*Calendars*, pp. 63-64), on the basis of a photographic copy of the hieratic text of the papyrus supplied to him by Gardiner, reads the six pairs of dates as follows:

ꜣbd II šmw 26 nfryt r ꜣbd III šmw 25
ꜣbd IIII šmw 25 nfryt r hꜣt-sp 31 ꜣbd I ꜣḫt 19
hꜣt-sp 31 ꜣbd II ꜣḫt 20 nfryt r ꜣbd III ꜣḫt 19
ꜣbd IIII ꜣḫt '19 or *18' nfryt r ꜣbd I prt 18*
ꜣbd II prt 18 nfryt r ꜣbd III prt 17
ꜣbd IIII prt 17 nfryt r ꜣbd I šmw 16."

Parker goes on to remark that Borchardt (and E. Meyer) believed that the intervening months began on the day after the second date in each recorded pair. But because Borchardt misread the second date in the second pair as day 20 rather than the correct day 19, he produced a sequence of days of 30, 29, 31, 29, 30, 29, 30, 29, 30, 29, 30, which included the impossible number 31. But with the correct reading of 30 instead of 31, a sequence of 30, 29, 30, 30, 30, 29, 30, 29, 30, 29, 30, would be produced according to this interpretation (cf. Fig. III.83b). But a second interpretation of the dates was suggested by G.H. Wheeler, namely that the second date in each of the six recorded pairs was repeated as the first date in each of the six unrecorded pairs. Then with this assumption and the corrected "19" instead of "20" in the second pair of recorded dates, Parker deduced the following sequence of days in successive months of the whole lunar year under study: 29, 30, 29, 31, 29, 30, 29, 30, 29, 30, 29 (cf. Fig. III.83c). But he noticed again an offending 31-day month. Consequently he decided to use only the specified first days of each of the six recorded months. These would, of course, be the first days of each lunar month. Using these six days and three other Middle Kingdom lunar days found by Borchardt, and employing the late 25-year lunar cycle which he (Parker) had reconstructed (see Document III.9), he was able, he thought, not only to date to July 17, 1872 B.C. Julian, the earliest Sothic rising date extant in Egyptian records, namely that of Year 7 of Sesostris III's reign (see Document III.10 below), but also to assign the lunar dates here discussed in this Document III.1 to Years 30-31 of Amenemhet III's reign. However, on the basis of additional lunar dates, this assignment was rejected by Krauss in favor of Borchardt's belief that the year 30-31 was rather that of Sesostris III's reign (see R.K. Krauss, "Sothis, Elephantine und die altägyptische Chronologie," *Göttinger Miszellen*, Heft 50 [1981], p. 74). Parker was able to calculate the first days of those 12 lunar months of that lunar year, showing that they agreed in ten out of the twelve days with those given in our document and supplemented by the intervening lunar first days deduced with the second interpretation, namely that the first day of each unrecorded month ought to be the same as the last day of each preceding recorded month. I need add further that Parker, in reducing the impossible 31-day duration to a 30-day duration, produced further changes in the durations of some of the other months deduced from the list of given days as noticed in Fig. III.83d. In any case, the lunar year so deduced probably consisted of 355 days.

Document III.1

Feast Lists in the Old and Middle Kingdoms

The Old Kingdom

I. [Cf. Fig. III.77, Tomb of Ptahshepses, late Dynasty 5.]

A boon which the king gives, and a boon [which] Anubis, who is in front of the divine booth and who is upon his mountain, [grants] that there may be [made] invocation-offerings of bread and beer for him (Ptahshepses) on every feast day. Ptahshepses.[1]

II. [Cf. Fig. III.78, C. 9, M. p. 130, Tomb of an earlier Ptahshepses, middle of Dynasty 5.[2] The numbers in parentheses here and in other lists from the Old Kingdom indicate the positions of the feasts in the canonical ordering of the feasts. Hence we have presented this list of C. 9 as an early example of the so-called canonical list.]

A boon that Anubis, who is in front of the divine booth, who is Lord of the Holy (or Free) Land on his Mountain, and who is "the One in the Embalming Place (*imy-wt*),"[3] gives that there be [made] invocation-offerings of bread and beer for him (Ptahshepses) at (1) the Feast of the Opening of the Year *(wp rnpt)*, (2) the Feast of Thoth *(ḏḥwtyt)*, (3) the Feast of the Head (or First) of the Year *(tp rnpt)*,

(4) the Feast of Wag (or Exultation, i.e., *wꜣg*), (5) the Feast of Seker *(ḥb skr)*, (6) the Great Feast *(ḥb wr)*, (7) the Feast of the Burning *(rkḥ)*, (8) the Feast of the Going Forth of Min *(prt mn)*, (9) the Feast of Sadj *(śꜣḏ)*, ... (10) the Feast of the [Head (or First) of the] Month (i.e., of the Day of the First Appearance of the Crescent?) *(ꜣbd)*,[4] (11) the Feast of the [Head (or First?) of the] Half-Month *([tp] śmdt)*, and (12) every feast every day *(ḥb nb rᶜ [*or *hrw] nb)*....

III. [p. 348, D.60 (Hetepherakhti), Dynasty 5; cf. note 2 for the regular feast list in his tomb; our text here is on a round alabaster offering table in Fig. III.79.][5]

A boon the king gives and Anubis gives, [i.e., Anubis] who is in front of the divine booth, that invocation-offerings of bread and beer be [made for the deceased at] (1) the Feast of the Opening of the Year, (3) the Feast of the Head (or First) of the Year, (4) the Feast of Wag (or Exultation), (2) the Feast of Thoth, (10) the Feast of the Head (or First) of the Month, (11) the Feast of the Head (or First) of the Half-Month, (12) [and] every feast every day on behalf of the Judge belonging to Nekhen and Prophet of Maat, Hetepherakhti.

IV. [Testament of Nekankh, Prophet, i.e., Priest of Hathor, dividing the services of that priesthood and its land-benefice among his children according to the months of the year, Time of Weserkaf, Dynasty 5; text in Sethe, *Urkunden*, I, pp. 24-28, from which I have inserted line numbers, cf. Fig. III.81a; also see the translation in J.H. Breasted, *Ancient Records of Egypt*, Vol. 1 (New York, 1908), pp. 101-06.]

[1] Steward of the Palace, [2] Governor of the New Towns, [3] Chief Prophet of [4] Hathor, Mistress of Royenet, [5] confidant of the King, Nekankh; [6] his wife and lady revered [7] by Hathor, Hedjethekenu.

[8] He makes a decree for his children to be priests of Hathor, Mistress of Royenet:

[9] "These are the prophets whom I made of my children from the endowment, to be priests of Hathor. [9a] Two parcels of land [expressed] in arouras were conveyed by the majesty of King Menkaure to these [i.e., such] prophets to be priests therewith."

[Attested by] [10] the King's confidant and steward of the Palace, Nekankh; his wife, the king's confidant, Hedjethekenu; and her children,

[Assignments][6]

[11] Hedjethekenu, King's confidant and revered lady [with female portrait]; [period of] epagomenal (?) days and month I of Akhet; land [support], 5 arouras.

[12] Henhathor, Scribe of the King's records [with male portrait]; month II of Akhet; land, 5 arouras.

[13] Hathorshepses, a priest [with male portrait]; month III of Akhet; land, 5 arouras.

[14] Nessuhathoryakhet [with male portrait]; month IV of Akhet; land, 5 arouras.

[15] Hathorshepses [with male portrait]; month I of Peret; land, 5 arouras.

[16] Webkauhathor [with male portrait]; month II of Peret; land, 5 arouras.

[17] Qaisuthathor [with male portrait]; month III of Peret; land, 5 arouras.

[18] Khabauhathor [with male portrait]; month IV of Peret; land, 5 arouras.

[19] Khentisuhathor (*portrait disappeared*); [month I of Shemu]; land, 5 arouras.

[20] Royenet (*portrait disappeared*); [month II of Shemu]; land, 5 arouras.

[21] *(This line is vacant except for a superfluous addition at end:)* land, 5 arouras.

[22] --meat, his tenth of all that is paid [into] the temple, beside the rations of bread and bear. Prophet Henhathor [portrait of male with libation vessel]; month III of Shemu; land, 5 arouras.

[23]-[24] Mortuary priest Mer-[rekhet?] [with male portrait] and mortuary priest Kashere [also a male]; month IV of Shemu; land, 5 arouras.

[The Decree of King Weserkaf for Nekankh]

[25] It was the majesty of King Weserkaf who commanded that I should be priest of Hathor, Mistress of Royenet. As for everything paid into the temple, it was I who was to be priest over everything that came into the temple. [26] Now it is these my children who shall act as priests of Hathor, Mistress of Royenet, as I myself did, while I travel to the beautiful west, ...in charge of these my children.

[The Mortuary Priesthood of Khenuka]

[27] Now it is these priests who make the invocation-offerings of bread and beer to the king's confidant, Khenuka, his father, his mother, his children, and all his progeny.

[Then follows in lines 28-40 a list of monthly assignments to the children of Nekankh that is essentially the same as the assignments given to them as priests of Hathor. But missing in this list is the name of the first priest, the daughter Hedjethekenu. In this list, the monthly periods appear in the first column and the priests' names in the second. The land assignments of 5 arouras per priest are here missing. Perhaps the third column of the earlier list was meant to apply also to this list. Recall that in the original establishment of these endowments two parcels of land were specified for the two endowments. If the amount of each parcel land was the same, then the total amount of land was 120 arouras.]

[The Decree concerning the Khenuka Mortuary Priesthood]

[41] Now it is these my children who make the invocation-offerings of bread and beer to the king's confidant, Khenuka, his father, his mother, and all his progeny, at the Feast of Wag, the Feast of Thoth, and on every feast day.

The Middle Kingdom

I. [Sethe, *Urkunden*, VII, pp. 29-30, Tomb of Khnumhotep II at Beni Hasan, Dynasty 12; see Fig. III.82.]

I commanded invocation-offerings of bread and beer, oxen and fowl at every feast in the necropolis *(ẖrt-nṯt)* at (1) the Feast of the First of the Year *(tp rnpt)*, (2) the Feast of the Opening of the Year *(wp rnpt)*, (3) the Feast of the Great Year *(rnpt ꜥꜣḫ̆t)*, (4) the Feast of the Small Year *(rnpt nḏst)*, (5) the Feast of the End of the Year *(ꜥrḳ rnpt)*, (6) the Feast of the Great Festival *(ḥb wr)*, (7) the Feast of the Great

Burning *(rkḥ ꜥꜣḥ)*, (8) the Feast of the Small Burning *(rkḥ nḏs)*, (9) the Feast of the Five Epagomenal Days *(5 ḥrw rnpt)*, (10) the Feast of the Digging of the Sand *(šḏt šꜥ[y])*, (11) the 12 Feasts of the [First (or Head) of the] Month *(ꜣbd)*, (12) the 12 Feasts of the [First (or Head) of the] Half-month *(śmdt)*, and (13) every festival of the dwellers on the good land (i.e., those who are living) and of the dwellers on the mountain (i.e., those who are dead).

II. [A papyrus fragment from Illahun (Berlin Museum, Papyrus 10056, verso) including six alternate monthly periods (demarked by the names of the overseers of priestly phyla) during which the year's income for the temple priest Heremsaf was to be paid, but the actual amount of the income each period (or in totality) is not recorded; see Fig. III.83a and the Introduction to Document III.1, notes 3-4:]

[1] Reckoning of... and... for one year.

[2] Amount *(rḫt)*[7] of the income of six months for the temple scribe Heremsaf, year 30 [to year 31 of the reign of Amenemhet III (?) or, more likely, of that of Sesostris III (?)].

[3] The setting forth of this amount *(rḫt)*.

[4] The [lunar] month of the overseer of the phylum Maketen's[8] son, ...seneb *(mꜥktn ...snb)* which begins [civil] day 26 of month II of Shemu [of year 30] and extends up to day 25 of month III of Shemu.

[5] [The (lunar) month of the overseer of the phylum, ...'s] son, Senwosert, from day 25 of month IV of Shemu up to day 19[9] of month I of Akhet, year 21 (!, 31).[10]

[6] [The (lunar) month of the overseer of the

phylum, ...'s] son, [...], in year 31, from day 20 (?)[11] of month II of Akhet up to day 19 of month III of Akhet.

[7] [The (lunar) month of the overseer of the phylum,] Herhernakht's son, Herwernakht, from day 19 (or 18?)[12] of month IV of Akhet up to day 18 of month I of Peret.

[8] [The (lunar) month of the overseer of the phylum,] Senebi's son, Khakheperreseneb, from day 18 of month II of Peret up to day 17 of month III of Peret.

[9] [The (lunar) month of the overseer of the phylum,] Senwosert's son, ...ankh, from day 17 of month IV of Peret up to day 16 of month I of Shemu.

[10] The sum [of income] ....; the remainder of ....(?). *{Income numbers are missing throughout.}*

Notes to Document III.1

1. This is the general formula *ḥtp di nsw* that usually precedes (with small variations) the list of festivals or feasts at which offerings are to be made on behalf of the deceased. In reality the general formula which ends on "every feast day" is simply giving a single phrase to stand for all of the 12 entries of the canonical list of feasts given in our next text from tomb C 9. I have followed (with only minor changes) the translation of A. Gardiner, *Egyptian Grammar*, 3rd ed. rev. (London, 1973), p. 171. I give the text itself in Fig. III.77.

2. The following offering-prayer *(ḥtp di 'Inpw)* contains the standard list of twelve entries specifying the Old Kingdom feasts to be celebrated by invocation-offerings on behalf of the deceased. Other Old Kingdom examples may be found in Mariette, *Les Mastabas de l'Ancien Empire*, at the pages listed below, with dates taken from B. Porter and R.L.B. Moss, *Topographical Bibliography of Ancient Egyptian Hieroglyphic Texts, Reliefs and Paintings*, Vol. III2, revised by J. Mālek (Oxford, 1978-79, 1981); see pp. 911-12 for an indexed list of the tombs at Saqqara from Mariette.

p. 116, C 3 (Ptahuser), Middle Dynasty 5 or later, entry 12 reads "the Feast of Every Year" *(ḥb n rnpt nbt)*;

p. 149, C 18 (Tjenty), Middle Dynasty 5 or later, has entries 1-6, plus "every feast forever" *(ḥb nb ḏt)*;

p. 152, C 21 (Kay), Dynasties 5-6, has all 12 entries;

p. 154, C 22 (Remeryptah), Dynasty 5 or later, has entries 1-5, 10-12 (with entry 11 twice);

p. 195, D 10 (Tepemankh), End of Dynasty 5 or Dynasty 6, has entries 1-6, and 12;

p. 203, D 12 (Niankhsekhmet), Time of Sahure, Dynasty 5, has entries 1-4, 6-12; a second list has 1-4, 6-7, and 12;

pp. 214-15, D 16 (Ankhmaaka), Time of Niuserre, or later, has entries 1-5;

p. 230, D 19 (Kai), Middle Dynasty 5 or later, has all 12 entries;

pp. 247-49, D 23 (Kaemnefert), Time of Niuserre or later, has feasts 1-6+part of 7 (horizontally); and 1-3 twice (vertically) on p. 247; 4-7 twice (vertically) on p. 248; and 8, 10, 9 twice (vertically) on p. 249;

p. 250, D 24 (Nimaatptah), Time of Neferirkare or later, has feasts 1-9, 12;

p. 259, D 28 (Senedjemib), Time of Niuserre or later, has feasts 1-12 (but with text intruding between 11 and 12);

p. 268, D 38 (Washptah), Time of Neferirkare, has feasts 2-6, 10-12, with "Feast of the Firsts of the Years *(tpw rnpwt)*" between 5 and 6, and 12 simply given as "every feast in every course of eternity *(ḥb nb m ꜥwt nbt ḏt)*;"

p. 278, D 39 (Kapure), Time of Isesi or later, has feasts 1-12, with "in the course of eternity" added to 12; on p. 279, the inscription has feasts 1-8, 10-11;

pp. 307-09, D 47 (Nenkheftka), Time of Sahure or later, has feasts 1-7, 10-12 twice in vertical columns;

p. 311, D 48 (Nikaankh), Time of Neferirkare or later, has feasts 1-8, 10, 9, 12;

p. 336, D 59 (Duahap), Dynasty 6, has feasts 1-4, and "every feast" as 12;

p. 341, D 60 (Hetepherakhti), Time of Niuserre or later, has feasts 1-8; see also the list given in Document III.1, which was arranged in a circle on an offering table;

pp. 355-56, D 62 (Ptahhotep), Time of Isesi, has feasts 1-5, 8-9;

p. 366, D 67 (Nakhtsas), Time of Niusserre or later, has feasts 1-6, 12.

There are other lists that do not follow the regular order:

p. 164, C 27 (Kaihap), Dynasty 5, has only feasts 10, 11, 1, and 3 (Parker in a slip incorrectly says "2"), arranged in a circle on offering table, together with his name (see Fig. III.80);

pp. 319-20, D 52 (Senenuankh), Time of Sahure or later, or Dynasty 6, has feasts 4, 2, 12 *(ḫb nb)*, twice;

p. 349, D 61 (Duaenre), Middle Dynasty 5 or later, has feasts 3, 1, 2, 12;

pp. 368-69, D 69 (Shedabd), Dynasty 6, or 1st intermediate period, has feasts 1, 4, 2, 3, 5-7, 10-11.

Without giving their details, I note further examples in Sethe's *Urkunden* I, pp. 72, 121, 165, 177, 217, and 252. All lists are incomplete in some respect; some are not in the conventional order.

In addition, I mention, without details, some more inscriptions from the Old Kingdom transcribed by Schäfer in the collection of the Berlin Museum's *Ägyptische Inschriften*, Vol. 1, pp. 8, 32, 39, 40, 50, 62, 63, 66, and Vol. 2, p. 102. None of the lists has all twelve entries of the canonical list, and many are out of order. The most interesting divergence in order is the placing of the Feast of Wag as the first feast in four of these lists (see pp. 32, 40 [twice], and 50). This divergence (in addition to being implied in the Old Kingdom inscription from Nekankh's tomb given in Document III.1, IV, line 41) also occurs in Spell 936 of the *Coffin Texts* (and thus in the Middle Kingdom): Ed. of A. de Buck, Vol. 7, p. 137: "A boon which the king gives and [also] Anubis .... Lord of invocation-offerings at the Wag-feast, at the Thoth-feast, at the Seker-feast, at the Feast of the First of the Year, at the Feast of the Opening of the Year, and at [all] good feasts for Osiris on behalf of the honored [deceased]...." Thus the order of these festivals is 4, 2, 5, 3, 1, and 12. Note that 12 is here restricted to the feasts of Osiris, instead of being quite general as in the Old Kingdom. Another list of feasts from the Middle Kingdom that also begins with the Wag-feast appears in an inscription of the "Great Openers of the Way," i.e., Princes of Abydos under Sesostris I and Amenemhet II (K. Sethe, *Ägyptische Lesestücke: Texte des Mittleren Reiches*, reprint [Hildesheim / Zurich / New York, 1983], p. 73, lines 17-18). The feasts are those of Wag, Thoth, Haker, Opening of the Year, Ferrying of the God, The Burning, First of the Year, The [First of the] Month, The Half-month, Seker, and Sadj.

3. In this translation I have clarified an ambiguous reading of

one of Anubis' epithets in the first line of the offering text (Fig. III.78, line 1). I assume *imy wt*, i.e., "He who is in the embalming place." This interprets the bird given by Mariette, which looks something like an ibis with a short beak, as a quail chick ( ), which is the correct glyph for this epithet of Anubis.

4. Needless to say, celebrations on the day of the first appearance of the crescent, usually the second day of the month, and on the day of the full moon in the middle of the month (celebrations apparently indicated by feasts 10 and 11 of the canonical list) surely reveal divisions of the month based on the waxing and waning of the moon and hence are persistent remnants of the lunar calendar (see Document III.6 for a discussion of these lunar days). On the other hand, as I have suggested in the introduction, such lunar celebrations were no doubt simply taken over into the civil calendar in its month of three weeks of ten days each, and hence we cannot argue from these celebrations that the *whole* list of feasts is a chronological one derived from the old lunar calendar.

5. See also the circular offering-inscription for Kaihap, Dynasty V, mentioned near the end of note 2 (cf. Fig. III.80).

6. Breasted, in his *Ancient Records of Egypt*, Vol. 1, p. 102, note b, indicates the structure of the table of assignments as follows (cf. Fig. III.81a for Sethe's depiction of it): "In the original, the names are arranged in perpendicular columns, and a figure of the person named is depicted below each name [except those given in lines 19 and 20]. I have added the sex of the person in each case [i.e., males in all cases but the first]. The first column is therefore the priests, the second (double) column is the time of service for each priest, and the third is the amount of land from which each draws the necessary income." Needless to say, the priestess in the first monthly slot is the daughter (and perhaps the eldest child) and not the mother of the same name.

7. The word "rekhet" *(rẖt)* usually means "knowledge" but it can also mean "number" or "amount" or "account." See F. Ll. Griffith, *Hieratic Papyri from Kahun and Gurob*, 2 vols. (London, 1898), Volume of Plates: no. 21, lines 3 and 30-31; Volume of Text: pp. 52 and 54.

8. Gardiner, *Egyptian Grammar*, p. 66 (Sect. 85) remarks that in Dynasty 12 the father's name comes first.

9. Read by Borchardt as "20," but this was rejected by Gardiner, Erichsen, and Parker in favor of "19" (*Calendars*, pp. 65

and 82, note 8). This corrected number establishes the month as 30 days rather than as 31, the latter figure being impossible. All six recorded months therefore would be 30-day months and all intervening months 29 days (except for one which is a 30-day month), if the first interpretation mentioned in note 5 of the Introduction to Document III.1 were followed. See also the affect of the correction if we assume the second interpretation discussed in the same note.

10. This lunar month that extends from day 25 of month IV of Shemu (i.e., the last month of year 30) to day 19 of I Akhet (i.e., the first month of year 31) includes the five epagomenal days inserted between day 30 of month IV of Shemu and the first day of month I of Akhet. This information assures us that monthly watches of the priestly phyla, which were reckoned in lunar months, were designated in the universally used civil calendar.

11. Borchardt regards "20" as uncertain, but Parker indicates that this number is perfectly clear (*Calendars*, p. 82, note 9).

12. Parker, *ibid.*, p. 82, n. 10: "The 10 is clear but the units [to be added to it] are in lacuna. Any other reading is excluded by the preceding and following dates."

Document III.2: Introduction

The Ebers Calendar

In about 1862 the American dealer Edwin Smith, who lived in Luxor at that time, temporarily came into possession of (but not ownership of) the hieratic medical papyrus that now bears the name Papyrus Ebers, Smith probably acting as an agent for the Egyptian owner. This papyrus was subsequently (1873) sold by the Egyptian owner (without any role in the transaction being played by Smith) to the Egyptologist Georg Ebers, who published its hieratic text in facsimile in 1875.[1] I shall have much to say about this papyrus in the subsequent volume when I treat of Egyptian medicine, but at this point I confine my discussion to the so-called Ebers Calendar consisting of thirteen lines of hieratic text that occupy the verso of the first column of the Ebers Papyrus.

In my account of the Ebers Calendar in Chapter III, I noted that Egyptologists (until the time of Borchardt in 1935) usually accepted it as a table correlating a Sothic calendar year with a specific civil year. In fact, this was the view expressed by Heinrich Brugsch when, five years before its publication by Ebers, he published in 1870[2] a German version of the calendar which was based on a tracing of its hieratic text that was provided by an Egyptologist from the University of Heidelberg (and later in 1872 its professor) August Eisenlohr, who had himself received a tracing of the calendar from Edwin Smith in February, 1870. In that same year

Eisenlohr, noting that Brugsch had published his version of the calendar without informing him (Eisenlohr), sent to Lepsius (who was the editor of the *ZÄS*) an article on the calendar which included the hieratic text of it. But in publishing the article, Lepsius substituted his own better version of the hieratic text which was based on a tracing of it provided to Lepsius by E. Naville (Naville's tracing having been made from the papyrus in 1868, when Smith alerted him to the calendar's existence). Eisenlohr accepted it as a Sothic-Civil double calender and added some erroneous views on the age of the text.[3]

But most noteworthy in the early acceptance of the Ebers Calendar as a correlation of a Sothic and a civil calendar for a given civil year was an article by Richard Lepsius in 1875, the very year of Ebers' publication of the papyrus.[4] The most interesting part of the article suggested that the purpose of the calendar was to provide a doctor using it at the time when the medical text was extracted or copied from a much older work, with a scheme to find when the medicines prescribed in the medical text for seasonal periods ought to be given in the civil year, which with a length of only 365 days had wandered from coincidence with the Sothic seasonal year of 365 1/4 days.

Subsequently, in 1885, the Viennese Egyptologist Jakob Krall[5] determined that the king's name given in the cartouche in line 1 was Djeserkare, the prenomen or throne-name of Amenhotep I, a reading that Ebers himself had proposed as early as 1873, not long after he acquired the papyrus. And so the double calendar was quite solidly dated to the ninth year of that monarch (ca. 1525-1517 B.C. in the chronology based on the heliacal rising of Sirius at Thebes, or ca. 1544-1537 in the chronology determined at Memphis or Heliopolis). Thus, to most Egyptologists it seemed certain, on the evidence of the second line of the calendar (see Figs. III.10 and III.11), that the year in column one was a specific fixed Sothic year which began with "The Feast of New Year's Day" at the time of the [heliacal] rising of Sothis (i.e., Sirius) on the 9th day of the third month of the season of Shemu in civil Year 9 of Amenhotep I's reign.

The view of the Ebers calendar as a double calendar correlating a specific but idealized Sothic year with a given civil year was canonized by Eduard Meyer in his enormously influential *Ägyptische Chronologie* (Berlin, 1904), pp. 46-51 (see my Fig. III.11 with its hieroglyphic transcription of the text of Ebers' calendar). I should add, however, that Meyer (see pp. 31-38) in no way believed that a formal Sothic fixed calendar determined by the regular addition every four years of a sixth epagomenal day was continuously kept throughout the whole dynastic period (a point I shall return to in Document III.10 when I discuss the full range of evidence concerning coordinated Sothic-Civil dates). In his initial discussion of the Ebers calendar Meyer noted certain difficulties.[6] We can follow his discussion by referring to my Fig. III.11. The first difficulty concerned the fact that the scribe has added what seems to be a mark of repetition in each of the remaining eleven lines directly under the phrase at the end of the second line ("The going out of Sothis"), though it is obviously impossible to have the helical rising on the 9th day of every month of the civil year. Though Meyer says that this is not understandable, he adds that one is inclined to take this for a scribal error. The second difficulty is that the epagmomenal days were omitted from the reckoning in the civil calendar, as is evidenced by the fact that the first day (or feast day) of each month of the Sothic year fell on the 9th day of all the successive months of the civil year in the Ebers calendar, though one would have expected in line 4 the first day of the Sothic month of Menkhet to fall on the fourth day rather than on the ninth. This does not trouble Meyer, for the epagomenal days "here as elsewhere are not counted as part of the 'year'." He cites as an example the astronomical ceiling of the Ramesseum (see my Fig. III.2). I have discussed the question of the meaning of *rnpt* and its use in the civil and lunar calendars in the Introduction to Document III.1.

Finally Meyer mentions what seemed to him a more puzzling problem. This was the supposed difficulty of having every month of the Sothic year ordered one month later than the similarly named month of the civil year, i.e., the name of the festival or first day of the first month of the Sothic year appeared to be the same as the later name of the last month of the civil year, the name of the festival beginning the second month of the Sothic year the same as the later name of the first month of the civil year, the name of festival beginning the third month of the Sothic year the same as that of the second month of the civil year, and so on, so that the name of the festival beginning the twelfth month of the Sothic year was the same as that of the eleventh month of the civil year. He simply concluded that he was unable to solve what he thought to be the discrepancy inherent in these conflicting orders.

Alan Gardiner, in an early paper,[7] addresses this last difficulty, and gathers considerable evidence to show that in fact at least four (and possibly six) of the entries (or their equivalents) in the first column of the Ebers Calendar (i.e., those for the Sothic year) were indeed feast days, whose dates of celebration could be independently shown from sources other than the Ebers Calendar to have fallen on the first days of the months that *followed* the months to which the gods of the feast days gave their names. But he still believed (without evidence) that *originally* those eponymous feast days were the first days of their homonymous months and that some kind of shift of one month in the ranks of the twelve months took place historically to give the common order of months that we see from the New Kingdom onward. But he admitted grave difficulties connected with the circumstances of the historical shift from the positions of the Old Order to those of the New Order, and indeed it was clear to him that the old and the new positions overlapped and thus existed at the same time in the Ramesside period.

Meyer welcomed Gardiner's account in the 1908-additions to his *Chronologie,*[8] and he summarized the sources for the Old and New Orders (see Fig. III.6b). Accordingly, he produced a complicated ad hoc scheme to explain the shift from the old positions (represented by the Ebers calendar) and those of the new (apparent in the Ramesseum), a scheme based on a completely unsupported view that the early Egyptians identified the summer solstice with the month of Mesore in its alternative name *Wp rnpt* (New Year's Day). The use of *Wp rnpt* as the name for IV Shemu (i.e., Mesore) was first discovered by Brugsch in his early paper of 1870 (*op. cit.* in note 2, pp. 109-10), and, as is evident from Figs. III.10 and III.11, *Wp rnpt* in the Ebers Calendar occupied the position of the beginning of the year when Sirius would rise heliacally. Meyer's scheme went on to suggest that the closeness of the summer solstice to the rising of Sirius could have been noticed as early as 4241 B.C., when, according to Meyer, the ancient Egyptians began to consider a calendar at Memphis. Meyer assigns 19 July Jul. for the latter event (the rising of Sirius) and 25 July Jul. for the former event (the summer solstice). He notes further that both events fell on 19 July Jul. in the 35th century and that in 2781 B.C., at the beginning of the second Sothic period, the solstice fell on 13 July Jul. and the rising of Sirius on 19 July, both events still, for all practical considerations, coinciding. These observations, Meyer believed, could have given rise to the identification of Mesore (i.e., the Birth of Re) with the New Year's Day. But the two events separated from each other significantly as the centuries went by since the year from solstice to solstice was shorter than the Sothic year (see Chapter Three above, note no. 5). Hence it was necessary, Meyer reasoned, that the month beginning with the feast of the birth of Re, i.e., Mesore, be shifted to its later canonical position as the last month in the preceding year, and simultaneously that the positions of all the other eleven months be shifted to their later canonical positions one month earlier, i.e., one rank above their original positions.

But this neat ad hoc explanation (deprived as it was of any hard evidence) was successfully torpedoed by Kurt Sethe. In the first place Sethe,[9] after going over all the evidence presented by Gardiner and Meyer, emphasized as centrally important the fact that (1) the twelve entries in the Ebers Calender were in fact *feast days and not months*, and (2) that Meyer had gone off track in thinking that in early times the 12 months must have appeared in an order which can be called the Old

Order, i.e., an order in which the months were each one rank above their positions in the calendar of months of a New Order, an order which began to be used (though sparingly) from the time of the New Kingdom onward and which became canonical by the Ptolemaic period. Instead Sethe held that the month names were applied to the thirty-day periods that preceded and culminated with the gods' festivals. Thus feast days and the months are not to be confused, as they tended to be by both Gardiner and Meyer. Though there are evidences of some shifting of feast days, there is no evidence at all of the general one-month shift of such days assumed by Meyer. Hence there was no need to look for a so-called Old Order of months differing from the later generally accepted New Order.

After this presentation, Sethe then proceeds to dispose of Meyer's proposal that the celebration of the Birth of Re (i.e., Mesore) was conceived of as occurring at the summer solstice in the late 5th millennium B.C. when that event was close in time to the helical rising of Sothis. He found absolutely no evidence of this cosmological conception hypothesized by Meyer and indeed he only uncovered evidence that would link Egyptian views of the Birth of Re with the winter solstice.

Sethe's arguments were fortified vigorously and with clarity by Raymond Weill in 1926,[10] but his account need not be summarized here since it differs little from Sethe's.

We have now reached Borchardt's radically distinctive interpretation of the Ebers Calendar, which, as I have already noted in Chapter 2, assumes that the first column contains lunar months and the hieratic "9" at the end of the second column (Borchardt calls it the third column) in each line, understood by earlier authors as the "ninth day" of the respective months of the civil year, ought to be translated in each of the twelve lines as "New Moon Day" of each of the twelve lunar months (see Fig. III.12).[11] But, as I have already noted in my earlier account of the Ebers Calendar in Chapter 3, there is no paleographical evidence for the use of the "9"-sign for "New Moon Day." Hence Borchardt's solution of the irregularities in the Ebers Calendar must be rejected. I have also rejected (and shall not repeat here) Richard Parker's opinion that the Ebers Calendar represents a correlation between a schematized lunar calendar (with the first entry being the festival of the rising of Sirius in the eleventh month of the year) and the corresponding days of civil years 9-10 of Amenhotep's reign.

Thus we are left with the view which I outlined in Chapter 3, namely, that the Ebers Calendar is an ad hoc correlation of (1) twelve feast days (30 days apart) marking a fixed Sothic year beginning with the Feast of New Year's Day determined by the helical rising of Sirius with (2) the corresponding days of the civil year extending from III Shemu 9 in civil year 9 of Amenhotep I's reign to II Shemu 9 in civil year 10 of that reign.

As is usually the custom, in my translation of this document I have included bracketed phrases which, I believe, complete the intention of the author. The reader will find it useful to consult Fig. III.10 and Fig. III.11, as well as the various studies quoted in the notes to this introduction. The Arabic numerals in parentheses refer to the lines of the hieratic text.

Notes to the Introduction to Document III.2

1. G.M. Ebers, *Papyros Ebers, das hermetische Buch über die Arzeneimittel der alten Ägypter in hieratischer Schrift.... Mit hieroglyphisch-lateinischem Glossar von Ludwig Stern,* Vol. 1 (Leipzig, 1875), verso of the first column (see Fig. III.10).

Incidentally, Ebers in an article on the papyrus published not long after his acquisition of it ["Papyrus Ebers," *ZÄS*, Vol. 11 (1873), p. 41, n. 4 (full article, pp. 41-46)], rather harshly treats Smith's role in the dissemination of the calendar, suggesting that Smith was claiming the papyrus as his own and in his possession, while, according to Ebers, it never was: "Mr. Smith in Luqsor gab unserem Papyrus, auf dessen Rücken sich das erwähnte Kalenderfragment befindet, für den seinen aus, *obgleich er ihn niemals besessen* hat. Noch, nachdem ich den Papyrus von dem *wahren* Eigenthümer ohne Mr. Smith's Wissen erstanden, erklärte er Herrn Professor Lauth aus München, der Besitzer des Schriftstückes zu sein, von dem es ihm Copieen zu erlangen gelungen war. Diesem letzteren zeigte er sogar ein Blättchen des Originals, da sich schon als unbeschädigte, vollständige Rolle seit mehreren Tagen in meiner Hand befand. Daraus geht hervor, dass er thatsächlich ein kleines medicinisches Papyrusfragment von ähnlicher Schreibung, wie die des meinen besitzt, welches er benutzte, um in etwaigen Käufern den Gedanken nicht aufkommen zu lassen, dass ein anderer als er der Eigenthümer des grossen medicinischen Papyrus sein könne. Ich darf diese Vermuthung kühnlich aussprechen, da Mr. Smith meinem verehrten Collegen und Freunde Prof. Eisenlohr und mir selbst erzählte, neben dem grossen einen kleinen medicinischen Papyrus zu besitzen. Bei meinem Aufenthalte in Theben 1869 konnte ich leider die Smith'schen Copieen nicht studiren, weil ich damals von einem bedenklichen Augenübel heimgesucht war; 1873 gelang es mir, bei einem längeren Aufenthalte in Theben den wahren Besitzer des Papyrus aufzufinden und so eines der ehrwürdigsten Denkmäler des ägyptischen Alterthums für Deutschland zu werben." J. H. Breasted gives a measured and reasonable defense of Edward Smith in his *The Edwin Smith Surgical Papyrus*, Vol. 1 (Chicago, 1930), pp. 20-25. In one point Ebers seems definitely to be in error, for Eisenlohr in his article of 1870 (see below, note 3, p. 165) said that when he returned to Thebes in February of 1870, he sought out Smith, and the latter showed him his collection of antiquities and "two papyri of medical content, one of which contained over 100 pages and the other 19 pages." Surely, these are the papyri now called the Papyrus Ebers and the Edwin Smith Papyrus. So, though he apparently did not own the Papyrus Ebers, Smith certainly appears to have had it in his possession in early 1870. I suspect that one cause of Ebers' resentment toward Smith was the fact that the articles on the Calendar published in 1870 before he (Ebers) had acquired the papyrus, and which I discuss here in the notes, designate the great medical papyrus as the Papyrus Smith. And we see how quickly (actually in the same year in which he obtained the papyrus) Ebers named the papyrus after himself!

2. H. Brugsch, "Ein neues Sothis-Datum," *ZÄS*, Vol. 8 (1870), pp. 110-11: "Die erste Reihe der Monate, welche in dem Original durch die eponymischen Bezeichnungen der 12 Monate ausgedrückt sind, enthält die Aufeinanderfolge derselben im festen Jahre, in welchem der Sothis-Stern in der Nacht vom letzten Schalttag zum 1. Thoth aufging.... Die zweite Reihe der Monate unseres Textes zeigt uns die Stellung des beweglichen Jahres zum festen. Es erhellt daraus, dass im Jahre 3 [! 9] der Regierung des Königs N. N. [! Amenhotep I] der Sothisaufgang am 3. [! 9.] Epiphi Statt fand, dass also dieser Tag mit dem 5. Schalttag des festen Jahres zusammenfiel. Die dem Epiphi folgenden Monate sind sämmtlich durch ein regelmässiges Intervall von 30 Tagen von einander getrennt." (Full article, pp. 108-11.) I should point out, however, that Brugsch misread (or took the misreading from Eisenlohr of) the number "9" in the hieratic text as "3," though Smith seemed to have already decided that the number was "9," since the learned student of hieratic texts C. W. Goodwin tells us in "Notes on the calendar in Mr. Smith's papyrus," *ZÄS*, Vol. 11 (1873), p. 107 (full article, pp. 107-09), that, when Smith communicated the calendar text to him in 1864, he had pointed out to him (Goodwin) that the correct reading was "9." Indeed Goodwin confirmed this by citing specific examples of the sign where it must be read as "9" (pp. 107-08). To return to Brugsch's article, we note further that he believed that the King whose name is mentioned was probaby a king during the next Sothic period (i.e., 1460 years later than Amenhotep I, who turned out to be the king mentioned in the

rubric). Furthermore, he followed the anachronistic custom of early Egyptologists by expressing civil months, which are given in the documents in terms of the numbered months of the three seasons, by the later month names (without month numbers and seasons). Such names were only rarely used before the twenty-fifth dynasty. See above, Chapter Three, note 4.

3. A. Eisenlohr, "Der doppelte Kalender des Herrn Smith," *ZÄS*, Vol. 8 (1870), pp. 165-67. He also has the error of reading the hieratic sign for "9" as "3." He was unable to interpret the hieratic signs within the cartouche (indicated in the text by embracing parentheses), which were later to be interpreted as Amenhotep I's prenomen or throne-name, i.e., Djeserkare (see note 5 below). Indeed Eisenlohr assumed rather tentatively that the pharaoh in question was one of the Ptolemaic kings and that the year of his reign in which the double calendar fell was about 117-14 B.C. Jul., though he did add: "or 1460 Jul. years earlier, i.e., 1577-1542 (! 1574?)." However he could not identify the signs in the cartouche with the names of Ptolemy Soter II Philometor, or indeed with any earlier monarch.

4. R. Lepsius, "Über den Kalendar des Papyrus Ebers und die Geschichtlichkeit der ältesten Nachrichten," *ZÄS*, Vol. 13 (1875), pp. 145-57. Lepsius reports and accepts the views of Smith, Ebers, Goodwin, and others that the hieratic number sign which had first been read by Brugsch and Eisenlohr as "3" (and by himself as "6" in "Einige Bemerkungen über denselben Papyrus Smith," *ZÄS*, Vol. 8 [1870], pp. 167-70) was in fact "9." The reading of the pharaoh's name in the cartouche in the rubric he still considered as uncertain.

The reader will also find interesting the confirmation by Lepsius of his belief in the keeping of a Sothic fixed seasonal calendar by the ancient Egyptians, while he explains the possible use of the Ebers calendar by a physician reading the medical papyrus (pp. 149-50):

"Es ist jetzt allgemein anerkannt, was früher lange zeit hindurch zähen Widerspruch fand, dass die Aegypter von Alters her neben dem beweglichen auch das feste Jahr und die unmittelbar daraus hervorgehende vierjährige Schaltperiode, nicht nur in Allgemeinen, sondern kalendarisch genau kannten. Man rechnete zwar nicht im gemeinen Leben danach; aber die Priester führten die genaue Sothis-Rechnung fort von einer bestimmten Epoche an, nämlich von dem Anfange der laufenden Sothisperiode, die ihrerseits eben so nothwendig und unmittelbar aus der fortschreitenden vierjährigen Verschiebung der beiden Kalender sich ergab, weil 4 mal 365 = 1460 (Jahren) ist. Jeder gebildete Mann musste den jedesmaligen Stand des festen Kalenders kennen, ja jeder Feldbauer musste seine agrarischen Geschäfte danach regeln, und konnte sich auch über den genauen Tag desselben leicht unterrichten, oder ihn selbst berechnen. Ebenso bedurfte der 'Prophet' des festen Kalenders, um die allgemeinen und die Tempelfeste, die theils nach dem beweglichen Jahre theils nach den Jahreszeiten angesetzt waren, näher zu bestimmen; ebenso natürlich der 'Horoskop' bei seinen astronomischen und chronologischen Beschäftigungen; ebenso auch jede andre Klasse nach ihrem Berufe, und so auch der Arzt.

"Wenn es nun in unserm Papyrus T. 61, 4. heisst, das eine gewisse Augensalbe in den Monaten Phamenoth und Pharmuthi gebraucht werden soll, während eine andre das ganze Jahr hindurch dienen soll, und 61, 14 dass ein Mittel um das Gesicht zu kräftigen in den Monaten Tybi und Mechir zu gebrauchen ist, so ist es klar, dass sich diese Gebrauchsvorschriften auf die Jahreszeiten bezogen, in welche diese Monate fielen; sie konnten nicht an die genannten Monate des Wandeljahrs in der Weise gebunden sein, dass sie mit ihnen allmählig das ganze Jahr durchwanderten..."

He goes on to say (and the discussion is somewhat confusing, as Meyer later pointed out) that the author of the medical tract uses the months for the original prescriptions that are to be taken in given seasons, but in order to find when the seasonally oriented months occur during a given civil year, we must know when in that civil year the helical rising of Sirius takes place and start our Sothic year from that point in the civil year, lining up the two calendars month by month. Needless to say, if this is indeed the way the calendar was to be used by doctors at the time of the copying of the medical papyrus, it could also be used at a later time by noting when the rising of Sothis took place, adjusting the months to fit the time elapse. This, in brief, is how a doctor may correctly use the seasonal prescriptions in any given civil year, or so Lepsius quite reasonably believed.

5. J. Krall, "Der Kalender des Papyrus Ebers," *Recueil de travaux relatifs à la philologie et à l'archéologie égyptiennes et assyriennes*, Vol. 6 (1885), pp. 57-63, and particularly p. 61.

For Ebers' earlier suggestion that the time of the calendar at the earliest was that of Amenhotep I, whose prenomen he identified as Re sor ka [=Djeserkare], see his article mentioned in note 1, "Papyrus Ebers," *ZÄS*, Vol. 11 (1873), p. 41:...der Stil der einzelnen Lettern [und auch die Zahlen] sowie der Gebrauch gewisser in späteren Texten kaum mehr vorkommender Gruppen bestimmen mich, und mein vortrefflicher Freund and College Ludwig Stern theilte diese Meinung, seine Enstehung in die ersten Jahrhunderte des neuen Reichs zu verlegen. An die Ptolemäerzeit wird keiner denken, der den Papyrus selbst gesehen hat; dagegen spricht auch für sein hohes Alter ein schon früher nach Copieen bekannt gewordenes Königsschild, das sich doch wohl am ersten mit dem Vornamen Amenhotep I...Ra sor ka [=Djeserkare] zusammenstellen lässt." However he made no mention of this suggestion in the introduction to his edition two years later.

6. Meyer, *Ägyptische Chronologie*, pp. 46-48:

"Im übrigen aber bietet der Kalender der Interpretation Schwierigkeiten, die zu lösen bisher nicht gelungen ist. Der Kalender folgt nebenstehend [cf. Fig.III.11].

"Wie man sieht, wird der 9. Epiphi als Tag des Sothisaufgangs bezeichnet; wie es aber zu verstehen ist, dass diese Bermerkung durch den die Wiederholung anzeigenden Punkt auch zu allen folgenden Daten gesetzt ist, ist völlig unverständlich, und man wird geneigt sein, das für ein Versehen des Schreibers zu halten. Nun ist aber auch von jeden der fölgenden elf Monate der 9. Tag genannt mit Übergehung der 5 Epagomenen, die hier so wenig wie sonst irgendwo zum 'Jahre' gerechnet sind (ebenso fehlen sie z. B. im Deckenbild der Ramesseums...[Fig. III.21]); und vor ihnen stehen die Namen der Monatsgottheiten oder Monatsfeste, aus denen die meisten der späteren uns geläufigen ägyptischen Monatsnamen hervorgegangen sind. Man würde diese also zunächst für Bezeichnungen der Monate eines festen, mit dem Siriusaufgang beginnenden Jahres, und den Kalender für einen Doppelkalender des festen und des Wandeljahres halten. Aber der erste von ihnen, Techi, steht weder beim Siriusfest, neben dem wir ihn erwarten müssten, noch etwa beim Thoth des bürgerlichen Jahres, sondern zwischen beiden, neben dem 9. Mesori, und die Epagomenen sind hier auch nicht berücksichtigt. Nach ihrer Folge würde man den Gott des Mesori, Re^{c} Harmachis, neben dem 9. Epiphi erwarten müssen; statt seiner steht hier das uns bekannte 'Neujahrsfest'. Und nun hat Brugsch nachgewiesen, dass dies Zeichen in ptolemäischer Zeit in der That als Aequivalent des Mesori gebraucht wird. Auf der anderen Seite aber kann man das

'Neujahrsfest des Siriusjahrs' [hieroglyphs] von dem danebenstehenden 'Aufgang des Sirius' unmöglich trennen. Wir stehen hier vor lauter Räthseln, die wir als solche anerkennen müssen, die zu lösen ich aber gänzlich ausser Stande bin."

7. A. H. Gardiner, "Mesore as first month of the Egyptian Year," *ZÄS*, Vol. 43 (1906), pp. 136-44: "It is necessary first of all to summarize the results hitherto reached. As is well-known, the Egyptians of the Pharaonic periods did not employ month-names in dating their monuments, but numbered the months in reference to a division of the year into the three seasons of Inundation, Winter and Harvest. This cumbrous system was retained by the native Egyptians, with characteristic conservatism, down as far as the Coptic period. The practice of dating the months by names of their own seems to have become usual among the foreigners dwelling in Egypt in the Persian period, as it is found in the Aramaic papyri. In the Greek documents of the Ptolemaic and Roman ages the month-names are in regular employment, and here Thoth corresponds to the first month of the year...[i.e., I Akhet] in the native mode of writing.

"The names used by the Greeks for the Egyptian months betray their native origin at a glance, and at an early stage of Egyptological science Champollion was able to show the connection of a few of them with the names accompanying the tutelary deities of the months recorded on the ceiling of the Ramesseum. The proof that at a still earlier date the Egyptians possessed, though without using them for dating, designations for the various months that are practically equivalent to month-names, was given by the Calendar of the Ebers medical papyrus, where most of the months [of the first column] are called in the same way as in the Ramesseum. A comparison of the three lists [i.e., that of the Greek month names, that of those on the Ramesseum ceiling, and that of those names on the Ebers papyrus] shows only four names (those of Athyr, Pharmouthi, Pakhon and Epiphi) that are common to all; to these may be added Khoiak, which occurs in the Ebers and Greek lists, but is represented in the Ramesseum by the name of the goddess Sekhmet.... The relative position of these five months to one another remains the same in all the lists, so that it is clear that the original scheme of months was never altered as a whole, but that such changes as it underwent are due to the substitution, at different times, of new names for old ones. It is the merit of Brugsch to have recognized that the month-names are derived from festivals celebrated in, and considered typical of, the months called after them. This affords, in several instances, an explanation for the change of name. Festivals that had fallen into disuse or insignificance ceased to be looked upon as the characteristic feasts of the months, and were replaced by others of greater popularity....

"Having pointed out that the month-names were derived from festivals, Brugsch sought to find evidence in the case of each particular month-festival that it was actually celebrated in the month to which it gave its name. Unfortunately the material here proved defective, and even with the help of the contradictory and late festival calendars in the Ptolemaic and Roman temples he was unable to find much support for his thesis. In one case indeed a surprising contradiction revealed itself:... [in each of two tombs of the 18th dynasty] the feast of...[Rnwty] (Pharmouthi) is dated as having occurred on the first day of the first month of Harvest (... [i.e., Tepy Shemu]), that is to say on the first day of the month called by the Greeks Pakhon, Pharmouthi being for them the preceding month...[i.e., IV Peret].

"As a matter of fact this anomaly is far from being an isolated one...[several others being listed; cf. the data given in Figs. III.6a and III.6b] ....

"Ignoring the last two instances as too dubious to serve as arguments in the discussion, we have yet sound evidence of various dates for the fact that the feasts of Athyr, Pharmouthi, Epiphi and Mesore were in early times celebrated, not in the months of the usual notation...[i.e., III Akhet, IV Peret, III Shemu, IV Shemu], where from the Greek testimony we should expect to find them, but at the beginning of the following months. Applying the argument used by Brugsch that the names of the months are derived from characteristic festivals celebrated in them, we now reach the conclusion that the earlier *name* of...[IV Akhet] was not Khoiak but Athyr, that of ...[I Shemu] not Khons (Pakhon) but *Rnwtt* (Pharmouthi), that of ...[IV Shemu] not Mesore but Epiphi, and that of ...[I Akḥet] not *Tḫj* (Thoth) but Mesore. The striking feature here is that the month-names retain the same relative position to one another, but in reference to the beginning of the year in the notation of the civil calendar they have in the course of ages receded one place backwards. If now we recall the inference made at the beginning of this article that the original scheme of month-names remained rigid and unaltered as a whole, whatever changes might be found in the individual names, it follows that what has been found true of four month-names must be true of the rest, and that all twelve month-names stood in early times one place ahead of their later position.

"The inference would be a rash one if it were not thoroughly confirmed by the Ebers Calendar, which has hitherto so puzzled its commentators. All are agreed that the object of this was to compare the months of a fixed year beginning from the rising of Sothis with the months of the shifting civil year....

"....The discrepancy of the place of the month-names in early and later times is [however] a very grave difficulty. It could on the one hand be accounted for by the assumption that a certain number of days were at a given moment intercalated in the civil calendar of seasons and months, so that the monthly festivals, if they were celebrated, without any break of continuity, at the traditional intervals from one [an]other, fell back into the preceding calendar months. Such an assumption would naturally vitiate the whole of our chronology. Happily there is another alternative, namely that the festivals, and not the months of the civil calendar, were transferred, as a body, from their original place. But the difficulty is to find a motive for such a proceeding....

"We have hitherto spoken of the displacement of the month-names as evidenced by the Greek testimony only: and it has been pointed out that as late as the reign of Rameses XI the feast of Epiphi was still held on the first (or second) day of the fourth summer month, while under Rameses IX the first day of the year could still be named Mesore. Yet there is conclusive evidence in the Ramesseum list, dating from the earlier period of Rameses II, that the Greek position of the month-names was already then accepted, at least theoretically. Thus in point of date the system according to which Mesore was the first month of the year and that according to which the latter was called Thoth overlap."

8. E. Meyer, *Nachträge zur ägyptischen Chronologie* (Berlin, 1908). For his suggested identification of the summer solstice with *Wp-rnpt* and Mesore, see pp. 10-13:

"Der Fund Gardiner"s hat uns, im Anschluss an die alte Entdeckung von Brugsch, nun noch eine dritte Bedeutung des Ausdrucks *(ḫeb) wepet ronpet* kennen gelehrt, in der er im Kalender des Papyrus Ebers gebraucht wird, nämlich als Bezeichnung des Monatsfestes *Mesure*^{c} (Mesore) 'Geburt des Re^{c}', das ursprünglich den ersten, später den letzten Monat des Jahres bezeichnet. Dem entspricht es, dass der Gott Re^{c}-Ḫor-achuti, der

'Sonnengott am Horizonte', im Ramesseum und im Kalender von Edfu der Schutzgott des Monats Mesore ist.

"Und nun ergiebt sich eine neue Seltsamkeit: die Verschiebung der Monatsfeste und der aus ihnen hervorgegangenen Monatsnamen ist nicht etwa in einem bestimmten Moment eingetreten, so dass die eine Bezeichnung von der anderen abgelöst würde, sondern beide stehen wenigstens im Neuen Reich neben einander. Denn während im Papyrus Ebers wie in den von Gardiner besprochenen Texten der 20. Dynastie die alte Ordnung herrscht, zeigt schon weit mehr als ein Jahrhundert vor den letzteren das astronomische Deckengemälde des Ramesseums die jüngere Ordnung, die den späteren Monatsnamen zu Grunde liegt....

"....Die Geburt der Sonne, mit der das ägyptische Jahr beginnt, muss natürlich mit der Sonnebahn in Beiziehung stehen, und kann bei einem im Hochsommer beginnenden Jahr nur die Sommersonnenwende sein, die ja auch in vielen griechischen Kalendern den Jahresanfang bezeichnet. Somit bestätigt sich die oft ausgesprochene Annahme, dass der ägyptische Kalender nicht nur den Sirius, sondern auch den Sonnenlauf berücksichtigt. Es war das um so eher möglich, da nicht nur das Siriusjahr von 365 1/4 Tag und das wahre Sonnenjahr fast gleich lang sind, sondern auch der Siriusaufgang und die Sommersonnenwende zur Zeit der Enstehung des ägyptischen Kalenders nahezu zusammenfielen. Im Jahre 4241 v. Chr. fiel der Siriusaufgang in Memphis auf den 19. Juli jul., die Sommersonnenwende auf den 25. Juli jul., also nur 6 Tage später, was für die Praxis kaum in Betracht kommt. Die Aegypter konnten daher zu Ende des 5. Jahrtausends sehr wohl des Glaubens sein, dass der Siriusaufgang mit dem Solstitium zusammensfalle, und das Geburtsfest des Re^{c} mit dem wahren Neujahr zusammen feiern. Im Laufe der folgenden Jahrhunderte rückten beide Punkte astronomisch immer näher an einander, im 35. Jahrhundert fielen beide auf den 19. Juli, und auch beim Beginn der zweiten Sothisperiode, im Jahr 2781 v. Chr., wo die Sonnenwende am 13. Juli eintrat, fielen sie für die Praxis noch nahezu zusammen. Die Verbindung von Siriusaufgang, Neujahrstag und Geburtsfest des Re^{c} (Mesore), die sich so für das Idealjahr ergab, ist dann, wie alles andere, von diesem auf sein unvollkommenes Abbild, das bürgerliche Wandeljahr, übertragen.

"Aber in den folgenden Jahrhunderten entfernt sich die Sommersonnenwende immer weiter vom Siriusneujahr. Zu Beginn der dritten Sothisperiode, 1321 v. Chr., fällt sie bereits auf den 1.

Juli jul., 18 Tage vor das Neujahrsfest, also mitten in den letzten Monat des Idealjahres. Diese Vershiebung konnte nicht unbemerkt bleiben; und so erklärt es sich, dass im Neuen Reich neben die auf dem Siriusjahr beruhende Gleichung

wepet ronpet = Geburt des Re^{c} (Mesore) =
Siriusaufgang = erster Echetmonat I [i.e., I Akhet]

die neue, dem jetzigen Stand der Sonnenwende entsprechende Gleichung

wepet ronpet = Geburt des Re^{c} (Mesore) =
vierter Somumonat XII [i.e., IV Shemu]

getreten ist. Diesen Stand giebt das Deckengemälde des Ramesseums wieder. Dies Gemälde stellt eben das ideale Normaljahr (das feste Sothisjahr) nach dem Stande der Sonne zum Siriusaufgang in der Zeit Rameses' II. dar. Damals fiel die 'Geburt des Re^{c}' in den letzten Monat des Siriusjahrs, und Re^{c} Ḥor-achuti musste daher aals Schutzpatron dieses Monats (XII) erscheinen; 'Isis-Sothis...' dagegen blieb selbstverständlich in Verbindung mit dem I. Monat. Aber dieser musste jetzt, da Re^{c} in den XII. Monat gerückt war, mit dem Techifest, das bisher den II. Monat bezeichnete, verbunden werden, und in derselben Weise verschoben sich im Idealjahr alle weiteren Schutzpatrone und Feste um eine Stelle, bis zum Epiphi- (Epet-) Fest hinab, das aus dem XII. in den XI. Monat rückte."

9. K. Sethe, "Die Zeitrechnung der alten Aegypter im Verhältnis zu der der andern Völker," *Nachrichten von der Königlichen Gesellschaft der Wissenschaften zu Göttingen, Philologisch-historische Klasse aus dem Jahre 1920*, II. Abt. pp. 30-39: "Die 12 Monate des Kalenderjahres, die offiziell in den aegyptischen Texten zu allen Zeiten bis ans Ende des Heidentums nur mit Ordnungsziffern als 1., 2., 3., 4. Monat einer der 3 Jahreszeiten bezeichnet werden, z.B. 'Monat 3 der Winterjahreszeit,' haben in den griechiscen Texten ihre bestimmten Namen, die nach gewissen Gottheiten, Festen u. dergl. benannt sind: 1. Thoth. 2. Phaopi. 3. Athyr. 4. Choiak. 5. Tybi. 6. Mechir. 7. Phamenoth. 8. Pharmuthi. 9. Pachon. 10 Payni. 11. Epiphi. 12. Mesore....

"Auf den aegyptischen Denkmälern lässt sich eine entsprechende, mit den Monaten des Jahres verknüpfte Namenreihe, die zum grossen Teil schon dieselben Elemente enthält, in der aber einige Namen noch durch ältere, von ihnen erst später ersetzte Vorgänger vertreten sind, bis in den Amfang des Neuen

Reiches zurückverfolgen. Das älteste Vorkommen enthält... Korrespondenzkalender des Papyrus Ebers aus dem 9. Jahr Amenophis I (1540 v. Chr.). Dort erscheinen die Namen nicht, wie man bisher meist annahm, als Monatnamen, sondern als Namen von Festen; sie bezeichnen die damals auf den 9. Tag der bürgerlichen Kalendermonate fallenden Monastanfänge des mit Siriusaufgange beginnenden festen Normaljahres. Als Beizeichnungen bestimmter monatlicher Festtage, und zwar anscheinend des bürgerlichen Wandeljahres, kehren sie, wie Gardiner gezeight hat, auch einzeln in gewissen geschäftlichen Schriftstücken der 19/20 Dynastie in Verbindung mit dem entsprechenden Kalenderdatum wieder...." Then after summarizing Gardiner's evidence showing that feast days following on the first days of the months following the months to which the feast days gave their names, Sethe declares (p. 32):

"Wir haben hier also den unzweifelhaft recht merkwürdigen Tathestand, dass die Feste, die den Monate den Namen gegeben haben, selbst gar nicht in diesen Monaten lagen. Man had deshalb an eine wirkliche Verschiebung der Feste oder ihrer Namen gedacht, und zwar hat man, in der Voraussetzung, dass die namengebenden Feste am Anfang der nach ihnen benannten Monate gestanden haben müssten, eine Verschiebung um einen ganzen Monat angenommen. Ed. Meyer hat dafür auch eine sinnreiche, auf den ersten Blick geradezu bestechende Erklärung gegeben. Er will diese Verschiebung auf die Präzession det Tage- und Nachtgleichen zurückfähren. Durch diese habe sich auch die Sommersonnenwende, auf die sich die Bezeichnung 'Geburt der Sonne' augenscheinlich beziehen müsse, verschoben. Ursprünglich bei Begründung des Kalender auf den Neujahrstag gelegt (19 Juli), dem die Sonnenwende damals sehr nahe lag (25 Juli), sei das Fest seitdem auf diesen Tage liegen geblieben, bis es schliesslich zu einer gegebenen Zeit den veränderten tatsächlichen Verhältnissen entsprechend um einen Monat vorgeschoben wurde, die ganze Reihe der andern Monatsfeste, soweit sie alt waren, nach sic ziehend.

"Diese auf den Namen Mesore als die mutmassliche Bezeichnung eines Sonnenstandes zugeschnittene Erklärung passt aber zu den übrigen Festbezeichnungen nicht recht. Denn für diese wäre nach allem, was wir über aegyptische Fest wissen und was oben gerade wieder für eine Reihe von Fällen evident festgestellt werden konnte, keine solche Verschiebung im Kalender zu erwarten. Und tatsächlich lässt sich denn auch für mehrere der oben gennanten Monatsfeste zeigen, dass sie auch noch in der spätesten Zeit ihren alten Platz auf dem 1. Tage der Monats, der dem nach ihnen benannten Monat folgte behalten haben , wie sie ihn zur Zeit Amenophis I, und in der 19/20 Dynastie in den Gardiner'schen Schriftstücken einnahmen." After studying in some detail shifts of certain feast days, Sethe summarizes as follows (p. 35):

"....Die Monate sind, wenigstens in der älteren Schicht der Namen, nicht nach dem Feste, mit dem sie selbst begannen, sondern nach dem, durch das sie beschlossen wurden, benannt worden. Die Namen bezeichnen sie also als den Monat, der zu dem betr. Feste führt, als seine Vorbereitung, die mit dem eigentlichen Festtage ihre Krönung findet." Finally, the reader should note Sethe's account of the Egyptian identification of the "Birth of the Sun" with the winter solstice rather than with the summer solstice (see pp. 37-38):

"Es ist oben schon darauf hingewiesen worden, dass der Aegypter der älteren Zeit den Tageslauf der Sonne mit einen Menschenleben vergleicht (S. 29). Die Sonne wird am Morgen als Knäblein geboren, wächst bis zum Mittag zum Manne heran, altert dann und sinkt Abends als Greis ins Grab. Mit einer solchen Auffassung verträgt es sich eigentlich nicht, denselben Vergleich auch auf den Jahreslauf der Sonne anzuwenden, wie das in der Beizeichnung 'Geburt der Sonne' für den Neujahrstag der Fall zu sein scheint. Es findet sich denn auch sonst nirgends im aegyptischen Schrifttum der älteren Zeiten eine Spur eines solchen Vergleiches. Das aber ist klar: fände sie sich, so könnte dieser Vergleich naturgemäss nur in dem Sinne durchgeführt sein, das die 'Geburt der Sonne' die Wintersonnenwende bedeute, nicht aber die Sommersonnenwende; denn diese könnte doch nur die Mittagshöhe im Leben der Sonne darstellen, von der sie alsbald absteigt, um in Herbst zu altern und am kürzesten Tage des Jahres zu sterben."

Sethe quotes later passages from Macrobius (Sat. I 18) and Plutarch (Is. et Osir. 65). So the final conclusion of Sethe's is that no shifting of the order of the months, as posited by Meyer, took place.

10. R. Weill, *Bases, méthodes et résultats de la chronologie égyptienne* (Paris, 1926), pp. 112-26.

11. L. Borchardt, *Quellen und Forschungen zur Zeitbestimmung der ägyptischen Geschichte*, Vol. 2: *Die Mittel zur zeitlichen Festlegung von Punkten der ägyptischen Geschichte und ihre Anwendung* (Kairo, 1935), pp. 19-20. He says concerning the

four columns of the Calendar: "Keine der vier Spalten ist ohne Unregelmässigkeiten, die noch der endgültigen Erklärung harren. Die meisten sind in Spalte 3 zu finden, daher mag mit dieser begonnen werden.

"Diese Spalte ist dafür, dass sie angeblich nur den Monatstag angibt, merkwürdig weit von Spalt 2 getrennt. Monatstage pflegen sonst dicht hinter den Kalendermonat geschrieben zu werden. Die 9, die hier geschrieben ist, ist eigentlich nicht das Zeichen für den betreffenden Monatstag, sondern das gewöhnliche Zahlzeichen. Endlich ist die Zeitspanne zwischen Zeile 2 — bisher 9.12.W. gelesen — und Zeile 3 — bisher 9.1.W. gelesen — wegen der Shalttage zwischen 12. und 1. Monat des Wandeljahres um 5 Tage länger als die übrigen Zeitspannen von Zeile zu Zeile, trotzdem die erste Spalte der 'Monatsfeste' Zeile 2 — *tḫy* — und Zeile 3 — *mnḫ.t* — den üblichen Abstand von einem Monat erwarten lässt.

"Die erste Unregelmässigkeit, der Spaltenabstand, deutet darauf hin, dass die 'Neuen' eben nicht den Monatstag bedeutet, sondern etwas anderes.

"Die zweite Unregelmässigkeit, die Schreibung der 'Monatstages', deutet auf dasselbe.

"Die dritte, die falsche Zeitspanne zwischen Zeile 2 und Zeile 3, macht die Lesung "Neuen" unmöglich.

"Nun wird in den Berliner Illahun-Papyrus an 3 stellen mit zeitchen für 9 *'pśḏ'* der erste Mondtag, der Neumondtag *'pśd(n)tjw'* geschrieben, in einem der dortigen Beispiele wird sogar vielleicht nur *'pśḏ'* geschrieben. Es kann also nicht bestritten werden, das die 'Neuen' in Spalte 3 des Ebers-Kalenders auch nur *pśḏ* 'Neumontag', 'erster Montag' zu lesen sind. Damit wird auch die Schwierigkeit der zu grossen Zeitspanne zwischen Zeile 2 and 3 aus der Welt geschafft. Es ist also in keinem der 12 Monate ein fester Tag angegeben, sonder stets, wie das z. B. auch im Kalender von Medinet Habu sowie in Edfu verkommt, nur: der Neumond in jedem dieser Monate, d.h. gleiche Zeitspannen von rd. 29 1/2 Tagen, oder kalendarisch ausgedrückt, von abwechselnd 29 und 30 Tagen.

"Danach ist nun auch Spalte 1 etwas anders aufzufassen als dies bisher meist geschehen. Es handelt sich hier nicht um Monatsfeste oder Namen von Monastgottheiten, sondern um die von diesen abgeleiteten Namen der Mondmonate, die den jeweils ihnen gegenüber gesetzten Neumonden (Spalte 3) in den Kalendermonaten (Spalte 2) beginnen."

Document III.2

The Ebers Calendar

(1) Year 9 under the Majesty of the King of Upper and Lower Egypt Djeserkare (=Amenhotep I).[1]

(2) The Feast of the Opening of the Year[2] (i.e., of New Year's Day). III Shemu 9 (i.e., Day 9 of the third month of the third season Shemu in civil year 9 of Amenhotep I).[3] The Going Forth of Sothis (i.e., the heliacal rising of Sirius).

(3) [The Feast of] Tekhy (i.e., Drunkenness, culmination of celebrations for Thoth?). IV Shemu 9 (i.e., Day 9 of the fourth month of Shemu in civil year 9).

(4) [The Feast of] Menkhet (i.e., Clothing). I Akhet 9 (i.e., Day 9 of the first month of the first season Akhet in civil year 10 of Amenhotep I).[4]

(5) [The Feast of] Hathor. II Akhet 9 (i.e., Day 9 of the second month of Akhet in civil year 10).

(6) [The Feast of] Kaherka. III Akhet 9 (i.e., Day 9 of the third month of Akhet in civil year 10).

(7) [The Feast of] Shefbedet. IV Akhet 9 (i.e., Day 9 of the fourth month of Akhet in civil year 10).

(8) [The Feast of] Rekeh [wer] (i.e., the Greater Burning). I. Peret 9 (i.e., Day 9 of the first month of the second season Peret in civil year 10).

(9) [The Feast of] Rekeh [nedjes] (i.e., the Lesser Burning). II Peret 9 (i.e., Day 9 of the second month of Peret in civil year 10).

(10) [The Feast of] Renutet. III Peret 9 (i.e., Day 9 of the third month of Peret in civil year 10).

(11) [The Feast of] Khonsu. IV Peret 9 (i.e., Day 9 of the fourth month of Peret in civil year 10).

(12) [The Feast of] Khenet khety. I Shemu 9 (i.e., Day 9 of the first month of the third season Shemu in civil year 10).

(13) [The Feast of] Ipet hemet. II Shemu 9 (i.e., Day 9 of the second month of Shemu).

Notes to Document III.2

1. This line is written in red.

2. For the phonetic transcriptions of all the feast days, see Fig. III.12.

3. I have here used the system of civil year dates followed throughout this volume, i.e., month in roman numerals, name of season, day number in arabic numerals.

4. In using day "9" here and in all of the entries, it is evident, as I have said in the Introduction to this document, that the five epagomenal days between IV Shemu 30 and I Akhet 1 have been neglected by the author of the calendar. In this respect, the Ebers Calendar resembles water clocks, as I have indicated in my discussion of water clocks in Chapter 3.

Document III.3: Introduction

The Astronomical Ceiling of the Secret Tomb of Senmut

In Chapter Three I discussed the astronomical ceiling of the secret tomb of Senmut, the Steward of the Estate of Amun under Queen Hatshepsut and her right-hand man during the building of her remarkable temple at Deir el-Bahri in Western Thebes, in two contexts: (1) the use of the ceiling by Richard Parker in his less than successful effort to reconstruct the Old Lunar Calendar[1] and (2) its importance as the oldest extensive representation of the ancient Egyptian celestial diagram.[2] I shall not repeat here the general description contained in those earlier discussions, except to note once more (1) that we are discussing the astronomical ceiling of Theban tomb no. 353 (the so-called secret tomb of Senmut) rather than tomb no. 71 (Senmut's completed tomb), (2) that the tomb (and thus the ceiling decorations) dates from about 1473 B.C., and (3) that the secret tomb was discovered during the excavations of the Metropolitan Museum in 1925-27 A.D.

Here I shall simply translate (or phonetically transliterate when I am unable to translate) the hieroglyphic writing of the various parts or elements of the astronomical ceiling. I shall start with the Southern Panel (as shown in the top of Fig. III.4) and proceed column by column and register by register. From this we shall obtain a detailed description of the entries devoted to the 36 "weekly" (i.e., 10-day) decans originally used to determine 12 nighttime hours in the course of the 12 months or first 360 days of the civil year. And following this, the succeeding columns of the panel have entries pertaining to 6 of the 12 triangle decans used to supplement the main 36 decans and to mark the nighttime hours during the five epagomenal days at the end of the year, as well as entries embracing four of the five planets known to the ancient Egyptians (the exterior planets are in the columns before the epagomenal decans and the interior planets follow after those decans).

I have already exerted some effort to characterize the sundry entries on the Southern Panel in Chapter Three above, stressing the difficulty of identifying these star-decans with any certainty. We reiterate the earlier conclusions that the decanal stars occupied a band that seems to be roughly parallel to and south of the ecliptic, that several of the decanal stars are parts of or are near to our constellation of Orion, and that the yearly decanal columns in the celestial diagram culminate in that of Sirius. And hence the reader can turn immediately to the document. I further remind the reader that the graphical and textual material beneath the names of the decans and planets reflects the intimate relationship that existed during Pharaonic times between the constellations, stars, and planets named and the divinities associated with them. That relationship in the astronomical domain bears out my general conclusion in Chapter Two of Volume One, that no physical science independent of religious thought existed in Ancient Egypt.

Following the presentation of the Southern Panel

the remainder of Document III.3 gives the elements of the Northern Panel (see Fig. III.4, bottom). As in the case of the Southern Panel, I have already briefly described the main elements of the Northern Panel in Chapter Three: (1) the centrally located northern constellations, (2) the flanking month-circles (eight on the right and four on the left), and (3) below them flanking deities that have some relationship with some protective lunar day-gods but seem to have no particular astronomical significance in the context of the celestial diagram. Hence we can once more turn directly to the document.

Between the border stars of the two panels (again see Fig. III.4) are five horizontal lines of prayer for the deceased, the middle one of which gives the partial titularly of Hatshepsut. Since they are of no astronomical importance I refrain from translating them.

This document seems to have played an important role in the history of the celestial diagram, as I have also pointed out in Chapter III above, since it not only contains the oldest more or less complete copy of the diagram extant but is the earliest example of the most important family of monuments containing that diagram.

The works on which I have depended in preparing this document have been mentioned in passing in the general description of the ceiling in Chapter Three. But the most important of them should be noted here:

H.E. Winlock, "The Egyptian Expedition 1925-1927," *Section II of the Bulletin of the Metropolitan Museum of Arts* (February, 1928), pp. 32-44, full article pp. 3-58.

G. Roeder, "Eine neue Darstellung des gestirnten Himmels in Ägypten aus der Zeit um 1500 v. Chr.," *Das Weltall*, 28. Jahrgang (1928), Heft 1, pp. 1-5.

A. Pogo, "The Astronomical Ceiling-decoration in

the Tomb of Senmut (XVIIIth Dynasty)," *Isis*, Vol. 14 (1930), pp. 301-25.

O. Neugebauer and Parker, *Egyptian Astronomical Texts*, Vol. 3 (Providence and London, 1969), pp. 10-12, 105-18.

S. Schott, "Erster Teil: Die altägyptischen Dekane," in W. Gundel, *Dekane und Dekansternbilder, Ein Beitrag zur Geschichte der Sternbilder der Kulturvölker* (Glückstadt and Hamburg, 1936), pp. 1-21.

W.M.F. Petrie, *Wisdom of the Egyptians* (London, 1940); I have reported a few of Petrie's suggested identifications of decans, though most of them are not now credited; I have also reproduced as my Figs. III.84a-b his effort to represent the whole Egyptian decanal belt and principal constellations.

In my translation I have added column numbers and decan numbers, as well as numbers designating the northern constellations, starting at the top with the Big Dipper and proceeding more or less in clockwise direction through the rest of them. I have continually spoken of a decan or deity as "determined by" one star. By this I of course mean that the star glyph is a determinative for the decan's name or for the deity's name.

Notes to the Introduction of Document III.3

1. Chapter Three, note 15.
2. See the section "Astronomical Ceilings" in Chap. 3.

Document III.3

The Astronomical Ceiling of the Secret Tomb of Senmut

[The Southern Panel, Fig. III.4, top]

[The First Thirty-five (! Thirty-six) Decans in Columns 1-29]

[Col. 1, from right] [Decan 1: The Star called] The Forerunner of Kenmet *(tpy-ᶜ knmt)*.[1] [Determined by] one star. [Deities:] Hapy and Imseti [written together and determined by] two stars. [Below and covering cols. 1-2: a declining string of] 5 stars [with] a sixth star [immediately below in col. 2. Below at the bottom of cols. 1-2, another declining string of] 5 stars.

[Col. 2] [Decan 3:] The One under the Tail (or Buttocks) of Kenmet *(ẖry ḫpd knmt)*. [Determined by] one star. [Inserted below Decan 3 is Decan 2:][2] Kenmet *(knmt)* [Determined by] one star. [Deity:] Isis [with strings of declining stars below, as indicated in the text of col. 1].

[Col. 3] [Decan 4:] The Beginning of Djat *(ḥꜣt ḏꜣt)*. [Determined by] one star. [Decan 5:] The End of Djat *(pḥwy ḏꜣt)*. [Determined by] one star.[3] [Deity:] Duamutef [for Decan 4] and The Children of Horus [for Decan 5].[4] [The latter children are determined by] 4 stars.

[Col. 4] [Decan 6:] The Upper Tjemat *(ṯmꜣt ḥrt)*.[5] [Determined by] one star. [Decan 7:] Lower Tjemat *(ṯmꜣt ẖrt).* [Determined by] one star. [Deity:] Duamutef [for Decan 6?] and the Children of Horus [for Decan 7]. [The latter is determined by] 4 stars.

[Col. 5] [Decan 8:] The Two Wesha-birds *(wšꜣti)*.[6a] [Determined by] one star. [Decan 9:] The Two Pregnant Ones *(bkꜣti)*. [Determined by] one star.[6b] [Deity:] Duamutef [for Decan 8] and Hapy[7] [for Decan 9]. [Determined by] two stars [presumably one for each god].

[Col. 6] [Decan 10:] The Forerunner of Khenett *(tpy-ᶜ ḫntt).* [Determined by] one star. [Decan 11:] Upper Khenett *(ḫntt ḥrt).* [Determined by] one star.[8] [Deity:] Horus [followed by] three stars.

[Col. 7] [Decan 12:] Lower Khenett *(ḫntt ẖrt).* [Determined by] one star [and plural sign]. [Deity:] Seth. [Followed below by] three stars [covering cols. 7-8]. [At bottom of cols. 7-12 is the constellation] Ship [with] 5 stars [horizontally above it, and] 4 stars [vertically arranged from the prow downward, and] one star [more horizontally next to the first of the vertically arranged stars]. [In the middle of the Ship is] one star [and behind that star is another] star. [And see star descriptions under col. 13.]

[Col. 8] [Decan 13:] The Red One (Star?) of Khenett *(ṯms n ḫntt).* [Determined by] one star. [Deity:] Horus. [Ship and stars as noted for col. 7.]

[Col. 9] [Decan 14:] The Sapty of Khenuy *(sꜣpty ḫnwy)*.[9] [Determined by] one star. [Deities:] Isis and Nephthys.

[Col. 10] [Decan 15:] The [Star in the] Middle of the Ship *(ḥry-ib wiꜣ).* [Determined by] one star. [Deity:] Seth (or Horus?)[10].

[Col. 11] [Decan 16:] The Guides *(sšmw)*.[11] [Determined by] one star. [Deity:] Seth.

[Col. 12] [Decan 17:] The Kenmu [Stars] *(knmw)* [Determined by] one star [and a plural sign].[12] [Deity:] Children of Horus.

[Col. 13] [Decan 18:] The Forerunner of the [Star of the] Half-month *(tpy-ᶜ smd).* [Determined by] a star surmounted by a half crescent. [Deity:] Horus [determined by] one small star [followed by] 13 larger stars [irregularly placed in the vertical direction and some spilling over into cols. 11-12 on the right and col. 14 on the left].

[Col. 14] [Decan 19:] The [Star of the] Half-month *(smd).* [Determined by] one star. [Deity:] Hapy [determined by] one star surmounted by a half crescent (cf. note 7). [Below in cols. 14-16, the constellation of] The Sheep (or Ram) [with] one star [above] and one star [below the hind-part of the Sheep].

[Col. 15] [Decan 20:] The [Principal Star of the] Sheep *(sỉt).* [Determined by] one star. [Deity:] Isis [followed by the expression:] the 3rd star cluster *(3-nwt ẖt).*[13]

[Col. 16] [Decan 21:] The Two-children (a star-doublet?) of the Sheep *(sꜣwy sỉt).* [Determined by] one star. [Deity:] Duamutef.

[Col. 17] [Decan 22:] The One under the Buttocks of the Sheep *(ẖry ḫpd srt).* [Determined by] one star. [Deity:] Qebehsenuf [followed vertically by] 4 stars [with a space between the 3rd and the 4th].

[Col. 18] [Decan 23:] The Forerunner of the Two Spirits *(tpy-ᶜ ꜣḫ(wy)).* [Determined by] one star. [Deity:] Duamutef [followed vertically by] 4 stars [with a space between the 3rd and the 4th].

[Col. 19] [Decan 24:] The Two Spirits *(ꜣḫwy).* [Determined by] two stars. [Deities:] Duamutef and Qebehsenuf [determined by] two stars.

[Col. 20] [Decan 25:] The Two Souls *(bꜣwy).* [Determined by] two stars. [Deities:] Hapy and Imseti. [Bottom of the column:] 7 stars.

[Col. 21] [Decan 26:] The Upper Khentu *(ḫntw ḥr(w))* [and Decan 27:] The Lower Khentu *(ḫntw ẖrw).* [Together determined by] three stars. [Deities:] The Children of Horus. [Somewhat below and occupying cols. 21-22 is the legend:] the 4th star cluster *(4-nwt ẖt).* [Immediately below this, covering cols. 21-23, is an] egg-shaped figure [roughly depicted within] 4 stars. [Below this, and still covering cols. 21-23, is a declining line of] 7 stars. [Then below that in col. 21 are] two stars [and the legend:] the 4th star cluster (?). [Next to this in col. 22 is a vertical line of] 3 stars.[14]

[Col. 22] [Decan 28:] Qed *(ḳd).* [Decan 29:] The children of Qed *(sꜣwy ḳd).* [Together (?) determined by] two stars. [Deities:] Hapy [for Decan 28] and Qebeh[senuf] [for Decan 29]. [As noted above, in col. 22 there are] three stars [below the declining line of stars found in cols. 21-23].

[Col. 23] [Decan 30:] The Thousands *(ḫꜣw)* [Followed immediately by] The Children of [plus space, and below the horizontal line:] Horus [thus representing the relevant deities]. [Immediately below is written:] The 5th star Cluster *(5-nwt ẖt).*

[Col. 24-28 give some erroneous and misplaced readings, all related to the constellation of Orion *(sꜣḥ),* which I initially repeat here but subsequently note at the end of these columns the suggested order and readings of Neugebauer and Parker, *op. cit.*, Vol. 3, pp. 112-15.]

[Above the horizontal line running from col. 24 through col. 28:]

[Col. 24] The One below the Arm of Orion *(ẖry*

rmn s'ḥ). [And no stars.] [Deities:] Children of Horus. [Determined by] two stars.

[Col. 25] [Repeated:] The One below the Arm of Orion *(ẖry rmn s'ḥ).* [Deity:] Osiris [and no star].

[Col. 26] [No decan, but the following is included, perhaps as a collective enumeration of the stars related to Orion and called "the 6th star cluster":] Two large stars [on the first level]. 8 small circles or stars [on the next two levels].[15] 4 large stars [on next two levels].[16]

[Col. 27] Aret *(ʿrt).* [Determined by] one star. [Deity:] The Eye of Horus. [No star].

[Col. 28] The One above the Arm of Orion *(ḥry rmn s'ḥ).* [Determined by] one star. [Deity:] The Children of Horus.

[Below the horizontal line from col. 24-28:]

[The figure of] Orion standing in a ship (Fig. III.4) with a was-scepter in the left hand and an ankh-sign in the right. [Around the ship are the following stars:] 3 large stars [above the whole figure] with a small star to the right of the third one, [a vertical line of] 6 small stars [to the right of the was-scepter], one star [on the bow] and one star [on the stern].

[The reconstruction by Neugebauer and Parker of Decans 31-35:]

[Decan 31:] Aret *(ʿrt).* [Determined by] one star. [Deity:] The Eye of Horus. [Determined by] one star. (Cf. col. 27 above.)

[Decan 32:] The One above the Arm of Orion *(ḥry rmn s'ḥ).* [Determined by] one star. [Deity:] The Children of Horus. (Cf. col. 28 above.)

[Decan 33:] The One below the Arm of Orion *(ẖry rmn s'ḥ).* [Determined by] one star. [Deities:] The Children of Horus. (Cf. col. 24 above.)

[Decan 34:] The Arm of Orion *(rmn sꜣḥ)*. [Determined by] one star. [Deity:] The Eye of Horus (cf. col. 25 above). [Followed by the group of stars that is not a single decan but is called by Neugebauer and Parker "the constellation [[of Orion]]." Cf. col. 26 above.]

[Decan 35:] Orion *(sꜣḥ)*. [Determined by] one star. [Deity:] Osiris. [Missing as a separate decan in the Senmut list.]

[Returning to the Senmut ceiling, go to:]

[Col. 29] [Decan 36:] Isis-Sothis (i.e., Sirius). [Immediately below is a standing] goddess [holding a was-scepter]. [Determined by] one star. [At the bottom of the column is] Isis-Sothis standing in a boat. [She is wearing a crown with double plume and a feather like that of Maat, surmounted by a solar disk, her right hand near the rear of the crown and left hand clutching an ankh-sign and a flat headed staff.]

[Exterior Planets in Cols. 30 and 31]

[Col. 30] [At the right hand side of the column is the Horus name of Hatshepsut] Female Horus: The One who is Powerful of Kas.[17] [Exterior planet Jupiter:] *Horus who bounds the Two Lands* is his name; Southern Star of the Sky *(Ḥr tꜣš tꜣwy rnf rwy pt n sbꜣ)*.[18] [The latter epithet is determined by] one star. [Depicted at the bottom of the column is the falcon-headed] Horus [surmounted by] a star [and] standing in a boat.

[Col. 31] [At the right hand side of the column is once more the Horus name of Hatshepsut] Female Horus: The One who is Powerful of Kas.[17] [Exterior planet Saturn:] *Mut (Horus) Bull of the Sky* is his name.[19] The Eastern Star which crosses the Sky.[20] [The latter epithet is determined by] one star. [Again, depicted at the bottom of the column is the falcon-headed] Horus [surmounted by] a star [and] standing in a boat.

[Epagomenal Decans in Columns 32-36]

[Col. 32] [Additional decan 1:] The Two Tortoises *(šṯwy)*. [No stars.] [Deities] Hapy and Duamutef. [No stars.] [Depicted at the bottom of the column are] two tortoises.

[Col. 33] [Add. decan 2:] Neseru *(nsrw)*. [Below:] It is a cluster *(ẖt pw)*.[21] [Determined, however, by] one star.

[Col. 34] [Add. decan 3:] Shespet *(šspt)*. [Determined by] one star. [Deities:] Eyes of Horus [followed by] three stars.

[Col. 35] [Add. decan 5:] Abshes *(ᶜbšs)*. (!, Probably should be Sebshesen, i.e., *sbšsn*). [Deity:] Horus. [Inserted below but probably meant to be before add. decan 5 is add. decan 4:] Hepdes *(hpds)*. (!Perhaps should be Ipedes, i.e., *ỉpds*.) [Followed by] three stars.[22]

[Col. 36] [Add. decan 6:] The Honored of God *(nṯr wỉš)* [No stars.] [Deity:] Duamutef.

[The Interior Planets in cols. 37 and 38]

[Col. 37] Mercury *(Sbg)* [Determined by] one star. [Deity:] Seth.

[Col. 38, split in two at the top] [On the top right:] [Venus:] The crosser *(ḏỉ)* [Deities:] Bah *(bḫ)* [on the left top] and Osiris [on next level below]. [Depicted below:] Heron [surmounted by] a star.

[The Northern Panel, Fig. III.4, bottom]

[The Northern Constellations, Fig. III.4, center][23]

[For the constellations with bracketed numbers, see Fig. III.66, where the Senmut constellations are rearranged.]

[0] The Adze[24] (or Foreleg, i.e., the Big Dipper) *(msẖtyw)*. [Depicted here as a bull with an oval body from which tiny legs protrude, as well as a tail of three circles (stars); the third is encircled in red and from it are two diverging vertical lines that proceed to the bottom of the panel.[25] The name Meskhetyu is written in glyphs within the oval body, with three stars above it.]

[1] [The Goddess] Serqet *(srḳt)*. [Here above and parallel to Meskhetyu and depicted as a goddess having a red solar disk on her head and her name in hieroglyphs above the disk, i.e., to its left.]

[2] [The ram-headed God] Anu [below and parallel to Meskhetyu. Apparently he is in the act of spearing the bull. His name in hieroglyphs is to the left of the spear.][26]

[3] Isis-Djamet Festival of the Sky *(ꜣst-ḏꜣmet ḥb pt)*. [This is the puzzling legend over the constellation Hippopotamus which bears a crocodile on her back.[27]] [The hippopotamus has in her right paw] a mooring post and [in her left paw] a crocodile standing on his tail.

[4] The crocodile [lunging at Man] [with legend above Crocodile:] "Restful of Feet" *(ḥtp-rdwy)*.[28]

[5] Man [standing with upraised arms, perhaps to spear Crocodile, though the spear is missing].

[6] Lion [depicted with crocodile tail, and with a

legend above:] "Divine Lion, who is between them" [i.e., the crocodile above and the crocodile below].[29]

[7a] Crocodile with curved tail, "The Plunderer" *(ḫ(ꜣ)ḳw)*, [with epithet preceding the crocodile, which is very much larger than the glyphs of the epithet].

[7] [Another] Crocodile with curved tail. "The Gatherer" *(sꜣḳ)*.[30]

[Flanking the Northern Constellations on both sides are] 12 [monthly] circles, each divided into 24 sectors.[31] [To the right of the Northern Constellations on the first level, the 4 circles (i.e., months) in the season of Akhet, beginning from the right, have above them the following feast-names:[32]] (1) Tekhy *(tḫy)*, (2) Menkhet *(mnḫt)*, (3) The sky together with her stars: Hathor *(pt ḥnᶜ ꜣḫꜣḫw ḥt-ḥr)*, (4) Ka-[her]-Ka *(kꜣ-[ḥr]-kꜣ)*. [Then on the left side on levels 1 and 2, the 4 month circles of the season of Peret have over them these names:] (1) Shefbedet *(šf bdt)*, (2) Rekeh [wer] *(rkḥ [wr])*[33], [3] Rekeh [nadjes] *(rkḥ [nḏś])*, [4] Renutet *(rnwtt)*. [Finally, on the right side, level 2, beginning from the constellations, are the 4 month circles of the season of Shemu with the following names:] (1) Khensu *(ḫnśw)*, (2) Khent-Khety *(ḫnt-ḫty)*, (3) Ipet-hemet *(ipt ḥmt)*, (4) Wep-renpet *(wp rnpt)*.

[On level 3 are the depictions of 16 protective gods (eleven of which can be identified as gods of feasts of the lunar month listed in Document III.6 and marked with asterisks below); the 9 on the right side face left toward the back of the constellation of Hippopotamus and the 7 on the left side face the back side of the constellation of the lunging crocodile. There is in addition a small half-sized figure of a man (without name) in back of the lunging crocodile that appears to belong neither to the protective gods nor to the constellations. All but one of the gods bear disks on their heads. All of the disks are colored red except for the first two on the right side, namely Isis and Imseti. Above the figures are the names of the deities:[34]]

[On the right side following Hippopotamus:] Isis *(ꜣst)*[35], Imseti* *(ỉmstỉ)*, Hapy* *(ḥpy)*, Duamutef* *(dwꜣ-mut.f)*, Qebehsenuf* *(ḳbḥ-snw.f)*, Manitef* *(mꜣ-n-ỉt.f)*, Irendjetef* *(ỉr-n-ḏt.f)*, Irrenefdjesef* *(ỉr-rn.f-ḏs.f)*, Haqu *(hꜣḳw)*;

[And on the left side following the lunging crocodile:] Iremawa* *(ỉr-m-ꜥwꜣ)*, Teknu* *(tknw)*, Shedkheru* *(šd-ḫrw)*, Nehes* *(nhs)*, Aanera *(ꜥꜣ-nr)*, Imysehnetjer *(ỉmy-sḥ-nṯr)*, Herhekenu *(hr-ḥknw)*

Notes to Document III.3

1. W.M.F. Petrie, *Wisdom of the Egyptians* (London, 1940), pp. 16-17, comments, "[Decan] 1. 'Beginning of *Kanemut*,' [Decan] 2 *Kanemut* under side of *Kanemut*; this 'cow of Mut' is unknown, but later it was altered to taking 1 as the Nile Tortoise from the example of Cancer (gr.), and [Decans] 2,3, were Kenem or Kenmem the vine, and the scattered group of stars would easily outline a vine, starting from Cor Hydrae." This transcription and translation of 'Kenmet' is not generally supported. We do not know for sure what the shape of the Kenmet constellation was, but it appears to have been a string of 5 stars, with a prominent star below.

2. The insertion of Decan 2 after Decan 3 presumably resulted from the scribe's careless copying of 3 first.

3. The second star below is the initial glyph for the name Duamutef.

4. Note that the falcon indicating Horus is here replaced by the sign for "way" *(wꜣt)*, as on Middle Kingdom coffins (see Gardiner, *Egyptian Grammar*, sign N-31, p. 489).

5. Tjemat was translated "mat" by Schott, but this is refuted by Neugebauer and Parker, *Egyptian Astronomical Texts*, Vol. 1, p. 24. Concerning Decans 6 and 7, Petrie, *Wisdom*, p. 17, says: "6 'Upper *Themat*,' 7 lower *Themat*. Later versions seem to explain

this by the variant *Demat*: a wing with which the stars would agree, as drawn [see my Fig. III.84b, taken from Petrie's drawing, covering decades XVII and XVIII]."

6a. Translated by Roeder, "Eine neue Darstellung," p. 1, as "Die beiden Töchter." That is "The Two Daughters." But in any case note that there is only one star as a determinative.

6b. Translated by Roeder, *ibid.*, as "Die (beiden?) Schwangere(n?)." That is, "The (Two ?) Pregnant Ones (?)." A second star below is not part of the determinative but is rather the first glyph of Duamutef, the God's name which follows.

7. Represented by two birds (apparently geese).

8. Concerning "Khenett," Neugebauer and Parker, *Egyptian Astronomical Texts*, Vol. 1, p. 24, say "It is probable that the *ḫntt* group of stars is part of the large constellation of the ship [given below cols. 7-12]." If this suggestion is correct, then Decan 9 would have been a star rising before the Ship constellation.

9. Perhaps to be translated as "The Edge of the Reeds," and as such is a reference to the fact that ships are often pictorially represented in patches of reeds.

10. This is almost surely Seth in the Senmut list, but in later versions it may be Horus.

11. In a later version we find the glyph for "vine" or "vineyard." With this in mind, Petrie, *Wisdom*, p. 17, says: "16 'Wine press,' fairly outlined by stars in Sagittarius."

12. Sometimes both *-w* and the plural sign are missing in later lists.

13. The star under the Sheep's hind-part and two of the irregularly placed stars on the right described in col. 13, could constitute the 3rd star cluster mentioned here in col. 15 (not, however, as a decan). But notice that the star under the hind-part of the Sheep is Decan 22, q.v.

14. Concerning all of this collection of stars and groups of stars, see Neugebauer and Parker, *Egyptian Astronomical Texts*, Vol. 3, p. 111.

15. Perhaps an error for a plural sign of 9 circles.

16. If the small stars are a mistake for a plural sign (as suggested in the preceding note), then this whole collection would be one consisting of only six stars.

17. Since this column (30) and next (31) contain planets under the protection of Horus, it must have seemed proper to the artist to include Hatshepsut's Horus name ("The One who is Powerful of

Kas") as the living Horus in each of them.

18. In my translation I have accepted the textual emendation suggested by Neugebauer and Parker, *Egyptian Astronomical Texts*, Vol. 3, p. 177. The text actually has what I have included in parentheses.

19. The usual name is, of course, "Horus Bull of the Sky." Perhaps the error arose from the common epithet for Horus as "Bull of his Mother" *(kꜣ mwt.f)*. Needless to say, in this case where Hatshepsut is a female Horus, "bull" is hardly appropriate.

20. Once more the translation reflects the discussion of Neugebauer and Parker *(ibid.)*, Vol. 3, p. 178.

21. Neugebauer and Parker, *ibid.*, Vol. 3, p. 117: "'It is a cluster,' should mean that the decan is comprised of a number of stars. Confirmatory evidence is to be found in the Seti I A family, where *nsrw* has 4 stars."

22. For a detailed discussion of the two decans in col. 35, see Neugebauer and Parker, *ibid.*, Vol. 3, p. 117.

23. For comments on these constellations, see Chapter Three, notes 139-147, and the text in the chapter to which they refer.

24. See Wainwright's article cited in Chapter Three, n. 139.

25. Pogo suggests that the three stars are respectively Delta, Epsilon, and Zeta Ursae Majoris, and thus Zeta is the one encircled in red. See the text in Chapter Three to which note 139 refers. He further comments that "around 3000 B.C. Zeta was the only 2nd magnitude star within ten degrees of the pole."

26. The identification of Anu as Cygnus by Wainwright should probably be rejected as I remarked above note 143 in Chapter Three.

27. For a discussion of the possible interpretations of the legend, see Neugebauer and Parker, *Egyptian Astronomical Texts*, Vol. 3, pp. 189-90. Does "Isis" refer to the Hippopotamus and "Djamet" to the crocodile? Does "Festival of the Sky" refer to a feast that was meant to mark one of the sectors on the nearest month-circle (which we mention below)? As I noted above in the text of Chapter Three, over note 142, Pogo believed that the Hippopotamus constellation was formed from some stars of Draco (see Fig. III.64a).

28. I have corrected the erroneous reading *tdwy* to *rdwy*. See Neugebauer and Parker, *Egyptian Astronomical Texts*, Vol. 3, p. 194. As they note from Gardiner, "Restful of feet" is an epithet of the Crocodile God Sobek.

29. The given reading on the Senmut ceiling is *nṯr rw nt ỉmysny*. But in the text above I have accepted the corrections (namely, *rw nṯry nty ỉmytw)* and translation ("Divine Lion, who is between them") suggested by Neugebauer and Parker, *Egyptian Astronomical Texts*, Vol. 3, pp. 192-93.

30. The word "saq" with the meaning of "gathering" or "collecting" often has the crocodile with curved tail as its determinative (*Wb*, Vol. 4, p. 25). It also appears as the name of a ferryman in the Otherworld (*ibid.*, p. 26; cf. Budge, *Dictionary*, p. 639, col. b).

31. I suggested in Chapter Three, note 148, that the sectors were meant to include in one half of each circle the transits of twelve decans on the first of each month and in the other half of each circle the twelve decans on the 16th day of its month. I also asked in note 27 to this document whether the expression "Festival of the Sky" included with the name of the Hippopotamus constellation was some vestige of one of the missing legends on the sectors of one of the monthly circles. But I admit that I do not see how this would fit into my reasonable suggestion that the sectors were meant to contain biweekly risings for each month.

32. Compare the twelve feast-day names that yield the names of months with those discussed in Documents III.1 and Document III.2, where a number of them have been rendered into English. However these festivals originated, whether in seasonal or lunar calendars, they are given here as the months of the three seasons of the civil year, which seasons were almost certainly rooted in tropical, seasonal events and not in lunar phenomena. Note, however, that the epagomenal days are not represented. In this respect the ceiling duplicates the celestial diagram given on water clocks, as I have said in Chapter Three.

33. On the Senmut ceiling the two feasts of Rekeh (the Burning) are not distinguished; and so I have added the adjectives "wer (great)" and "nedjes (small)" to identify them.

34. For a list of the protective-gods and their locations on the various copies of the celestial diagram, see Neugebauer and Parker, *Egyptian Astronomical Texts*, Vol. 3, p. 195.

35. Isis may in fact refer to the Hippopotamus and may not be one of the day-gods.

Document III.4: Introduction

The Vaulted Ceiling of Hall K in the Tomb of Seti I

Of the many depictions of the celestial diagram we shall add to our treatment of that found in Senmut's tomb only one more, that on the ceiling of the sepulchral Hall K in the beautiful tomb of Seti I (1306-1290 B.C.)[1] in the Valley of the Kings (see Figs. III.65a-c). The hall containing this ceiling is more than "one hundred and fifty feet within the limestone wall of the valley and about seventy-five feet below the level of the valley floor."[2] As we remarked in Chapter Three in the text which note 149 supports, the diagram given on Seti I's ceiling is the prototype of a family of depictions of rising decans that Neugebauer and Parker designate as the third family. The over-all organization and main features of the South-Western and North-Eastern parts of the ceiling (equivalent to the Southern and Northern panels of the Senmut tomb), have been described briefly in Chapter III but will emerge in greater detail in Document III.4 itself. Note that I shall continue to use the terms Northern and Southern Panels in Document III.4, since these are the parts of the sky being represented. Suffice to say now, we are no more able to identify surely the various stars that comprise the decanal belt of Seti's I depiction than we were in our examination of the similar arrangement in Senmut's representation. That is to say, we can identify with certainty only Sirius (Sothis) and we know in general that several of the decans before it are related in some fashion to Orion (without identifying with exactness any of the individual stars). The same is true for the other part of the ceiling containing the so-called Northern Constellations grouped about the Big Dipper. Only the Dipper itself is identified with absolute certainty. I have reported a number of guesses concerning both the decans and the northern constellations in Chapter Three, notes 66, 134, 139-40, 143-47, 149-153 and the text which refers to those notes. I shall not repeat them here.

The following works served me well in preparing Document III.4:

L. S. Bull, "An Ancient Egyptian Astronomical Ceiling-Decoration," *Bulletin of the Metropolitan Museum of Art,* Vol. 18 (1923), pp. 283-86. Contains reproductions of Harry Burton's photographs of the astronomical ceiling in the Sepulchral Hall K of Seti's tomb, commissioned by the Metropolitan Museum of Art. These are rather small reproductions and are inferior to the reproductions of the same photographs in the works of Neugebauer-Parker and Hornung mentioned below.

G. Roeder, "Eine neue Darstellung des gesternten Himmels in Ägypten aus der Zeit um 1500 v. Chr.," *Das Weltall,* 28. Jahrgang (1928), pp. 1-5. Though primarily about the Senmut ceiling, its consideration of the decans of the Southern Panel is also helpful for translating the decan list in Document III.4.

A. Pogo, "The Astronomical Ceiling-decoration on the Tomb of Senmut," *Isis,* Vol. 14 (1930), passim, pp. 301-25, and see the charts on pp. 318-19, comparing the decanologies of Senmut I, Seti I, and of the Ramesseum.

O. Neugebauer and R.A. Parker, *Egyptian Astronomical Texts*, Vol. 3 (Providence and London, 1969). The celestial diagram in Seti's tomb is named as Seti I C, and it is established as the prototype of a family. See pages 14-16, 129-33.

E. Hornung, *The Tomb of Pharaoh Seti I...photographed by Harry Burton* (Zurich and Munich, 1991). This includes the whole collection of Burton's superb black and white photographs of the tomb. For the photographs of the astronomical ceiling, see pp. 224-25, 236-41. See also the color photo of the northern constellations on p. 264.

The form of my English translation and transliteration of the text in the Seti diagram is much the same as that employed in Document III.3, and of course depends crucially on the diagrams noted in Document III.4. Note that the photograph of Fig. III.65b shows, in white on the left, the area from which parts of the ceiling have fallen, while Fig. III.65c, drawn by Lepsius before those parts had fallen, shows the southern panel intact.

Notes to the Introduction of Document III.4

1. Hornung, *The Tomb of Pharaoh Seti I*, p. 9, adopts a somewhat different chronology, which dates Seti I's reign as 1293-1279 B.C.

2. Bull, "An Ancient Egyptian Astronomical Ceiling Decoration," p. 283.

Document III.4

The Vaulted Ceiling of Hall K in the Tomb of Seti I

[Southern Panel, Figs. III.65b and III.65c, top]

[The Main Decans for the Twelve Months in Columns 1-23]

[Col. 1, from right] [Decan 1:] The Forerunner of Kenmut (! Kenmet?) *(tp-^{c} knmwt).* [Determined by] one star [in the second register]. [Deity in third register:] Geb. [Followed by] two stars. [At the bottom of the third register is] a human-headed god standing on the base-platform.

[Col. 2] [Decan 2:] Kenmut (! Kenmet?)[1] *(knmwt).* [Determined by] one star [in the second register]. [Deity in third register:] Ba [the ram]. [Followed by] two stars. [At the bottom of the third register is] a ram-headed god [presumably Ba] standing on the base-platform.

[Col. 3] [Decan 3:] The Carrier of the Tail [of Kenmut] *(ẖry ẖpd).* [Determined by] one star [in the second register]. [Deity in the third register:] The One Pre-eminent in Khas (i.e., Sekhmet). [No stars.] [At the bottom of the third register is] a lion-headed goddess [presumably Sekhmet] standing on the base-platform.

[Col. 4] [Decan 4:] The Beginning of Djat *(ḥꜣt ḏꜣt).* [Determined by] one star [in the second register].

[Deity in the third register:] Isis. [No star.] [At the bottom of the third register is] human-headed Isis with her glyph on her head and standing on the base-platform.

[Col. 5] [Decan 5:] The End of Djat *(pḫwty ḏꜣt)* [Determined by] one star [in the second register]. [Deity in the third register:] The Lady of Aphroditopolis (i.e., Hathor of Kom Ishqaw). [No star.] [At the bottom of the third register is] a cow-headed Hathor with a disk between the horns on her head and standing on the base-platform,

[Col. 6] [Decan 6:] The Upper Tjemat *(ṯmꜣt ḥrt)*. [Determined by] one star [in the second register]. [Deities in the third register:] Imseti and Hapy. [No stars.] [At the bottom of the third register are, respectively,] the baboon-headed Hapy and the falcon-headed[2] Qebehsenuf standing on the base-platform.

[Col. 7] [Decan 7:] Lower Tjemat *(ṯmꜣt ẖrt)*. [Determined by] one star [in the second register]. [Deity in the third register:] Qebehsenuf. [No star.] [At the bottom of the third register is] the jackal-headed Duamutef [standing out of position] on the base-platform.

[Col. 8] [Decan 8:] The Two Wesha-Birds *(wšꜣti)*. [Determined by] one star [in the second register]. [Deity in the third register:] Duamutef. [No star]. [At the bottom of the third register are, respectively,] the falcon-headed Qebehsenuf and the jackal-headed Duamutef standing on the base-platform.

[Col. 9] [Decan 9:] The Two Pregnant Ones *(bkꜣti)*. [Determined by] one star [in the second register]. [Deities in the third register:] Duamutef and Qebehsenuf [followed by a] single star. [At the bottom of the third register are, respectively,] a falcon-headed Qebehsenuf and a jackal-headed Duamutef.

[Col. 10] [Decan 10:] The Forerunner of Khenett *(tpy-ᶜ ḫntt)*. [Determined by] one star [in the second register]. [Deities in the third register:] Duamutef and Hapy [followed by] two stars. [At the bottom of the third register are, respectively,] a falcon-headed Qebehsenuf and a jackal-headed Duamutef.

[Col. 11] [Decan 11:] Upper Khenett *(ḫntt ḥrt)*. [Decan 12:] Lower Khenett *(ḫntt ẖrt)*.[3] [Decan 13:] The Red One (Star?) of Khenett *(ṯms n ḫntt)*. [The three decans are determined by] three stars [in the second register]. [Deities in the third register:] Horus, Seth, Horus [written together on the same line as *ḥr stš ḥr*]. [In the middle of the third register are] three levels of three stars (i.e., nine stars) [possibly 3 for each decan].[4] [At the bottom of the third register are, respectively,] a falcon-headed god (Horus) jackal-headed god (Seth), and a falcon-headed god (Horus), [all standing on the base-platform].

[Col. 12] [Includes decans connected with the Ship Constellation.] [Decan 14:] The Sapty of Khenuy *(sꜣpty ḫnwy)*.[5] [Decan 15:] The [Star in the] Middle of the Ship *(ḥry-ib wiꜣ)*. [Deity 16:] The Guides *(sšmw)*. [The three decans are determined by] three stars [in the second register]. [Deities in third register, written all together on a single line:] Isis and Nephthys [for Decan 14], Seth [for Decan 15], and Horus [for Decan 16]. [At the bottom of the third register are, respectively,] the gods Isis, Nephthys, Seth, and Horus [standing in] The Ship[6] [with a line of] six stars [over their heads].

[Col. 13] [Decan 17:] The Kenmu [Stars] *(knmw)* [Determined by] one star [after the name in the first register].[7] [No star in the second register.] [Deities in the third register comprise the names of all four sons of Horus, which are written together in a single line:] Imseti, Hapy, Duamutef, and Qebehsenuf. [In the middle of the third register are] nine stars (i.e., three in each of three levels).[8] [At the bottom of the third register are the gods] Imseti. Hapy, Duamutef, and Qebehsenuf [all standing on the base-platform].

[Col. 14] [Decan 19, out of place at the top of the column:] The [Star of the] Half-month *(smd* [written *smdt* as in Decan 18]). [Decan 18, out of place below Decan 19:] The Forerunner of the [Star of the] Half-month *(tpy-ᶜ smdt).* [These two decans determined by] two stars [in the second register]. [Deities named in the third register:] Hapy with 3 stars and Horus with one star. [At the bottom of the third register are] two falcon-headed gods [standing on the base-platform, the first of which should rather be a baboon-headed god for Hapy].

[Col. 15] [Decan 20:] The [Principal Star of the] Sheep *(srt).* [Determined in the first register by] a figure of the Sheep, with 4 stars above its back and 3 stars on its stomach. [A] star [is added in the second register]. [Deity in the third register:] Isis. [At the bottom of the third register is a figure of the goddess] Isis [standing on the base-platform].

[Col. 16] [Decan 21:] The Two-children (a star-doublet?) of the Sheep *(sʾwy srt).* [Determined by] one star [in the second register]. [Deities named in the third register:] Duamutef and Qebehsenuf [determined by] two stars. [At the bottom of the third register are depicted three gods:] a jackal-headed god [for Duamutef] and two falcon-headed gods [for Qebehsenuf?] [all standing on the base-platform].

[Col. 17] [Decan 22:] The One under the Buttocks of

the Sheep *(ẖry ḫpd srt)*. [Determined by] one star [in the second register]. [Deity named in the third register:] Qebehsenuf [followed by] one star. [At the bottom of the third register are depicted] a jackal-headed god [for Qebehsenuf] and an extraneous falcon-headed god [both standing on the base-platform].

[Col. 18] [Decan 23: The Forerunner of the Two Spirits *(tpy-ᶜ ꜣḫwy)* should be here, but is missing.[9] Instead the diagram jumps to] [Decan 24:] The Two Spirits *(ꜣḫwy)*. [Determined by] one star [in the second register]. [Deities named in the third register:] Duamutef and Qebehsenuf. [Determined by] two stars. [At the bottom of the third register are depicted] a jackal-headed god [for Duamutef] and a falcon-headed god [for Qebehsenuf] [both standing on the base-platform].

[Col. 19] [Decan 25:] The Two Souls *(bꜣwy)*. [Determined by] one star [in the second register]. [Deities named in the third register:] Imseti and Hapy. [Determined by] one star. [At the bottom of the third register are depicted] a human-headed god [for Imseti] and a baboon-headed god (Hapy) [both standing on the base platform].

[Col. 20] [Decan 26, out of place below Decan 28 but after Decan 27][10] [Decan 28:] The Lower Khentu *(ḫnt(w) ẖrw)*. [Decan 27:] The Star between Upper and Lower Khentu *(ḥry-ib ḫntw)*. [Decan 26:] The Upper Khentu *(ḫnt(w) ḥr(w))*. [Decan 29:] Qed *(ḳd)*.[11] [Determined in the second register by] two stars.[12] [Deities named at the top of the third register:] Horus three times [no doubt for Decans 26-28]. [Below the triple Horus is the egg-shaped figure which was included in cols. 20-23 of the Senmut ceiling. Here it is outlined by] 12 [or more] stars. [Beneath the figure is the statement:] the 4th cluster of stars. [Finally, at the bottom of the third register are depicted:] three falcon-headed gods [which are for the three Horus gods indicated above] [and which stand on the base-platform].

[Col. 21] [Decan 30:[13]] The children of Qed *(s'wy ḳd)*. [Determined by] four stars [in the second register]. [Deities named in the third register in three vertical lines of which the first includes:] Imseti and Hapy [followed below, but horizontally, by] two stars, [the second] Qebehsenuf and Duamutef [followed vertically by] two stars, [and the third] Qebehsenuf and Hapy [followed vertically by] two stars.[14] [Below are depicted two lines of gods, of which the first is probably the four children of Horus, with] the jackal-headed god Duamutef [immediately visible, and the second is a pair of gods, namely] the falcon-headed Qebehsenuf and the baboon-headed Hapy.[15] [And all the gods depicted are standing on the base-platform.]

[Col. 22] [In the first register are written vertically the following decans.] [Decan 32:] Aret *(ᶜrt)*. [This is followed below by a floral insertion sign and Decan 31, which had been omitted.] [Decan 31:] The Thousands *(ḫ'w)* [Decan 34:] The [Star above the] Upper Arm [of Orion] *(rmn ḥr[y]w)*. [Then below Decan 34 is inserted the omitted Decan 33, namely] The Column of Orion *(iwn s'ḥ)*. [Decan 35:] The Ear of Orion *(msḏr s'ḥ)* [Decan 36:] The [Star below] the Lower Arm of Orion *(rmn ẖry s'ḥ)* [Decan 37:] The [Star near the] Hand of Orion *(ᶜ s'ḥ)* [Decan 38:] Orion *(s'ḥ)*. [At the bottom of the first register these eight decans have a horizontal row of] seven (!) stars [as determinatives].[16]

[Then in the second register of col. 22 there is another horizontal row of stars, but this time one of] eight stars [presumably as a corrected set determining the eight decans].

[At the top of the third register are the names of the gods, the first two vertical sets from the right being the Children of Horus:] Imseti, Hapy, Duamutef, and Qebehsenuf [no doubt for the omitted Decan 31]. [Then follow in succession:] Eye of Horus, Horus, Eye of Horus, Children of Horus, Eye of Horus, Eye of Horus, and Osiris [respectively for Decans 32-38]. [Then at the bottom of the third register, beginning at the left, we see depicted a bank of] the four Children of Horus with stars above their heads [and all standing on the base-platform]. [To the left is depicted] Orion standing in a boat [with head looking back at Isis[-Sothis] in col. 23][17]. [He has a number of stars about his body that seem to conform approximately to the names of Decans 32-38.]

[Col. 23] [Decan 39:] Sothis *(spdt)*. [Nothing in the second register.] [In the third register the deity is named] Isis [determined by] one star. [Below the name is depicted] Isis standing in a boat [wearing a crown with double plume and a Maat-like feather, her left hand clutching an ankh-sign and a flat-headed staff, her right hand raised near the feather and with the star determining her name directly over the crown].

[Exterior Planets in Columns 24-26 in Figs. 65a, top, and 65b]

[Col. 24] The Southern Star *(sbꜣ rsy)* [i.e., Jupiter]. [Nothing in the second register.] [Depicted at the bottom of the column is the falcon-headed] Horus [surmounted by] a star [and] standing in a boat.

[Col. 25] [The planet Saturn:] The Western (! Eastern?) Star which crosses the Sky *(sbꜣ imnty dꜣ pt)*.

His name is Horus Bull of the Sky *(ḥr-kꜣ-pt rn.f)*. [Deity's name in the second register:] Horus. [Depicted at the bottom of the column is the falcon-headed] Horus [surmounted by] a star [and] standing in a boat.

[Col. 26] [The Planet Mars:] The Eastern Star of the Sky *(sbꜣ iꜣbty pt)*. His name is Horus of the Horizon *(ḥr-ꜣḫty rn.f)*. He travels backwards *(ddsk.f m ḫtḫt)*. [Deity's name in the second register:] Re. [Nothing in the short third register.] [Depicted in the fourth register at the bottom of the column is the falcon-headed] Horus [surmounted by] a star [and] standing in a boat.

[Epagomenal Decans in Columns 27-33]

[Cols. 27-28] [Though two columns, these are for one additional decan named in col. 28:] The Two Tortoises *(štwy)*. [Determined by] two tortoises and one star. [There is nothing in the first three registers of col. 27 and the second and third registers of col. 28.] [Deities named in fourth register of cols. 27 and 28 respectively:] Duamutef and Hapy. [Depicted at bottom of cols. 27 and 28 respectively:] the jackal-headed god Duamutef and the baboon-headed Hapy each with a star on his head [and both standing on the base-platform].

[Col. 29] [Add. decan 2:] Neseru *(nsrw)* [Determined by] one star. [Nothing in the second register.] [Deity named in third register:] Imseti. [In the fourth register is depicted] the human-headed god Imseti with a star over his head [and standing on the base-platform].

[Col. 30] [Add. decan 3:] Shespet *(šspt)*. [Deities:] Eyes of Horus [with the two eyes in the second register and Horus in register 3]. [In the fourth register is depicted] the falcon-headed god Horus with 5 stars over his head [and standing on the base-platform].

[Col. 31] [Add. decan 4:] Ipesedj *(ipsḏ)*. [Determined by a] solar disk and one star. [Nothing in the second register.] [Deity named] Horus [in the third register]. [In the fourth register is depicted] the falcon-headed god Horus with 2 stars over his head [and standing on the base-platform].

[Col. 32] [Add. decan 5:] Sebshesen *(sbšsn)*. [Determined by] one star. [Nothing in the second register.] [Deity named] Horus [in the third register]. [In the fourth register is depicted] the falcon-headed god Horus with 2 stars over his head [and standing on the base-platform].

[Col. 33] [Add. decan 6:] The Honored of God *(nṯr wšʾ)*. [Deity name] Duamutef [in third register]. [In the fourth register is depicted] the jackal-headed god Duamutef with 1 star over his head [and standing on the base-platform].

[The Interior Planets in Columns 34 and 35]

[Col. 34] Mercury *(sbg)*. [Determined by] one star. [Nothing in the second and third registers.] [Deity:] Seth [in the fourth register]. [Followed by] two stars. [Below is depicted] the long-eared animal-headed god Seth [standing on the base-platform.]

[Col. 35] [Venus:] The star which crosses *(sbʾ ḏʾ)* [Deity named:] Osiris [in the second register]. [Nothing in the third register.] [Depicted in the fourth register] Heron surmounted by a star [and standing on the base-platform].

[The Northern Panel, Fig. III.65b, bottom]

[The Northern Constellations]

[For the constellations with bracketed numbers, see Fig. III.69.]

[1] The Adze (or Foreleg, i.e., the Big Dipper). [Above is its name] Meskhetyu *(msẖtyw)*. [Depicted here as a striding bull on a rope-like platform with a hump in the rope-platform between the fore- and after-legs.][18]

[2] An *(ᶜn)*, the falcon-headed god [which appears to be supporting the platform of Meskhetyu. Its name appears to its left.] Five stars appear on An's body [i.e., on his left arm, torso, right arm and each ankle.]

[3] Man (or better, a god) with disk on his head. [He is holding reins which extend from the rear end of Meskhetyu to a mooring post held in the paws of Hippopotamus (5).] The mooring post rests on the base-platform. [Marking the constellation of Man are] four circles [for stars]: [one on his elbow, one on his shoulder, one on his left leg, and one on his right leg.]

[4] Falcon [with no name] on perch in front of Meskhetyu. [No stars indicated.]

[5] Hippopotamus [with legend:] *sꜣ-mwt.*[19] [She is marked by] eight stars [one on her shoulder and seven on the tail-like continuation of her headdress].

[6] Crocodile on Hippopotamus' back. [It has an indeterminate number of ovals and other marks on its body, which, however, do not seem to indicate stars from which the constellation was imagined.]

[7] The Lunging Crocodile with straight tail [who seems to be attacking Man (7a)]. [He has many small stars along his tail. He has no name or epithet.]

[7a] Man without name [probably spearing Crocodile (7), but the spear is missing].

[8] Lion on his haunches and outlined by stars, with a truncated epithet *imy-rw* (see Document III.3, note 29).

[9] Crocodile with bent tail and a corrupt epithet.[20]

[10] Serqet *(srḳt)* [the human-headed Scorpion Goddess who lies parallel to the base line and to the left of Meskhetyu]. [The hieroglyphic name is above the depiction.]

[Flanking both sides of the northern constellations are the protective gods, eleven of which (in fact, the same eleven as on the ceiling in Senmut's tomb; see the flanking gods with asterisks in Document III.3) are protective gods of days of the lunar month (for the full list of 30 gods, see the lower halves of the columns in Fig. III.91a).[21] Their names appear above them in the first register. The nine deities on the right side are the same as those in Document III.3, where I have listed the names. On the left side there are eleven gods, the first seven of which are very similar to those on the Senmut ceiling (see in Document III.3). For the whole set of gods with some attention to the variant names, consult the discussion and chart given by Neugebauer and Parker, *Egyptian Astronomical Texts*, Vol. 3, pp. 194-99.]

Notes to Document III.4

1. Unlike the Senmut ceiling, that of Seti I has Decan 2 in the correct order in a column of its own. Notice the spelling *knmwt* here (and) in col. 1 instead of Senmut's *knmt*. If *knmwt* is correct, then this should be translated "The Ape" and Decan 1 as "The Forerunner of the Ape"; then Decan 3 would be "The Carrier of the Tail [of the Ape]."

2. The two standing gods are jammed together and the second one is hard to see. Its head looks like a falcon head rather than a human head (i.e., the head of Qebehsenuf rather than the head of Imseti as perhaps it was intended to be, as we might expect from the fact that Imseti's name appears above). On the other hand in the next column (the seventh) we see the figure of the jackal-headed Duamutef rather than the falcon-headed Qebehsenuf who is named in the eighth column, and so perhaps it simply indicates that Qebehsenuf, intended perhaps for the eighth column, has mistakenly been crowded into the seventh. Note further that from column eight through column eleven we have a series of alternating falcon-headed and jackal-head gods not fitting very well with the gods' names listed above them.

3. Decans 11 and 12 are written beside each other, and below them is written Decan 13, with *ẖntt* squeezed in before *ṯms.*

4. The nine stars may be a mistake for the super plural of nine circles or strokes often used in connection with the Ennead and occasionally elsewhere. See note 15 of Document III.3.

5. Consult note 9 of Document III.3.

6. The stern of The Ship intrudes somewhat into col. 13.

7. Perhaps a plural is intended. See Document III.3, Decan 17.

8. See note 4 above. It appears that the first two vertical rows of stars are aligned (but not exactly) with the first two sons, and that the last vertical row of three stars is approximately aligned between the third and fourth sons.

9. The omission of Decan 24 is probably explained by the fact that the entries in the second and third register in other copies of this family of decans are identical.

10. Col. 20 contains four decans numbered by Neugebauer and Parker as Decans 26-29 in *Egyptian Astronomical Texts,* Vol. 3, p. 131. These replace numbers 26-29 in Senmut's diagram (see Document III.3). It is Seti's Decan 28, "The Star between Upper and Lower Khentu," that has been added, or at least it is missing in the Senmut list.

11. There seems to be an extraneous after the *ẖnt(w)* which goes with Decan 28.

12. It is probable that the artist at first included only Decans 28 and 29. This would account for the appearance of only two stars. Later he seems to have added in somewhat confused fashion Decans 27 and 26.

13. Since an additional decan was added in col. 20, the decan

of col. 21 is numbered 30 instead of 29 as it was on the Senmut ceiling (see Document III.3).

14. The stars following the gods' names look like they may form some constellation.

15. As I have described it, the depicted gods represent all the gods whose names are mentioned above.

16. These seven stars were probably inserted before one or the other of the omitted decans 31 and 33 was added.

17. See Chapter Three, above note 150, for the significance of Orion's looking back at Isis-Sothis.

18. Here the Big Dipper is depicted as the full bull with a circle on the neck and one on the rump. No doubt both indicate key stars. There are in addition three stars in a vertical row from the top of the shoulder toward his under belly.

19. This legend seems to be a corruption of *ḫsꜣ-mwt*, which may be translated "the mother is fierce" (see Neugebauer and Parker, *Egyptian Astronomical Texts,* Vol. 3, p. 190). A possible translation of the phrase as given, i.e., *sꜣ-mwt*, is "the chapel of Mut." In *loc. cit.* Neugebauer and Parker note mythological texts that speak of a relationship between Hippopotamus and Meskhetyu.

20. See the discussion of the epithet in Neugebauer and Parker, *Egyptian Astronomical Texts*, Vol. 3, p. 192. It may be a corruption of the name *sꜣḳ* and the epithet for the lion in Document III.3 (see Fig. III.66, numbers [6] and [7a]). The depiction of the crocodile is larger than the glyphs of the epithet. Hence it is probably not a determinative.

21. It was Heinrich Brugsch, *Thesaurus inscriptionum aegyptiacarum*, I. Abt., pp. 52-53, who first linked the gods here in the Seti ceiling with those of the lunar month.

Document III.5: Introduction

Extracts from the Calendar of the Temple at Medina Habu in Western Thebes

The Temple of Amon-Re, built by Ramesses III (1194-63 B.C.) at Medina Habu in Western Thebes in the early years of the 12th century B.C., has on its entire southern wall a calendar consisting primarily of lists of offerings to be prepared for specified daily, monthly, and annual feasts. Though there is evidently a long history of such temple calendars, no doubt going back to the Old Kingdom or earlier, the earlier ones are missing or yield only fragments, while that at Medina Habu is quite complete and is surely the longest one extant, containing as it does over 1470 lines of hieroglyphs.[1]

In our examination of the calendar of Ramesses III, we are particularly interested in those monthly and annual feasts which are more traditional and which accordingly relate to our previous documents and to the discussions of calendars provoked by those documents. We are not, however, interested in the types and quantities of the various breads, cakes, and beer that are contained in the sundry offering-lists for the feasts included in the Medina Habu calendar, although the grain and beer measures and their fractions are of concern to us in Chapter IV when we discuss ancient

Egyptian applied mathematics.[2] Accordingly, I have not given the actual offering-items and their quantities, except in the case of the rather short list for the annual Feast of the [Heliacal] Rising of Sothis which can act as a model.

The numbers by which we refer to the feasts with their offerings and to the plates of photographs and their hieroglyphic transcriptions are those given in the volume prepared by the The Epigraphic Survey of the University of Chicago under its field director, H.H. Nelson: *The University of Chicago Oriental Institute Publications, Volume XXII: Medinet Habu—Volume III, Plates 131-192: The Calendar, the "Slaughterhouse," and Minor Records of Ramses III* (Chicago, 1934). In the preface to this volume Nelson succinctly describes the rather slight prior record of publication concerning the calendar.

Though my objective here is to present extracts from the calendar, some general remarks about the calendar as a whole might be helpful to the reader. The chief source of the calendar appears to have been a similar calendar on the southern wall of the Ramesseum, the mortuary temple of Ramesses II (1290-1224 B.C.). Thus, in large part, the lists of feasts of the later calendar (except those originated by Ramesses III himself) and their specific offerings seem to have been identical to those in the calendar of the Ramesseum, though to be sure we have only fragments of the latter. In the preface to *Medinet Habu—Volume III*, p. ix, Nelson describes the relationships between the two calendars:

> In Ptolemaic days the Ramesseum was a ruin and was used extensively as a source of building material. The late additions to the

> little Eighteenth Dynasty temple at Medinet Habu are largely built from stone derived from Ramses II's temple. ...today the fragments of the reliefs and inscriptions of Ramses II are plainly to be seen. In 1881 Duemichen published thirty of these fragments of the Ramesseum calendar. Plates 187-89 of this volume [i.e., *Medinet Habu–Volume III*] reproduce photographs of these stones, including all that Duemichen saw and some additional fragments which had not been uncovered in his day....
>
> A comparison of the Ramesseum material with the Medinet Habu Calendar will show at a glance that the former inscription was much more compact and probably occupied less lateral space than does the latter....
>
> While the Medinet Habu scribes copied the earlier calendar, even to its arrangement on the wall of the temple, they modified the forms of the signs in accordance with the calligraphy of their own day. Ramses III's inscription is distinctly of the Twentieth Dynasty, with all the deterioration of the signs characteristic of the period. This later copy is slovenly and unpleasing in comparison to the Ramesseum style....

While the temple at Karnak in Eastern Thebes contains calendars, they do not seem to be part of a systematic scheme like those of Ramesses II and Ramesses III. Most of the other mortuary temples in Western Thebes are not well enough preserved to yield adequate comparisons with lists of the temple at Medina Habu.

The arrangement of the calendar on the southern wall (see Figs. III.87a-c) is succinctly summarized by Mr. Nelson:[3]

> Because the floor of the building rises toward the rear and at the same time the level of the roof descends, the wall area is higher toward the front of the structure than it is at the back. In the adornment of such an edifice, the Egyptian artist dealt in straight lines and preferred to keep the various units of the decoration in squares or rectangles. He achieved this purpose with the Medinet Habu Calendar by placing along the upper and lower margins of the wall west of the second pylon long lines of inscriptions and reliefs that left him a free area between them approximately the same in height throughout its entire length. In this long space he inscribed the Calendar, or at least such part of it as could be accommodated therein, as though he had unrolled upon the wall a papyrus from the temple archives. Within the area thus arranged for the reception of the Calendar the scribe next laid out thirty-six rectangles by drawing at intervals two parallel lines, spaced close together, running from top to bottom of the area. Two of the sections thus marked off—the first and the ninth [see Plates 136 and 144 in *Medinet Habu Volume III*], counting from the rear of the temple—were reserved for reliefs, illustrations to the document, depicting the Pharaoh announcing to the Theban Triad [Amon-Re, Mut, and Khons] the institution of the Calendar and recounting his good deeds in

> their behalf .... Two more sections—the second and the third [see my Figs. III.88 and III.89]—are devoted to the king's speech to the gods (*in fact, the speech is addressed only to Amon-Re rather than to the whole Theban Triad*) and to the royal decree establishing new endowments .... The remaining thirty-two sections contain the lists of feasts and offerings that compose the body of the Calendar. Four more sections were inscribed between the pylons .... These four sections, added to those located west of the second pylon, bring the total for the entire document up to forty sections.
>
> The Calendar is divided into two parts, each introduced by one of the reliefs already referred to. The first part, which includes eight sections, deals with Ramses III's new creations, his temple, its equipment and organization, and his personal contributions to its endowment. The remaining thirty-two sections, which constitute the second part, deal (except for a very few lists which also record new endowments) with old established feasts and offerings which the king merely reaffirmed. .... Moreover, in the vocabulary, spelling, and calligraphy of the two parts there is a noticeable difference...

In my extracts I have included at the beginning some of the speech of Ramesses III to Amon-Re, recounting his deeds on behalf of the god. I do this to reinforce for the reader the relationship that exists between Egyptian views of eternity and everlastingness, which I described at some length in Volume One, and the establishment of the civil calendar. My version of this address follows, for the most part, the translation given by Nelson,[4] though I include some minor material at the beginning not translated by Nelson and a few phonetic transcriptions in parentheses.

Following the speech, I have given some lines of the decree instituting the calendar. As the reader will notice, the regnal year when the decree was inscribed on the wall is not given, though its month and day are (see Fig. III.89, beginning of line 53). The date of the decree itself is added (end of line 60): "Year 4, II Peret" (day not given). Following the decree are extracts from a number of titles of the offering-lists for the eight monthly feasts to be observed and for some of the annual feasts that make up most of the second part of the calendar, as is mentioned in the quotation from Nelson's monograph just given.

Initially, I want to stress that the title of each annual feast includes the date of its observance: i.e., the month number, the season name, and the number of the day of the month. *Thus the form of the annual feast-dates in the Medina Habu calendar is always that of dates in the civil calendar.* Furthermore, in the totals of Upper Egyptian and Lower Egyptian grains to be produced for the daily and monthly feasts a grand total is given in both cases for "one year and 5 days," that is, the civil year for which the supplies were to be provided. Hence the efforts by some earlier Egyptologists to establish the Medina Habu calendar and other such temple calendars as fixed Sothic calendars of 365 1/4 days themselves or at least as giving evidence of the existence of a separate but regularly maintained fixed Sothic calendar of festivals were surely in vain.[5]

It is also of interest that only eight of the thirty possible feasts of the lunar month listed in later temples at Edfu and Dendera were included in the Medina Habu calendar: the feasts for the following days: 29th, 30th, 1st, 2nd, 4th, 6th, 10th, and 15th. These are the feasts of the two possible days of last visibility of the moon's crescent in one month (depending on whether it is a 29-day month or a 30-day month) and some of those that lead up to and include the Feast of the Full Moon in the next month (cf. Document III.6).

As I have indicated in the text, the titles of the offering-lists of Ramesses III's calendar were always written in a vertical column, while the items of the offering-lists were written in horizontal lines to the right of the vertical column.

In the extract giving the offering-list for the annual Feast of the Rising of Sothis (i.e., the helical rising of Sirius), no number appears after ☉ to indicate the specific day of the month for the feast. As I have said, this is usually taken to mean that the first day is to be understood by the very appearance of the solar sign, but it probably should be interpreted that no particular day is indicated because the rising of Sothis was actually delayed one day in four years. Hence the appearance of the solar sign without a number actually meant that the feast was to be celebrated sometime in the first month of the season Akhet. Since the helical rising of Sirius that last took place in the civil year on the first day of the first month of Akhet (i.e., on New Year's Day) prior to the time of the building of the Medina Habu temple or its principal source, the Ramesseum, occurred in the quadrennium 1321-1318 B.C., the entry for this feast day in the Medina Habu calendar (which may have been copied from the

Ramesseum, some one hundred years earlier than the time of Ramesses III) may imply only that the celebration was scheduled, when this entry was originally prepared, for some day in the first month of the season Akhet when the rising would occur. The precise day would be determined by the particular year of its celebration, that year falling within the 120-year period extending from the beginning of the Sothic cycle in 1321-18 B.C., for, as I noted in Chapter Three above, the rising of Sirius is delayed in the civil calendar by one day every four years and so in order for it to be celebrated in the first month of Akhet (as this calendar indicates), the year would have to be in that 120-year period noted. It is somewhat amusing that the temple at Medina Habu, frequently designated in its inscriptions as a house of millions of years called "United with Eternity," should bother to specify the time of the rising of Sirius as occurring during the first month of the season Akhet only, since, of course, it would rise one day later every four years in the civil year through a whole Sothic period of 1460 years, and repeatedly so in succeeding Sothic periods. It could be, of course that the scribe did indeed mean that the feast was to occur on the first day of the first month of Akhet and that it was considered as fixed in the civil year regardless of the actual day of the rising of Sothis. But, in view of the later evidence given in Document III.10, I do not believe this to be so.

After the titles of the feast days, and to their right, are listed the offering items. The structure and content of these lists is succinctly described by Nelson.[6]

> The lists themselves are each divided into three parts: first, an itemized statement of foods that were prepared by cooking and in

the composition of which grain was used; second, a summary of the preceding, giving the number of units of different kinds of food listed and the quantity of grain needed for their preparation; third, a statement of miscellaneous offerings, edible and otherwise, for which no grain was required. ....

The items in the first part of each list are arranged in practically the same order throughout the Calendar and contain a certain minimum of objects, the number of which increases with the importance of the feast to which they are assigned. Thus for six of the regular monthly feasts the minimum list is specified. It consists of two sizes of *byt*-bread, one lot of *psn*–bread, one lot of white fruit bread, and one lot of beer, giving a total of 84 loaves of bread of various sorts and 15 jars of beer. This is a humble offering for a group of minor feasts which recurred at frequent intervals. On the other hand, for the more important of the monthly feasts this group contains as many as 28 different items and embraces a larger variety of objects and an increased range of sizes of the same object. ....

After the items of cooked foods in each list the scribe, as is usual in Egyptian documents, totaled up the units of various kinds which he had just given. In these totals he classified the foods under six heads: bread, cakes, sweets, beer, a second form of sweets known as *bnr-nḏꜣ*, and cereals or meal. ... After these totals the scribe gave the quantities of both Upper Egyptian grain and Lower Egyptian grain which were required. These totals are given in sacks, *ḫḳꜣt*-measures..., and fractions of the latter.

Coming now to the last part of the offering list, we find that it named a miscellaneous lot of objects that were not cooked or in the preparation of which no grain was required. Among these items certain offerings regularly occur in practically every list. These are *r*-geese, *śʿšꜣ*-fowl, wine, incense, fruit, and flowers. In the longer lists these are supplemented by other foods—meats, vegetables, fats, oils, and honey.

Now we may press on to the document itself. In each part of it I have indicated the plate, list, and line numbers from *Medinet Habu Volume III*, and, where pertinent, my figure-numbers. As I have said above, I have included the actual list of offerings only for the feast of the rising of Sothis (Sirius), so that the reader may see how such lists are organized. Incidentally, it ought to be noted that the list for this feast is a very modest one, indicating that it was not such an important feast in the Medina Habu calendar.

Notes to the Introduction of Document III.5

1. H. H. Nelson and U. Hölscher, *The Oriental Institute of the University of Chicago, Oriental Institute Communications, No. 18, Work in Western Thebes, 1931-33* (Chicago, 1934?), p. 2.

2. See the classical summary treatments of these measures in F.L. Griffith, "Notes on Egyptian Weights and Measures," *Proceedings of the Society of Biblical Archaeology*, Vol. 14 (1892),

403-50, Vol. 15 (1893), pp. 301-16, and in A. Gardiner, *Egyptian Grammar*, 3rd ed. (1973), pp. 191-200.

3. Nelson, *op. cit.* in note 1, pp. 4-8, which includes three figures. The first shows Ramesses III before the Theban Triad of Amon-Re, Mut, and Khons and the goddess Maat, who, we saw in Volume One, was the goddess of cosmic order. The second gives the text of the pharaoh's speech, and the third includes the text of the decree establishing the calendar and the list of new endowments set up by Ramesses III. Note that I use the conventional forms "you," "your," etc. instead of Nelson's "thee," "thy," etc.

4. *Ibid.*, pp. 12-15.

5. For example, see R. Weill, *Bases, méthodes et résultats de la chronologie égyptienne* (Paris, 1926), p. 128, where he asserts that the Medina Habu calendar fixes the feast of the rising of Sothis as the first day of Akhet, whereas we can see by looking at our document that it is probably only the first month of the season of Akhet that is specified and not the first day of that month, i.e., that the solar sign after Akhet without a following number is acting only as a determinative for Akhet [or much less likely as a consonantal complement for ḫ without the circle being filled in]. Cf. Weill, pp. 145, 155. See also J.H. Breasted, *Ancient Records of Egypt*, Vol. 4, p. 83 (note d, with earlier bibliographical references), and p. 84 (where he implies that the calendar indicates the rising of Sothis as being on New Year's Day). In fact the text in Brugsch, *Thesaurus*, II, p. 364, which Breasted cites on p. 83, note d, does not refer to the rising specifically as on the *first* day of the first month of Akhet but only as in "the first month of Akhet," with the circular sign following Akhet probably acting as a determinative of Akhet. We should notice that Nelson himself in the preface to the *Medinet Habu— Volume III*, p. viii, declares concerning the Feast of the Rising of Sothis: "The actual number of days occupied by these annual feasts, exclusive of special feasts of victory, is, according to the Calendar, sixty-nine, the first being the Feast of the Rising of Sothis on the first day of the first month of the year,..." He has obviously taken the common view that the solar sign by itself indicates the first day of the first month of Akhet. Hence those Egyptologists who thought that the annual, seasonal feasts were falling on the same days of the months in the mobile civil calendar of the temple of Medina Habu and those of other late temples as if the temple calendars were identical with a calendar based on a fixed Sothic year reached the conclusion that the priests maintained

a fixed Sothic year of festivals along side of the civil year.
6. Nelson, *Work in Western Thebes*, pp. 17-21.

Document III.5

Extracts from the Calendar of the Temple at Medinet Habu in Western Thebes

[Ramesses III's Address to Amon-Re: Plate 138, Lines 24-52 (Fig. III.88)]

/Line 24/ Address of the King of Upper and Lower Egypt, Lord of the Two Lands: Usermaatre Meriamon; Son of Re, Lord of Thrones: Ramesses, Prince of On, to his father Amon-Re, King of the Gods. "Behold, I know eternity *(nḥḥ)*, O my august father, and I am not ignorant of everlastingness *(ḏt)*.[1]

/Line 25/ "My heart is glad for I know your strength is more than that of the [other] gods, since it is you who has fashioned *(ms)*[2] their images and created their majesties [and] you have made the sky ....

....

/Line 28/ ..."I built for you my mansion of millions of years[3] in the city of Thebes *(wꜣst)*, the Eye of Re. I fashioned your august images dwelling within it, while the great Ennead are in splendid shrines /line 29/ in their sanctuaries.

....

/Line 30/ ..."I have made excellent his offerings and his ordinances and his ceremonies in accordance with the festal usages of the House of Ptah[4] [i.e., Ptah-Seker (or Ptah-Sokar)] in order to observe all the occasions of the year.

"I know that /line 31/ you have given the lands in order to supply his offering-loaves, O Min-Amon in your beautiful forms, so that he may appear at his [accustomed] times, according to what you desire.

....

/Line 32/ ..."I provide for you daily divine offerings and I establish the Feasts of the Sky at all their proper times .... /line 33/ .... I have made festive your regular offerings with bread and beer, while cattle and desert game are butchered in your slaughterhouse.

....

/Line 38/ ...."I clear the way for the lord of gods, Amon-Re, in his Feast of Millions of Years; for I am a feast-leader, pure of hands, offering great oblations /line 39/ before him who begat me.

....

/Line 51/ .... "May you do that which my majesty desires /line 52/ in making excellent my house. Then it shall endure as the heavens endure, with your majesty in its midst like the Horizon-Dweller,[5] for happy is my temple if you dwell in it to eternity, and it shall abide forever."

[The Decree Instituting the Calendar: Plate 140, lines 53-61, Fig. III.89]

/Line 53/ Year ___ , the first month of the season Shemu. The king appeared on his throne and acquired unto himself the adornments of his father in order to observe 1,300,000 festivals under his majesty,... /line 54/...Ramesses [III], Prince of On.

/Line 55/ His majesty has decreed offerings for his father, Amon-Re, King of the Gods, in Opet (i.e.,

Karnak), and for the fathers of the Ennead, and for the holy image [of the processional bark] of the King of Upper and Lower Egypt, Usermaatre Meriamon, in the mansion /line 56/ of millions of years [called] "Usermaatre Meriamon endures to eternity"; in the estate of Amun, as the regular offerings of every day, abiding and endowed forever and ever.

....

/Line 60/ .... As offerings to his father, Amon-Re, King of the Gods, by a decree of Year 4 [of Ramesses III], second month of the season Peret /line 61/ ....

[Ramesses III's Additions to the Endowments for Daily Services in the Temple: Plate 140, List 1, Lines 62-123]

[Omitted here in Document III.5.] [Also omitted are Lists 2-5 given on Plate 142 and List 6 (Daily Offerings at the Temple) on Plate 146 where we notice in line 259 that the total production of Upper and Lower Egyptian grain needed for the daily offerings is given for "one year and 5 days" (the whole civil year of 365 days), and the figure of "one year and 5 days" is continually repeated in the remaining lines of List 6. I have as well omitted List 16, which also contains daily needs for the temple but seems to be out of place on Plate 150 after the lists for regular monthly feasts.]

[Offerings for the Monthly *Feasts of the Sky*: Plates 148, 150, Lists 7-15]

[Lines 293-303:] /Line 293/ [In vertical columns is the introduction to the Feasts of the Sky, and the title to List 7:] Feasts of the Sky which shall occur. That which is offered to Amon-Re and to the image of the holy [processional] bark of the King of Upper and Lower Egypt, Usermaatre Meriamon, in the house of Usermaatre Meriamon, [called] "United in Eternity"; in the estate of Amun in Western Thebes, furnished every month....

[*Feast of the Attender (?)*, List 7:] /line 294/ Every day of the Feast of the Attender(?) (i.e., of the 29th day of the lunar month) which shall occur. That which is offered to Amon-Re, King of the Gods, as festal supplies this day. [Then follows, in horizontal lines 295-305, the short list no. 7 of offering-supplies.]

[*Feast of the Going Forth of Min*, List 8, lines 306-17:] /Line 306/ [In vertical column:] Every day of the Feast of the Going Forth of Min (i.e., of the Feast of the 30th day of the lunar month)[6] which shall occur. That which is offered to Amun, King of the Gods from the festival supplies this day. [Then follows, in horizontal lines 307-17, the short list no. 8 of offering-supplies.]

[*Feast of the First day of the Lunar Month*, list 9, lines 318-66:] /Line 318/ [In vertical column:] Every Day of the Feast of *psḏtyw*[7] (i.e., the first day of the lunar month) which shall occur. That which is offered to Amon-Re and to the holy image [of the bark] of Usermaatre Meriamon (i.e., the procession bark of the Ramesses III) as festal supplies this day. [Then follow, in horizontal lines 319-66, the exceedingly long list no. 9 of supplies, revealing the importance of this feast day.]

[*Feast of the Month*, list 10, lines 367-78:] /Line 367/ [In vertical column:] Every day of the Feast of the Month *(ꜣbd)*[8] (i.e., the second day of the lunar month) which shall occur. That which is offered to Amon-Re, King of the Gods, as festal supplies this day.

[Then follows, in horizontal lines 368-78, the short list no. 10 of supplies.]

[*Feast of the Going Forth of Sem* (i.e., the Sem-priest?), list 11, lines 379-90:] /Line 379/ [Vertical column:] Every day of the Feast of the Going Forth of Sem (i.e., of the feast of the 4th day of the lunar month) which shall occur. That which is offered to Amon-Re, King of the Gods, and [to the Ennead] with him, as festal supplies this day. [Then follows, in horizontal lines 360-90, the short list no. 11 of supplies.]

[*Feast of the Sixth Day of the Lunar Month*, list 12, lines 391-439] /Line 391/ [Vertical column:] Every day of the Feast of the Sixth Day which shall occur. That which is offered to Amon-Re, King of the Gods, and to the holy image [of the bark] of Usermaatre Meriamon as festal supplies this day. [Then follows, in horizontal lines 392-439, the very long list no. 12 of offerings, again indicating the importance of this feast day.]

[*Feast of the Tenth Day of the Lunar Month*, plate 150, list 13, lines 440-51] /Line 440/ [Vertical column:] Every day of the Feast of the Tenth Day of the Month which shall occur. That which is offered to Amon-Re and to his Ennead as festal supplies this day. [Then follows, in horizontal lines 441-51, the short list no. 13 of supplies.]

[*Feast of the Half-Month*, list 14, lines 452-63] /Line 452/ [Vertical column:] Every day of the Feast of the Half-month[9] (i.e., the Feast of the Full Moon, or of the 15th day) which shall occur. That which is offered to Amon-Re and to his Ennead as festal supplies this day. [Then follows, in horizontal lines 453-63, the short list no. 14 of supplies.]

/[List 15, lines 464-529] Total of all good and clean

bread, beer, cattle, and fowl, [at the Feasts of the Sky] which are offered on behalf of Amon-Re in the house of Usermaatre Meriamon in the estate of Amun as festal supplies this day.

....

/Line 495/ Grain to be produced for one month: That of Upper Egypt: sacks, 4; heqats, 1 + 1/2 + 1/20. That of Lower Egypt: sacks, 1; heqats, 1 + 1/5.

/Line 496/ Grain to be produced for the year and 5 days[10] (i.e., for the civil year of 365 days): That of Upper Egypt: sacks, 56; heqats, 1 + 1/2 + 1/10; That of Lower Egypt: sacks, 35; heqats 1 + 1/2.

/Line 497/ Total of grain [to be produced]: 92 sacks[, summing approximately the items in line 496].

[Lists 16-18 on Plate 150, namely, (16) of *Daily offerings to the Royal Amun-standard*, (17) of some *Supplementary annual donations from the Pharaoh's Treasury*, and (18) of some *Miscellaneous annual temple supplies*, are omitted here from Document III.5.]

[Annual Offering Feasts, Plates 152-167, Lists 19-67]

[*The Coronation Feast*, Plate 152, Lists 19-22, Lines 551-628:] /Line 551/ [Vertical columns:] First (or head) festival of the seasons. That which is offered to Amon-Re, King of the Gods, and to the sacred image [of the bark] of the King of Upper and Lower Egypt, Usermaatre Meriamon and to the Ennead in the house of millions of years of /line 552/ the King of Upper and Lower Egypt, Usermaatre Meriamon in the estate of Amun in Western Thebes.... /line 553/ The first month of the season Shemu, Day 26, the day of the coronation of the King of Upper and Lower Egypt, Usermaatre Meriamon. That which is offered to Amon-Re and to the sacred image [of the bark] of the King of Upper and Lower Egypt and to the Ennead in his house as the festal supplies this day.... [Then follow, in lines 554-628, long lists of offerings. Needless to say this was an extremely important feast for Ramesses III.]

[The *Feast of the Rising of Sothis*, plate 152, list 23, lines 629-45 (Fig. III.90):] /line 629/ [Vertical col.:] First month of the season Akhet, day of the Feast of the [Heliacal] Rising of Sothis.[11] That which is offered to Amon-Re, King of the Gods, and to the holy image [of the bark] of the King of Upper and Lower Egypt, Usermaatre Meriamon, with his Ennead, from the festival supplies this day.

[Then horizontally follow the items of the offering list:]

/Line 630/ *byt*-bread, a cooking *(pfśw)*[12] [of] 30 per *ḥḳꜣt*-measure, [of which offer] 15 loaves.

/Line 631/ *byt*-bread, a cooking [of] 40 per *ḥḳꜣt*-measure, [of which offer] 25 loaves.

/Line 632/ *psn*-bread, a cooking [of] 20 per *ḥḳꜣt*-measure, [of which offer] 40 loaves.

/Lines 633-35/ [Three more breads.]

/Line 636/ Beer, *wś(?)*-jars, a cooking [of] 5 per *ḥḳꜣt*-measure, [of which offer] 2 jars.

/Line 637/ Beer, *dn*-jars (! *ds*-jars), a cooking [of] 20 per *ḥḳꜣt*-measure, [of which offer] 5 jars.

/Line 638/ Total of various breads for the divine offerings: 112 loaves; beer, 7 *ḥnw*-jars.

/Lines 639-40/ [These lines note the quantities of the grain of Upper Egypt and that of lower which are to be produced for the offerings.]

/Line 641/ Cattle, 1; fruit, *dny*-baskets (or sacks), 2.

/Line 642/ *r*-geese, living, 1; ....-flowers, bunches, 5.

/Line 643/ *sꜥš*-fowl, 3 (or 5?); flowers, formal

bouquets, 4.

/Line 644/ Wine, *mn*-jars, 1; flowers for *ḥtp[t]*-bunches, 4.

/Line 645/ Incense, *dny*-sacks, 2.

[*Feast of the Eve of the Wag-Feast*, list no. 24, lines 646-66:] /line 646/ [Vertical column:] First month of Akhet, day 17[13], day of the Feast of the Eve of the Wag-Feast. That which is offered to Amon-Re, King of the Gods, and his Ennead, and to the holy image [of the bark] of the King of Upper and Lower Egypt, Usermaatre Meriamon, as festal supplies this day. [Then follows, in lines 647-66, the medium-length supply list no. 24. When coupled with the list for the next-day's Wag-Feast, it shows this to be a moderately important festival.]

[The *Feast of Wag*, plate 154, list no. 25, lines 667-85:] /line 667/ [Vertical column:] First month of Akhet, day 19 (!, *should be* 18), day of the Wag-Feast.[14] That which is offered to Amon-Re and to the holy image [of the bark] of the King of Upper and Lower Egypt, Usermaatre Meriamon, as festal supplies this day. [Then follows, in horizontal lines 668-85 supply list no. 25.]

[The *Feast of Thoth*, list no. 26, lines 686-704:] /line 686/ [Vertical column:] First month of Akhet, day 19, day of the Feast of Thoth.[15] That which is offered to Amon-Re, to the holy image [of the bark] of the King of Upper and Lower Egypt, and to the Ennead in his temple, as festal supplies this day. [Then follows, in horizontal lines 687-704, supply list no. 26.]

....

[The *Feast of Hathor*, plate 158, list 40, lines 917-31:] /917/ [Vertical column:] Fourth month of Akhet, day 1, day of the Feast of Hathor.[16] That which is offered to Amon-Re and his Ennead, and to the image [of the holy bark] of Usermaatre Meriamon, as festal supplies this day. [Then follows, in horizontal lines 918-31, supply list no. 40.]

....

[The *Feast of Opening the Window in the Shrine of Seker (Sokar)*, list 42, lines 943-53:] /line 943/ [Vertical column:] Third month of Akhet, day 20, day of the Feast of Opening the Window in the shrine [of Seker]. That which is offered to Seker[17] as festal supplies this day. [Then follows, in horizontal lines 944-53, supply list 42.]

....

[The *Feast of Proceeding in the Shrine of Seker*, list 44, lines 974-87:] /line 974/ [Vertical column:] Fourth month of Akhet, day 13, day of the Feast of Proceeding in the Shrine [of Seker].[18] That which is offered to Ptah-Seker (Ptah-Sokar) as festal supplies this day. [Then follows, in horizontal lines 975-87, supply list no. 44.]

[The *Feast of Placing Seker in their Midst*, list 45, lines 988-1002:] /line 988/ [Vertical column:] Fourth month of Akhet, day 24, day of the Feast of Placing Seker in their Midst.[19] That which is offered to Ptah-Seker (Ptah-Sokar) as festal supplies this day. [Then follows, in horizontal lines 989-1002, supply list no. 45.]

....

[The *Feast of Seker*, plate 160, list 47, lines 1025-1107:] /line 1025/ [Vertical column:] Third month of Akhet, day 26, day of the Feast of Seker.[20] That which is offered to Ptah-Seker-Osiris (Ptah-Sokar-Osiris) and to Nefertum ... within the house of Usermaatre Meriamon in the estate of Amon-Re in

Western Thebes as festal supplies this day.... [Then follows, in horizontal lines 1026-1107, the long list no. 47, indicating the great importance of this feast.]

....

[The *Feast of Nḫb-kꜣw*, plate 163, list 52, lines 1191-1222:][21] /line 1152/ [Vertical column:] First month of Peret, day 1, day of the Feast of *Nḫb-kꜣw* .... That which is offered to Amon-Re, King of the Gods and to his Ennead, as well as to the holy image [of the bark] of the King of Upper and Lower Egypt, Usermaatre Meriamon, as festal supplies this day .... [Then follow horizontal lines 1192-1222 specifying the offerings suffered in the erasure of this feast; see note 21.]

....

[The *Feast of the Procession of Min*, plate 167, list 66, lines 1430-50:][22] /1430/ [Vertical column:] First month of Shemu, day 11, day of the Feast of the Procession of Min to the Stairs ... at dawn. That which is offered to Amun and to the sacred image of the bark of Usermaatre Mariamon as festal supplies this day. [Then follows, in horizontal lines 1431-50, supply-list no. 66.]

Notes to Document III.5

1. These conform to the translations of *nḥḥ* and *ḏt* that I used throughout Volume One (see p. x).

2. For the use of *ms* for the making or fashioning of statues, see Vol. 1, p. 99.

3. "Mansion (or house) of millions of years" is the conventional expression for a mortuary temple and thus here refers to the temple at Medina Habu. See Vol. 1, p. 371.

4. Nelson, *Work in Western Thebes*, p. 9, n. 7: "Ramses III here states that at Medinet Habu he introduced, in connection with

the cult of Ptah-Sokar, the ritual and ceremonies long established at Memphis, the god's home."

5. The horizon-dweller is one who has been buried in the west and who has passed to the other world, i.e., the immortal dead. This includes the spirits, gods, and so on.

6. There is a feast day of the Procession of Min which is included in the canonical list of feast days found in the Old Kingdom as no. 8 (see Document III.1). But that one appears to be an annual feast rather than a monthly one; and so is equivalent to Feast No. 66 in the Medina Habu calendar (Plate 167, The Feast of Procession of Min). The monthly feast of Min is found listed as the last of the feast days of the lunar month (see below, Document III.6).

7. This seems to be written as *ḫn nb*, but what is surely intended is *pśḏtyw* with the added festival sign as a determinative instead of the basket sign *(nb)* which actually appears in the text. If Parker is correct in saying that the first day precedes the appearance of the crescent (*Calendars*, p. 13), which follows on the second day or later, then we should not translate *pśḏtyw* "Feast of the New Moon," as was conventionally done by Borchardt, Nelson, and others.

8. This is found in the canonical list of feasts given in Document III.1 (Feast no. 10), where it is also called the "Feast of the Head (or First) of the Month." Parker, *Calendars*, p. 11, calls it "new crescent day" to bring it in line with his view that the month starts on the day prior to the first crescent. However it is rendered, it is obviously the feast of the second day of the lunar month.

9. This feast is included among those of the Old Kingdom (see feast no. 11).

10. It may seem puzzling that the civil year is specified here, since all the monthly Feasts of the Sky by name are designated month-day feasts and not epagomenal-day feasts. But in fact these monthly Feasts of the Sky are lunar or celestial related festivals that fall on the days of the lunar months and thus they are determined by celestial phenomena. Accordingly, they must fall, from year to year, on different days of the civil year and so even on epagomenal days. Hence, since accounts and supply lists were kept in terms of the civil year, it was quite proper to use the whole 365 days of the civil year in reckoning the totals.

11. See my introduction to the document for a discussion of this feast and the reason why it is dated in the month of Akhet without any specification of the day of its celebration. The phrase "with his Ennead" later in the paragraph is partly destroyed and in fact may be out of position here. It probably should follow after "King of the Gods."

12. Concerning *pfśw* Nelson, *Work in Western Thebes*, p. 18, says: "*pfśw* is a word meaning 'a cooking'." Following Nelson's explanation, we can see that, if 30 loaves were made from 1 heqat of grain, only 1/2 heqat of grain would be required to produce the 15 loaves specified as the desired quantity of *byt*-bread. "Therefore, as is well known, the *pfśw* value of a loaf determined its size. Similarly the *pfśw* value of beer determined its strength."

13. The writing is partially obscured, but I Akhet 17 is the Eve of the Wag-Festival. See S. Schott, *Altägyptische Festdaten* (Wiesbaden, 1950), p. 961, reference 17.

14. The Wag-Feast appears in the old list of feasts as no. 4. See Document III.1. There I have translated it as "Exultation."

15. The Feast of Thoth appears in the old list of feasts as no. 2. See Document III.1.

16. The Feast of Hathor is noted in the Ebers Calendar as the beginning of the fourth month of the year. As a part of an ad hoc fixed Sothic year, this feast, like that of the rising of Sothis, ought to be celebrated in the civil year one day later every four years. However it seems to have become a feast attached to the first day of the fourth month of Akhet in the civil year of the Medina Habu calendar, unless this is a gratuitous addition of "day 1" by an unknowing scribe. One of the difficulties in making sure that "day 1" was originally intended is that the sun-sign often acts only as a determinative for the seasonal names and at other times (as most often in the Medina Habu calendar) it simultaneously acts as a determinative for the season and as an ideogram for the day of the season. Hence in the entry for the Feast of the Rising of Sothis, it is acting only as the determinative for Akhet while in this entry for the Feast of Hathor, where it is preceded by a single stroke, it appears to be acting simultaneously as a determinative for Akhet and as an ideograph for "day." Hence one wonders whether the scribe confronted in a copy by the sun-sign alone without any day number following it (and thus indicating an unspecified day of the month) decided that it must be the ideogram and so he placed the single stroke before it to make it "day 1." But that itself is odd

since ordinarily the vertical stroke indicating "1" would be written after the sun-sign. A somewhat similar situation exists in the entry for the Feast of *Nḫb-kꜣw*, which is our penultimate extract in this document (where, however, the single vertical stroke in conventional fashion follows rather than precedes the sun-sign), though the identity of this last feast with that of *kꜣ-ḫr-kꜣ* in the Ebers calendar is not absolutely certain. However, it seems better to stick to what is in the text and conclude that what were fixed feasts in the calendar of Ebers in the course of time became fixed feasts in the calendar of Medina Habu (and other calendars). Such, however, was not the case of the Feast of the Rising of Sothis, which was apparently celebrated in the civil years on the day of its appearance (see Schott, *Altägyptische Festdaten*, p. 960). It was also obviously not true in the case of the feasts tied to lunar phenomena which are included in the Medina Habu calendar among the so-called Feasts of the Sky.

17. A Feast of Seker is found in the old list of feasts. See Document III.1, no. 5. But that feast is no doubt the important feast mentioned in list 47 below.

18. See note 17.

19. See note 17. There is confusion on the writing of "Seker." Instead of writing the "r" of Seker the artist writes the similarly shaped "eye" glyph, which is the first glyph of Osiris. Accordingly he added the "seat" glyph, which is the second glyph of Osiris. But perhaps he meant to write "Ptah-Seker-Osiris," which is a known syncretic god (see Vol. 1, p. 268) and merely omitted the "r." This is supported by the almost certain reading of "Ptah-Seker-Osiris" in line 1025 of list 47.

20. This is by far the most important of the four feasts for Seker and, almost certainly, is the old feast mentioned in note 17 above.

21. Nelson says (*Work in Western Thebes*, p. 59): "Directly following the Sokar feasts comes a palimpsest, in which a new feast to celebrate the king's victory of year 11 over the Meshwesh was superimposed upon the Feast of *Nḫb-kꜣw*. It is interesting that Ramses III should have chosen to erase the list for the Feast of *Nḫb-kꜣw* to make room for that of his new feast. The former was a fairly important celebration; and, of course, the erasure of the list does not indicate that its observance was abandoned. It came on the first day of the first month of the second season...."

Despite Parker's objections (*The Calendars of Ancient Egypt*,

p. 58), I believe that Gardiner may well have been correct in identifying the Feast of *Nḫb-kꜣw* with the Feast of *kꜣ-ḫr-kꜣ* given as the feast marking the first day of the fifth month of the Ebers calendar. Hence its status in the civil calendar of Medina Habu ought to have been like that of the Feast of the Rising of Sothis, coming one day later every four years. However, it seems, in the civil calendar of Medina Habu (and other temple calendars), to have become attached to the first day of the first month of Peret (i.e., the fifth month of the civil year) just as it was apparently attached to the first day of the fifth month of the Ebers calendar, based on an ad hoc fixed Sothic calendar. That it was a fixed feast of I Peret 1 in the civil year of other temple calendars is indicated in the evidence given by Schott, *Altägyptische Festdaten*, p. 973, references 82-86. But see also my note 16 above.

22. This is probably identical with the Feast of the Going Forth (or Procession) of Min in the Old Kingdom lists (feast no. 8).

Document III.6: Introduction

The Names of the 30 [Feast-Days] of the [Lunar] Month[s]

This document reveals the late expanded list of 30 monthly lunar feasts, i.e., the names of the days of a 29-day month and that of the additional day of a 30-day month, months of both lengths having been recorded in the lunar calendar described in the Illahun Temple records of the Middle Kingdom (see Document III.1, "Middle Kingdom," Section II and Figs. III.83b and III.83c). In the canonical list of feasts that goes back at least to the Old Kingdom only three or possibly four feast days of the lunar month were mentioned (see Document III.1 above: feasts 9, 10, and 11, with the possible addition of 8, which latter however is probably an annual rather than a monthly feast). Then in the astronomical ceilings of the tomb of Senmut (ca. 1473 B.C.) and that of Seti I (ca. 1306-1290 B.C.) the protective gods of eleven lunar feast days (along with nine or ten others) were indicated, i.e., those for the feasts of the following lunar days: the 4th, 5th, 6th, 7th, 8th, 9th, 10th, 15th, 13th, 16th, and 30th, and the names of those feasts of the protective gods are mentioned in the notes to those numbered days below in Document III.6. Some one-hundred years later, there were given in the Medina Habu calendar eight monthly lunar feasts under their primary names and accompanied by lists of the offerings to be supplied at their celebration. These were feasts for the following days of the lunar month in the following order: 29th, 30th, 1st, 2nd, 4th, 6th, 10th, and 15th. And then skipping to Ptolemaic and Roman times, in the temples of Edfu and Dendera, the full list of feasts for each of the thirty days of the longer lunar month were presented. These are the days included in this document.

We have discussed in Chapter Three the importance of these lists of thirty lunar feasts in giving hints as to the structure of the lunar month. The earliest dominant view among the Egyptologists was that of Lepsius who in 1849 had "no doubt that it (the lunar year [and hence the lunar month]) began with a new moon."[1] We have already mentioned that Richard Parker's treatment of the subject, a century later, abandoned that earlier view and went far toward establishing, as the first day of the Egyptian lunar month, the day of first invisibility of the crescent after the moon of the previous month has waned. Thus, as I note below, he believed that the day embracing the mean conjunction of moon and sun with the earth constituted the first day of the Egyptian lunar month. Parker was by no means the first to suggest this. As Parker's brief historical account indicates, Brugsch, Mahler, Sethe, and Borchardt held this view with varying degrees of consistency and certainty.[2] But surely none of them presented the evidence for it so cogently as did Parker. His evidence was both religious and astronomical in nature. Let me recall that evidence.

The first important bit of evidence was an inscription discovered and published by Brugsch first in 1862 and again in 1864.[3] It can be translated as follows: "He (Khons, the God of the Moon) is conceived on the Feast of *psḏntyw* (i.e., on the first day of the lunar month); he is born on the Feast of the Month (i.e., on the Feast of the Second Day of the Month); he comes to maturity on the Feast of the Half-Month (i.e., on the Feast of the Full Moon or Fifteenth Day of the Month)." This seems quite good evidence that the Egyptians conceived that first day of the month was the day of the invisibility of the first crescent, that the second was the day of first visibility of the new crescent, and that the fifteenth day was the day of the full moon. And indeed Brugsch drew the conclusion that Parker was later to develop more fully, namely, that the month started with the first day of invisibility of the crescent, and the second day marked "the first visible apparition of the lunar disk." He notes that a "host of religious texts" supports his interpretation of this passage.

Parker points to an earlier passage (Middle Kingdom) in the *Coffin Texts* that is similar to the passage quoted by Brugsch:[4] "O Souls of Hermopolis, I know what is small *(šrt)* on [the Feast of] the Month (i.e., the second day of the lunar month) and what is great on the Feast of the Half-Month (i.e., the 15th day of the month), it is Thoth (i.e., the moon)."

The most important part of Parker's argument for the identification of the first day of the month with the day of crescent invisibility lies in his actual calculations, the results of which I here report:[5]

> .... In the latter part of this chapter I have had occasion to make sixty-five calculations of conjunctions with their accompanying mornings of crescent invisibility and evenings of new crescent visibility. In forty-six cases (70 per cent) the crescent was visible on the evening of the day *(ꜣbd)* after that on which

there was no lunar visibility in the morning *(psḏntyw)*. In the other nineteen cases (30 per cent), new crescent was first visible on the third day of the month, *mśpr*, which I have termed "'arrival' day."

.... In Figure 10 [not given here; see Parker's work] is diagrammed the astronomical situation at the beginning of the month. Just before dawn on either the 29th or the 30th day, the last crescent is still to be seen. On the following morning it cannot be seen and the new month begins with *psḏntyw*. Just after sunset on the following day, *ꜣbd*, the new crescent is visible in seven months out of ten. In the other three months it is first seen on *mśpr*.

.... When new crescent can be seen on *ꜣbd*, there is a period of 60 hours from old crescent to new. The mean time of conjunction is thus at noon on *psḏntyw*. If we take 17 hours as the minimum time that must elapse from last visibility to conjunction and from conjunction to new visibility..., we have a possible range of time, during which conjunction may occur, of about 26 hours. When the number of hours required for visibility increases to 30, or the hours required plus the number of hours after noon on *psḏntyw* to time of conjunction total 30 or more, then new crescent is delayed until *mśpr*. In that event the time from last to new crescent is 84 hours, mean conjunction is at midnight on *psḏntyw*, and the possible range is as shown [in Fig. 10; not given here; again see Parker's work]. It is instructive to note that most conjunctions fall on *psḏntyw* (so in fifty-seven of my sixty-five calculations), a few on the day before (seven out of sixty-five), and still less (one out of sixty-five) on the following day. Thus Brugsch, Mahler, and Sethe were roughly correct in their theory.... Their error was in failing to associate *psḏntyw* with an *observable* phenomenon.

.... On the basis of Figure 10 [not given here; see Parker's work] and the time required from conjunction to full moon (13.73-15.80 days...), it is possible to diagram the situation for that time of month. Figure 11 [not given here; see Parker's work] shows the possibilities. When new crescent is visible on *ꜣbd*, then mean full moon is just at the end of *smdt* with a possible range before and after of some 72 hours. When new crescent is delayed until *mspr*, then full moon is also delayed to the beginning of the night on *mspr sn-nw*, "second 'arrival' day," to my mind a deliberate and meaningful choice of name. In no case does full moon ever occur earlier than the night of the 14th or later than the night of the 17th; both these days bear the same name, *siꜣw*, and while I am unable to translate this, I cannot believe that it is lacking in significance.

In presenting the days of the lunar month I have followed the table compiled by Brugsch (Fig. III.91a) that was prepared primarily from inscriptions in the Temples of Edfu and Dendera (see also Brugsch's separate list and German translation, Fig. III.91b).

Notes to the Introduction of Document III.6

1. R. Lepsius, *Die Chronologie der Aegypter. Einteilung und erster Theil. Kritik der Quellen* (Berlin, 1849), p. 157.

2. R.A. Parker, *Calendars of Ancient Egypt*, pp. 9-10, with the accompanying notes.

3. Published by Brugsch in his *Recueil de monuments égyptiens*, I (Leipzig, 1862), Pl. XXXVIII, 2, and his *Matériaux pour servir la construction du calendrier des Égyptiens* (Leipzig, 1864), pp. 59-60.

4. Parker, *Calendars of Ancient Egypt*, p. 12. See A. de Buck, *The Egyptian Coffin Texts*, II (Chicago, 1938), Spell 156, pp. 322-24. Cf. the somewhat confused rendering of R.O. Faulkner, *The Ancient Egyptian Coffin Texts*, Vol. 1 (Westminster, England, 1973), pp. 134-35.

5. Parker, *Calendars of Ancient Egypt*, p. 13; see pp. 13-23 for Parker's detailed report of the calculations by which he supports his view that the ancient Egyptians used the first day of the invisibility of the old crescent before dawn as the first day of the new lunar month. See also Parker's spirited defense of his views, "The Beginning of the Lunar Month in Ancient Egypt," *Journal of Near Eastern Studies*, Vol. 29 (1970), pp. 217-20.

Document III.6

The Names of the 30 [Feast-Days] of [Lunar] Month[s]

[1] Feast of Psedjentyu *(psḏntyw)* (i.e., Feast of the First Day of the Month).[1]

[2] Feast of the Month *(ꜣbd)* (i.e., Feast of the 2nd Day of the Month; perhaps "new crescent day").[2]

[3] Feast of the First Mesper (or Arrival I) *(mśpr tp)*[3] (i.e., Feast of the 3rd Day of the Month).

[4] Feast of the Going Forth of Semet (or the Sem-priest) *(prt śmt* [or] *śm)*[4] (i.e., Feast of the 4th Day of the Month).

[5] Feast of the Offerings on the Altar *(iḫt ḥr ḫꜣwt)* (i.e., Feast of the 5th Day of the Month).[5]

[6] Feast of Senet *(śnt)* (i.e., Feast of the 6th Day of the Month).[6]

[7] Feast of the First Quarter Part [of the Month] (*(dnit)* (i.e., Feast of the 7th Day of the Month).[7]

[8] Feast of the Tep *(tp)* (i.e., Feast of the 8th Day of the Month).[8]

[9] Feast of Kap *(kꜣp)* (i.e., Feast of the 9th Day of the Month).[9]

[10] Feast of Sif *(śif)* (i.e., Feast of the 10th Day of the Month).[10]

[11] Feast of Setet *(śtt)* (i.e., Feast of the 11th Day of the Month).[11]

[12] Feast of the 12th Day of the Month (reading uncertain).[12]

[13] Feast of Maa setjy *(mꜣꜣ šṯy)* (i.e., Feast of the 13th Day of the Month).[13]

[14] Feast of Recognition [I] *(šiꜣw)* (i.e., Feast of the 14th Day of the Month).[14]

[15] Feast of Half-Month Day *(šmdt)* (i.e., Feast of the 15th Day of the Month, Full Moon Day).[15]

[16] Feast of Mesper II (Arrival II) *(mšpr II)* (i.e., Feast of the 16th Day of the Month).[16]

[17] Feast of Recognition [II] *(šiꜣw)* or *(sꜣ)* (i.e., Feast of the 17th Day of the Month).[17]

[18] Feast of the Moon *(iꜥḥ)* (i.e., Feast of the 18th Day of the Month).[18]

[19] Feast of Hearing His Words *(šḏm mdw.f)* (i.e., Feast of the 19th Day of the Month).[19]

[20] Feast of Choice *(štp)* (i.e., Feast of the 20th Day of the Month).[20]

[21] Feast of Ornaments *(ꜥprw)* (i.e., Feast of the 21st Day of the Month).[21]

[22] Feast of the Back of Sepedet (Sothis?) *(pḥ spdt)* (i.e., Feast of the 22nd Day of the Month).[22]

[23] Feast of the Second Part [of the Second Half of the Month, i.e., the Last Quarter of the Month] *(dniꜣt* II) (i.e., Feast of the 23rd Day of the Month).[23]

[24] Feast of the Shadows (?) *(ḳnḥw)* (i.e., Feast of the 24th Day of the Month).[24]

[25] Feast of Emitting Light *(stw* or *štt)* (i.e., Feast of the 25th Day of the Month).[25]

[26] Feast of the Going Forth *(prt)* (i.e., Feast of the 26th Day of the Month).[26]

[27] Feast of Wesheb *(wšb)* (i.e., Feast of the 27th Day of the Month).[27]

[28] Feast of the Jubilee of Nut *(ḥb-šd nwt)* (i.e., Feast of the 28th Day of the Month).[28]

[29] Feast of the Attender (?) *(ꜥḥꜥ nṯr ?)* (i.e., Feast

of the 29th Day of the Month).[29]

[30] Feast of the Going Forth of Min *(prt mn)* (i.e., Feast of the 30th Day of the Month).[30]

Notes to Document III.6

1. There is no general agreement on the meaning of *psḏntyw*. For a discussion of its possible meaning with references to the literature, see Parker, *The Calendars of Ancient Egypt*, p. 12. It was one of the two most important monthly Sky Feasts at Medina Habu (see Document III.5). It was also called from the name of its protective god "The Feast of Thoth" (see Fig. III.91a, lower col. 1). This feast is not to be confused with the regular annual Feast of Thoth listed in the Old Kingdom and at Medina Habu in the 19th dynasty. The alternate names for the feasts in terms of their protective Gods here and throughout the 30 entries are given in the lower columns of Fig. III.91a and generally translated in Brugsch, *Thesaurus inscriptionum*, pp. 49-51. The god of this day appears with 10 other protective deities among the 11 gods known to be protective gods flanking the northern constellations on the astronomical ceilings of the tombs of Senmut, Seti I, and others. I shall give only the names of these lunar day-gods here in the notes so that they may be compared with the day-god list at the end of Document III.3. For the remainder, the reader may, as I have suggested, consult Figs. III.91a and the above-mentioned translations in Brugsch's *Thesaurus*. Finally, as I noted in the introduction to this document, this first day of the month was thought by Parker and others earlier to coincide with the first day of invisibility of the waning crescent.

2. The translation or rather epithet given this feast by Parker is "new crescent day." Called simply "Feast of the Month," it was listed as one of the feasts in the Old Kingdom offering-lists (see Document III.1, feast no. 10). It also appears as the fourth feast of the Feasts of the Sky in the Medina Habu calendar (Document III.5). Further, it was named from its protective god "Feast of Horus, Protector of his Father" (see Fig. III.91a, lower col. 2).

3. The translation by Parker as "'arrival' day" was based on his belief that its title was predicated on the fact that often first crescent appeared not on the second but rather on the third day of the lunar month, as I have noted in discussing Parker's calculations

in the introduction to this document. The alternate name of this feast from its deity was "Feast of the Day of Osiris" (see Fig. III.91a, lower col. 3).

4. The phrase "*śm*-priest" is that of Parker. The feast's alternate name from its deity was "Feast of Imset" (see Fig. III.91a, lower col. 4). The god is found among the protective gods on the astronomical ceilings (see Document III.3).

5. The feast's alternate name from its protective deity was "Feast of Hapy" (see Fig. III.91a, lower col. 5). The god appeared among the gods flanking the northern constellations in astronomical ceilings (see Document III.3).

6. Its alternate name from its protective god was "Feast of Duamutef" (see Fig, III.91a, lower col. 6). Its god was among the gods flanking the northern constellations in the astronomical ceilings (see Document III.3). As is evident from its offering-list, this was an extremely important feast at Medina Habu as befitting the eve of the first quarter day (see Document III.5). For various views of the name and character of this Feast including the author's own, see G.H. Hughes, "The Sixth Day of the Lunar Month and the Demotic Word for 'Cult Guild'," *Mitteilungen des Deutschen Archäologischen Instituts Abteilung Kairo*, Vol. 16, II.Teil (1958), pp. 147-60.

7. Its alternate name from its protective god was "Feast of Qebehsenuf" (see Fig. III.91a, lower col. 7). Its god is present among the gods flanking the northern constellations in the astronomical ceilings (see Document III.3). As the feast of the first quarter, it draws its name from an astronomical phenomenon.

8. Its alternate name from its protective god was "Feast of [the One] Observing his Father" (Fig. III.91a, lower col. 8). This god was among the gods flanking the northern constellations in astronomical ceilings (see Document III.3).

9. Its alternate name from its protective god was "Feast of [the One who] Created his Eternity" (Fig. 91a, lower col. 9).

10. Its alternate name from its protective god was "Feast of [the One who] Created his Own Name" (Fig. III.91a, col. 10). For its earlier usage, see Documents III.3. Note its place as the penultimate feast among the monthly Feasts of the Sky at Medina Habu; see Document III.5.

11. For the alternate name of this feast from its protective god, see Fig. III.91a, lower col. 11.

12. For the alternate name of this feast from its protective

god, see Fig. III.91a, lower col. 12.

13. The alternate name of this feast from its protective god was "Feast of Teknu" (Fig. III.91a, lower col. 13). For its earlier usage, see Document III.3.

14. For the alternate name of this feast from its protective god, see Fig. III.91a, lower col. 14. This "Feast of Recognition" on the 14th day of the month has some unknown relationship with the similarly named feast of the 17th day. I suspect that it refers to the "recognition" on this day that the waxing moon is almost complete, i.e., when the full moon will occur on the fifteenth day (see note 17 below), or even in rare instances when the full moon occurs on the 14th itself. See the long passage on phases of the moon I have quoted from Parker's *Calendars* in the introduction to this document.

15. Needless to say, this day, the one on which the full moon was most likely to occur, was half-way between the day of the first quarter (no. 7) and the day of the last quarter (no. 23). The alternate name of this feast from its protective god was "Feast of [the god] Irmawa" (Fig. III.91a, lower col. 15). For its early usage, see Document III.3. This was the last of the Feasts of the Sky celebrated at Medina Habu (see Document III.5), in view of the fact that the Feasts of Day 29 and Day 30 were specified before the feasts of the following month and no more Feasts of the Sky were celebrated in that month.

16. This day was translated "second 'arrival' day" by Parker. I have given his explanation of why this day would be so-named in the quotation from his *Calendars of Ancient Egypt* in the introduction to this document. The alternate name of this feast from its protective god was "[Feast of the God] Who Pronounces his [Own] Words (or Name)" (see Fig. III.91a, lower col. 16). For its early usage in a different form, see Document III.3.

17. For the alternate name of this feast from its protective god, see Fig. III.91a, lower col. 17. See note 14.

18. For the alternate name of this feast from its protective god, see Fig. III.91a, lower col. 18.

19. *Ibid.*, lower col. 19.

20. *Ibid.*, lower col. 20.

21. *Ibid.*, lower col. 21.

22. *Ibid.*, lower col. 22.

23. *Ibid.*, lower col. 23.

24. *Ibid.*, lower col. 24.

25. *Ibid.*, lower col. 25.

26. *Ibid.*, lower col. 26.

27. *Ibid.*, lower col. 27.

28. *Ibid.*, lower col. 28.

29. *Ibid.*, lower col. 29. The Feast of this day appeared as the first of the Feasts of the Sky on the calendar of Medina Habu. In 29-day months, the final crescent appeared on this day.

30. The alternate name of this feast from its protective god was "Feast of Nehes" (see Fig. III.91a, lower col. 30). This god was included among the gods flanking the northern constellations in astronomical ceilings (see Document III.3). This feast was included as the second Feast of the Sky in the calendar of Medina Habu. It was the last day of a 30-day lunar month, i.e., it was the day of final crescent of such a month.

Document III.7

A Table of the Lengths of the Daylight and of the Nighttime at Monthly Intervals, i.e., on the First of each Month (from Cairo Museum Papyrus No. 86637)

The extant copy of this short document (see Figs. III.58a and III.58b) dates from the twelfth century B.C., but it was composed earlier, in the period between 1400 and 1250 B.C. It is discussed and given in its entirety in Chapter Three, pages 98-101.

Document III.8

A Table of the Lengths of the Daylight and of the Nighttime at Semimonthly Intervals, i.e., on the First and Fifteenth of each Month.

The stone fragments of this plaque (see Fig. III.59) were discovered at Tanis and are now in the Cairo Museum. Though the date of the copy is not known, the table itself probably dates from the time of Necho II in Dynasty 26. This short document is discussed and given in its entirety in Chapter Three, pages 101-06.

Document III.9: Introduction

An Egyptian 25-Year Lunar Cycle

Much of the discussion in Chapter Three of the 25-year lunar cycle edited by Neugebauer and Volten[1] need not be repeated here. But a few words concerning its initial discovery and editing seem appropriate.

The text appears in Papyrus Carlsberg 9 (D 7) and was written down in or after year 7 of the reign of Antoninus, i.e., 144-45 A.D. It came from Tebtunis in the Fayyum. In the Neugebauer-Volten edition of 1938 the crucial part of the text in Column II, lines 1-20 and Column III, lines 1-8, was established as a lunar cycle of 25 years, in which the normal year is 12 months and the "great year" is 13 months.

In Column I we first find a list of regnal years of five emperors, the regnal year most likely translating the well established expression *ḫꜣt-sp*. The emperors' names were enclosed in cartouche rings which reminded the reader that in Egypt they were not only emperors of Rome but pharaohs of Egypt as well. The regnal years in this list are the initial years of past 25-year lunar cycles and include Tiberius 6 (19 A.D), Vespasian 1 (69 A.D.), Domitian 14 (94 A.D.), Hadrian 3 (119 A.D.), and Antoninus 7 (144 A.D.). It is obvious that the year Claudius 4 (44 A.D.) was omitted from the list. It should have occupied the line between actual lines 1 and 2, i.e., between Tiberius 6 and Vespasian 1.

Following this list of the beginnings of cycles in

lines 7-12 the names of the twelve zodiacal constellations are given, two per line. Line 6 no doubt made some reference to the constellations that followed but only the words "the sky" *(tꜣ pt)* remain at the end of the line.

Lines 13-17 in Column I contain 25 numbers needed in the cycle described later in columns II and III. These numbers are the day numbers of month II, season I, of the beginning of the second lunar month in each of the 25 years of the lunar cycle. These numbers are generated by beginning with 1 and, assuming a module of 30 (determined by the unchanging length of the month in the civil year), continually subtracting 11 except from numbers 28, 4, 10, 16, and 22 from which 10 is subtracted to produce 18, 24, 30, 6, and 12. As Neugebauer and Parker point out, "the scheme is periodic" because, if we subtract 11 from the last number 12, the result is once more 1, and thus the cycle of numbers will be repeated.[2]

Moving to Column II, lines 1-20, and Column III, lines 1-8, we find the data thought sufficient to reveal the "procedure of enumerating the 25 years of the moon in order to make them known," as we are told in line II, 1. The given numbers are tabulated in Fig. III.8a; they are those above and to the left of the gnomon formed by the double lines. The numbers below and to the right of the gnomon are those deducible from the given numbers. The formula for generating the proper numbers is that implicit in the numbers given in lines 13-17. Parker further extended the procedures to construct the full 25-year cycle (see Fig. III.9) in Chapter II of his *The Calendars of Ancient Egypt*. Later he and Neugebauer described the procedure of going from the numbers given for a limited number of even-numbered months (the 2nd, 4th, 6th, 8th, 10th, and 12th months of the first year and the 2nd and the 4th months only of the remaining 24 years) to the numbers for all the even-numbered months and then from them to the full 25-year cycle of numbers for all the remaining odd-numbered months; and they followed this discussion by a succinct characterization of the cycle:[3]

> It is easy to restore all day numbers in this 25-year cycle for all even-numbered months by following the pattern of the first four lines (...[see Fig. III.8a]). The dates in each line decrease by 1. The transition from [Month] XII to the next I ordinarily requires a lowering of the day numbers by 5+1 because of the 5 epagomenal days with the exception of the five years 4, 9, 14, 19, and 24 where the dates are lowered only by 5. Since by these rules Year 25 XII 7 will be followed by II 1 we see that the whole scheme is strictly periodic with a period of 25 years.
>
> Ordinarily the interval between two consecutive dates is 59 days with the exception of the above-mentioned five years where the interval is 60 days. Hence we are dealing here with lunar months ordinarily 29 1/2 days long with an occasional insertion of two 30-day months.
>
> Normally each year contains 12 lunar months. But whenever the day numbers increase numerically the interval is not two months but three, e.g., from II,1 to IV,30 in year 1 or from XII,4 to II,28 in the transition from year 3 to year 4. In such cases the interval amounts to 59 + 30 = 89 days. Such

> years with 13 lunar months are the nine cycle years numbered
>
> 1 3 6 9 12 14 17 20 23.
>
> The remaining 16 years are ordinary years with 12 lunar months. Consequently our scheme contains
>
> $$9 \cdot 13 + 16 \cdot 12 = 309$$
>
> lunar month in 25 Egyptian years .... [and] 25·365 [days] = 9125 days. Therefore our scheme is based on the relation
>
> 309 lunations = 9125 days = 25. Eg. years
>
> which is well known in ancient astronomy.[4]

The last part of our document, Column III, lines 9-21, has two sub-columns. The first, in lines 9-17, gives the 9 Great Years of the 25-year cycle (those with 13 months): 1st, 3rd, [6th], 9th, 1[2]th, 1[4]th, 1[7]th, 20th, and 23rd, with the statement in lines 18-19: "These are the 9 great years under the 25 years [of the moon]." The second lists 12 of the remaining 16 "small" or ordinary 12-month years, years 21, 22, 24, and 25 presumably appearing in another column.

Finally, I remind the reader, as I have already said in Chapter III (the text over notes 27-34), that we owe to Richard Parker (1) the determination of the approximate date of the introduction of this cycle in ca. 357 B.C. and this because of (2) his demonstration that, when the cycle was first constructed for the Egyptian civil year, the first day of the new lunar month was considered to be the first day of invisibility of the waning crescent of the old. The original editors of the document, and Parker as well, believed that the cycle's structure and form and the absence in it of contemporary Hellenistic astronomy seemed to stamp it as an Egyptian invention, though the fourth century was a period of foreign intervention in Egypt. Even if the underlying knowledge of the existence of a 25-year cycle was imported, the expression of it in terms of the Egyptian civil year and its complete use of Egyptian terminology show that it was thoroughly adapted to Egyptian procedures by the Egyptian astronomer-priests. The very idea of schematizing calendars was itself inherent in the invention and use of the Egyptian civil calendar in the beginning of the third millennium and the use of a lunar calendar tied to that civil calendar in the Middle Kingdom (see Document III.1) is another example of a simple numerical schematization. This Egyptian 25-year cycle was adapted by the Ptolemies to their lunar calendar, in which, however, the lunar month began with the first visibility of the new crescent.[5]

In my translation I have followed the column and line numbers of the Neugebauer-Volten text, which is the text from which the translation was made (with some attention, however, to the phonetic transcription of Parker in Neugebauer and Parker, *Egyptian Astronomical Texts*, Vol. 3, pp. 220-21). The German translation accompanying the original edition and the various comments in the Neugebauer and Parker volume were useful. The columns are designated by capital Roman numerals, the lines by Arabic numerals inserted between slants. Square brackets used within the text itself contain conjectured but illegible or missing readings. Much of that bracketed material is clearly evident from the numerical procedures that underlie the cycle and I shall only occasionally call attention to this in the notes to the document. Angle brackets include additions that appear to be certain. I have generally followed the editors or Parker in my use of brackets.

Notes to the Introduction to Document III.9

1. O. Neugebauer and A. Volten, "Untersuchungen zur antiken Astronomie IV," *Quellen und Studien zur Geschichte der Mathematik, Astronomie, und Physik,* Abt. B, Vol. 4 (1938), pp. 401-02; full monograph, pp. 383-406.

2. Neugebauer and Parker, *Egyptian Astronomical Texts*, Vol. 3 (Text), p. 223.

3. *Ibid.*, pp. 223-24.

4. Neugebauer and Parker in *ibid.*, p. 224, n. 1, add the following: "It is quoted, e.g., by Ptolemy in the *Almagest* VI, 2 and underlies many numerological constructions in Greek astrology (cf., e.g., Bouché-Leclercq, *AG*, p. 410)."

5. A. E. Samuel, *Ptolemaic Chronology* (Munich, 1962), pp. 54-74.

Document III.9

An Egyptian 25-Year Lunar Cycle

[I] [Preceding 25-year Lunar Cycles [1]]

/1/ [Year] 6 of the Pharaoh-Emperor[2] Tiberius, l.p.h. (life, prosperity, and health), is Year 1 of a lunar cycle *(iʿḫ try)* (*lit.*, a moon period).

/2/ [Year 1] of the Emperor Vespasian, l.p.h., is year 1 of a lunar cycle.

/3/ [Year] 13 (*!, should be* 14) of the Emperor Domitian is year 1 of a lunar cycle.

/4/ [Year] 4 (*!, should be* 3) of the Emperor Hadrian is year 1 of a lunar cycle.

/5/ [Year 7] of the Emperor Antoninus is year 1 of a lunar cycle.

[Signs of the Zodiac]

/6/ [.........] of the sky
/7/ [The Lion] The Maiden
/8/ [The] Balance The Scorpion
/9/ [The Ar]cher The [Face] of the Goat
/10/ [The] Water[bearer] The [Fis]h
/11/ [The] Ram The [Bull]
/12/ [The] Twins The [Cra]b

[Day Numbers for Month II of Season I in years 1 to 25 of the 25-year Lunar Cycle[3]]

/13/	[1]	[20]	9	28	[18]	7
/14/		[26	1]5	4	23(!24)	1<3>
/15/		[2]	[2]1	10	30[4]	19
/16/		8	27	16	6	25
/17/		1<4>	3	22	12	

[Col. II] /1/ Here is the procedure of enumerating the 25 years of the moon in order to make them known.[5]

/2/ Year *(ḥ͗t sp)* 1, Month I of Akhet, Day *(sw)*; Month II of Akhet, Day 1; Month III of Akhet, Day; Month IV of Akhet, Last Day (i.e., Day 30);

/3/ Month I of Peret, Day; Month II of Peret, Day 29; Month III of Peret, [Day]; Month IV of Peret, Day 2[8].

/4/ Month I of Shemu; Month II of Shemu, Day 29; Month III of Shemu; Month IV of Shemu, Day 26.

/5/ Year 2, Month I of Akhet, Day; Month II of Akhet, Day 20; Month III of Akhet, Day; Month IV of Akhet, Day 19.

/6/ Year 3, Month I of Akhet, Day; Month II of Akhet, Day 9; Month III of Akhet, Day; Month IV of Akhet, Day 8.

/7/ Year 4, Month I of Akhet, Day; Month II of Akhet, Day 28; Month III of Akhet, Day; Month IV of Akhet, Day 2[7].

/8/ Year [5], Month I of Akhet, Day; Month II of Akhet, Day 18; Month III of Akhet, Day; Month IV of Akhet, Day 17.

/9/ Year 6, Month I of Akhet, Day; Month II of Akhet, Day 7; Month III of Akhet, Day; Month IV of Akhet, Day 5.

/10/ Year 7, Month I of Akhet, Day; Month II of Akhet, Day 26; Month III of Akhet, Day; Month IV of Akhet, Day 5 (! 25).

/11/ Year 8, Month I of Akhet, Day; Month II of Akhet, Day 15; Month III of Akhet, Day; Month IV of Akhet, Day 14.

/12/ Year 9, Month I of Akhet, [Day]; Month II of Akhet, [Day] 4; Month III of Akhet, Day; Month IV of Akhet, Day 3.

/13/ Year 10, Month I of Akhet, Day; Month II of Akhet, Day 24; Month III of Akhet, Day; Month IV of Akhet, Day 23.

/14/ Year 11, Month I of Akhet, Day; Month II of Akhet, Day 1[3]; Month III of Akhet, Day; Month IV of Akhet, Day 12.

/15/ Year 12, Month I of Akhet, [Day]; Month II of Akhet, Day [2; Month III of Akhet, Day; Month IV of Akhet, Day 1].

/16/ Year 13, Month I of Akhet, [Day]; Month II of Akhet, Day [21; Month III of Akhet, Day; Month IV of Akhet, Day 20].

/17/ Year 1[4], Month I of Akhet, Day; Month II of Akhet, [Day 10; Month III of Akhet, Day; Month IV of Akhet, Day 9].

/18/ Year 15, Month I of Akhet, Day; Month II of Akhet, [last] day; [Month III of Akhet, Day; Month IV of Akhet, Day 29].

/19/ Year 16, Month I of Akhet, Day; Month II of Akhet, [Day 19; Month III of Akhet, Day; Month IV of Akhet, Day 18].

/20/ Year 17, Month I of Akhet, Day; Month II of

Akhet, [Day 8; Month III of Akhet, Day; Month IV of Akhet, Day 7].

[Col. III]

/1/ [Year 18, Month I of Akhet, Day; Month II of Akhet, Day 27; Month III of Akhet, Day; Month IV of Akhet, Day 26.]

/2/ [Year 19, Month I of Akhet, Day; Month II of Akhet, Day 16; Month III of Akhet, Day; Month IV of Akhet, Day 15.]

/3/ [Year 20, Month I of Akhet, Day; Month II of Akhet, Day 6; Month III of Akhet, Day; Month IV of Akhet, Day 5.]

/4/ [Year 21, Month I of Akhet], Day; [Month II of Akhet, Day 25; Month III of Akhet, Day; Month IV of Akhet, Day 24].

/5/ [Year 22, Month I] of Akhet, Day; [Month II of Akhet, Day 14; Month III of Akhet, Day; Month IV of Akhet, Day 13].

/6/ [Year 23, Month I] of Akhet, Day; Month II of Akhet, [Day 3; Month III of Akhet, Day; Month IV of Akhet, Day 2].

/7/ [Year 24, Month I] of Akhet, Day; [Month II] of Akhet, Day [22; Month III of Akhet, Day; Month IV of Akhet, Day 21].

/8/ [Year 25, Month I] of Akhet, Day; Month II of Akhet, Day 12; Month III of Akhet, Day; [Month IV of Akhet, Day 1].

[Great Years, left; Small Years, right][6]

/9/ [Year 1 of the moon] is a great yr.
—Yr. 2 of the moon [is a small yr.].
/10/ [Yr. 3 of the moon] is a great yr.

—Yr. 4 of the moon [is a small yr.].
/11/ Yr. [6 of the moon] is a great yr.
—Yr. 5 of the moon [is a small yr.].
/12/ Yr. 9 [of the moon] is a great yr.
—Yr. 7 of the moon [is a small yr.].
/13/ Yr. 1[2 of the moo]n is a great yr.
—Yr. 8 of the moon [is a small yr.].
/14/ Yr. 1[4 of] the moon is a great yr.
—Yr. 10 [of the moon is a small yr.].
/15/ Yr. 1[7] of the moon is a great yr.
—Yr. 11 [of the moon is a small yr.].
/16/ Yr. 20 of the moon is a great yr.
—Yr. 13 of the moo[n is a small yr.].
/17/ Yr. 2[3] of the moon is a great yr.
—Yr. 15 of the moo[n is a small yr.].
/18/ These are the 9 great years under the 25.
—Yr. 16 [of the moon is a small yr.].
/19/ *(cont. from line 18, left)* Yrs. of the moon.
—Yr. 18 of the moo[n is a small yr.].
/20/ Yr. 19 *(delete?)*.
—Yr. 19 of the moon [is a small yr.].
/21/ *[blank on the left]*
—....*[small years 21, 22, 24, 25*
were in another column now missing?]

Notes to Document III.9

1. Between the first and the second numbered lines the scribe has probably omitted a line concerning an intervening 25-year lunar cycle: "Year 4 of the Emperor Claudius is the first year of a 25-year lunar cycle," since there is no good reason that year should be omitted in a list of preceding cycles. Year 4 of Claudius equals 44 A.D.

2. I translate here the cartouche ring itself by the expression "Pharaoh-Emperor." But from this point on, I simplify it by simply

saying "Emperor."

3. As Neugebauer and Parker note, the restorations and emendations in this section are required by the cycle's pattern.

4. The text has *sw* *ᶜrḳy*, i.e., "the last day," for which we may substitute "30."

5. Again I note that the emendations and corrections throughout the text of II,1-III,8 are required by the cycle's pattern.

6. Enough remains of the repetitive lines to ensure the correctness of the restorations in the remainder of the text. As I have pointed out in the Document, the four additional normal, "small" years needed to complete the right-hand column are missing and were presumably added in another, missing fragment.

Document III.10: Introduction

The Heliacal Rising of Sirius and Sothic Periods in Ancient Egypt

I have treated the so-called fixed Sothic year in some detail in Chapter Three, and in the course of that discussion I have mentioned the three earliest recordings of the helical rising of Sirius in terms of the Egyptian civil year: (1) in the reign of Sesostris III, year 7, IV Peret 16; (2) in the reign of Amenhotep I, year 9, III Shemu 9; (3) in the reign of Tuthmosis III, year ?, III Shemu 28. To them I have added extracts (4) and (5) found respectively in the reign of Ptolemy III: II Shemu 1 (238 B.C. Julian, recorded in the Decree of Canopus promulgated in that year), and in the reign of Ptolemy IV: II Shemu 6 (found on a monument from Aswan). Lastly, (6) I have included a reference to a heliacal rising of Sothis on the 1st day of Thoth (=I Akhet 1) in the Julian year 139 A.D., which is given in Censorinus' *De die natali*, where he also noted that this day was the beginning of a new Sothic period. To each of the six extracts is appended the location of its text and a short commentary with appropriate notes.

I should say once again that I am not primarily concerned in my treatment of ancient Egyptian calendars, clocks, and astronomy with the detailed astronomical efforts by modern scholars to determine Egyptian chronology, important as that subject is to Egyptology, but instead I am concentrating on the nature of Egyptian scientific knowledge and the procedures to acquire that knowledge, the latter being germane to the procedures adopted by modern chronologers. I should note, however, that even after almost one hundred and ninety years of chronological study of the calendaric data recorded by Egyptians, *precise* determinations of their historical Julian dates remain quite difficult and the results vary considerably according to the underlying assumptions made by the chronologists. This is shown in the comments and notes for the first three extracts in Document III.10.

Still, in this introduction to that document a few remarks are in order to inform the reader of the possible techniques used by the ancient Egyptians to determine the day of the heliacal rising of Sirius during any given civil year. In brief, these techniques are quite simple and involve one or more of the following: (1) continued observation of Sirius as the brightest star in the sky (with a possible observational error of about plus or minus one day in the reporting of a rising), (2) a probable hit-rate of about 80% of correctly observed lunar observations of first or last crescent-visibilities that can be useful for dating reported Sothic risings, (3) a prior, simple calculation by the Egyptians of the rising of Sothis, a calculation based on the fairly accurate assumption (at least for Pharaonic times) that the rising would occur one day later in the civil year after every four civil years. Following discussions by Meyer, Borchardt, and others, we may, by shorthand, call the techniques of (1) and (2) "observational" and that of (3) "cyclical" or perhaps better "cyclically schematized." (See note 7 below for Borchardt's preliminary discussion.)

The fact of the delay of the heliacal rising of Sirius

quadrennium after quadrennium was surely detected in the first years, or at least decades following upon the adoption of the day of the rising of Sirius as the first day of the civil year. Indeed Krauss goes even further and suggests that knowledge of the progressive delay of Sirius' rising was a planned part of the firm establishment of the civil calendar of 365 days.[1] But actually we do not know when (or even if) the ancient Egyptians before the Ptolemaic period proceeded from the fact of delay to the conclusions (1) that the progressive retardation would continue indefinitely at that same, or approximately same, rate for over fourteen and one half centuries until once more the heliacal rising of Sirius would fall on the first day of the civil year and (2) that the whole process of nearly uniform retardation would be repeated again and again, thus defining the historic era or period of 1461 Egyptian civil years, the so-called Sothic period. There is no discussion of such a historic period by the ancient Egyptians themselves, though we might expect such a discussion at least after the end of the first Sothic period in ca. 1318 B.C.

In fact, the first explicit historical or descriptive interest in the approximately one-day delay after four years of the rising of Sothis and in the role of the so-called Sothic period and its relationship to the Egyptian calendar began with the Macedonians and the Greeks. The former were conquerors who established the Ptolemaic dynasty. As rulers they were forced to face the prevailing Egyptian calendar with (1) its strengths, such as its uniformity and its long use, and (2) its deficiency, i.e., its non-conformity with the seasons. It was apparently that seasonal incompatibility which dictated the attempted reform of the civil calendar during the reign of Ptolemy III. This is illustrated by my extract from the Decree of Canopus given in Document III.10. This extract shows that the Egyptian priests had kept track of the progressive delay of the rising of Sirius over a long enough time that, by means of that decree, they could order the reform of the civil year and thus eliminate that delay by adding a sixth epagomenal day to the end of every fourth year.

The Ptolemaic rulers yielded to the Romans (see the Chronology in Volume One of my study). And Augustus reformed the calendar in Egypt in a manner like that suggested in the Decree of Canopus, producing the so-called Alexandrine calendar inaugurated in 26/25 B.C. Julian with its first day of Thoth rendered fixed on 25 Epiphi, i.e., III Shemu 25, of the then current Egyptian civil year (which was thus made equivalent in terms of the Julian calendar to August 29) by the addition of a sixth epagomenal day.

Meanwhile many Greek and Roman authors mentioned the Egyptian civil calendar because of its unvarying length and long history and often used it to construct their astronomical tables.[2] They also noted (1) the delay of about one day in the rising of Sirius after every period of four civil years and (2) that delay's production of the Sothic period of 1461 civil years. Their accounts (not detailed here)[3] were probably based, at least in part, on Egyptian sources, perhaps both oral and written ones. While I shall not quote those various classical authors, I do include at the end of Document III.10 pertinent passages of Censorinus' *De die natali*, for these passages show several important things: (1) The Sothic period is defined as one of 1461 civil years. (2) Despite the introduction by Augustus in 26/25 B.C. of the Alexandrine (Roman) calendar mentioned above (but not by Censorinus), the old Egyptian civil year was probably still in use to some extent in 239 A.D.; this latter point is deduced from the fact that Censorinus speaks of Sirius' rising in 239 A.D. (the year in which he is writing) as being on vii kalends July (June 25) and he couples this statement with the observation that therefore Sirius arose on the first day of Thoth 100 years earlier, i.e., on the xiii kalends August (July 20), 139 A.D. Julian. (3) This latter day was the first day of a new Great Year (Sothic period), which was equivalent to saying that the heliacal rising of Sirius was once more synchronous with the first day of the old Egyptian civil year. (5) Two other historical eras, those of Philip and Nabonassar (written as Nabonnazar), take their beginnings from the first day of that month "whose name among the Egyptians is Thoth," thus intimating that the usage of the Egyptian civil year was widespread.

In my general account of the so-called Sothic year in Chapter Three, I alluded to the extraordinary conclusion of Meyer that the rising of Sothis came to be fixed by the Egyptians through calculation rather than by observation as it was no doubt initially determined.[4] He based this conclusion predominantly on these considerations: (1) the persistent use of the remarkable civil year of 12 months of 30 days each plus five epagomenal days in an uninterrupted and unreformed way from the time that the rising of Sothis was used to determine the first day of that year; (2) clear evidence existed, in the form of the recording of double dates of Sothic risings (i.e., their dates in specific Egyptian civil years), that the ancient Egyptian priests kept track of the march of the rising of Sothis through the civil year in the course of long periods of time; and

(3) late evidence that the Egyptian priests knew that the progress of the Sothic rising through the days of the civil year could be halted (almost completely) by adding a sixth epagomenal day to the year every fourth year, though Meyer insisted that no such step was taken or attempted until the failed reform of the Decree of Canopus in 238 B.C. One of Meyer's most enthusiastic supporters was Raymond Weill, who reviewed and extended Meyer's conclusions with what he believed was further supporting evidence of the use by the Egyptians of a fixed Sothic year of festivals in the temples from at least the New Kingdom onward.[5] To be sure, in a 1928 supplement to his study, which he called *Compléments*,[6] Weill was able to discuss questions of the actual observation of Sirius in terms of the latitude of the place of observation and the possible range of measurements in degrees of the so-called "arc of vision," as presented by Borchardt and P.V. Neugebauer in their attempt to judge the actual dates of the heliacal risings of Sirius mentioned in the ancient Egyptian documents. In fact, Meyer's conclusion that the Julian dates of the risings of Sirius mentioned in the documents were easily determinable if one assumed that the priests at some early time simply held that the progressive delay in Sirius' rising was one day after each period of four civil years, and so from that time forward they used that mechanism (though it was not entirely accurate) to announce ahead the days when the festivals of Sothis were to be held. In our first extract in Document III.10, such an announcement of the Sothic rising some 22 days in advance to the Lector Priest at Illahun (El-Lahun) in the twelfth dynasty seemed to confirm, Meyer thought, his belief that this was the way in which the rising of Sothis was able to

be announced ahead of time, namely by the simple calculation we have mentioned, which conclusion, as we shall see, was vigorously attacked by Borchardt (see note 7).

Such a scheme as that proposed by Meyer had as a corollary a continuous set of Great Years, i.e., Sothic cycles, each one of 1461 civil years, the beginnings and ends of those periods being determined backwards from the conjunction of the Sothic rising with the first day of the civil year in 139 A.D. Julian, that date being the single specifically cited coincidence of the rising of Sirius and the first day of the Egyptian civil year with a designated day, month, and year of the Julian calendar.

However, despite the fact that the lengths of the Sothic year and the Julian year were very nearly the same during the Pharaonic times, they were not exactly the same. In the period 2776-1318 B.C. Julian there was one civil year in which the delay of Sothic rising was not one day after four years but was one day after three years, and in the period 1318 B.C.-139 A.D. there were three years with the exceptional delay of one day after three years. Hence Borchardt in 1917 used these facts (along with the additional calculation that there was one exceptional delay of one day after five years during the period 4236-2776 B.C., which he believed —surely incorrectly—to be the first Sothic period in which the civil calendar was used in Egypt) as his initial argument of those presented to reject Meyer's simplistic view of the determinations of risings by cyclic calculation.[7]

Borchardt joined to that argument his opinion that the announcement at Illahun of a rising of Sirius 22 days before the event showed that observation rather than calculation lay at the base of the prediction, since the Lector priest at the temple who received the announcement or prediction would have been too accomplished a calculator and determiner of festival dates to need a date so simply calculable to be reckoned for him (which appears to me to be an extraordinarily weak argument in view of the obviously bureaucratic organization of Egyptian temples). Hence, Borchardt concluded, after additional speculation, that the Egyptians kept tables of the transits of Sirius and other important stars (like the Ramesside charts we give in Document III.14) which would have been helpful for predictive announcements like that at Illahun. He therefore reasoned that the view holding that a so-called cyclical technique was used by the Egyptians ought to be given up and replaced by one that affirms that actual observations underlay the Egyptian reports of Sothic risings. So, according to Borchardt and his successors in the next generation, the view that the Egyptians used the easy cyclical method as their technique for reporting Sothic risings in terms of the civil year, should be abandoned (see n. 7, end).

Hence Borchardt's discussion appears to have been the favored point of departure for the modern dating techniques (described with tables by P.V. Neugebauer), and used by Ludwig Borchardt, W.F. Edgerton, Richard Parker, E. Hornung, and almost all other students of Egyptian chronology, as is evident in the references given in the notes to Document III.10. But it should also be noted that R.K. Krauss in 1981 nevertheless believed that a schematic Sothic Year (based on the assumption of the one-day delay after every four civil years) was more probably used in the Sothic rising reported in the Ebers Calendar (see note 4 to Document

III.10).

It is also evident, above all in the investigations of Borchardt, Parker, and Krauss (see the notes to the first two extracts in Document III.10), that reports of lunar dates in temple papyri (and particularly those of the Middle Kingdom) have played a crucial role in the diverse efforts to establish the historical dates of Sothic risings (see note 3 to Document III.10).

The perusal of both the astronomical and the schematic methods as valid techniques among the Egyptians in reporting specific Sothic dates in terms of the civil year has opened the gates to quite different and varying assumptions by modern scholars of the latitude of the place or places of ancient Egyptian observations and to discussions of the uncertainties of the meaning and actual values of the *arcus visionis* (the arcal altitude of Sirius above the sun at the time of Sirius' sighting) that lay behind the specific observation by the Egyptian of the heliacal risings of Sirius. So the fixing of actual Julian dates from existing ancient evidence becomes very difficult indeed, as Hornung so succinctly noted in his general discussion of the four factors to be used in attempts to establish sound dates.[8]

With these general comments and caveats behind us, we may now proceed to Document III.10 itself. The English translations of each of the extracts making up this document are my own.

Notes to the Introduction of Document III.10

1. R.K. Krauss, "Sothis, Elephantine und die altägyptologischen Chronologie," *Göttinger Miszellen. Beiträge zur ägyptischen Diskussion*, Heft 50 (1981), pp. 76-77: "5. *Einführung des*

365tägigen Kalenders. — Nach älterer ägyptologischer Auffassung ist der 365tägige Kalender und die ihm inhärierende relative Verschiebung des Sothisaufgangs das Resultat einer groben Fehlbestimmung der Länge des Sonnenjahres: 365d statt ca. 365.25d. Nach [O.] Neugebauer bzw. Parker soll das 365tägige Jahr zuerst als 'gemitteltes Niljahr' bzw. als 'schematisches Mondjahr' eingeführt worden sein. Die Funktionsweisen dieser hypothetischen Jahrformen sollen die Erkenntnis der relativen Verschiebung des Sothisaufgangs in einem 365tägigen Kalender verhindert haben. Beide Modells sind m.E. in sich widersprüchliche und nicht praktikabel. Ich sehe hier ein Scheinproblem und begründe in meiner Dissertation den Vorschlag, dass der 365tägige Sonnen- bzw. Sothiskalender mit seiner inhärenten Verschiebung gegenüber dem wahren Sonnen- bzw. Sothisjahr überlegt und planmässig eingeführt worde." Full article, pp. 71-80. For Krauss' dissertation, see note 3 to Document III.10.

2. E. Meyer, *Ägyptische Chronologie*, (Berlin, 1904), p. 3, n. 1, gives examples of Greek usage of the Egyptian civil year.

3. Still useful for the treatment of these classical authors with quotations from their texts is the detailed study by R. Lespius, *Die Chronologie der Aegypter* (Berlin, 1849), pp. 148-96.

4. Meyer, *Ägyptische Chronologie*, pp. 17-38, and his *Nachträge zur Ägyptischen Chronologie* (Berlin, 1908), passim.

5. R. Weill, *Bases, méthodes et résultats de la chronologie égyptienne* (Paris, 1926), Chaps. VI-IX.

6. R. Weill, *Bases, méthodes et résultats de la chronologie égyptienne, Compléments* (Paris, 1928), Chap. IV.

7. L. Borchardt, *Quellen und Forschungen zur Zeitbestimmung der ägyptischen Geschichte*, Vol. 1: *Die Annalen und die zeitliche Festlegung des alten Reiches der ägyptischen Geschichte*, (Berlin, 1917), pp. 56-58: "Wenn im Jahre 139 n. Chr. der Frühaufgang der Hundssterns so fiel, dass in diesem Jahre eine neue Hundssternperiode begann, so ist dasselbe Ereignis nach astronomischer Berechnung einmal 1456 (jul.) Jahre = 1457 ägyptische Wandeljahre, dann wieder 1458 jul. Jahre = 1459 ägyptische Wandeljahre und endlich 1460 jul. Jahre = 1461 ägyptische Wandeljahre, früher eingetreten und zu beobachten gewesen, also in der uns hier interessierenden Zeit im Jahre 4236 v. Chr, wobei ein Fehler von 2 Jahren möglich ist. Bei Annahme einer zyklisch berechneten Periode soll aber der nach diesem Ereignis eingerichtete Kalender im Jahre 4242 v. Chr. eingeführt

worden sein, also in einem Jahre, in dem der Frühaufgang des Hundssterns noch auf den 4. Schalttag fiel. Die Annahme, man könne selbst eine theoretisch zyklisch gedachte Periode vor ihrem tatsächlichen astronomischen Anfang einführen, ist ausgeschlossen. Also ist der Hundssternperiode bezw. die Verschiebung des Frühaufgangs des Hundssterns durch das Wandeljahr hindurch von den Ägyptern nicht zyklisch berechnet, sondern durch Beobachtung ermittelt worden.

Dagegen spricht auch nicht die von. Ed. Meyer angeführte Stelle des Dekrets von Kanopus, in der ausdrücklich gesagt wird, dass der Hundssternfrühaufgang sich alle 4 Jahre um einen Tag verschiebt. In der zeit von 4236 bis 2776 v. Chr. (1461 ägypt. Wandeljahre) ist nämlich die Verschiebung nur einmal 5-jährig und in der von 2776 bis 1318 v. Chr. (1459 ägypt. Wandeljahre) einmal 3-jährig. Erst in der folgenden Periode, 1318 v. Chr. bis 139 n. Chr. (1457 ägypt. Wandeljahre), liegen drei 3-jährige Verschiebungen. Sonst sind die Verschiebungen stets 4-jährig. Wenn in den fast 4000 Jahren, die seit Einführung des Kalenders damals bereits verflossen waren, höchstens 4 Abweichungen von der Regel beobachtet worden waren, von denen sich 2 sogar wieder aufgehoben hatten, so konnte das Dekret von Kanopus ganz gut sagen, dass der Hundssternfrühaufgang sich alle 4 Jahre um einen Tag verschiebe.

Der andere Grund, den Ed. Meyer gegen die Beobachtung anführt, ist aber geradezu ein Beweis dafür. Da in der Anweisung an den Vorlesepriester des Tempels bei Illahun der Frühaufgang des Hundssterns bereits 22 Tage vorhergesagt wird, so kann er nur aus dem Kalender berechnet sein, wird behauptet. Nein, wenn er aus dem Kalender berechnet worden wäre, wäre die ganze Voranzeige nicht nötig gewesen. Der Tempelschreiber, der viel Rechnungen mit grössen Zahlen schrieb und die Feste so genau im Tagebuch verzeichnete, wird doch wohl bis 4 haben zählen können, um auch ohne Voranzeige den Hundssterntag richtig zu bestimmen, wenn die Verschiebubg zyklisch, alle 4 Jahre regelmässig um einen Tag, bestimmt worden sein sollte. Sie muss also anderweitig bestimmt worden sein, was leicht aus irgendeiner Berobachtung etwa einer Frühkulmination oder eines Frühuntergangs durch Hinzurechnung einer gleichfalls leicht nach zwei bis drei Beobachtungen auszuzählenden Reihe von Tagen geschehen kann. Für Beobachtungen von Kulminationen zu jeder Nachtstunde sind uns Tabellen aus dem neuen Reich enthalten, vielleicht haben wir auch eine Tabelle über die Zeitabstände besonders wichtiger Stellungen wie Frühaufgang, Frühhöchststand und Frühuntergang der Hauptgestirne gleichfalls aus dem neuen Reich. Die Unterlagen zu der hier gemeinten Vorherberechnung des Hundssternfrühaufgangs auf Grund einer Beobachtung hatten also die alten Ägypter, es ist daher durchaus möglich, dass ein Astronom in einem der grossen Tempel, nachdem er den etwa im September (jul.) stattfindenden Frühhöchststand des Hundssterns beobachtet hatte, durch einfaches Hinzuzählen einer ein für alle Mal bekannten Anzahl von Tagen genau den im Juli (jul.) des kommenden Jahres stattfindenden Frühaufgang auf den Tag vorherbestimmen konnte. Man braucht auch gar nicht einmal anzunehmen, dass der Frühaufgang des Hundssterns von seinem Frühhöchststand aus vorher bestimmt wurde; es würde sich gewiss auch ein auffallender Stern finden lassen, der etwa gerade die bewussten 22 Tage vor dem Frühaufgang des Hundssterns in leicht zu beobachtender Stellung sich befindet, so dass von da ab die Tage gezählt sein könnten. Jedenfalls ist dieses Verfahren der Vorherbestimmung ausserst einfach und auch mit den Kenntnissen und Hilfsmitteln der alten Ägypter leicht ausführbar. Diese Vorherbestimmung wurde dann den Tempeln des ganzen Landes mitgeteilt. Nur so wird die bei Illahun gefundene Mitteilung erklärlich." In this passage Borchardt made significant use of the article by Theodor v. Oppolzer, "Über die Länge des Siriusjahres und der Sothisperiode," *Sitzungsberichte der Kaiserlichen Akademie der Wissenschaften. Mathematisch- naturwissenschaftliche Classe*, Vol. 90, II. Abt. (1884; publ. Wien, 1885), pp. 557-84. A more detailed treatment of the same topic, referring to the tables published by P.V. Neugebauer, *Astronomische Chronologie*, 2 vols. (Berlin, 1929): Vol. 1, pp. 159-61; Vol. 2, Taf. E 58, et seq., was given by Borchardt, in his *Quellen und Forschungen zur Zeitbestimmung der ägyptischen Geschichte*, Vol. 2: *Die Mittel zur zeitlichen Festlegung von Punkten der ägyptischen Geschichte und ihre Anwendung* (Cairo, 1935), pp. 10-15. He dismisses, on p. 12, the cyclical computation of the Sothic risings "to the scrap-heap (zum alten Eisen)," and it is evident that Egyptologists for the most part have acceded to that dismissal.

8. *Untersuchungen zur Chronologie und Geschichte des Neuen Reiches* (Wiesbaden, 1964), pp. 15-21. Hornung discusses succinctly and accurately the four factors he believes must be considered in efforts to determine the historical dates of the Sothic risings

recorded by the Egyptians, though stressing only the date given in the Ebers Calendar. A more detailed account was given in Borchardt's *Die Mittel*, pp. 10-35, mentioned at the end of the preceding note. For Christian Leitz's more recent effort to use astronomical data other than the conventional data discussed here in connection with the various extracts comprising Document III.10, see above, Chapter III, note 49.

Document III.10

The Heliacal Rising of Sirius and Sothic Periods in Ancient Egypt

1. The Heliacal Rising of Sirius in Year 7 of the Reign of Sesostris III on IV Peret 16

[See Fig. III.92a, in hieroglyphic transcription; cf. Fig. III.92b, the hieratic text from Papyrus Berlin 10012, 18-21, for the predicted rising of Sothis, i.e., for the lines beginning with "The Prince..." and ending with "... announcement-board of the Temple."]

Year 7 [of the reign of Sesostris III], Month III [of Season] Peret, Day 25 .... The Prince and Overseer of the Temple Nebkaure has said to the Chief Lector Priest Pepyhotep: "You should know that the Going Forth (i.e. heliacal rising) of Sothis takes place on [Month] IV [of Season] Peret, Day 16.... You might wish to inform (?) the lay-priests of the Temple of the city called 'Mighty is Sesostris the Justified' and [of the Temple] of Anubis and of [that of] the Crocodile-god. .... And let this letter be produced for the announcement-board of the Temple."

....

Year 7 [of the same reign], Month IV, [of Season] Peret, Day 17 ... Receipts [from the] Festival Offerings of the Going Forth of Sothis:... Loaves, assorted, 200; beer, jars, 60....

[This apparent prediction has caused considerable discussion, as I noted in the introduction to this document. Meyer and those who followed him believed that this prediction showed conclusively that the priests predicted, on the basis of the simple calculation of the appearance of Sirius one day later after every four years, that the rising was to be celebrated on the 16th day of Month IV of Peret and that the account of receipts (tallied on the 17th) indicated that the feast had taken place on the preceding day. But the opponents were convinced that this proved just the opposite, namely that the rising was determined by observation each year. The details of Borchardt's reasoning have been given in note 7 to the introduction.

[The general opinion accepted for a generation or so after Borchardt was that of the opponents, i.e., that the actual historic dates of the risings of Sirius reported in Egyptian documents must be determined by considering the factors affecting actual observations, namely those of the probable latitude and of the specific *arcus visionis* at the time of observation. This has been discussed in Chapter Three, where I have included the essential results given by Edgerton for both this rising during the reign of Sesostris III and the next rising discussed, namely that of the ninth year of the reign of Amenhotep I. As I have suggested in the introduction to this present document, there are difficulties in determining the latitudes and the conditions of observations that have produced sharp differences in modern calculations, even when there is some supplementary evidence reporting lunar years.

[Concerning this first extract I point first to the detailed calculation of the Sesostris date by L.

Borchardt, who settled on -1874, July 18 (Julian).[1]]

[Then I once again remind the reader of Edgerton's calculations, which I quoted in the body of Chapter Three (the text to which note 48 is the reference), where he concluded that the date of the heliacal rising of Sirius is 1870 B.C. ± ca. 6 years.

[Parker's calculation of the rising of Sothis as taking place precisely in 1872 B.C. (Julian), by using much of the supplementary lunar data available,[2] caused considerable stir and almost universal acceptance until the publication of R.K. Krauss' thesis. Indeed the favorable reception of Parker's dating convinced chronologists that they could date much of Dynasty 12 (down into the reign of Sesostris III) with complete accuracy, though W. Barta arrived at a Sothic rising date of 1875 B.C. by accepting most of the conditions set down in Parker's treatment, while deciding against Parker that lunar dates B and D (identified below in note 3 where all of the Illahun lunar dates are listed) fell in Amenemhet III's reign rather than in Sesostris III's reign. (See note 3 below, where the reference to Barta's determination has been added, and where that determination, along with Parker's, is rejected by Krauss.)

[The widespread acceptance of Parker's date of 1872 in the generation after the publication of his *The Calendars of Ancient Egypt* in 1950 is illustrated by my quotation of Hayes's statement in *The Cambridge Ancient History* in 1970 (see note 49 to Chapter Three) and is reflected in the Chronology of the Middle Kingdom which I took from Baines and Målek and presented in the Appendixes to Volume I, like them putting asterisks before those dates considered as definite. Unfortunately, when that part of the volume was prepared, I did not have available to me the very important work of Krauss, which, at the very least, made Parker's dating uncertain.

[Krauss, by means of a more penetrating investigation of the 17 lunar dates from Illahun that can be tied to the Sothic rising (two more than those referenced by Parker), reached a radically different conclusion, namely that the year of the Sothic rising reported for Sesostris III's reign was 1836 B.C. and that the place of observation was Elephantine.[3] His general estimate was that the dates of the Middle Kingdom should be put some 40 years later than the dates generally accepted as the result of Parker's and Barta's investigations. Finally it should be noted that in the tables found in the Book of Nut, there is a reference to a rising of Sothis in IV Peret 16, which is apparently the very Middle Kingdom rising being discussed here and which allows us to date the origin of those tables to the 7th year of the reign of Sesostris III (see the Introduction to Document III.12 below).]

2. The Helical Rising of Sirius on III Shemu 9 in the 9th year of the Reign of Amenhotep I

(1) Year 9 under the Majesty of the King of Upper and Lower Egypt Djeserkare (=Amenhotep I).

(2) The Feast of the Opening of the Year [lies on] III Shemu 9 (i.e., Day 9 of the third month of the third season Shemu in civil year 9 of Amenhotep I). The Going Forth of Sothis (i.e., the heliacal rising of Sirius).

[This text is taken from my translation of the Ebers Calendar in Document III.2 above. As I mentioned at some length in Chapter Three above, this rising of

Sothis was calculated by Edgerton as taking place within the period between -1543 and -1536 Julian if the observation was made at Heliopolis or between -1525 and -1518 if it was made at Thebes (see the quotation in Chapter Three above note 50). Concluding his argument, Edgerton noted: "Other localities in Egypt would yield yet other results. Sais as an observation point would yield earlier years than Heliopolis, and Assuan would yield later years than Thebes. No one believes that the observations were made at Sais or at Assuan, but this belief rests on purely historical considerations, which have nothing to do with astronomy." But as a matter of fact, Krauss seriously presented the case, based on his fixing of the Sothic rising of Sesostris as 1836 B.C., that the rising noted in the Ebers Calendar was probably at Elephantine (Aswan) in 1506 B.C., though it depends on assuming that it was the lunar year that was being related to the civil year in the Ebers calender, and thus that III Shemu 9 of Year 9 of Amenhotep I was the first day of the initial lunar month, a view I suggested earlier was incorrect.[4]]

3. The Heliacal Rising of Sirius on III Shemu 28 in an Unknown Year in the Reign of Tuthmosis III

Month III of the Season of Shemu, Day 28, the Day of the Feast of the Going Forth of Sothis. That which is to be offered this day....

[This is followed by a list of offerings, including beef, fowl, and other items like those given in the offerings lists at Medina Habu (see Document III.5). This Elephantine text is given by H. Brugsch, *Thesaurus inscriptionum*, 2. Abtheilung, p. 363, text "a" (see my

Fig. III.93; cf. also K. Sethe, ed., *Urkunden der 18. Dynastie*, Vol. 3, p. 827). We are not certain to which regnal year of Tuthmosis III this belongs. On the basis of assuming Heliopolis (lat. 30.1°) as the place of observation and assuming a range of values for the *arcus visionis* from 8.5° through 9.5°, Borchardt[5] calculates that the Julian year of the observation lies ("with very great probability") in the period from year -1464 to year -1461. Incidentally, there is a reference to the culmination of Sirius at the beginning of the night in the 12th table of the Ramesside Star Clock, the table concerning II Peret 1-15. By using the terminology of the Book of Nut we can deduce a rising of about III Shemu 26, which Parker would date about 1472 B.C., and this is close to the date mentioned here (see below, Introduction to Document III.14).]

4. The Decree of Canopus of March 7, 238 B.C., Ninth Year of the Reign of Ptolemy III

[This decree resulted from a convocation at Canopus (not far from Alexandria) on the 17th of Tybi in the 9th year of the reign of Ptolemy III. It brought together the chief priests, prophets, those who enter the inner shrine for the robing of the gods, feather-bearers, sacred scribes and the rest of the priests who came together from the temples throughout the land. After praising Ptolemy and his queen Bernice for continually performing many great benefits to the gods, their temples, and the sacred bulls and increasing the honors of the gods, the Assembly of priests decided, among other things, to honor Ptolemy and Bernice with two additional festivals in the course of which they ordered what later turned out to be an unsuccessful inauguration of a fixed Calendar of 365 1/4 days by means of adding a sixth epagomenal day to the conventional five of the civil year. The pertinent extract follows the texts of Lepsius and Sethe (the line numbers are those of Fig. III.94a; cf. Fig. III.94b).]

..../16/ And since, according to an act of a decree published earlier, there is celebrated a feast /17/ to the Benefactor Gods in all temples in the course of each month on the 5th day, on the 9th day, and on the 25th day, and [since] there is also celebrated a feast to the great gods as a general national feast (*lit.,* a great procession [of celebrants] about) the land (i.e., Egypt) at its season during the year, let there also be celebrated a great national feast at its time during the year for the King of Upper and Lower Egypt, Ptolemy, Living Forever and Beloved of Ptah, /18/ and Queen Bernice, Benefactor Gods, in the temples of the Two Lands (Upper and Lower Egypt), that is, in the whole kingdom, [this feast to be celebrated on] the day of the [heliacal] rising *(prt)* of the goddess Sothis (*spdt*), [i.e., the star Sirius] (a feast) called the "Feast of the Opening of the Year" *(wp rnpt)* as so named in the Writings of the House of Life, which corresponds [now] in Year 9 to Day 1, Second Month of Shemu (*Gr.*, the first day of the month of Payni), and in it are celebrated the Feast of the Opening of the Year as well as the [simple] Feast of Bast and the Great National Feast of Bast, and it is also the time of /19/ gathering all the fruits (i.e., crops) of the land, and the time of the rising of the Nile. But if it happens that the Feast of the Rising of Sothis (Sirius) changes to another day after every 4 years, the day of observing it shall not be changed but it shall be celebrated on Day 1, Second

Month of Shemu, just as it was in year 9. /20/ And this festival shall be celebrated for 5 days.... And in order that the seasons may all correspond to the ordinances of heaven at this /21/ time and so that feasts [originally] celebrated in the land in Peret (i.e., the winter) shall not be celebrated at some time in Shemu (i.e., the summer) as the result of the displacement of the Feast of Sothis one day [later] every 4 years, [and in order that it not happen] that other feasts [originally] celebrated in Shemu come to be celebrated in Peret in the future as has happened /22/ in earlier times and would now happen if the year *(rnpt)* remains 360 days plus the 5 [epagomenal] days customarily (?) added to them at the end, [let it now be decreed] that henceforth there shall be added to the 5 epagomenal days one day (i.e., a sixth day) before the New Year for a Feast of the Benefactor Gods. [This shall be done] so that it will be known to all people that what was a bit defective in the dispositions of the seasons and the /23/ year has been corrected and that the opinions concerning the laws *(hpw)* on the science *(rḫ)* of the courses *(m^{c}tnw)* has been completed ....

....

/36/.... This decree shall be written out by the councilors of the temple, by the heads of the temple, and by the scribes of the temple and inscribed on a stela of stone or bronze in the script of the House of Life (i.e., hieroglyphics), in book writing (i.e., demotic), and in the writing of the Lords of the North (i.e., Greek) (cf. Fig. III.94c), and it shall be placed in the Hall of the People /37/ in temples of the first order, temples of the second order, and temples of the third order, so that all the people may know of the honor rendered by the priests of the temples of Egypt to the

Benefactor Gods and their children, as is fitting to do.

[I have translated this extract from the hieroglyphic version published by Richard Lepsius (see Fig. III.94a) and that of K. Sethe (see Fig. III.94b and note 9 below). So far as we know, the decree was the first historical effort to establish a Sothic fixed year by adding a sixth epagomenal day at the end of each quadrennium of the Egyptian civil year. Hence the history of its discovery and the preparation of the texts of its three versions (hieroglyphic, demotic, and Greek) are of some interest. There are two extended copies and two fragments of this inscription.

[The first two publications of the Hieroglyphic and Greek versions from the Stela of Tanis, with discussion and German translations (but without any mention of the existence of the demotic version on the right hand edge of the stela), were those of (1) S.L. Reinisch and E.R. Roesler, *Die zweisprachige Inschrift von Tanis* (Vienna, 1866) and (2) Richard Lepsius, *Das bilingue Dekret von Kanopus in der Originalgrösse mit Übersetzung und Erklärung beider Texte* (Berlin, 1866). These authors were all involved in the discovery of the stela at Tanis in 1866 and there was some acrimony over who should be regarded as the true discoverers, since a French engineer (M. Gambard) working for the Suez Canal Company had first told Reinisch and Roesler and later Lepsius of a block at Tanis with a Greek inscription, and all three of these authors were in Tanis together in 1866.[6] The publications are very much alike but Lepsius' work seems to me to be the more useful one, at least so far as the treatment of the hieroglyphic version is concerned. According to Reinisch and Roesler, the stela measured 2.22 m. long, .78 m. wide, and .33 m. thick. The hieroglyphic text consisted of 37 lines; the Greek, below it, of 76 lines; and the demotic, on the right hand edge, of 75 lines. The stela is in the Cairo Museum with the number 22187.[7]

[The first work to consider the demotic version on the right edge of the Tanis stela was that of É. Révillout, *Chrestomathie démotique. Études égyptienne, Livraisons 13-16* (Paris 1880). The decree is given on pp. 125-76 (my extract on calendars is on pp. 146-57). It contains the Greek text, a French translation, and the demotic text in three parallel columns. Errors and comments are given on pp. 448-54.

[P. Pierret, *Le Décret trilingue de Canope* (Paris 1881). This work presents an interlinear translation of the hieroglyphic text, preceded by a quite useful synoptic translation of the three versions in three columns; in the translation of the demotic text he uses the work of Révillout.

[Not long after this, a second copy of the Decree of Canopus was discovered by G. Maspero in 1881 at Kom-el-Hisn. It is virtually complete, having all three versions, though a fragment is missing from the top and along the right side of the stela (see Fig. III.94c).[8] It was quickly used to reedit or improve the three versions of the decree.[9]

[In addition to the two more or less complete copies of the decree from Tanis and Kom-el-Hisn, there are at least two fragments, one of which gives an interesting variant to the Greek version and the other of which is completely useless.[10] Neither is of interest to our treatment of Sothic risings.

[As the second most important document inscribed simultaneously in Egyptian hieroglyphics, Egyptian demotic, and Greek, the Decree of Canopus continues to day to be used by historians of Ancient Egypt and of the Ptolemies.[11]]

5. The Helical Rising of Sirius during the Reign of Ptolemy IV

[Fig. III.1] /col. 1/ "Hail to you, Isis-Sothis .../col. 2/...Lady (?) of 14 [centuries?] and mistress of 16 [what?], who has followed her dwelling place (i.e., been advancing through the civil year up to now?) for 730 years, 3 months, 3 days, and 3 hours."

[See the hieroglyphic text in Fig. III.1. The references to 14 and 16 are puzzling, though, as I suggest (along with Weill, *Bases*, p. 58, n. 3) the "14" is possibly a reference to 14 centuries (a rounding out of the 1460 years of the Sothic period?). I am stumped by the number "16." If it is interpreted as "ten[s of years], six [of them]" rather than "16," as Weill, *ibid.*, thought possible, it would then be 60 years and thus complete the Sothic period, but I have never seen this type of numerical representation in Egyptian texts and hence I think Weill's suggestion is highly unlikely. The second part of the statement is also not easily understood but I am more confident of my interpretation of it as the recording of the position of Sirius' rising on a specific day in a datable year of the reign of Ptolemy IV. I believe it has not been so recognized because it is presented in numbers that are mixed measures. Let me explain my phrase "mixed measures." By using quotation marks around the given measures of the inscription, the meaning of the whole extract can be interpreted as follows: In order to reach the position the rising of Sothis has in the civil year of the Ptolemaic inscription, she has travelled "730 years" (i.e., half of the supposed 1460 years needed for the star to rise on each of the days of a civil year) to delay her rising half-way through the civil year (i.e., 182 1/2 days), and 360 more years to delay her rising another "3 months," and 12 more years to delay her rising a further "3 days," and one-half more year to delay her rising by "3 hours." And so the mixture of measures I mentioned is that the first one is the number of years it took for Sirius' rising to be delayed a full half of the civil year, while the next three numbers count the additional months, days and hours of further delay in the appearance of Sothis in the current civil year (but without any mention of the numbers of years to produce the remaining days of delay). Thus at the time of the Aswan inscription 730+360+12+1/2 (i.e., 1102 1/2 years) had passed since the time when the helical rising of Sirius took place on the first day of the civil year. If we take that time of the first day as occurring in the quadrennium 1321-18 B.C. Julian, then the date of the inscription should lie in the quadrennium 219-216 B.C. Julian. This latter quadrennium lies within the reign of Ptolemy IV (221-205 B.C.). And indeed if we examine the cartouches found in the Aswan inscription (Fig. III.1, cols. 3-4), we see that they are the cartouches of Ptolemy IV Philopator. From all of this we may then conclude that the Egyptians at the times of the Ptolemaic period (and perhaps long before that time) had been keeping yearly track of the rising of Sirius, perhaps by the assumption that the rising was delayed one day after every four civil years even if they had not created a formal, fixed calendar by adding a sixth epagomenal day, as they attempted unsuccessfully to do in 238 B.C. during the prior reign of Ptolemy III.

Incidentally, this is the first text in the Egyptian language to mention (though by indirection) the specific length of the Sothic period. I say "by indirection" since we are given first the number for half of a Sothic period (730 years) before recording the continuing progress of Sirius' rising through the civil year until the year of the inscription.

[But we can come to an even more specific date of the year of the rising given by this inscription without using any assumption of the dates of the quadrennium in which the Sothic period itself began if we note that according to the text (interpreting the numbers in the manner I proposed above), the rising of Sothis has taken place on the 276th day of the civil year (i.e., 182 1/2 + 90 + 3 + 1/2 days into the civil year). This equates the day of rising to II Shemu 6 of the civil year. Now notice that in the Decree of Canopus of 238 B.C. (presented in the extracts just above the present one), the heliacal rising of Sirius was said to have taken place on the first day of Payni, i.e., II Shemu 1, in that year. Hence 20 years later (at the rate of one day delay per four years) the rising would take place five days later on II Shemu 6, the day deduced from the inscription. Therefore the rising ought to be dated 218 B.C. by this reckoning. Again we note that this falls within the reign of Ptolemy IV whom the cartouches confirm to be the monarch responsible for the inscription].

6. Extracts from the *De die natali* of Censorinus on the Sothic Period and the return of the rising of Sirius to the First Day of Thoth (i.e., I Akhet 1) in 139 A.D. (Calendar Julian)

[Chapter 18, Section 10:] [After describing "Great

Years" or cycles involving the moon, Censorinus says:] But the moon does not pertain to the Great Year (*annus magnus*) of the Egyptians, which in Greek is called *"kunikon"* and which we call in Latin the "Caniucular" [year, i.e., the Year of the Dog-Star (Sirius)] because its beginning is taken when, on the first day of the month which is called Thoth by the Egyptians, the Dog-star rises [heliacally]. For their whole civil year has 365 days without any intercalation. And so a quadrennium for them is about one day shorter than a quadrennium in the natural [year]. Thus it happens that in the 1461st [civil] year it is returned to the same beginning point. This [Great] Year is also called the Solar Year *(heliakos)* by some and the "Year of god" by others....

[Chapter 21, Sections 9-10:] For just as among our ancestors and the Egyptians certain [different] eras (*anni*) are utilized in their writings, like that which they call the "Years of *Nabonnazar*" (i.e., the Era of Nabonnazar), because they (i.e., the years of the Era) begin from the first year of his reign. We are now in the 986th year of that era. Then comes the "Years (Era) of Philip," whose years are numbered from the death of Alexander the Great, and if they are extended to our time, they embrace 562 years. But the beginnings of these eras are always taken from the first day of that month whose name among the Egyptians is "Thoth," which day in the current year was vii kal. July (= June 25) and which 100 years ago [in 139 A.D. Julian] under the second consulate of the emperor Antoninus the Pius and that of Bruttius Praesens fell on xiii calends August (July 20), at which time the Dog-star was accustomed to rise in Egypt. Thus we see that we are today in the 100th year of that Great Year (Era) which, as has been said above, is called "Solar

Year," "Year of the Dog-star," and the "Year of god."

[I have translated the Teubner Latin text of N. Sallmann, *Censorini de die natali liber ad Q. Caerellium* (Leipzig, 1983), Chap. 18, p. 43, lines 2-9; Chap. 21, p. 52, line 18 – p. 53, line 11. I have already indicated in the Introduction to Document III.10, the importance of this extract for the treatment of the Sothic risings (and particularly for the rising of 139 A.D. when, according to Censorinus, the first day of the Egyptian month Thoth fell on July 20, the day when the Dog-star Sirius customarily rose in Egypt), and so, he points out, at the time he was writing, "we are in the 100th year of that Great Year," i.e., the 100th year of a Sothic period.]

Notes to Document III.10

1. L. Borchardt, *Quellen und Forschungen zur Zeitsbestimmung der ägyptischen Geschichte*, Bd. 2: *Die Mittel zur zeitlichen Festlegung von punkten der ägyptischen Geschichte und ihre Anwendung* (Cairo, 1935), pp. 29-32.

2. R.A. Parker, *The Calendars of Ancient Egypt* (Chicago, 1950), pp. 65-66.

3. R.K. Krauss, *Probleme des altägyptischen Kalenders und der Chronologie des Mittleren und Neuen Reiches in Ägypten* (Diss. Berlin, 1981), pp. 86-127, and also his "Sothis, Elephantine und die altägyptische Chronologie," *Göttinger Miszellen. Beiträge zur ägyptischen Diskussion*, Heft 50 (1981), pp. 71-80, and see particularly pp. 74-75 for a brief summary:

"Wesentlich günstiger ist die Situation beim Kahundatum (7. Jahr Sesostris III.), da es mit insgesamt 17 Monddaten verknüpft ist. Es handelt sich dabei um gegebene oder erschliessbare Mondmonatstage, die entweder aus der Zeit Sesostris III. oder Amenemhets III. stammen; andere Könige kommen nicht in Frage.

A P.Berlin 10090 Jahr 3 S.III. oder A.III.
B P.Ber. 10062 Jahr 29 (ditto).

C P.Ber. 10006 Jahr 36 (ditto).
(12 Daten) D P. Ber. 10056 Jahr 30/31 (ditto).

....

E P. Ber. 10009 Jahr 5 Sesostris III.
F P. Ber. 10248 Jahr 14 (ditto).

[N.B. In the list of lunar dates I have abbreviated "Berlin" as "Ber." and "S.III. oder A.III." as "(ditto)" in all but the first line.]

"Nur die daten A,B,C,D wurden bisher von Parker und Barta ausgewertet. Nach eindeutigen astronomischen Kriterien lassen sich die Monddaten A und C einerseits. B und D andererseits als in je eine Regierung gehörend erkennen. Die Zuweisung an Sesostris III. oder Amenemhet III. muss dabei offen bleiben. Durch die von mir herangezogenen Daten E und F kann die Zuordnungsfrage eindeutig geklärt werden: B,D,E und F sind daten Sesostris III. und A,C sind solche Amenemhets III.

"Nach der Anordnung A,C: Sesostris III., B,D: Amenemhet III. und unter Voraussetzung eines unterägyptischen Sothisbezugsortes, sowie einer astronomischen Kalenderform, bestimmte Parker 1872 v. Chr. als Jahr des Kahundatums. Da die Monddaten anders anzuordnen sind, entfällt dieser Ansatz.

"Barta [W. Barta, "Die Chronologie der 12. Dynastie nach den Angaben des Turiner Königspapyrus, *Studien zur Altägyptischen Kultur*, Vol. 7 (1979), pp. 1-9] argumentierte für die Anordnung B,D: Sesostris III. und A,C: Amenemhet III. Unter sonst gleichen Bedingungen wie bei Parkers Ansatz, folgt daraus die Datierung des Kahundatums auf 1875 v. Chr. Gegen diese Lösung spricht, das dabei nur 5 von 17 überlieferten Monddaten zur Berechnung stimmen. Anstelle dieser Übereinstimmung von weniger als 25% wäre eine solche von ca. 80% zu erwarten (s.u.). Die 12 abweichenden Daten sind alle verfrüht, statt in einer wahrscheinlichen Weise auf Verfrühungen und Verspätungen verteilt zu sein; hinzu kommt, dass in zwei Fällen extrem unwahrscheinliche Abweichungen von 2^d zwischen überlieferten und berechneten Monddaten vorliegen. Der Ansatz des Kahundatums auf 1875 v. Chr. kann demnach nicht richtig sein. Mit den Ansätzen 1872 und 1875 v. Chr. entfällt prinzipiell die Möglichkeit eines auf die Gegend von Memphis bezogenen astronomischen Sothiskalenders.

"Wenn Bezugsort und Kalendarart offen sind, dann kommt für das Sothisdatum von Kahun das Intervall von ca. 1880 bis 1830

v. Chr. in Frage. In diesem Zeitraum ist das Jahr 1836 v. Chr. die einzige Ansatzmöglichkeit für das mit den Kahun-Monddaten gekoppelte Sothisdatum. Bei diesem Ansatz stimmen 15 der 17 überlieferten Monddaten zur Berechnung, was dem bei einem korrekten Ansatz zu erwartenden Anteil von ca. 80% entspricht (s.u.). Zu dieser Lösung gehört ein auf Elephantine bezogener schematischer Sothiskalender."

We should note that Krauss, not always but often, follows the older custom of speaking of the Sothic and lunar dates from Illahun as the "Kahundatum" and "Kahun-Monddaten," though the designation of "Illahun" as "Kahun" (or "el-Kahun") was simply an error of Petrie's, as has long been recognized.

4. Krauss, "Sothis, Elephantine und die altägyptische Chronologie," p. 73: "*Fixierung von Kahun- und Ebersdatum.* — Die Festlegung dieser beiden aus dem MR und NR stammenden auswertbaren Sothisdaten auf bestimmte Jahre ist infolge ihrer Verkoppelung mit Monddaten möglich. Beim Ebersdatum nehmen Hornung und Parker bekanntlich Koinzidenz mit einem 1. Mondmonatstag an. Diese Annahme glaube ich durch eine geeignete Analyse der Eberskalenders absichern zu können.

"Für das Ebersdatum kommen zunächst die Jahre von ca. 1550 bis 1500 v. Chr. in Frage, wenn der Sothisbezugsort irgendwo zwischen Elephantine und Mittelmeer lag und der Sothiskalender astronomisch oder schematisch gehandhabt wurde. Nach Hornungs sicherem Ergebnis, dass der Bezugsort des Ebersdatums nicht in ÜÄ lag, lässt sich dieses Intervall auf ca. 1525 bis 1500 v. Chr. reduzieren. In diesem Zeitraum koinzidiert nur im Jahre 1506 v. Chr. ein 1. Mondmonatstag mit III šmw 9 (Kalendertag des Ebersdatums). Dieser Lösung entspricht eindeutig ein schematischer Sothiskalender mit Elephantine als Bezugsort.

"Daneben kommt ein Ansatz auf 1517 v. Chr. in Frage, bei dem allerdings mit einem Fehler im Monddatum zu rechnen wäre. (p. 74:) Dieser Lösung würde ein astronomischer Sothiskalender mit Bezugsort Elephantine entsprechen. Bei diesen beiden möglichen Ansätzen für das mit einem Monddatum gekoppelte Ebersdatum ergibt sich eindeutig Elephantine als Bezugsort; als historisch wahrscheinlicher Form erweist sich der schematische Sothis-kalender....

(p. 75:) "Die Fixierung des Kahunsdatums auf 1836 v. Chr. findet ihre Entsprechung im obigen Ansatz des Ebersdatums auf 1506 v. Chr. Zwischen diesen beiden Sothisdaten liegen $330^a \pm 2^a$

(Abstand *IIII prt 16 - III šmw 9)*, wenn dazwischen keine Kalenderregulierung stattfand. Zugunsten einer solchen Regulierung lassen sich keine Argumente beibringen, wohl aber Gründe die dagegen sprechen."

5. *Die Mittel zur zeitlichen Festlegung*, pp. 18-19.

6. For some details of the dispute over the discovery, see R. Lepsius, "Entdeckung eines bilinguen Dekretes durch Lepsius," *ZÄS*, Vol. 4 (1866), pp. 29-34, and the introduction to the publication of Reinisch and Roesler mentioned earlier.

7. Ahmed Bey Kamal, *Stèles ptolémaiques et romaines*, Vol. 1 (Cairo, 1905), p. 183.

8. *Ibid.*, Vol. 1, pp. 182-83, and Vol. 2, Plates LIX-LXI. This stela has the Cairo Museum number 22186.

9. The copy of the decree from Kem-el-Hisn was used by H. Brugsch, *Thesaurus inscriptionum Aegyptorum*, Abt. 6 (Leipzig, 1891), pp. 1554-78, to present the hieroglyphic and demotic texts running in parallel lines. Also note that K. Sethe, *Hieroglyphische Urkunden der griechisch-römischen Zeit* (Leipzig, 1904), Vol. 2, pp. 124-54, employed it for variant readings to his hieroglyphic text of the decree. Earlier E. Miller, "Découverté d'un nouvel exemplaire du Décret de Canope," *Journal des savants*, Avril, 1883, pp. 214-29, and plate opposite p. 240, on the basis of photographs sent to him by Maspero, had produced the text of the Greek version (with a French translation), which latter he thought to be superior to the text on the Tanis stela. Finally, I note that the copy from Kom-el-Hisn served W.N. Groff for his *Les deux versions démotiques du Décret du Canope* (Paris, 1888).

10. For these fragments, see F. Daumas, *Les moyens d'expression du Grec et de l'Égyptien comparé dans les Décrets de Canope et de Memphis*, in *Supplements, Annales du Service des Antiquité de l'Égypte*, Cahier No. 16 (Cairo, 1952), pp. 4-6. For the more important fragment from el-Kab, consult A. Bayoumi and O. Guéraud, "Un nouvel exemplaire du Décret de Canope," *ASAE*, Vol. 46, pp. 373-82 and Plate LXXXI, and A. Bernand, *De Thèbes à Syène* (Paris, 1989), no. 37, pp. 48-49.

11. There are numerous other studies of the Decree of Canopus beyond those I have mentioned, but they are too many for us to list here. Still, we can alert the reader to three titles. For the Greek text, see W. Dittenberger, *Orientis graeci inscriptiones selectae*, Vol. 1 (Leipzig, 1903), pp. 91-110. Also noteworthy is the Greek text of the decree and an interesting commentary on it by J.F.

Mahaffy, *The Empire of the Ptolemies* (London and New York, 1895), pp. 226-39. Finally the reader should note that E.A. Wallis Budge gathered together much useful material (including some of the versions of the decree I have mentioned here and in the notes) in his *The Decree of Canopus* (London, 1904).

Document III.11: Introduction

The Decanal Clock on the Lid of the Coffin of Meshet

In the section of Chapter Three entitled "Decanal Clocks" I have described the organization, structure, and purpose of a generalized decanal clock (see Fig. III.13). This description was based on incomplete examples found on the lids of twelve coffins dating from Dynasties Nine or Ten through Twelve (ca. 2154-1783).[1] It illustrates how the twelve nighttime hours were determined by the risings and nightly progression of 36 hour-decans (primarily 12 stars or star groupings per 10-day "week," or decade), supplemented by 12 so-called epagomenal or triangular decans, the first 11 of which shared with the last 11 main decans the task of determining nighttime hours through the last 11 decades and all 12 of which marked the hours of night during the pentad of the 5 epagomenal days.

The document at hand, Document III.11, presents an actual (nearly complete) decanal clock inscribed on the inside of the lid of the inside coffin of a ninth or tenth dynasty noble from Asyut called Meshet — 𓅓𓊃𓐍𓏏 *(msḫt)*. The wooden coffin of Meshet was discovered at Asyut in 1893 and was briefly described, with some attention to its astronomical significance, in 1900 by George Daressy.[2] He compared the decans on the lid to those found in several later decanal lists that appeared in celestial diagrams. The text of the decanal clock on the lid of Meshet's coffin (which by that time had been moved to the new Cairo Museum as No. 28118) was published by Pierre Lacau in 1906,[3] though he made no effort to translate or analyze it. In Fig. III.86 I have reproduced Lacau's text of the decanal columns, his text being presented in columns from left to right rather than from right to left as they are on the lid. Note also that Lacau did not add the vertical strings of stars which divide the columns and individually act as determinatives of the decans (see note 4).

The first serious astronomical attempt to treat of the genre of coffin decanal clocks was that of A. Pogo, "Calendars on coffin lids from Asyut (Second half of the third millennium)," *Isis*, Vol. 17 (1932), pp. 4-24, and particularly pp. 11-14 for Meshet's lid. He called these clocks "calendars" and his analysis starts from that premise, and so leaves much to be desired, since he did not properly understand the function of the so-called triangular or epagomenal decans.

It was not until the treatment of O. Neugebauer and R.A. Parker, *Egyptian Astronomical Texts*, Vol. 1, passim (pp. 4-5 for Meshet's lid), that decanal clocks were adequately understood. My document, though rendered from Lacau's text, assumes the discussion of Neugebauer and Parker as a point of departure. My truncated references in the notes to the document include their corrections of Lacau's text of Meshet's clock. They are primarily drawn from Vol. 1, pp. 4-5; also useful was their commentary, pp. 23-27.

If we examine the exterior surface of the lid of Meshet's interior coffin, we see from the initial paragraph of Document III.11 that Meshet was designated in one place as "Count; Seal-bearer of the

King of Lower Egypt; Sole Companion [of the King]; Overseer of the Priests of Wepwawet, Lord of Asyut;" and in another place as "Hereditary Prince; Count; Seal-bearer of the King of Lower Egypt; Sole Companion [of the King]; Overseer of the Priests of Anubis, Lord of Ra-qerret." Other slight variations and shortening of these largely honorific titles appear elsewhere on the exterior of the lid.[4]

Proceeding to the interior of the lid, one notes first the diagonal arrangement of the decans specified in the document. The columnar arrangement of the decans that make up the clock is schematized in Fig. III.85, where the columns proceed from right to left (as they do on the lid) and where numbers and letters are inserted in the place of the glyphs that designate (1) the hour-decans in columns 1-36 used for marking twelve nighttime hours during each of the 36 decades of the year (including some omissions and errors as indicated in the document below), and (2) those in column 40 used for the twelve decans that mark the twelve hours of the night during the 5 epagomenal days (again with omissions, i.e., with E unreadable, J and K interchanged, and boxes for H and L not available). Columns 37-39 were meant to contain the names of the first 36 decans (again with the deficiencies and errors reported in the document). At the bottom of column 40 (rows 11 and 12) we read: "Total of those (i.e., decans) in [their] place[s], gods of the sky: 36." This is a reference to the canonical list of the 36 decans which successively rise through the hours in the 36 decades.

We also note in Fig. III.86, in the top row (T) above the decans (rearranged, as I have said, by Lacau so that the columns proceed from right to left), the division of the civil year into 36 "decades" (10-day periods) covering the four months not named but numbered successively (1-4) for each of the three seasons, with each of the three seasons being named along with the month numbers (i.e., in columns 1, 4, 7, 10, 13, 16, 19, 22, 25, 28, 31, 34). The decades for each of the twelve months are entitled "first," "middle," and "final." In the top row (T) over column 40 nothing is written. Perhaps this box ought to have included some name or indication of the epagomenal pentad, such as "the 5 days upon the year." Or, as I suggested in my general account, in the original clocks, column 37 had the twelve epagomenal decans (with some such title). But when the Meshet clock was prepared, the convention was to omit such an hour list and rather to list in columns 37-40 the names of all 48 decans (36 main and 12 triangular or epagomenal decans) without any decade identification in a superior row.

Since I have already explained in the general account of Chapter Three how the diagonal chart of the decans was supposed to function as a nighttime clock, I shall not repeat that explanation here. I do note however that columns V and R appear here as on other coffin lids, and their texts need no more discussion than that present in the document.

Notes to the Introduction of Document III.11

1. These are the twelve diagonal clocks described by Neugebauer and Parker, *Egyptian Astronomical Texts*, Vol. 1, pp. 4-21. See also the coffin lid of Nakhet (no doubt from Asyut), with its diagonal decanal clock, which was acquired recently by the Pelizaeus Museum in Hildesheim, numbered T 17 and described by R. Hannig in the new catalogue edited by A. Eggebrecht, *Suche*

nach Unsterblichkeit: Totenkult und Jenseitsglaube im Alten Ägypten (Hildesheim, 1990), pp. 58-61.

2. G. Daressy, "Une ancienne liste des décans égyptiens," *ASAE*, Vol. 1 (1900), pp. 79-90.

3. P. Lacau, *Catalogue général des antiquités égyptiennes du Musée du Caire: Nos. 28087-28126: Sarcophages antérieurs au novel empire*, Vol. 2 (Caire, 1906), pp. 101-28, and Pl. IX. The photographs of the lid printed as Plates 1 and 2 by Neugebauer and Parker, *Egyptian Astronomical Texts*, Vol. 1, are too poor to be readily used. However, we can note from these plates the strings of stars that separate the columns of decans.

4. Lacau, *ibid.*, pp. 101-04, passim.

Document III.11

The Decanal Clock on the Lid of the Coffin of Meshet

[Lacau's text of the exterior of the lid, as referred to in note 4 of the Introduction to Document III.11, includes titles of Meshet, among which are:] Count; Seal-bearer of the King of Lower Egypt; Sole Companion [of the King]; Overseer of the Priests of Wepwawet, Lord of Asyut...Hereditary Prince; Count; Seal-bearer of the King of Lower Egypt; Sole Companion [of the King]; Overseer of the Priests of Anubis, Lord of Ra-qerret.

[Interior lid, top row (T) (see Fig. III.85 and Fig. III.86):]

[Col. 1:] First [decade of I Akhet].
[Col. 2:] [Middle decade].
[Col. 3:] Final decade.
[Col. 4:] First decade of II Akhet.
[Col. 5:] Middle [decade].
[Col. 6:] [Final decade].
[Col. 7:] First decade of III Akhet.
[Col. 8:] Middle decade.
[Col. 9:] Final decade.
[Col. 10:] First decade of IV Akhet.
[Col. 11:] Middle decade.
[Col. 12:] Final decade.
[Col. 13:] First decade of I Peret.

[Col. 14:] Middle decade.
[Col. 15:] Final decade.
[Col. 16:] First decade of II Peret.
[Col. 17:] Middle decade.
[Col. 18:] Final decade.
[Col. 19:] [First decade III Peret.]
[Col. 20:] [Middle decade.]
[Col. 21:] Final decade.
[Col. 22:] First [decade] of IV [Peret].
[Col. 23:] Middle decade.
[Col. 24:] Final decade.
[Col. 25:] First decade of I Shemu.
[Col. 26:] Middle decade.
[Col. 27:] Final decade.
[Col. 28:] First decade of II Shemu.
[Col. 29:] Middle decade.
[Col. 30:] Final decade.
[Col. 31:] First decade of III Shemu.
[Col. 32:] Middle decade.
[Col. 33:] Final decade.
[Col. 34:] First decade of IV Shemu.
[Col. 35:] Middle decade.
[Col. 36:] Final decade.

[The Grid of Diagonally Arranged Decans]

[The twelve hourly decans in col. 1:] 1. The Upper Tjemat *(ṯmꜣt ḥrt)*.[1] 2. Lower Tjemat *(ṯmꜣt ẖrt)*. 3. Weshat bekat *(wš[ꜣt] bkꜣt)*.[2] 4. Ipedjes *(ỉpḏś)*.[3] 5. Sebshesen *(sbšśn)*.[4] 6. Upper Khenett *(ḫntt ḥrt)*. 7. Lower Khenett *(ḫntt ẖrt)*. 8. The Red One (Star?) of Khenett *(ṯmś n ḫntt)*.[5] 9. Qedty *(ḳdty)*. 10. The Two Fish *(ḫnwy)*. 11. The [Star in the] Middle of the Ship *(ḥry-ỉb wỉꜣ)*. 12. Crew (?).

[The hourly decans of col. 2:] Decans 2-12 [whose names are specified in col. 1 plus Decan] 13. The Kenmu [stars] *(k[n]mw)*.

[The hourly decans of col. 3:] Decans 3-13 [plus Decan] 14. The Smed of the Sheep *(śmd srt)*.

[The hourly decans of col. 4:] Decans 4-14 [plus Decan] 15. The Sheep *(srt)*.

[The hourly decans of col. 5:] Decans 5-15 [plus Decan] 16. The Children of the Sheep *(sꜣwy srt)*.

[The hourly decans of col. 6:] Decans 6-16 [plus Decan] 17. The One under the Buttocks of the Sheep *(ẖry ḫpd srt)*.

[The hourly decans of col. 7:] Decans 7-17 [plus Decan] 18. The Forerunner of the Two Spirits *(tpy-ꜥ ꜣḫ[wy])*.

[The hourly decans of col. 8:] Decans 8-18 [plus Decan] 20. The Follower of the Two Spirits *(imy-ḫt ꜣḫwy)*. [Decan 20 is out of place here; Decan 19 obviously should precede it.]

[The hourly decans of col. 9:] Decans 9-18[6] [and] 20 [plus Decan 19]. The Two Spirits *(ꜣḫwy)*. [This should have been in col. 8 instead of Decan 20. The two decans have been interchanged on the Meshet-lid in the diagonal lists, but are in the proper order in the list of decans in col. 38 below.]

[The hourly decans of col. 10:] Decans 10-18, 20-19 [plus Decan] 22. Qed *(ḳd)*. [Decan 21. The Two Souls *(bꜣwy)* is missing in the diagonal chart on this lid.[7]]

[The hourly decans of col. 11:] Decans 11-18, 20-19, 22 [plus Decan] 23. The Thousands *(ḫꜣw)* [a star-cluster?].

[The hourly decans of col. 12:] Decans 12-18, 20-19, 22-23, [plus Decan] 24. Aryt *(ꜥryt)*.[8]

[The hourly decans of col. 13:] Decans 13-18, 20-19,

22-24 [plus Decan] 25: The One under Aryt *(ẖry ʿryt).*

[The hourly decans of col. 14:] Decans 14-18, 20-19, 22-25 [plus Decan] 26. The Upper Arm [of Orion] *(rmn ḥry).*

[The hourly decans of col. 15:] Decans 15-18, 20-19, 22-26 [plus Decan] 28. The Abut-scepter [of Orion] *(ʿbwt).* [Note that in the diagonal chart, Decan 27. The Lower Arm [of Orion] *(rmn ẖry)* is missing, though it is given in the list of decans, col. 39, top row (T).]

[The hourly decans of col. 16:] Decans 16-18, 20-19, 22-26, 28 [plus Decan] 29. The One under the Leg (or Lower Leg) [of Orion] *(ẖrt wʿrt).*

[The hourly decans of col. 17:] Decans 17-18, 20-19, 22-26, 28-29 [plus Decan] 30. Forerunner of Sothis (i.e., Sirius) *(tpy-ʿ śpd).*

[The hourly decans of col. 18:] Decans 18, 20-19, 22-26, 28-30 [plus Decan] 31. Sothis (i.e., Sirius) *(spd).*[9]

[The hourly decans of col. 19:] Decans 20-19, 22-26, 28, 29[10], 30-31 [plus Decan] 32. Kenmut *(knmwt).*[11]

[The hourly decans of col. 20:] Decans 19, 22-26, 28-32 [plus Decan] 33. The Children of Kenmut *(sꜣwy knmwt).*

[The hourly decans of col. 21:] Decans 22-26, 28-33 [plus Decan] 34. The One under the Tail (or Buttocks) of Kenmut *(ẖry ẖpd knmwt).*

[The hourly decans of col. 22:] Decans 23-26, 28-34 [plus Decan] 35. The Beginning of the Thousands *(ḥꜣt ḫꜣw).*

[The hourly decans of col. 23:] Decans 24-26, 28-35 [plus Decan] 36. The End of the Thousands *(pḥwy ḫꜣw).*

[The hourly decans of col. 24:] Decans 25-26, 28-36, 1 [repeated].[12]

[The hourly decans of col. 25:] Decans 26, 28-36, 1-2 (see again note 12).

[The hourly decans of col. 26:] Decans 28-36, 1-2 [plus Decan] A. Southern Smed *(śmd rśy)*.[13]

[The hourly decans of col. 27:] Decans 29-36, 1-2, A [plus Decan] B. Northern Smed *(śmd mḥty)*.

[The hourly decans of col. 28:] Decans 30-36, 1-2, A-B [plus Decan] C. The God who Crosses the Sky *(nṯr d͟ꜣ pt)*.[14]

[The hourly decans of col. 29:] Decans 31-36, 1-2, A-C [plus Decan] D. The Lower Arm [of Orion] *(rmn ẖry)*. [Equivalent to Decan 27 in col. 39, top row (T).]

[The hourly decans of col. 30:] Decans 32-36, 1-2, A-D [plus Decan] E. The Thousands [No. II] *(ḫꜣw)*. [Not equivalent to Decan 23.][15]

[The hourly decans of col. 31:] Decans 33-36, 1-2, A-E [plus Decan] F. The Forerunner of Sothis (i.e., Sirius) *(tpy-ᶜ špd)*. [Equivalent to Decan 30.]

[The hourly decans of col. 32:] Decans 34-36, 1-2, A-F [plus Decan] G. The Follower of Sothis (i.e., Sirius) *(ỉmy-ḫt špd)*.

[The hourly decans of col. 33:] Decans 35-36, 1-2, A-G [plus Decan] H. The Two Spirits *(ꜣḫwy)*. [Equivalent in writing to Decan 19.]

[The hourly decans of col. 34:] Decan 36, 1-2, A-H [plus Decan] J. [Letter "I" not used] *(ḫꜣw)*. [Equivalent to Decan 23.]

[The hourly decans of col. 35:] Decans 1-2. A-J [plus Decan] K. The God who Crosses the Sky *(nṯr d͟ꜣ pt)*.[16]

[The hourly decans of col. 36:] Decans 2, A-K [plus Decan] M. The End of Sabu *(pḥwy sꜣbw)*. [Note: Missing here is Decan L. Sabu *(sꜣbw)*, which is found in two other coffins and should be there, with Decan M saved for the last hour of the epagomenal pentad. Decan L is also missing in the list of col. 40.]

[Col. 37 lists the names of] Decans 1-13.

[Col. 38 lists the names of] Decans 14-20, 19 [repeated, and 21 is missing], 22-26.

[Col. 39 lists the names of] Decan 27 [followed by a blank box], and [then] Decans 28-36, 1, 4 (instead of 2).

[Col. 40, following blank row T, lists in rows 1-10 the names of epagomenal-triangular] Decans A-D, [E], F-G, K-J, and M. [Note: Decans H and L are omitted and Decans J and K are interchanged.] [Rows 11-12:] Total of those who are ⸢in their places⸣, [i.e.,] the gods of the sky: 36.[17]

[The Horizontal Stripe R between Rows 6 and 7, with Invocation Offerings, Lacau's Text, p. 110][18]

A good boon which is given to Re, Lord of the Sky in all of his Places, [for] Invocation offerings of bread and beer, oxen and fowl, on behalf of the Count and Overseer of Priests Meshet; a boon which is given to Meskhetyu (the Foreleg or Big Dipper) in the northern sky; a boon which is given to Nut; a boon which is given to Orion in the southern sky; a boon which is given to Smed in the southern sky and to Smed in the northern sky; a boon which is given to the God who Crosses the Sky (a star in Orion?) and to the Upper Arm [of Orion]; a boon which is given to Sothis (i.e., Sirius) and to the Follower of Sothis; a boon which is given to the Two Spirits and to the Follower of the Two Spirits; a boon which is given to the Beginning of the Thousands and to the End of the Thousands; a boon which is given to Upper Khenett and to Lower Khenett [for] Invocation Offerings on the behalf of Count Meshet.

[The Vertical Stripe V: cf. Fig. III.85 for its position between Columns 18 and 19, Fig. III.17 for an enlargement from another lid, and the text of Lacau, p. 110]

[The goddess] Nut [figured as holding up the sky, with the epithet behind her:] the one who carries.

Meskhetyu (the Foreleg or Big Dipper) [figured and outlined with] seven stars [and bearing the epithet behind it:] Meskhetyu in the northern sky.

Sah (Orion) [figured with name above his head and looking backward toward Meskhetyu and with right arm lifted and grasping an *ankh*-sign and left arm lower, carrying a *w's*-staff. Inscribed before him:] ... in the southern sky [and behind him:] Orion, turn your face backward so that you may see [the deceased] Osiris [Meshet].

Sothis [with glyph:] Δ [on her head. She faces Orion and has a *w's*-scepter in her right hand and an *ankh*-sign in her left hand.] [To the left is the statement:] Sothis, she gives life to [the Count and Overseer of Priests Meshet *(thus in Lacau's text, but I am unable to read the photograph)]*.

Notes to Document III.11

1. Consult the lists of decans present on the southern panels of the celestial diagrams in Documents III.3 and III.4, where I have added comments concerning possible translations and other matters. The list in this decanal clock of Document III.11 starts with Decan 6 of the lists in those later documents, i.e., Decan 1 here is Decan 6 there. There is considerable variation between the decans that appear in the coffin clocks and those in the celestial diagrams. The

reader may see detailed differences by comparing the list of Fig. III.18, or, even better, the extensive chart of the decans used in the coffin clocks given in Neugebauer and Parker, *Egyptian Astronomical Texts,* Vol. 1, Plates 26-29, with the lists of decans from the astronomical monuments given in Vol. 3, Chap. 2, of the same work.

2. There seems to be a coalescing of two decans: numbers 8 and 9 in Document III.3, q.v.

3. Compare Document III.3, col. 35, add. decan 4, and Document III.4, Col. 31, add. decan 4.

4. Compare Document III.3, col. 35, add. decan 4, and Document III.4, col. 32, add. decan 5.

5. As Neugebauer and Parker note in Vol. 1, p. 4, this is the way the decan is written in col. 37, row 8. Here it is written *ḫntt ṯmś*.

6. In row ten of col. 9, we find that instead of two *ꜣḫ*-birds there is one *ꜣḫ*-bird and one *bꜣ*-bird (see Neugebauer and Parker, Vol. 1, p. 4).

7. *Ibid.*, Neugebauer and Parker would justly amend the spirit-birds of the misplaced Decan 19 in row 12 of col. 9 (and in all the succeeding columns through col. 20) to be the missing soul-birds of Decan 21. Hence in this estimate it would be Decan 19 instead of Decan 21 which is missing on the diagonal chart. But in conformity with the text of Lacau, Neugebauer and Parker, in their version of the schematized chart of the clock, indicate that it is Decan 21 which is missing. In the list of decans in col. 38, we find in order Decans 19, 20, and 19 again, giving support to the view that the author mistakenly omitted Decan 19 in his main chart and accordingly read the soul-birds as spirit-birds in col. 38 as well as in col. 9 and the columns following it.

8. In the celestial diagram of Senmut this decan is written *ꜥrt* (see Document III.3, col. 22). At any rate, its translation is not known for sure.

9. Lacau gives the glyph Δ by itself for Decan 31. Actually, as Neugebauer and Parker had already noted, the name *śpd(t)* is actually written out in full [hieroglyphs] Δ. NP omit the loaf-glyph [loaf-glyph], but it seems to me to be there. However the determinative stands alone for the name in its diagonal progression through cols. 19-29, as Neugebauer and Parker further stated. According to Lacau's reading, it also stands alone for the decan in column 39, row 5, but

I am unable to read the photograph.

10. This decan appears in col. 19, row 9 (and along the diagonal through the succeeding eight columns) with the words reversed: *w^c^rt h̲rt*, as Neugebauer and Parker have noted.

11. See Document III.4, cols. 1-2 and note 1, for the similar spelling of the name.

12. Since the scribe left out two decans (Decans 21 and 27), and he knows that there ought to be 36 main decans, he repeats Decan 1 here and Decan 2 in the next column, and, of course, each one along its diagonal through the succeeding columns.

13. The Smed of Decans A and B is no doubt the Smed of Decan 14.

14. Neugebauer and Parker, Vol. 1, pp. 25, 110, identify this god as "surely Orion," and thus the decan as some part of Orion. As Neugebauer and Parker note (*ibid.,* p. 5), the scribe, here and through col. 32, wrote out *pt* including the determinative, while Lacau merely uses the determinative as an ideogram in his text.

15. Neugebauer and Parker comment (*ibid.*, p. 25): "The writing [of Decan E] marks this as a different *ḫ'w* from [Decan] No. 23."

16. Neugebauer and Parker write (*ibid.*, p. 26): "This is [Decan] C in another hour." Cf. *ibid.*, p. 110.

17. I follow Neugebauer and Parker, *ibid.,* p. 5.

18. I follow the translation in Neugebauer and Parker, *ibid.*, p. 27 with only minor emendations.

Document III.12: Introduction

The Book of Nut that Accompanies the Figure of Nut in her Arched Celestial Position, Together with Extracts from a Commentary on It

The so-called Book of Nut (designated by Neugebauer and Parker as "The Cosmology of Seti I and Ramses IV") in fact is rather a diverse and often ill-connected set of mythological and astronomical sentences that describes the motion of the sun and the stars throughout the year and lists the decans twice: the first list being of those usually found indicating hours by their risings and the second (incomplete as to the names of the decans) being of those indicating the hours by their meridian culminations and appearing as part of a rather extensive table of the times of their crucial locations during their yearly course. Though the earliest copy of this collection of texts as a whole dates from the reign of Seti I (ca. 1306-1290), an entry in the decanal table states that the rising (i.e., heliacal rising) of the star of Sothis (Sirius) took place on IIII Peret 16 (see Document III.12, U35). This immediately suggests that the date of the rising in the table was none other than IIII Peret 16 of year 7 of Sesostris III, a well attested rising (see Document III.10, Section 1), which, if true, would stamp the table of decanal courses as originating at least as early as the Middle Kingdom.

What is more, it would also indicate that by that time a shift to using meridian transits of decans instead of horizon risings to mark the nighttime hours had taken place, since the table is obviously based on such meridian transits.

The extant sources that yield the text of the Book of Nut are four in number. Two (**S** and **R**) are collections of hieroglyphic inscriptions from sarcophagus chambers and they represent, so far as we know, the pristine version of the text. The other two (**P** and **Pa**) are papyri from Carlsberg that contain most of the text of **S** and **R** in hieratic, along with translations and commentary in demotic that help us understand the original text. The text of **S** and **R** is the focus of Document III.12. I have included in brackets some of the more important readings or comments from **P** and **Pa**.

First I shall briefly describe the four copies, depending significantly on the excellent textual studies of H.O. Lange and O. Neugebauer, *Papyrus Carlsberg No. I. Ein hieratisch-demotischer kosmologischer Text* (Copenhagen, 1940) and Neugebauer and Parker, *Egyptian Astronomical Texts*, Vol. 1, Chapter 2.

S = Text around the Figure of Nut on the West Half of the Roof of the Sarcophagus Chamber of the Cenotaph of Seti I, which is behind the Temple of Seti I at Abydos (see the drawing of Walter Emery here given as Fig. III.95a, reproduced from H. Frankfort, *The Cenotaph of Seti I at Abydos*, and the drawing in Fig. III.95b, modified by Neugebauer and Parker from the drawing given by Lange and Neugebauer). Above all, see the very clear photographs of A.M. Calverley reproduced as Plates 30-32 by Neugebauer and Parker

(Plate 33 gives the Dramatic Text of Document III.13). The Cenotaph figure, as I have implied, has the earliest extant copy of the Book of Nut and dates from Dynasty 19. The hieroglyphic texts in both **S** and **R** (with the quotations from the texts in **P** and **Pa**) are given in hand copy by Neugebauer and Parker as Plates 44-51 (=my Fig. III.95c). The hieroglyphic material from **S** and both the hieratic quotations from the Book of Nut transcribed into hieroglyphics and the demotic commentary transcribed into phonetic letters and signs from **P** were earlier given by Lange and Neugebauer, pp. 1*-31*; see below. English translations of **S** and **R** were published in Neugebauer and Parker (with accompanying references to textual readings and a running interpretive commentary). A German translation of the **S-R** text, with concise and informative notes, is found in E. Hornung, *Zwei ramessidische Königsgräber: Ramses IV. und Ramses VII.* (Mainz am Rhein, 1990), pp. 91-93.

R = Text around the Figure of Nut on the South Half of the Ceiling of Hall K in the Tomb of Ramesses IV, Valley of the Kings, Western Thebes. It dates from Dynasty 20. In contradistinction to what the text says, Nut's head is here directed toward the east. The locations of the various textual inscriptions are indicated in Fig. III.96a. An early drawing of the Nut picture by Brugsch is shown in Fig. III.96b, and an even earlier drawing by Champollion in Fig. III.96c. The varying colors of the hieroglyphs and other parts of the picture are described by Neugebauer and Parker (Vol. 1, p. 36). An interesting and lengthy essay on the colors of the hieroglyphs by E. Staehelin is included in the above-mentioned volume of Hornung (pp. 101-22). The serious student of the text will wish to examine the photographs of Charles Nims, taken with blue and yellow filters, which appear as Plates 34-35 in the Neugebauer and Parker volume and as Tafeln 68-71 in the volume of Hornung noted above. I have already mentioned the German translation of Hornung in the paragraph above, as well as the hand copies of the text in Fig. III.95c. Earlier bibliography is briefly presented by Neugebauer and Parker (Vol. 1, p. 36), the most significant being that of Brugsch, *Thesaurus inscriptionum aegyptiacarum,* (Graz, 1968, unaltered repr. of Leipzig ed. of 1883-91), pp. 167-79.

P = Papyrus Carlsberg No. I. This papyrus contains, as I have said, (1) the original textual statements in hieratic writing, (2) their translation into demotic (often with more than one effort at translating or restating each individual textual passage), and (3) commentary in demotic. For the most part the Commentary is superficial and not particularly clarifying, but on occasion, as in the treatment of the cycle of periods of the decans, it is useful and makes the original rather abbreviated text clearer. Furthermore, it helps to give some sense of wholeness and even purpose to the document, as I believe will be evident even in the syncopated form I have presented it here. The fact that the textual statements themselves are in hieratic is perhaps evidence that the original text itself was known to the commentator from a now unknown papyrus source or sources rather than from inscriptional sources (written in hieroglyphs) like **S** and **R**, though much of the hieratic textual quotation in **P** is quite close to our extant inscriptional copies.

P itself is thought to be from about the same date as the other papyri in the Carlsberg collection, i.e., 144

A.D. or later. There are seven columns (four recto and three verso). The first three and much of the fourth (i.e., I,6-IV,34) contain the mainly astronomical texts that constitute Document III.12. The remaining part of the fourth and the last three columns contain mythological texts that make up a commentary on the first part of the so-called Dramatic Text, extracts from which, comprising the first 13 and 2/3 vertical lines of **S**, form Document III.13.

The picture of Nut probably was placed before the text but is missing, as is some of the beginning of the text. The whole text of **P** was edited by Lange and Neugebauer, in their above-noted work, on the basis of photographs of the seven columns with line numbers added (see their Tafeln I-VII). Its hieratic parts, i.e., the passages taken from the Book of Nut and the titles of books the commentator used or thought pertinent to the text at hand, were presented by the editors in hieroglyphic transcription (the passages from the Book of Nut being compared by the editors to the similar passages in **S**). The demotic translations and commentary were rendered by the editors in phonetic transcription. The whole of **P** was translated into German with further notes and commentary.

In the later study of this document by Neugebauer and Parker mentioned above, a translation into English of the Commentary with interlaced translations of the original texts of **S** and **R** was also given, while new photographs appeared as their Plates 36-42. As the result of studying these plates and a completely new examination of the whole text, they rearranged the order of presentation of the texts, made some changes in the readings of the text with consequent changes in their translation.

Pa = P. Carlsberg Ia. This is incomplete, having four columns only (the Dramatic Text of Document III.13 being missing). It was probably written by the scribe of **P**. For details of its relationship to the latter copy and its contributions to the text, see Neugebauer and Parker, *Egyptian Astronomical Texts*, Vol. 1, pp. 37, 88-93.

It is evident that Document III.12 in its broad form that includes the original text of the Book of Nut and the Commentary represents something rather distinct in Egyptian astronomy. For it highlights and delineates in a unique way the common theory of the duration of decanal stars in their annual courses that lay behind the tabular listing for each of the decanal stars of (1) the time of the completions of the 10-day period in which its meridian culmination served as the marker of the first hour, of (2) the time of its descent into the Dat or Duat, and of (3) the time of its subsequent heliacal rising. This theory was based originally, I presume, on the long familiarity of the ancient Egyptians with the annual helical risings of Sirius. That is to say, the course of Sirius served as a model for the selection of the remaining 35 decans that obeyed similar periods.

This major contribution along with the general cosmological features are described with clarity in both the original edition of Lange and Neugebauer (pp. 65-75) and in the general summary of the text and its Commentary by Neugebauer and Parker. I repeat here much of that latter summary, which, it will be noted, follows the division into Chapters A-G initiated by Lange and Neugebauer and retained by Neugebauer and Parker (the letters so employed for chapter designations of the various parts of **P** and **Pa** are not to be confused with the capital letters used to indicate the locations of the textual statements on the figures in **S** and **R**):[1]

> Chapter A...refers to the sun and the Nut picture as a whole. With Chapter B the description of particular inscriptions begins, starting with Text A (Cf. Fig. 20 [=my Fig. III.95b]) .... The sun, depicted as a scarabaeus [beetle], is placed "under her thigh" in **S** and **R**. The sun originates or rises in far distant regions in the Southeast, the country of Punt. Perhaps this is the direction from which the sun is supposed to rise when it is farthest away from Egypt at the time of the winter solstice. It is difficult to say whether these two introductory chapters are concerned with the primeval origin of the sun or with its daily rising.
>
> Undoubtedly the latter is meant in Chapter C, but again in a new mythological form: the sun rises "upwards from the Duat", that is from the other world. With it are supposed to rise the decans *knmt* and *štꜣ*. We have in this the first occurrence of the important insight that appearance and disappearance of sun and stars are related phenomena, here expressed by their common wanderings through the Duat. This representation abruptly shifts in the second part of Chapter C to the myth of the birth of the sun as a scarabaeus. Towards the end we return again to the description of the sun leaving the Duat. Here even a specific moment is mentioned at which the withdrawal from the Duat towards

the earth begins, namely the "9th hour" which is called *sḫtp.n.s.* Finally Re[c] goes forth as a child of increasing strength.

Chapter D deals with a different subject, the outermost limits of the sky. There total darkness prevails, and the sun never reaches these regions. Thus Nut, with sun and stars, belongs, so to speak, to an inner part of the world, which is surrounded by the primeval waters as boundaries in all directions.

Chapter E returns to sun and stars. It is the central astronomical section in which the relation between the sun and the decans, their appearance and disappearance in the course of a year are explained. It corresponds to the lists of decans and dates which occupy in **S** and **R**, the middle part of the picture, to the right and to the left of Shu, "the prince whose hand[s] is [are] under the sky" [i.e., Nut].

The most important information which comes from this chapter is the fact that the decans indicate the hours of the night no longer by their successive rising [as in the diagonal charts on the coffin lids of the 9 and 10th, and occasionally the 11th and 12th dynasties; see Document III.11] but by their culmination or transit. The star of the "first" *(tpt)* hour is the decan which has completed its 10 days as first hour star and is seen in the meridian at the beginning of the night, that is, sometimes (some time?) after sunset. From then on a simple scheme controls the succession of phenomena. It takes 90 days ("in the west") after finishing as first hour star before a decan becomes enclosed *(šn)* by the Duat. At that time the decan is setting right after sunset and thus begins its period of invisibility, which is assumed to last 70 days. Reappearance from the Duat is called "birth" *(ms)*. From then on the decan is visible for a longer period each night, but it takes 80 days "in the east" before the decan really does "work", i.e., indicates an hour by its culmination. Since culmination at the end of work there have now elapsed 90 + 70 + 80 = 240 days.[2] Because the whole circle of sidereal phenomena is schematically assumed to be 360 days long, we now have 120 days left for the "working" of a decan. At first its culmination indicates the 12th hour, ten days later the 11th, and so on until it stops working after 120 days, having indicated at last the first hour.

This is the scheme on which the list of dates in **S** and **R** is based and which is commented upon in **P**. Its strongly schematic character is unmistakable....

With Chapter F we return again to a more mythological part of the text. Sunset is explained as entering the Duat by the way of the mouth of Nut and the same holds for the decans which are invisible. The end of the chapter seems to refer to the sunrise at which Re[c] appears again as a youth, similar to his first appearance (Chapters B and C).... At the beginning as well as here (Chapter G) a description is given of outer regions which lie in total darkness beyond the travel of the sun.

> In the west these regions are inhabited by birds, human headed with the speech of men. [These are the "Soul" Birds.]

The conflicting details of the sun's motion in the texts of **S** and **R** and in the Commentary are puzzling, but I shall not discuss them here. In this regard the reader is urged to read note 24 of Document III.12.

However, at this point I should remark about the overall system of decans deducible from Texts U and V and the Commentary on them, all a part of Document III.12. This system of decanal stars, though showing the close attention the Egyptian stargazers gave to the nightly rising, culmination, and setting of stars, reveals, as a careful reader of the document will see, how far removed it was from the later decanal system of the Greeks, which identified decans with 10 degrees of angular distance on a circumferential belt. In fact, the distribution of the 36 decans in the Egyptian system in a manner such that at any time there were 9 visible decanal stars in the western sky, 12 in the middle of the sky, 8 in the eastern sky, and 7 invisible in the Duat, showed the lack of any concept of angular measures by degrees of apparent star movements. Hence, despite its quasi-quantitative look and its use of a schematic, numerical system of time measurement, it remained essentially a grossly empirical system that floundered on its overly simple schematization and its inadequate unit of measurement of celestial movements.

One other interesting feature of the Commentary in **P** is its citation of several books or writings that are for the most part unavailable. They include the following (some citations noted here are not in my extracts of Document III.12): (1) *The Description of the Movements of the Stars* [= *The (Book) bnn*? — or is the first title in a collection of works of which *bnn* and some of the following works are parts?], **P**, Col. I, line 14; II,40-41; III,15,17; (2) *The (Book) [called] Protection of the Bed*, **P**, I,20; (3) *The (Book) of Seeing the Sun Disk*, **P**, I,26; (4) *The (Book) gꜣbt*, **P**, II,11; (5) *The (Book) Five Days on the Year*, **P**, II,12; (6) *The (Book) of the Sky*, **P**, II,19; (7) *The (Book) ḫr*, **P**, II,21; (8) *The (Book) šn idnw*, **P**, II,37; (9) *The (Book) sf*, **P**, II,42; (10) *The (Book) sḫn-spdt*, **P**, III,5; and (11) *The (Book) šd*, **P**, III,37-38. Works (1), **P**, VII, 20-21; (7) **P**, V,32, 34-35, and (9) **P**, VII,25; plus a further work, *The (Book) iꜣt*, **P**, V,32, 34-35, VI,15, 20 are mentioned in the columns and lines of copy **P** of the Commentary to the Dramatic Text of Document III.13; for the references included in my limited treatment of the Commentary in Document III.13, see notes 4, 12, 15, and 16 to that document.

Notice that the first work of those listed appears to be one mentioned in the catalogue of the library of the Temple of Edfu (see the Introduction to Document III.18 below and Volume One of my work, pp. 45-46).

My English translation constituting Document III.12 owes much to the prior translations into German and English mentioned above. Though it consists mainly of renderings of the hieroglyphic texts of **S** and **R** (quoted under the alphabetical letters locating the texts on Fig. III.95b and given in hand copy in Fig. III.95c), I have added from **P** and **Pa** enough extracts from the Commentary (abbreviated as "Com." and enclosed in special brackets { and }) to give a feeling for the whole. In quoting these extracts from **P**, I have used the column number (a Roman numeral identifying which of the papyrus columns is being quoted) and a line number (an Arabic numeral referring to the relevant line of the column). To save space I have not included the

Chapter letters assigned to the various parts of the Commentary and mentioned above in the quotation from Neugebauer and Parker, but the reader can find the correlation of the column numbers and chapter letters in the latter work. In quoting the extracts from the Commentary, I use dots not only to indicate that the text is missing or obscured but also for my omission of parts of the text that can be read but which I deem not to be especially useful for understanding the meaning of the original text of the Book of Nut.

In the notes to the document I have used the abbreviation LN for readings from the work of Lange and Neugebauer and NP for those from that of Neugebauer and Parker. Observe that I have given the translations of the texts A-Z and Aa to Ll in alphabetic order as they are given in the texts of LN and NP (but not uniformly so in the translation of the Commentary by NP, who insert them where they belong in the Commentary since they are following its order). In presenting the translation, I have depended mainly on **S** without attempting to note omissions and variant readings from **R** though I have occasionally taken the latter into account, and most of them are referred to by NP. In giving the names of the decans in the texts of U I have not repeated my earlier guesses as to their proper translation. The reader may consult my earlier efforts in Documents III.3, III.4 and III.11. No doubt under the influence of earlier Egyptologists and to be consistent with my quotations from Neugebauer and Parker, I have used the form "Duat" in my translation of this document, though I tended in my first volume to prefer "Dat," as Hornung does here in his German translation.

Notes to the Introduction of Document III.12

1. Neugebauer and Parker, *Egyptian Astronomical Texts*, Vol. 1, pp. 38 and 41.

[2. By a printer's error the text here actually reads "90 + 70 + 80 + 240 days." The final plus sign should, of course, be an equal sign, as I have given it.]

Document III.12

The Book of Nut that Accompanies the Figure of Nut in her Arched Celestial Position, Together with Extracts from a Commentary on It.

[Book of Nut, the Picture of Nut around which the book is organized; see Figs. III.95a and III.95b.]

{[Com., col. I: Introduction to the Picture of Nut; the original text being commented upon (if indeed there was one distinct from the statements in the remainder of the text) is not in **S** and **R**.]

/1/ This is ⌜the picture (*ṯky*)[1] on the papyrus⌝.[2] The female figure of this ⌜body⌝ [i.e., the one with her] head [in the] west /2/ [and her hind part in the eas]t, is the goddess (Nut) [i.e., who] is the Northern sky. It tells about her in the picture. .... /5/ .... He (Re) caused the hind part to be the beginning, i.e., the place of birth [of the sun and the stars], /6/ .... [Certain waters] are [beyond the limits] of the circumference of the sky. [They are those] in which Re never rises.[3] The upper part of the region (?) which is under the falcon[4] /7/ .... are .... (those regions?) of the circumference of the sky in which Re is accustomed to rise. This is what it (the text) says *(ḏd)*. /8/ .... in the water from which Re rises. The support which stands under the falcon (i.e., the text on the picture before the support?) /9/ .... (speaks of?) the surroundings of the circumference of [the] sky, that is to say, the waters which I have mentioned. /10/ .... they (i.e., the stars?) wander in the sky, i.e., they go around [just as][5] he (Re) has gone around. /11/ .... a star travels in accordance with the existing order of the travelling of the stars,[6] that is to say (of those) which rise in the sky /12/ .... those which rise in the sky, i.e., all (of them), all [these] which rise in the sky, /13/ that is to say, ⌜these which are on⌝ the figure of the Lady (Nut).

[Com., col. I] /14/ ⌜The texts which are on the picture⌝ follow those which *The Description of the Movements of the Stars* provides; it is *The [Book] Benn (bnn)*. /15/ [The text which is between the falcon and the vulture.][7]}

[Book of Nut; see Fig. III.95c, Text A:] This god (Re) exists on her (Nut's) southeastern side behind Pun[t].[8] [He] exists [there] before the lighting of the sky (i.e., before dawn).

{[Com., col. I] /19/ He is Re when he rises from the water at dawn. He is a falcon (i.e., Horus).... /20/ ⌜coming from⌝ Nun, says *The (Book) Protection of the Bed* ....}

[Bk. Nut, Text B:] Nekhbet ....[9] [Text C:] She flies there before the Eye.[10] [Text D:] The Kenmet decan lives (i.e., rises) as well as the Ab and Sheta decans, which means that Horus [also] lives.

{[Com. col. I] /37/.... The place of rising

which Kenmet fixes (*lit.* makes, *[ỉr]*), and also which Ab and /38/ [Sheta] fix, is the place of rising which Re fixes.[11]}

[Bk. Nut, Text E:] He opens his <ball of clay> *(nḫp(t).f)*.[12] He swims in his [morning-]redness (i.e., rosy dawn)

{[Com., col. I] /39/ .... It opens to the sky, that is to say, the place from which Re rises out of the Duat, that from which he rises daily .... /41/.... [there] he .... [is] in the form of Khepr̲er (i.e., the Scarabaeus Beetle), and (then?) he assumes the form of the sun disk.}

[Bk. Nut, Text F:] He is purified in the arms of his father Osiris. His father lives and his father is glorified when he has put <himself> under him. [Text G:] The redness <comes> after birth.[13]

{[Com., col. II] /1/.... The redness comes after birth. It is <in> the color which arises in the sun disk at dawn that he, that is to say, Re, produces /2/ his rays upon the earth in the designated color. Look at the picture ....}

[Text H:] He enters as this Scarabaeus (i.e., as the one on the left of the picture beneath the upper leg or thigh of Nut). He comes into existence as he came into existence the first time, [i.e.,] in Primeval Time.[14] [Text J:] The majesty of this God (Re) goes forth (or, went forth?) from the Duat, and in the Mesqet (*msḳt*, i.e., easterly region?) the stars go forth (or, went forth?) with him in his Primeval Time. He is (or, was?) reared in the Mesqet-region. He was glorified in the arms of his father Osiris at Ta-wer (Abydos?)[15] in the First Time of his antiquity. He is accustomed to withdraw to the ⌜sky⌝ in the hour Sehetepnes (i.e., the ninth hour).[16] He is strong .... [when he appears in the

(Two?) Lands (i.e., Egypt?)].[17]

> {[Com., col. II] /9/ .... It happens that he (Re) withdraws /10/ toward mankind from the Duat in the hour of Sehetepnes, which is the ninth hour of the night. This is what it (the text) means. It happens that /11/ *The (Book) Gabet (gꜣbt) (i.e., Gbt, The Sky?)* .... names the eighth gate of the Duat; the people who are in it enter .... /12/ ....see *(The Book) The Five Days on the Year.*[18] In the hour of /13/ *sḥtp.n.s*, the ninth hour of the night, that is to say, the one in which he rises .... /14/ His existence (i.e., the essential character of his existence), which is his glowing, is strong .... /15/ It exists in the Lands, that is to say, the glowing .... The place where he causes it to exist is not named.[19]}

[Bk. Nut, Text K:] He (Re) comes into being when his heart (i.e., his force) comes into being. He sees Geb (i.e., the earth) when [first] rising. When he is a child [later] in the morning he is the sun disk and he goes forth (i.e., rises further). [Text L:][20] That [region] which is over (beyond) the sky is in total darkness. [Its] boundaries are unknown in the direction of the south, [the north,] the west, and the east. .... These (residents) are established in the primeval water like weary ones. There is no light (or, rising) for the Soul *(bꜣ)*.... [there]. Its realm *(tꜣ)* is not known *(rḫ)* to south, north, west, and east by the gods and the spirits.[21] There is no injury *(ḫḏt)* [or, light *(ḫḏḏ)*] there.[22] Moreover every place there [in the Abyss] is devoid of sky (i.e., the region where the sun and stars travel), [while] the entire Duat is devoid of lands (i.e., the countries of the living?).[23] [Text M:] The majesty of this god (Re) goes forth from her (Nut's) hind part. [Text N:] He proceeds to the earth, risen and born. [Text O:] He rises afterwards. He opens the thighs of his mother Nut. He withdraws from (or, departs toward ?) the sky. [Text P:] Eastern horizon. [Text Q:] Western horizon. [Text R:] Evening.[24]

[Text S_1:] The Decan Sebshesen *(sbšsn).* [Text S_2:] The Decan Tjes areq *(ṯs ᶜrḳ).* [Text S_3:] The Decan Waret kheret *(wᶜrt ẖrt).*[25]

[The Texts labeled T_1 and T_3 on Fig. III.95b constitute a list of decans designated by NP as the Seti I A family of decans, with the names of the deities of the decans and all of the decanal figures omitted. Below the determinative star of each decan further stars are shown as circles. The list was derived from a star clock in which the hours were marked by the rising of the decans.[26] Since I have given detailed examples of other decanal lists originally used for that purpose in Documents III.3, III.4, and III.11, I shall not list these decans here but instead refer the reader to NP, Volume Three, p. 14, and pp. 118-28, where the decans are listed in phonetic transcription and in their original hieroglyphic form detailed description of them is given.]

[Text T_2:] The doing *(ir)* in the first month of Akhet, according as (or, when) Sothis rises (or, has risen).[27]

> {[Com., col. II] /36/ .... "All these stars begin /37/ in the sky in the first month of Akhet when Sothis rises," <says> *(The Book) Shen idnu (šn ỉdnw)* ...... /40/ It is [also] the beginning of the year ⌜with⌝ Re. He has made no alteration in the path of the stars. *The (Book) Benn* says: "Sothis, there are 18 [decanal] stars after her and 18 [decanal] stars

before her."}[28]

[The Tables of Decanal Days][29]

[Text X:] Register *(ᶜt?)* of the Decan Kenmet *(knmt)*.[30] 1. [Text U₁:] [End of marking] First (hour): Month III of Season Akhet, Day 16; Enclosed by the Duat (acronical setting): II Peret 26; Birth (helical rising): I Shemu 6.[31]

2. [Text U2:] The Decan Khery Kheped Kenmet *(ẖry ḫpd knmt)*. First h.: IIII Akhet 6; Enc. Duat: III Peret 6; Birth: I Shemu 16.

3. [Text U3:] The Decan Hat Djat *(ḫ'''t ḏ'''t)*. First h.: IIII Akhet 16; Enc. Duat: III Peret 16; Birth: I Shemu 26.

4. [Text U4:] The Decan Pehuy Djat *(pḫwy ḏ'''t)*. First h.: IIII Akhet 26; Enc. Duat: III Peret 26; Birth: II Shemu 6.

{[Com., col. II] /41/ .... Behold /42/ the way of calling every one of them (i.e., the decans). The last part of the text, i.e., [that including] these five stars from Kenmet to Dem (!, *tm'''t ḫrt ẖrt*),[32] is according to *(The Book) Sef*. They are called (i.e., finish their work at the ends of successive 10-day intervals from) III Akhet 26 (to) I Peret 6.

[The Commentator's Explication of the Days of the Fourth Decan (as a Model for All Decans): (1) the Day of the End of its Working as the Marker of the First Nighttime Hour, (2) the Day of its Entering the Duat (Setting just after Sunset), and (3) the Day of its Birth (Rising just before Sunrise):]

[Com., col. II] /44/ [Regarding the Decan]

"*pḫwy ḏꜣ(t)*. First (hour): I Peret 6." At I Peret 6, that is to say: I Peret 6 it stops working. [Com., col. III] /1/ It happens that on IIII Akhet 26 it begins to do work (as the marker of the first hour), which is to say, that it works ten days (at that task) according to *(The Book) Sef*, which is to say (in its words:) "a star dies (sets) and a star is born (rises) every ten days /2/ throughout the year."

[Com., col. III] /3/ [Regarding] "Enclosed by the Duat, IIII Peret 6." That is to say, on III[I] Peret 6 it knows the Duat [in order to go into] it. ⌜Drawing near⌝ from /6/I Peret 6, which is the day it stops doing the work it accomplishes, ⟨to⟩ [II]II Peret 6, is done in the west. /5/ They are the three months which *The (Book) Sekhen-Sepedet* has named. The stars pass them (i.e., the three months) in the west after doing work.

[Com. col. III] /6/ [Regarding] "Birth on II Shemu 16." That is to say, on [II] Shemu 16 it rises in the sky from the Duat. ⌜Drawing near⌝ from IIII Peret 6, /7/ which (is) the day it sets, to II Shemu 16, which is the day of rising, [a span] amounting ⟨to⟩ those 70 days which it passes in /8/ the Duat. It rises on II Shemu 16. It passes 80 days in the [east] before it does work. It passes 120 (days) /9/ ⟨doing⟩ work in the middle of the sky. It passes the 10 days which we mentioned above in the course of them (i.e., as the last part of the 120 days). [(Then, as said,) it passes three] months [in the] west...}

5. [Bk. Nut, Text U5:] The Decan Temat heret

kheret *(tmꜣt ḥrt ẖrt)*. First h.:I Peret 6; Enc. Duat: IIII Peret 6; Birth: II Shemu 16.

6. [Text U_6:] The Decan Weshati bekati *(wšꜣti bkꜣti)*. First h.: I Peret 16; Enc. Duat: IIII Peret 16; Birth: II Shemu 26.

7a. and 7b. [Texts U_{7a} and U_{7b} concern the same decan:] Ipedjes *(ipḏs)*. First h.: I Peret 26; Enc. Duat: IIII Peret 26; Birth: III Shemu 6.

8a. and 8b. [Texts U_{8a} and $_{8b}$ concern the same decan:] The Decan Sebshesen *(sbššn)*.[33] First h.: II Peret 6; Enc. Duat: I Shemu 6; Birth: III Shemu 16.

9. [Text U_9:] [The Decan Tepy-a Khenett *(tpy-ᶜ ḫntt)*.] First h.: II Peret 16; Enc. Duat: I Shemu 16; Birth: III Shemu 26.

10. [Text U_{10}:] [The Decan Khenett heret *(ḫntt ḥrt)*.] First h.: II Peret 26; Enc. Duat: I Shemu 26; Birth: IIII Shemu 6.

11. [Text U_{11}:] [The Decan Khenett kheret *(ḫntt ẖrt)*.] First h.: III Peret 6; Enc. Duat: II Shemu 6; Birth: IIII Shemu 16.

12. [Text U_{12}:] [The Decan Tjemes en Khenett *(ṯms n ḫntt)*.] First h.: III Peret 16; Enc. Duat: II Shemu 16; Birth: IIII Shemu 26.

13. [Text U_{13}:] [The Decan Septy Khenuy *(spty ḫnwy)*.] First h.: III Peret 26: Enc. Duat: II Shemu 26; Birth: I Akhet 6.

14. [Text U_{14}:] [The Decan Hery-ib Wia *(ḥry-ib wiꜣ)*.] First h.: IIII Peret 6; Enc. Duat: III Shemu 6; Birth: I Akhet 16.

15. [Text U_{15}:] [The Decan Seshmu *(sšmw)*.] First h.: IIII Peret 16; Enc. Duat: III Shemu 16; Birth: I Akhet 26.

16. [Text U_{16}:] [The Decan Kenem *(knm)*.] First h.: IIII Peret 26; Enc. Duat: III Shemu 26; Birth: II Akhet 6.

17. [Text U_{17}:] [The Decan Tepy-a Semed *(tpy-ᶜ smd).*] First h.: I Shemu 6; Enc. Duat: IIII Shemu 6; Birth: II Akhet 16.

18. [Text U_{18}:] [The Decan Semed *(smd).*] First h.: I Shemu 16; Enc. Duat: IIII Shemu 16; Birth: II Akhet 26.

19. [Text U_{19}:] [The Decan Seret *(srt).*] First h.: I Shemu 26; Enc. Duat: III Shemu 26; Birth: III Akhet 6.

20. [Text U_{20}:] [The Decan Sawy Seret *(sꜣwy srt).*] First h.: II Shemu 6; Enc. Duat: I Akhet 6; Birth: III Akhet 16.

21. [Text U_{21}:] [The Decan Khery Kheped Seret *(ẖry ḫpd srt).*] First h.: II Shemu 16; Enc. Duat: I Akhet 16; Birth: III Akhet 26.

22. [Text U_{22}:] [The Decan Tepy-a Akhuy *(tpy-ᶜ ꜣḫwy).*] First h.: II Shemu 26; Enc. Duat: I Akhet 26; Birth: IIII Akhet 6.

23. [Text U_{23}:] [The Decan Akhuy *(ꜣḫwy).*] First h.: III Shemu 6; Enc. Duat: II Akhet 6; Birth: IIII Akhet 16.

24. [Text U_{24}:] [The Decan Tepy-a Bawy *(tpy-ᶜ bꜣwy).*] First h.: III Shemu 16; Enc. Duat: II Akhet 16; Birth: IIII Akhet 16.

25. [Text U_{25}:] [The Decan Bawy *(bꜣwy).*] First h.: III Shemu 26; Enc. Duat: II Akhet 26; Birth: I Peret 16.

26. [Text U_{26}:] [The Decan Khentu heru *(ḫntw ḥrw).*] First h.: IIII Shemu 6; Enc. Duat: III Akhet 6; Birth: I Peret 16.

27. [Text U_{27}:] [The Decan Khentu kheru *(ḫntw ẖrw).*] First h.: IIII Shemu 16; Enc. Duat: III Akhet 16; Birth: I Peret 26.

28. [Text U_{28}:] [The Decan Sawy Qed *(sꜣwy ḳd).*] First h.: IIII Shemu 26; Enc. Duat: III Akhet 26; Birth: II Peret 6.

29. [Text U_{29}:] [The Decan Kheru *(ḫrw).*] First h.: I

Akhet 6; Enc. Duat: IIII Akhet 6; Birth: II Peret 16.

30. [Text U30:] [The Decan Aryt *(ᶜryt)*.] First h.: I Akhet 16; Enc. Duat: IIII Akhet 16; Birth: II Peret 26.

31. [Text U31:] [The Decan Remen hery *(rmn ḫry)*.] First h.: I Akhet 26; Enc. Duat: IIII Akhet 26; Birth: IIII Peret 6.

32. [Text U32:] The Decan Tjes arq *(ṯs ᶜrḳ)*.[34] First h.: II Akhet 6; Enc. Duat: I Peret 6; Birth: III Peret 16.

33. [Text U33:] The Decan Waret *(wᶜrt)*.[35] First h.: II Akhet 16; Enc. Duat: I Peret 16; Birth: III Peret 26.

34. [Text U34:] [The Dean Tepy-a Sepedet *(tpy-ᶜ spdt)*.] First h.: II Akhet 26; Enc. Duat: I Peret 26; Birth: IIII Peret 6.

35. [Text U35:] [The Decan Sepedet *(spdt)*.] First h.: III Akhet 6; Enc. Duat: II Peret 6; Birth: IIII Peret 16.

36. [Text U36:] The Decan Shetu *(štw)*. First h.: III Akhet 16; Enc. Duat: II Peret 16; Birth: IIII Peret 26.[36]

[Distribution of the 29 Decanal Stars Visible in the Night at one time][37]

[Bk. Nut, Text V:] There are 9 (stars) between the star which determines the first hour and the star which will be enclosed by the Duat.[38] And there are 20 stars between the star of birth and the star which determines the first hour. The total then is 29 (stars) which live and work in the sky. One dies and one lives every 10 days. (Or to put the distribution in another way:) there are 29 stars through the breadth of the sky between the star of birth and the star which will be enclosed by the Duat.

> {[Com. col. III, lines 12-30 repeat the essential numbers of stars in the regions of the sky, adding up to 29, as they are given in Text V. But a couple of additional points of

interest are made: (1) the decan star of the first hour, i.e., the one measuring the first hour by its culmination, is also called "the star of evening" (line 13) and (2) the Commentator does not limit himself to numbering the visible decanal stars in the sky but locates them in the western, the middle, and eastern sky, and in addition he notes the 7 remaining invisible decanal stars that occupy positions in the Duat, producing the total of 36 decanal stars:] "/22/ .... the rest of the stars [which are seen] in the sky [daily] are [29 stars] .... /23/ .... west, 9, the middle of the sky, 12, the east, 8, ... /30/ .... which it stated before concerning the 29 stars. They are the stars which are in the sky. The other 7 stars .... they are in the Duat. (Hence these together make) the 36 stars."}

[Bk. Nut, Text W_1:] The Decan Shetu *(štw)* lives. [Text W_2:] The decan Kenmet lives.[39] [For Text X, see the entry before Text U_1.] [Text Y:] Sand.[40] [Text Z:] Houses of Pillars is the place where Re is (or, enters, **P**)

{[Com. col. III] /39/ .... the place in which Re sets, is its name.}

[Bk. Nut, Texts Aa and Cc (Cc is simply the end of the passage in Aa):] The majesty of this god (Re) enters (the Duat) [by her (Nut's) mouth (**P**)] in the <first> hour of the evening. He becomes glorious; he becomes beautiful in the arms of his father Osiris. He becomes purified there. The majesty of this god rests from (or, sets in?) living (or, life) in the Duat in her (Nut's) second hour in early night. The majesty of this god gives commands to the Westerners (i.e., the dead) and he makes plans [for them, **P**] in the Duat. The majesty of this god goes forth on earth again, having appeared in Upper Egypt. His strength is great again like the first time of his antiquity (i.e., at the time of creation). He comes into existence as the great god in Behdet (i.e., Edfu). The majesty of this god travels up to the boundaries of the heavenly basin, (i.e.,) her (Nut's) arms. He enters her ... in the night in the hour of midnight and he travels therein in darkness, these stars with him.

> {[Com. col. III] /40/ .... The god enters by her (Nut's) mouth in the hour *sḫtp.n.s* (!, but should be, the first hour) /41/ ⟨of⟩ evening, that is to say, by her mouth, the god enters in the third (!) hour of the evening. He is accustomed to go forth from her in the hour /42/ *sḫtp.n.s*. It is the ninth (hour) of the night. [col. IV] /1/ "The majesty of this god sets in life [in the Duat in her second hour in early night] .... /11/ "He flies and travels in the hour of midnight" .... }

[Bk. Nut, Text Bb:] The majesty of this god enters within the Duat by her (Nut's) mouth. Afterwards he goes forth and travels in it (the sky). These stars enter and go forth with him. They travel about to their places *(dmiw)*.

> {[Com. col. III] /31/ .... It is by her mouth, that is to say, the sky, that the god Re enters within the Duat. Look at the picture, the disk which is in her mouth .... /34/ .... With him these stars set and with him they rise. When their time of /35/ setting is complete, the period in which they are not customarily seen occurs .... /36/ .... While travelling ⟨to⟩ their places, the ones proper to them, these (stars)

set. It (i.e., each place) is the place /37/ in which they do work. It happens that some are accustomed to set in order to receive others for them (i.e., in their places). *(The Book) šd* (says): "<Some> come at night. Others /38/ come at their hour when a star has gone into its places."}

[Bk. Nut, Text Cc. See Aa.] [Text Dd:] Total darkness, the heavenly waters (or, marshes) of the gods, is the place from which the birds come. They are from her (Nut's) northwestern side up to her northeastern (or possibly, southeastern?) side, which opens to the Duat, which is on her northern side. Her hind part is in the east, her head in the west.

{[Com. col. IV] /27/ that is to say, the birds which come to Egypt.}

[Bk. Nut, Text Ee:] These birds have human faces and are bird-shaped.[41] One speaks to the other with the speech of men. Now after they come to eat herbage and feed in Egypt, they alight under the bright rays of the sky. They appear in their bird-forms. [Text Ff:] The nests which are in the heavenly waters (or, marshes). [Text Gg:] Her (Nut's) head is in the western horizon, her mouth in the west. [Text Hh: This refers to two cartouches or ovals, the first of which is empty while the other contains three birds and thus appears to indicate a nest.] [Text Jj:] Her western position (or, her right arm) is on the northwestern side, [her eastern (or, left arm)] is on the southeastern [side]. [Text Kk:] Resting from living in the Duat. [Text Ll:] Going forth purified from evil.

Notes to Document III.12

1. NP's reading; LN had *rkj*.

2. This reference to the "picture on the papyrus" is a sure indication that the papyrus (or whatever it was copied from) had a Nut picture very much like the ones in **S** and **R**. There is a confused reference in line 4, which I have not attempted to render which seems to suggest that Geb, the Earth-god, was also present at the bottom of the picture (perhaps as in Fig. II.2b in my Vol. I, or as in Fig. III.97). Geb might have been intentionally eliminated in order to have enough room on the picture to add the decanal tables (U) which I mentioned in the Introduction to the document and which are of course a crucial part of the document.

3. Presumably these are the primitive waters of Nun that surround the Maat, and out of which the ordered cosmos was fashioned by the creator god (see Volume One, Chapter Two, pp. 264-65 and passim).

4. There is a falcon on its supporting pillar to the left of the vulture at the left of the Nut picture in **S** (see Plate 32 in NP). The reference could also be considered as invoking the common idea that the course of the sun's motion is symbolized by the flight of the falcon with his wings spread. Indeed the sun itself, when leaving Nut for its descent into the Duat, is shown as winged in the Nut picture (see Fig. III.95a, right).

5. Perhaps NP's suggestion of "after" is better than "just as."

6. LN interpret line 11 as a reference to a work entitled *The Order of the Wanderings of the Stars*. Such a work is mentioned in the temple library of Edfu (see my Volume One, p. 46, item 29): "The Governing of the Periodic Returns of the Stars." The work is unambiguously referred to in line 14 of this text where its shortened name *pꜣ bnn* is also given.

7. Lines /14/ and /15/ are given as part of Chapter A of the commentary by LN, while NP begin Chapter B with them. A possible inference from NP's arrangement and the text itself of these lines is that the Commentator had access to the text as inscribed on the picture in a funerary monument like **S** or **R** as well as to a continuously written text in a papyrus.

8. For the land of Punt or Pewenet ("the land of myrrh and

incense") see Volume I, p. 343. Apparently Punt was thought to be the place where the sun resided before rising, especially at the time of the winter solstice.

9. Thus in **S**. **R** has "Nekhbet the White One of Nekhen." Hornung translates the latter as "Nekhbet die Leuchtende," i.e., "Nekhbet the Shining One."

10. For the fortunes of the Eye in the creation legend, see NP, Vol, 1, pp. 592-93. The "She" mentioned here in Text C may be Tefenet, who along with Shu brought back Re's Eye to him. Or, as NP note (p. 45), Hathor and Nekhbet "seem to be participants." Indeed Nekhbet is mentioned in Text B and thus may be the "She" in this text. "But the goddess who is most intimately associated with Punt and God's Land is Hathor.... In our text [here] there may be an assimilation of the two goddesses *(ibid.)*."

11. This passage suggests the close relationship between the rising of the decans and the sun.

12. The reference is to Re as the Scarabaeus-beetle, pushing his mud ball in front of him. See the small scarabaeus on the left side of the Nut picture above H (Fig. III.95b) and Volume 1, p. 266 together with Fig. II.1. **R** has *wbꜣ.f m ḥtp.f*, "He opens as he sets."

13. **R** has "his birth."

14. For the concept "first time" see Vol. 1, pp. 369-70.

15. So translates Hornung, p. 92. NP Vol. 1, p. 50, note to line 7 says: "This is not the word for the nome of Abydos but rather that of *Wb. V*, pp. 230-31, 'larbord, portside' of a ship, in the extended meaning of 'left' and then 'east.'"

16. See NP's long discussion of the sun's motion quoted in note 24 below. It includes remarks on the meaning of "the ninth hour." See also Hornung, p. 95, n. 15.

17. The bracketed material is suggested by the reading in **P**. The ambiguity of the latter reading is that two "land-signs" are given and so suggest the rendering "The Two Lands," a conventional designation of "Egypt," as the place where the full vigor of the sun occurs, but then these two signs are followed by three strokes indicating the plural rather than the dual. Hence my question mark after the "Two." Hornung in his translation decides for "in den Beiden Ländern," i.e., Egypt.

18. See the reference to an extant work with such a title in the Introduction to Document III.1, n. 2.

19. This statement seems to tip the balance in favor of the view that "the lands" and not the "The Two Lands" is meant, for,

of course, "The Two Lands" signifies Egypt. See note 17 above.

20. The whole Text of L is reconstructed mainly on the basis of the quotations in **P**. See Hornung's work, p. 95, n. 21.

21. For the gods and the spirits, see Vol. 1, pp. 97, 344.

22. **S** and **R** have *ḥḏt*, but **P** has *ḥḏḏ*, i.e., "light."

23. I have translated **S** and **R** rather literally and perhaps not too clearly. The passage is attempting to delineate the Abyss from the region of the sky, and the entire Duat (i.e., the sky and underworld) from the region of the humanly inhabited lands.

24. Texts M through R are simply directional or locational labels that are tied essentially with the motion of the sun. Along with Text B, they are not quoted for commentary in **P** and **Pa**. NP add a superb essay on the sun's motion as revealed in the text and commentary at this point (pp. 81-83), which I now quote:

"We are now in a position to evaluate the apparently conflicting evidence of the Nut picture itself, the relevant texts already discussed in **P** and the present texts M, N, O, P, Q, and R. There are three depictions of the sun in the left half of the Nut picture, a sun-disc on her foot, a smaller sun-disc just to the right, on a line curving up to Nut's knee, with a small arc just above the wavy line of the earth on which Shu stands (see also text Y ...) and lastly a beetle on Nut's thigh apparently climbing upwards. In the right half of the picture the sun is shown but once, as a winged disc at Nut's mouth.

"Let us consider first the setting of the sun, as being perhaps the less complicated of the two events. Text Q tells us that Nut's mouth is the western horizon, text R that it is evening, and text Bb ... that it is by her mouth that the sun-god enters the Duat. Indeed in [the commentary III,31 to] Bb we are specifically instructed to look at the picture and the disc at Nut's mouth.

"So far all is clear sailing but [the commentary to] text Z (III, 39), located at the line of earth, just to the left of Nut's hands, tells us that this place, House of the Pillar by name, is where Re^{c} sets. The rather mixed-up text Aa now compounds the confusion. In [the commentary to it in] III, 40 we are told that the sun-god enters Nut's mouth in the hour of *sḥtp.n.s* in the evening which should of course mean the first hour to agree with texts R and Bb and [commentary] IV,1 and 11 of Aa. The demotic commentary to III, 40, however, informs us that what is meant is the *third* hour of the night, and moreover it recalls that it is in the ninth hour, which bears the same name, that the sun is accustomed to go forth from her (cf. also II, 9-10 of text J).

"There would appear to be a mingling of two versions of the sun's setting, in one of which he goes directly into the mouth of Nut in the first hour of the evening and another in which the setting takes place near Nut's hands and the actual entrance into the Duat through her mouth is later. There is additional evidence for the second version. Part of the shadow clock text which appears on the ceiling of Seti's cenotaph (Plate 32, 14-25) lists the twelve hours of the night and associates with them twelve parts of a female body, undoubtedly, as De Buck suggested, twelve stations in the body of Nut through or on which the sun passes during the night. These are:

"1. *ḏrt* hand 2. *spt* lip 3. *nḫ̥dt* tooth 4. *ḫ̥tt* throat 5. *šnbt* breast 6. ....*t* ...... 7. *mint* ...... 8. *mnḏr* gall bladder 9. *mẖtw.s* her intestines 10. *kꜣt* vulva 11. .... 12. *mnt* thigh

"We do not need the additional evidence of the unidentifiable words to see in the list a clear sequence but a sequence which, be it noted, requires the sun to spend the first two hours of the night actually outside the mouth of Nut. In the first hour he begins at her hand, in agreement with the location of text Z, in the next hour he is at her lip, and it is only in the third hour when, at her teeth, he can be said to be in her mouth, in accord with the commentary to III, 40.

"We have now to investigate the situation at rising and see if this mixture of versions holds true there as well. As is to be expected from the first version of the setting, the vulva of Nut is called 'Eastern horizon' in text P and it is to be presumed that the sun's birth takes place in the last hour of the night and that he has been in the Duat, the interior of Nut, for twelve hours. Exactly this do we find in the short form of the Book of Him Who Is in the Duat where we are told that in the twelfth hour the sun-god goes forth from the Duat and reposes in the bark of the day after he appears from the thighs of Nut. With this we must compare text O, the text closest to text P, where the same expression, 'opening the thighs of Nut,' occurs and where we are thus strongly tempted to take the last line of the text as stating: 'He departs toward the sky.' That we do not do so is because in text J, specifically II, 9, **R** has the very same sentence, *sw sḫ̥r.f r pt*, with **P** substituting *pꜥyt*, 'mankind', for *pt*, 'sky', and here it is certain that the movement of the sun is out of the Duat toward mankind but as it is still only the ninth hour of the night, the god is still in darkness and invisible to those on earth. If then in texts J and O 'sky' is the correct translation of *pt* what must be meant is Nut herself and the movement of the sun is away from her vulva towards the earth. And this movement is the subject of text N farther down on Nut's leg. Finally text M sums up the whole process and identifies the disc on her foot.

"It is clear that we have at least two versions of the rising of the sun as well. If we now refer II, 9-13 of text J and its commentary to the parts of Nut's body associated with the hours of the night which we have listed above we again find confirmatory evidence. The sun receives the order to start toward mankind in the ninth hour of the night. In this hour he begins in Nut's intestines and ends at her vulva and may thus be considered 'born' or 'risen.' In the tenth hour he moves from her vulva to an unknown place on her body intermediate between the vulva and the thigh and finally through the eleventh and twelfth hours past her thigh and down to her foot. All this time, although actually outside the body of Nut, the sun is not visible. But when he reaches the earth on which Nut's foot rests, we then see him in the eastern horizon at sunrise just making his appearance as a smaller disc (cf. II, 18-19) and later we see him transformed into a Scarabaeus once again on Nut's body but this time visible as he climbs the eastern sky. The texts which concern the smaller disc and the Scarabaeus, D, E, F, G, H, J, and K, bring in many other aspects of the rising of the sun. While there is an underlying unity of these there is also great diversity of expression. We do not propose to go further into that aspect of the problem."

The reader will, of course, find the line-number references to the Commentary mentioned here in this long quotation under the text letters (i.e., II, 9-13 under Text J; III,31 under Text Bb; and III, 39-40 under Text Aa, and so on).

25. As we see from Fig. III.95b these three decans labelled S_1, S_2, and S_3 are located on the body of Nut but below the list of decans labeled T. NP (Volume One, pp. 83-84) make a convincing case that they belong to the list of decan tables labeled U. Thus, as part of the tables of U, they are to be considered as decans used to mark nighttime hours by their meridian culminations.

26. Neugebauer and Parker, *Egyptian Astronomical Texts*, Vol. 1, p. 84: "The list of T would then represent the earlier and

superseded method of observing risings in the horizon [for marking the hours]." This is the authors' intention. I have added the bracketed phrase since it will be obvious from the tables of U below that, although the risings of the decans are given, their use as hour markers (that is when they "work") is based on their meridian culminations.

27. Presumably a reference to the decans labeled T on both sides of it. NP, Vol. 1, p. 54, have suggested that this might be taken as evidence of an ideal Sothic year, but they doubt this and believes that it is only a reference to the fact that in the time of Seti I the rising of Sirius fell in the first month of Akhet. While that is so, we have evidence that the decan list in T probably, and that in U more certainly, originated in the Middle Kingdom. Hence we would have to conclude that the statement in T2 is a general observation created by the person planning to add the Nut picture to the cenotaph of Seti I.

28. This curious statement (otherwise inexplicable on the basis of material in the original text or in the Commentary) may be explained by the position of the image of Sothis in the vertical V column of the lid of the coffin of Meshet (see NP, Vol. 1, p. 55, where they say rather ambiguously that there are 18 decans after the *decan* Sothis and 18 before, whereas in fact there are 18 decans after and 18 before the *image* of Sothis in column V). There, as usual in the star charts on coffin lids, the image of Sothis appeared at the bottom of column V with her back to the base line of the 12th hour (see NP, Vol. 1, Plate 1). There are 18 decans after column V (i.e., after Sothis) and 18 decans before V (i.e., before her). For a clear picture of the four images that are customary for column V on a coffin lid, see Figs. III.14 and III.17. I must admit that this is rather a thin explanation for the statement as given in the Book of Nut.

29. I have said in the Introduction to Document III.12 that each table contains three dates or days for points or stations of each decan in its annual course which is assumed schematically to be traversed in 360 days rather than 365 days of the civil year or, more accurately, the approximately 365 1/4 days of a Sothic year: (1) the day of the "first hour," i.e., the day when the decan completes its 10-day stint of marking the first hour; hence it is the day when the decan has completed all of its annual "work" as the marker of the nighttime hours through 120 days (10 days for each hour), it having worked its way from marking the 12th, 11th, 10th,

etc. through the 1st hour; (2) the day it is "enclosed by the Duat" i.e., the day the decan sets acronically 90 days from the day of the "first hour;" and finally (3) the day of its "birth," i.e., the day of its heliacal rising 70 days after setting into the Duat, from which time it passes 80 days in the eastern sky before it begins its work of measuring the 12th hour by its meridian culmination. It is a simple scheme that ignores the epagomenal days, and becomes quickly defective with regard to the actual annual motions, needless to say, more quickly than the system of the star charts on the coffins, in which decans were found to account for 5 epagomenal days, the so-called "triangular decans." I would suppose that the decans used for such a star clock would have been changed frequently.

This is obviously a grossly simple schematic device of marking nighttime hours and I have briefly indicated its unsatisfactory nature in Chapter Three and in the Introduction to this document. But its very simplicity makes it easy to calculate the "correct" entries for the tables, though they are often erroneously reported in both **S** and **R**. I have not reported the errors of the two sources here, since they can be readily found in the collection of tables reported by NP, Vol. 1, pp. 84-86 (and see also pp. 113-15).

It should be further noted that the names of only 11 of the 26 decans are given with the tables of **S** and **R**. But as, NP observe, those eleven allow us to identify the family of decans whose pivotal days are schematized in the tables. This family is designated as the Seti I B family. It is listed and studied carefully by NP in Vol. 3, pp. 133-40. All of this study permits us to include the names of all of the decans whose "days of finishing work as first hour, setting at sunset, and helical rising" are being tabulated in Document III.12; those decans taken from other copies of the list are here enclosed in square brackets.

30. NP (Vol. 1, p. 87) read the first word as *rmnt*, and indeed it does look like that. However, there is no great difference in the glyphs representing the two words. And surely it is possible that it is *ᶜt*, as I propose. This latter word can have the meaning of "writing" or "register" or the like. In either case, there is no known meaning of either word that seems to fit exactly what the context suggests, namely a word like "dates" or "schedule" (as NP suggest, without actually finding that usage), but I believe "register" is somewhat more fitting than any known reading of *rmnt*. Incidentally, the word here on the picture, whatever it means, actually has no determinative. If the word were *ᶜt* and meant

"writing" or "register," it ought to have had a determinative of a band of string or linen: [hieroglyph]. Along with NP, I believe that Text X refers to the table of dates for the Decan Kenmet as the first decan, and that the first word is titular in nature and probably is to be understood for all the decanal tables (or "registers," if you will). Hence I have given Text X out of alphabetic order before the tables of U to suggest its possible titular nature, and to introduce the name of the first decan.

31. I have expanded the entry for the first decan to make its meaning clear and abbreviated the terms in the entries for the remaining decans.

32. These first five decans are located below Shu on the Nut-picture (U_1 through U_5) in Fig. III.95b.

33. See Text S_1 and note 25 above.

34. See Text S_2 and note 25.

35. See Text S_3 and note 25.

36. As I remarked in the Introduction to this document, the day of the rising of Sothis in this part of the table may well refer to the rising of Sothis in the 7th year of the reign of Sesostris III and thus indicate a Middle Kingdom origin for the tables.

37. Having discussed the movements of the decanal stars in terms of dates and days, the Book of Nut now mentions the stellar grouping. That is to say, up to now the texts of the Book of Nut have traced the course of one star as it goes successively from one crucial date of a schematized 360-day year to the next, i.e., from the day of the end of its work in the sky to its setting 90 days later, then to that of its rising 70 days later, from there to the initial use of its culmination as a measurement of the 12th nighttime hour 80 days later, and finally to the last day when its culmination is used to measure the first hour. But in Text V we are told the number of the stars at one time (from a static viewpoint) between the separate stars that are at the crucial points mentioned earlier. The result is that at a single time there are 9 decanal stars visible in the western sky, 7 invisible in the Duat, 8 visible in the eastern sky, and 12 working, i.e., telling time, and visible in the middle sky. Now, in fact, the texts of V are simply interested in the 29 decanal stars that are visible in the nighttime sky, namely the 9 in the west plus the 20 stars that combine the 8 stars of the east ("living" but not "working," using the Egyptian terms) and the 12 stars "working" in the middle of the sky.

38. In my rendering of this text, I have changed the involuted style of Egyptian sentences which goes "As for such and such...., it is so and so" to simple declarative sentences to produce a style more attuned to English expression. Though this reduces somewhat the emphasis on the main consideration of the sentence, it surely does not change the meaning in any significant fashion.

39. In the Nut picture we see that the Decan Shetu is just above the horizon at position W1 and the name of the Decan Kenmet is below the horizon at position W2. As I noted above in the U tables, Kenmet is the first decan and Shetu is the 36th. In the phrases of the W texts the word "lives" means "rises" or "is revived" since the decans were "dead" in the Duat and come to life as they leave the Duat and enter the sky.

40. This is a label marking where Shu stands in the Nut picture (but this label is only in **S**).

41. This is the conventional depiction of Souls after death, i.e., with human faces or heads and bird-shaped bodies. See my Volume One, pp. 217-18.

Document III.13: Introduction

The Dramatic Text in the Cenotaph of Seti I

Somewhat to the left of the arched figure of Nut in the Sarcophagus chamber of the Cenotaph of Seti I at Abydos which served as a focus for the collection of texts which I have presented as Document III.12 there is another cosmological text related to that figure. It is a text comprising 46 vertical lines. For this text I have used the title Dramatic Text. This descriptive title was employed for the first time by A. de Buck in his partial translation and copy of the text published as Chapter IX and Plates LXXXIV and LXXXV of H. Frankfort's *The Cenotaph of Seti I at Abydos*, 2 vols. (London, 1933). Dr. de Buck's use of the designation Dramatic Text is made clear in his comments on the state of the text:[1]

> It is very unfortunate for us that a large part of this interesting text is utterly unintelligible. The first complete lines indeed tell us a clear, coherent story, but after a few lines the drift of the narrative is completely lost. The lacunae which interrupt the text at this point are the more serious because the subject is new to us and the text is written in an unfamiliar and partly enigmatic orthography. Moreover, as in the case of the worm-eaten original from which the text of

the Shabaka stone was copied [see my Vol. 1, p. 595], the ancient copyist of our document had a manuscript before him which was already at that time far from perfect. Small blank spaces corresponding to the lacunae of the ancient original are found in [Vol.] II, [lines] 15, 17, 23 [Plates LXXXIV-LXXXV; see my Fig. III.98a for the first two lacunae and the second of Frankfort's plates for the third], but towards the end [see lines 20-23, 29-31, and 34-46 of Plate LXXXV] the gaps seem to have been so long as to leave only fragments of legible text.

This state of affairs would not cause much regret if the text were a less important document, or belonged to a class of texts of which we have plenty of other examples; but it is particularly deplorable in the case of this text, which belongs to a literary *genre* of which only very few examples are known to us, viz., the so-called dramatic texts, which Sethe has recently analysed and explained in his book, *Dramatische Texte zu altägyptischen Mysterienspielen.* That our text bears the closest analogy to Sethe's material is clear at first sight; as regards the general plan it shows the same combination of narrative portions, explanations, and conversations, and in addition to this it uses many of the established phrases and peculiar words which are characteristic of this *genre* of literary works.

With regard to the date of the work it seems to me that there are two possibilities.

> The ambiguous character of the text, with its mixture of, partly at least, very ancient words and phrases and an orthography which often points to a later date, may be explained in two ways: either it is an originally ancient work in a modernized garb, or it is the more or less successful result of the archaising efforts of a later writer.

Unlike the case of the Book of Nut which I have given as Document III.12, there are extant only two sources for the Dramatic Text: the hieroglyphic inscriptions of **S** and the hieratic-demotic papyrus of Carlsberg comprising quotations from the text as well as commentary, and labeled **P** (both sources have been described in the Introduction to Document III.12). Because of the poor state of the earlier text in **S** (as the quotation from Dr. de Buck's account given above emphasized), in my translation I have placed some dependence on the textual passages found in **P**. I have not distinguished the differing readings from the two sources but have tried to make a consistent and coherent text from both. However, the reader can note such variants by referring to the text in Fig. III.98b and the translation of NP. Like LN, NP, and Hornung I have not gone beyond line 14 (of the 46 lines) since the commentary in **P** goes no farther and the state of the text in **S** makes any reconstruction grossly incomplete and hazardous (as Dr. de Buck's noble effort at translating the later material reveals). However, since the text of the part translated here is (with help from the textual passages in **P**) understandable for the most part, like Hornung,[2] I have confined to the notes any references in the Commentary that are not direct quotations of the text, a practice that diverges from the one I followed in Document III.12 where important quotations from the Commentary are inserted in the text itself. Needless to say, both LN and NP, who have the editing and translating of the Commentary as their principal objective, give full translations of the Commentary (texts and comments) and I have made extended use of these translations in the footnotes to the document. I have, however, not slavishly followed their literal styles but have tried to produce a smoother rendition. I add one caution. In view of the state of the text in both sources, my version of the Dramatic text, like all others, is not only incomplete in length, as I have said, but is in a number of places ambiguous and uncertain.

As I noted in the Introduction to Document III.12, the extract that makes up my Document III.13 lacks all but trivial astronomical detail. For example, nothing of the specificity found in the decanal tables U_1 to U_{36} of Document III.12 is evident in this document, and indeed the Dramatic Text is essentially mythological in character and content. This is particularly evident in the later lines not included in Document III.13 (see below, note 16 to that document). My main reason for including this short document in a volume devoted largely to technical detail is once more to underline the conclusions stressed in Chapter Two of Volume One: (1) such scientific knowledge that the ancient Egyptians acquired was presented integrally entwined with religion, myth, and magic, and (2) that knowledge has been transmitted to us almost exclusively in religious documents (see Volume One, p. 263 and passim).

I conclude this introduction with NP's short but just summary of the Dramatic Text (through two-thirds of line 14 of **S**), a summary devised by considering the original text in the light of the Commentary (and using the line numbers of **P**):[3]

We now turn to Part II, the so-called Dramatic Text which is concerned with the stars ([Com.] IV, 34-VII, 27). They rise (IV, 35-42) and set like the sun. The setting is discussed in a separate myth (IV, 43-V, 11) of Nut, seen as a mother-beast eating her piglets and quarreling about it with Geb. In V, 12-30 we revert to the simpler picture of stars entering the mouth of Nut. This refers now to the period of invisibility of 7 decans while 29 become visible during the night.

The stay in the Duat during 70 days is described in greater detail in V, 31-VI, 23 in the form of the embalming ritual and the freeing from impurity as a result. The stars are fish in the lake of the Duat, their tears become fish also. But eventually they withdraw to the sky. This seems to happen upon command of Geb (VI, 24-42) who is quarreling with Nut.

Thus the stars again become visible to man (VI, 13-VII, 27). The yearly circuit of the stars, which was interrupted by a period of invisibility, comes to a close. It finds its analogue in the circuit of the moon, which appears again, after invisibility, for 28 days in the sky. Thus the text concludes with a parallelism which may well have been the origin of the fundamental discovery that it is their relation to the sun which determines the yearly disappearance and reappearance of the stars as well as their visible rising and setting.

So much then for a brief characterization of our document.

Notes to the Introduction of Document III.13

1. *The Cenotaph of Set I*, Vol. 1, p. 82.
2. Hornung, *Zwei ramessidische Königsgräber: Ramses IV. und Ramses VII.*, pp. 93-94, 96, notes 59-80.
3. Neugebauer and Parker, *Egyptian Astronomical Texts*, Vol. 1, pp. 41-42.

Document III.13

The Dramatic Text in the Cenotaph of Seti I

/1/ These stars sail out at night to the limits of the sky outside of her (Nut); they shine and [accordingly] are seen. In the daytime they sail inside her, /2/ do not shine, and [hence] are not seen. They enter after (or, with)[1] this god (Re) and they go forth after (or, with) him. They travel with him on the Support of Shu (i.e., that which Shu supports, namely the arched body of Nut), and they settle in their places (in the night sky) after his majesty (Re) has set in the western horizon.[2] /3/ They enter into her (Nut's) mouth, in the position of her head [which is] in the west, and she eats them. Geb quarreled with Nut since he was angry because she eats them. And her name is called /4/ "Sow who eats her piglets" because she eats them. Her father Shu lifted her and supported her above himself saying, "Let Geb beware. Let him not quarrel with her because she has eaten /5/ their children, [for] they shall live and go forth to their places [again] from the place under her hind part in the east every day just as she gives birth to Re every day."[3] Not one /6/ of them has fallen since their birth.[4] The one (star) which goes to the earth dies and enters the Duat. It stops in the House of Geb 70 [days]. It is regenerated by loosing its impurity to the earth for 70 days.[5] There is no speaking the name /7/ of the one who is regenerated during the 7[0] days [since it has not achieved its identifiable or nameable character as a shining star until just after the end of the period when regeneration is complete].[6] Nor is the name "Living" applied to the one who is being regenerated [until all of its impurity falls to the earth] so that, like Sothis, it rises.[7] [Then] it is "pure" (regenerated) and ["lives" again].[8] /8/ Their heads (i.e., of the gods) are located in the east. Thus it happens: "one dies and another lives every ten days."[9]

These are the heads of the gods.[10] They celebrate the "First Feast *(ḥb tpy)*, (i.e., their birth day)" in the east.[11] One of them is given (back) its head, while its bones *(ḳsw)* fall /9/ to the earth, and (their) souls go forth on earth. Their tears fall and become fish. The life of a star develops (i.e., begins) in the lake.[12] It develops as a fish and goes forth (again) from the water. It flies upward to the sky out of /10/ the sea and out of its (previous) likeness (or, form) *(snn)*. This is the life (i.e., the rising) of a star. They (the stars) go forth from the Duat and they withdraw to the sky.[13] Then Geb became Prince of the Gods. Then Geb and Nut [fought]. He (Geb) commanded that they (the stars) show their heads /11/ in the east. Then a second time he, Geb, said to the gods (stars) "Fish out (i.e., find ?, or put an end to ?, or remove ?) your heads yourselves." Then Thoth commanded that they fish out their heads. Then they lived when their heads developed (i.e., took independent form as rising and shining entities?). Their bones (or, burials) developed /12/ like (those of) men.[14] Thus its (his ?, or perhaps, their) period in the Duat is appropriate for everything that has to be done (for regeneration).

The souls travel along on the inside of the sky at night. It happens that they withdraw to the /13/

boundaries of the sky by day without appearing to the sight. When it is seen by the living, it is (indeed) a star, a piglet of its mother, which makes its journey and shines forth in the sky in the hours of the night and which travels the /14/ sky to the end. This means its life is seen. The star which has gone forth and has been brought outside of her (Nut) proceeds as they do.[15] The moon of the second day is the feast of Horus.[16]

Notes to Document III.13

1. In this text, as in the Book of Nut, we note the intimate connection of the coursing of the stars with that of the sun (Re). NP in their translation, perhaps to stress this relationship, translate *m-ḫt* as "with." I prefer "after." But see the quotation from the Com. in note 2.

2. [Com. V:] "/12/ After /13/ Re sets in the western horizon, they enter into her (Nut's) mouth in the place [in] which Re sets, that is to say, the stars are with him (i.e., accompany him). These are the /14/ [2]9 (visible) stars which are on the (decanal) path."

3. [Com. IV:] "/44/ Geb has no quarrel with Nut for causing their children, i.e., the stars, to set. This is spoken of in *(The Book)*....." [Com. V:] "/8/ .... He raged against her... /9/ for causing her children to set (i.e., die as visible stars), which she did in order that she could give birth to them [again]."

4. [Com. V:] "/31/ 'Not one of them (the stars) falls'.... since (or, because) [she bore] them, /32/ i.e., the stars. There is no perishing of (even) one of them. [That] is what it (the text) means. It happens that it says [in] *(The Book) ḥr* .... /34/ It (the text) speaks about the time which is spoken of in *(The Book) ḥr*."

5. [Com. V:] "/37/ .... Its stay in the Duat 7 decades (of days, i.e., 70 days). /38/ .... It sheds the impurities in the 7 decades .... /39/ .... It is said: It is left [in] the Embalming-House for 70 days until .... (it is purified and rises ?)."

6. [Com. IV:] "/38/ .... The name of the one (i.e., the star) which has shed its impurity is pronounced (only) at its withdrawal /39/ from the House of Geb .... /40/ .... i.e., the Duat .... /41/ .... The name of the one (star) which sets is not customarily pronounced until it rises (again) /42/ from the Duat." One perhaps

sees evidence here of the Egyptian doctrine "to pronounce or speak after conceiving in the heart or mind of a creator god is to create or bring to life." This is extensively discussed in Chapter Two of Volume One (see especially pp. 307-12) and Document II.9.

7. [Com, V:] "/43/ .... It rises and comes into existence in the horizon like Sothis, i.e., all /44/ of them (the decanal stars). It means: She, Sothis, customarily spends 70 days in the Duat and (then) rises again." This is one more indication that ancient Egyptian astronomers thought of the annual course of Sothis (i.e., the star Sirius) as the model for selecting decanal stars to mark the nighttime hours.

8. Again we see evidence of the expressed analogy between the purification and rebirth procedure followed in the embalming of human beings and that used as an explanation for the setting and rising of the stars after a period of regeneration in the Duat (see also note 14 below). Of course, I must stress that the period of the stay of a star in the Duat like that of the human corpse was not to be one of death in the sense of "non-existence" but only in the sense of non-functionality or "rest" from "living," which is to say, it is a period in which it does not and cannot exercise its proper function, namely to shine and to be seen. Hence it cannot be called by its name.

9. [Com. VI:] "/2/ the setting of one (star) takes place and another rises every decade (i.e., 10 days)." "Head" in this mythological account becomes the term for a visible, shining, and rising star.

10. [Com. VI:] "/3/ .... these are the risings of the gods. Another version: these, i.e., Orion and Sothis, /4/ who are the first of the gods, ... customarily spend 70 days in the Duat <and (after this) they rise> again."

11. [Com. VI:] "/5/ .... The celebration of the First Feast .... is /6/ at their rising in the east."

12. [Com. VI:] "/14/ .... The life of a star begins in the lake (or, water). They are the water-forms, which are in /15/ the form of a fish, as *(The Book) ỉ't* makes it."

13. [Com. VI:] "/23/ .... They rise in the sky, becoming distant from the earth."

14. For my "bones" LN and NP have "Begräbnisse" and "<burials>" respectively, as they also have in their renderings of Com. VI: "/38/ .... The burials take place like (those of) men, i.e., they are similar to the burial-days which are /39/ for men today,

i.e., the 70 days they pass in the Embalming House before the utterance of words by them. /40/ That is their way of rest."

15. [Com. VII:] "/20/ ... like the journeys, i.e., *(The Book) bnn* says .... /21/ rise with the journeys they make."

16. [Com. VII:] "/23/ .... The moon of the second day is the feast of Horus, ... /24/ They are the appearances of the moon, those which it makes before other .... (i.e., before it makes its appearances every day of the rest of the month?). They are 28 /25/ when they fill the circuit [of 30 days]. It constitues *(The Book) sf* (i.e., this is all there is of it). I have not found more. /26/ It is its completion. /27/ It is its end." This is apparently a reference to the end of the treatment of the subject of lunar days in *The Book sf*. It clearly does not to refer to the end of the Dramatic Text, which continues another 32 1/3 lines in **S**. So far as we can tell from the gap-filled text of those lines in **S** and from the efforts of Dr. de Buck to translate it, the text continues to follow the mythologizing style of joining the activities of gods with simple references to heavenly phenomena. By far the greatest attention is given to the activities of Horus.

Document III.14: Introduction

The Ramesside Star Clock

In Documents III.11 and III.12 I have given texts that describe star clocks which were based respectively on (1) the risings of decanal stars and (2) the meridian culminations of decanal stars. Now in Document III.13 I present the so-called Ramesside star clock that was based on the transits of stars (different in all but three cases from the decanal stars used in the decanal star clocks) on the meridian line or on one of three parallel lines before or three parallel lines after the meridian. These additional lines are spaced from it at close but apparently equally distant increments, forming a grid with the 13 horizontal lines which successively represent the beginning of the night and the ends of the twelve hours into which the night has been divided. As we shall see, these star transits are given in 24 semimonthly tables, one for the first half of each month and one for the second half of each month throughout the civil year, with the transits on the epagomenal days probably assumed to be those recorded in the 24th table, as was apparently also the case in the hour-scale for the 12th month found on water clocks, as I have suggested in Chapter Three.

Imposed on or located below each grid was the figure of a seated man facing forward (i.e., the target figure), each vertical passing through a part of his body, to wit, his heart (the middle line, designated since the time of H. Schack-Schackenburg as line 0), or his right eye (line -1), right ear (line -2), or his right shoulder (line -3), or on the other side of the body, through the left eye (line +1), or the left ear (line +2), or the left shoulder (line +3). Hence on the charts "right" is on the left side of the chart and "left" is on its right side.

This transit-clock is ordinarily called The Ramesside Star Clock because all four extant copies are located in the tombs of Ramesside kings of the 20th dynasty, namely Ramesses VI (two copies), Ramesses VII (one), and Ramesses IX (one); all are briefly described below. But in fact these four copies are incomplete and often careless versions of a clock that dates from about 1470, as is deducible from the date II Peret 16 given as the culmination date of Sirius at the beginning of the night in Table 12.[1]

The nature of the Ramesside star tables was obscured by their early investigators. The founder of modern Egyptology, Jean François Champollion, who discovered the tables in 1829 in the tombs of Ramesses VI and Ramesses IX and published them in his great *Monuments*,[2] first thought them to be of an astrological nature indicating the influence of the positions of the stars during the nights of successive half-monthly periods throughout the year on the parts of the human body pictured in the figure imposed on or below each grid or chart representing the half-monthly spans.[3] This was denied by Lepsius, who presented an equally erroneous view of their nature, and was thoroughly refuted by Peter Le Page Renouf,[4] who was the first scholar to conclude that the 24 tables constituted a calendar of transits for the year (though without specifically designating it as a star clock). This explanation was presented neatly and succinctly in the following passages (referring to Fig. III.99a):[5]

> The Calendar, which is unfortunately imperfect in many parts, consisted of twenty-four columns, two being assigned to each month, or one to every 15 days. Each column contains thirteen entries, one for the beginning of the night, and one for each of the twelve hours. Throughout the Calendar a star occurs in one of seven positions, "the middle," the right eye, ear or shoulder, or the left eye, ear or shoulder. The position is not merely described in words, but [is] graphically indicated [see the figure on page 409 of Renouf's account, which is my Fig. III.99b]. The perpendicular line passing through each of the positions corresponds to the limb of a sitting figure, which is drawn underneath the diagram, and represented as facing the spectator. The line of "the middle" passes through its axis.
>
> If the text were Greek instead of Egyptian, there never would have been a doubt as to what was meant by a star being in "the middle." The verb *μεσοῦν*, "to be in the middle," when applied to sun, moon, or star, is equivalent to *μεσουρανεῖν* .... A star is in the middle of its course or in mid-heaven at the moment of its transit or *culmination*. The technical expression for this in the Egyptian Calendar now before us is ... *(r ᶜḳꜣ [ib])*, literally "in the middle." ....
>
> This explanation of the expression "the middle" is the key to the whole Calendar. As

the earth turns upon its axis in very nearly four minutes less than twenty-four hours, a star which today culminates at six o'clock will in fifteen days culminate very nearly at five, or it loses about an hour in position every fifteen days. Accordingly in our Calendar the *head of Sahu*, for instance, which culminates at the eleventh hour in the first column of the month Thoth [i.e., I Akhet], does so at the tenth hour in the second column of the same month, and the entries in each successive column imply the loss of an hour.

The entries, however, do not by any means always place a star in the same position which it held in the previous column. The *head of Sahu*, which was in the middle in the second column of Thoth, is on the right eye on the 1 Paopi [i.e., II Akhet 1] at the ninth hour, on the left eye on the 16 Paopi [i.e., II Akhet 16] at the eighth hour, and again in the middle both on the first and the sixteenth of Athyr [i.e., III Akhet 1 and 16], at the seventh and sixth hours respectively .... The conclusion which I draw from these facts is that "right eye," "left shoulder," and the like, signify certain relative *short* distances from the meridian; "left eye" being nearer to the meridian than "left ear," and this again less distant from the meridian than "left shoulder." Even this extreme distance from the meridian must have been *short*, for a star which is said to culminate at the twelfth hour on the first night of the month, and two hours later on the thirty-first night, cannot possibly be many degrees distant from the meridian at the eleventh hour of the sixteenth night. This is true, even upon the supposition that the hours of the Calendar may vary in length according to the season. It must, moreover, be remembered that in the climate [i.e., latitude] of Thebes the difference between the lengths of the days and nights is not so great as in the northern climates [i.e., latitudes], and that the difference between the twelfth parts of the longest and the shortest night in that latitude does not amount to many minutes.

The whole Calendar then, in my opinion, records nothing but real or approximate transits of stars. Once in the course of every fifteen nights, the observer appears to have noted down, at each successive hour, the name of the principal star which was either actually upon the meridian or close to it. We do not know how he determined his meridian, what instrument he used, or by what contrivance he limited the field of his observation. But he seems to have noted the passage of stars over seven different vertical lines. If a star were crossing the first line, beginning from the east, it was noted down as being on the left shoulder; if it were on the fourth line which represented the meridian, it was put down as in the middle; if on the fifth, it was "on the right eye" and so on.

This general view of the document is open to no serious objection that I am aware of. There are, however, difficulties to be encountered as soon as we endeavor to understand all the details. Part of these difficulties arise from the state of the text. We are not in possession of the original, or even of a copy intended for general perusal. Our copies were made inside tombs, and were never intended to be seen by mortal eye after the tomb was once closed. The Egyptian texts, which were made under these conditions, are always grossly inaccurate. The inaccuracy often arises from the ignorance or carelessness of the artist; but it is as often occasioned by the text being made subordinate to decorative effect. The two texts we possess [! now four] betray the most shocking confusion between the Egyptian signs for "right" and "left." .... The graphic indication of the position of stars is absolutely worthless in the tomb of Rameses (!) IX; in the tomb of Rameses VI some portions of this part of the work are carefully done, others most negligently. Some of the entries are manifestly made at the wrong hour ....

For a vast number of errors like these [listed in the lacuna and hence not given here] the original Calendar is not to be held responsible. But even this document no doubt may have contained very serious errors. It suggests many questions, which we have unfortunately no means of answering. Is it the work of one man or of several? Are all the entries made from direct observation, or have some of them been deduced from observations already made? Were all the observations corresponding to the entries made on the same night? How was time for each observation determined? ....

This is, however, the proper place to mention another interesting document .... There is in the British Museum a calcareous stone, No. 5635, upon which a note in hieratic character gives the names of certain persons who observed the transit of the *Star of the Waters* from the fifth Phamenoth [lit., III Peret 5] till the seventh Payni (!, should be Epiphi ?, i.e., III Shemu 7) of some year of an unknown king. There are thirteen entries altogether, and all are in the following form: .... "on the 13th Phamenoth – by the observation of Ken [i.e., *ḳn*] – the *Star of the Waters* in the middle." The observations recorded were made on the 5th, 6th, and 13th Phamenoth, the 7th, 9th, and 13th Pharmuthi, the 16th and 23rd Pachons, the 5th, 16th, and 21st Payni, and the 4th and 7th Epiphi. The names of the observers are Nebnefer, Pennub, Ken, Penamen, Nechtu, Het, Mes, Nebsemennu, Panebtmâ. No indication is given of the hours at which the observations were made, or of the name of the reigning king.

[Then follow the names of the constellations in which the transiting stars lie, here omitted since I give them below.]

Some of these constellations must have been of enormous extent [see Figs. III.84a and III.84b]. The entries for the whole of the first night of Epiphi are confined to stars belonging to the Hippopotamus and Necht [the Giant]. The head and rump of the Goose culminated at an hour's distance from each other, and Sahu [Orion] and the Lion have also two entries each in the nights when they are mentioned.

The Egyptian constellations of the northern sky, the *Thigh* (Great Bear) and the *Leg* (corresponding, I believe, to Cassiopeïa) do not appear at all in this Calendar, which probably contains only stars more closely approaching the equator. Two of them are known to us independently of this Calendar: Sahu is Orion and Sothis is Sirius.

From the acknowledged identity of Sothis and Sirius I endeavored, some years back, first of all to ascertain the date at which the Calendar was drawn up, and, secondly, to identify a certain number of the asterisms which it contains.

The method which I adopted was this: "Whatever may have been the length of the Egyptian hours of the night, the sixth undoubtedly corresponds to midnight." Now Sothis, that is Sirius, is said by the Calendar to be 'in the middle' at the sixth hour in the first column of the month Choiak, the fourth Egyptian month [i.e., IV Akhet]. The question, therefore, arises —in what year did Sirius culminate at midnight at Thebes within the first fifteen days of the Egyptian month Choiak? Through the very great kindness of the Astronomer Royal and of his First Assistant, Mr. Stone, to whom I am also indebted for a table of the approximate Right Ascensions of certain stars which I had specified, I am able to say this transit took place about year 1450 before Christ. "This inference of date," as the Astronomer Royal remarks, "is necessarily a very vague one .... but from the whole nature of the case a vague date is all that can be asked for. It is sufficient for us to know that the Calendar records observations in the fifteenth century before Christ, or thereabouts. It does not at all follow that the tomb of Rameses VI is of the same antiquity. The very same Calendar was found in the more recent tomb of Rameses IX, and it may have been inscribed on much earlier tombs. The decorators of those magnificent chambers did not think it necessary to alter the document in consequence of the changing positions of the heavenly bodies....

The approximate date of the Calendar being known, the next question is, what remarkable stars at that date culminated at the intervals before and after Sirius, which are assigned by the Calendar to its asterisms? And finding, for instance, that in 1450 B.C. the approximate Right Ascension of α Arietis was $23^h\ 5^m$, whilst that of Sirius was $4^h\ 11^m$ (the difference therefore being $5^h\ 6^m$), I have no hesitation in identifying α Arietis with the *Goose's head.* In the same manner I identify *Arit* as probably β Andromedae, the *Chu* (a group of stars [i.e., *ḫ'w*]) with the Pleïads, *Sârit* with α Tauri (Aldebaran), the *Lion* with part of our own constellation of the same name, the *Many Stars* with part of the Coma Berenices, the *Lute Bearer* with α Virginis: α

> Boötis and α Scorpionis are probably part of the Constellation *Menat.* Castor and Pollux, which at the present day come to the meridian about three-quarters of an hour after Sirius, seem at first sight to claim identity with the *Two Stars*, but their position in the sky with reference to Sirius was quite different at the time of our Calendar to what it is at present.
>
> [Then, after a few more remarks, follows Renouf's English translation of the tables (see Fig. III.99a), with a few additional comments.]

I have given this long extract from Le Page Renouf's account for two reasons. First, because it was the fundamental study of the Ramesside star clock that laid the basis for what is the now generally accepted opinion as to the nature of the observations that make up the contents of the clock's tables. Second, because it was so clearly presented. Renouf optimistically reported in his article (p. 401, n. 1) that his "explanation which I [i.e., Renouf] first published in the *Chronicle*, January 25, 1868, was promptly recognized by M. [François Joseph] Chabas as the true one" and that: "It has also been adopted by Dr. [Heinrich] Brugsch in his Dictionary, and not been controverted by any Egyptologist."

But in fact even before Renouf presented his fuller account in 1874, Friedrich Gensler published his *Die thebanischen Tafeln stündlicher Sternaufgänge* (Leipzig, 1872), which held that the stars in the tables indicated the hours by their risings, a view followed by Brugsch in 1883 in his *Thesaurus inscriptionum aegyptiacarum,* Abt. I, pp. 185-94 (despite his apparent earlier acceptance of Renouf's hypothesis) and by Gustav Bilfinger in his detailed and useful *Die Sterntafeln in den ägyptischen Königsgräbern von Bibân el Molûk* (Stuttgart, 1891), Section II, particularly pp. 7-19, which attempted to dispose of Renouf's hypothesis (see the reference to this work in note 7 below). But within the next decade Ludwig Borchardt (1899)[6] and Schack-Schackenburg[7] (who published his account in 1902 but finished it earlier in 1894) accepted the transit explanation of the document. Borchardt in his account proposed that a sighting instrument with its right-angled rule supporting a plumb line like the ones in the Berlin Museum (see Figs. III.20a and III.20b) was used to record the transits (see note 6). And Schack-Schackenburg hypothesized an instrument consisting of a frame with 7 equally separated vertical strings, each one used to observe transits of stars at or near the meridian in one or another of the positions assigned to a part of the body of the seated target figure (see Fig. III.99c and note 7). Although his analysis of the tables and their charts in terms of the instrument was ingenious, it was not convincing (again see note 7), and no such instrument as he proposes has been found. Borchardt in 1920 essentially repeated his earlier account of the star clock.[8]

The next and surely the most important of the succeeding investigations of the Ramesside star clock was that of Neugebauer and Parker, which occupied the whole second volume of their *Egyptian Astronomical Texts*. Not only did they establish a sounder text based on four sources rather than the two which the earlier authors had used but they brought up-to-date what information was available regarding the constellations and stars cited in the tables, and their graphic analysis of the star transits specified in the tables in relationship to the varying hour lengths from table to table was original if admittedly limited by (1) the generally poor state of the copies of the text, based as they were on copies that were made for decorative, funerary purposes and hence not easily correctable as was the case for papyri, (2) the crudeness of the sighting techniques, (3) the lack of adequate angular measures, and (4) our general ignorance of the identities of the stars involved in the document. For general and specific comments on all of these factors, see Chapter III above, the text over note 78, and the pages of their volume cited in note 7 to this Introduction.

Before listing the four copies of the clock, I should give here a few comments beyond those found in Chapter Three and in the first paragraphs of this Introduction in order to help the reader make his way through the document. First look at the Egyptian positional expressions used in the tables describing the bodily locations on the *target figure* through which the transit lines pass (see Figs. III.99b and III.99c):

ḥr ḳꜥḥ iꜣby ("on the right shoulder") = line -3;
ḥr msḏr iꜣby ("on the right ear") = line -2;
ḥr irt iꜣby ("on the right eye") = line -1;
r ꜥḳꜣ ib ("opposite the heart") = line 0; this is the line Renouf simply translates as "in the middle;"
ḥr irt wnmy ("on the left eye") = line +1;
ḥr msḏr wnmy ("on the left ear") = line +2;
ḥr ḳꜥḥ wnmy ("on the left shoulder") = line +3.

As I have said in the opening paragraph above, all but three of the stars used in the Ramesside star tables differ from the decanal stars found in the earlier star clocks based on star risings (i.e., those of coffin lids or, later, those on astronomical ceilings in tombs and temples) and other clocks based on the culminations of decanal stars (such as the clocks evident in the tables of decans found in the sarcophagus chamber of the Cenotaph of Seti I and the Tomb of Ramesses IV). The only Ramesside hour stars equivalent to, or closely related to, the Egyptian decanal stars are Sirius *(sbꜣ n spdt)*, the Star of the Thousands *(sbꜣ n ḫꜣw)* (simply called "Thousands" in the decanal lists), and a Star of Orion *(sbꜣ n sꜣḥ)*. Though one might have thought that the hour star *ꜥryt* was identical with the decanal star *ꜥrt*, it seems that they were not because "in the decans *ꜥrt* follows *ḫꜣw* and in the hour stars [of the Ramesside star clock] *ꜥryt* precedes, with the *ỉpd*-stars in between."[9]

As I have remarked in Chapter Three, the hour stars in the Ramesside clock are mostly prominent ones that are parts of popular constellations: the "Giant" (from which the remarkable number of 16 stars are used), the "Bird" (Petrie, following Renouf, calls it the "Goose") (from which four stars are used), "Orion" (two stars are related to it), "Sothis" (2 stars are mentioned in the tables, one being Sirius itself and the other the star called its "Predecessor"), the "Two Stars" (one being its "Predecessor" and the other the "Two Stars" itself), the "Stars of the Water," the "Lion" (one star is his "Head" and one his "Tail"), the "Many Stars" (presumably a cluster), the "Mooring Post" (6 of its stars are used), and the "Hippopotamus" (with 8 of its stars used). The hour stars themselves, with their names (in translation where possible), are of course given in the tables of the document below. Some of the guesses regarding their identity are given in the notes to the document. Despite the efforts of Renouf, Borchardt, and others, and except for the identification of the Star of Sothis with Sirius itself and the Star of Orion with some unknown star of the very familiar constellation of Orion, we cannot be sure of their identification with modern counterparts.[10] There are problems of location and terminology with identifying the constellations of the Lion, the Mooring Post, and the Hippopotamus, as given in the Ramesside star clock, with the similarly named constellations in the so-called Northern Constellations found in astronomical ceilings (see Document III.14, note 25). Again I remind the reader of Petrie's imaginative (but, so far as size is concerned, not realistically based) reconstruction of the celestial map, which depended in great part on the star lists in the Ramesside star clock (see Figs. III.84a and III.84b).

Now let us have a brief look at the four sources on which the tables of the Ramesside star clock given by Neugebauer and Parker in phonetic transliteration as well as my version of Document III.14 are based.[11]

A = Copy in the Tomb of Ramesses VI, Valley of the Kings: Hall E. This is one of the two copies of the Ramesside star clock found in the tomb of Ramesses VI (ca. 1151-43 B.C.). The other is copy **B** described below. The early depictions of the text of the star clock made by Champollion and Lepsius (see note 2 and see Figs. III.19a, III.19b, and III.19c) were made from copies **A** and **D** (the latter is also described below). Only small bits of Table 23 and no entries from Tables 13 and 24 are evident in **A**. In the preparation of Document III.14, I found indispensable the hand version of the hieroglyphic text of copy **A** and indeed of all of the copies of it given by Neugebauer and Parker in Plates 29-67 of their second volume (see note 11 below). Also of help were the photographs of all copies that are included as Plates 1-28. A short useful bibliography concerning copy A is given by the editors.

B = the second copy of the text found in Ramesses VI's Tomb, on the lower registers of the ceilings of corridors A and B. This copy was not used in the early publications of the text. Indeed most of the text in **B** is gone. Such fragments that do remain are recorded and treated by Neugebauer and Parker in the forms I just mentioned in describing copy **A**.

C = Copy in the Tomb of Ramesses VII, Valley of the Kings, north and south shoulders of the ceiling of Hall B. The confused and disorganized state of the tables in this copy is noted by Neugebauer and Parker.[12] The text used by Neugebauer and Parker consisted of "Photographs by Charles F. Nims, collated with the original in March, 1951." See their Voume 2, p. 3, and Plates 6-12, 14-28, and, of course, examine the references to copy **C** in their summary version of the hieroglyphic texts from all copies in Plates 29-67, passim.

D = Copy in the Tomb of Ramesses IX, Valley of the Kings, lower register to north and south of corridor B. This constituted the main source for Champollion's text of the tables in the *Notices* (with, of course, variant readings from copy **A**, see note 2). Again I remark on the indispensable plates from the second volume of Neugebauer and Parker.[13] The bibliography of works citing or studying copy **D** is essentially the same as that for copy **A**, and, as I have pointed out, the two copies were known from the very beginning of the publication and investigation of the Ramesside star clock.

Little need be said about my version of Document III.14 except that it makes use of the various preceding versions and above all the phonetic version of the tables given by Neugebauer and Parker and the various plates of photographs and summary hand copies that are included in their volume. Obviously when neither the meaning of the names of the stars nor their identity can be determined, I have been forced to settle for pronounceable representations of the hieroglyphs. Except for occasional remarks and comments in the notes to the document, I have left NP's full commentary on the shifting hour stars to the reader's perusal.

Notes to the Introduction of Document III.14

1. Neugebauer and Parker, *Ancient Egyptian Astronomical Texts*, Vol. 2, p. 9: "In Table 12, which concerns II *prt* 16, Sirius culminates at the beginning of the night. In the terminology of the ceiling of the cenotaph of Seti I [see Document III.13] the star has reached the 'end of work.' According to this scheme the date of 'birth' falls 160 days later, i.e., III *šmw* 26. Since 'birth' corresponds generally to helical rising we can say that III *šmw* 26 must be near the helical rising of Sirius. In the Seti cenotaph the 'birth' date of Sirius is IIII *prt* 16 or 100 days earlier. Consequently our new texts are about 400 years later than those of the cenotaph which leads to a year in the neighborhood of 1470 B.C., i.e., more than three centuries earlier than the preserved monuments. As a round date we shall use ... 1500 B.C." Consult also the independent confirmations on pp. 39 and 73 (which are tied in with NP's analysis of the seasonal nighttime hour lengths and the shift of the stars marking those hours from table to table and which use 1501-1500 B.C. as the year for which the tables were constructed). I further remind the reader of the approximate date of 1450 reported from the Astronomer Royal by Renouf (*op. cit.* in n. 4 below, p. 406) in the passage given in extenso later in this Introduction to the document. Also note that in Document III.10, Section 3, I quoted a passage giving a rising of Sirius for III Shemu 28 in some unknown year of the reign of Tuthmosis III. This is close to Parker's determination here of III Shemu 26. Indeed one can reasonably say that the independent report of

Sirius' rising in Tuthmosis III's reign is a confirmation of Renouf's now accepted hypothesis that the Ramesside star clock is based on meridian and close-meridian transits of stars rather than on risings.

2. J.F. Champollion, *Monuments de l'Égypte et de la Nubie*, Vol. 3 (Paris, 1845), Plates cclxxii (bis-quint). This text of the Ramesside star clock was published with no indication of the fact that it was prepared from sources in two tombs (those of Ramesses VI and Ramesses IX, and Ramesses VI is designated "Rhamses V"). But the text published in the separately printed work of Champollion, *Monuments de l'Égypte et de la Nubie. Notices descriptives conformes aux manuscripts autographes rédigés sur les lieux*, Vol. 2 (Paris, 1889), pp. 547-67 (Vol. 2 was written out in the hand of G. Maspero), has variant readings from the tomb of Ramesses VI clearly marked, with the main text taken from the tomb of Ramesses IX. In both texts the charts with the stars marked on the transit lines are missing, and hence my decision to include as Figures III.19a and III.19b the plates executed by Lepsius. The bold large hand of Maspero (see Fig. III.19c) makes the hieroglyphs of the second copy a pleasure to read, though the text as a whole must cede the palm to the relatively recent text of Neugebauer and Parker based on four copies, which we mention in the paragraphs of this Introduction devoted to the sources for Document III.14.

3. J.F. Champollion, *Lettres écrites d'Égypte et de Nubie en 1828 et 1829*, new ed. (Paris, 1868), pp. 197-98, describing these tables, says: "Ce sont des *tables des constellations et de leurs influences pour toutes les heures de chaque mois de l'année;* elles sont ainsi conçues:

"Mois de Tôbi, la dernière moitié. — *Orion* domine et influe sur l'oreille gauche.

"Heure 1re, la constellation d'*Orion* (influe) sur le bras gauche.

"Heure 2e, la constellation de *Sirius* (influe) sur le coeur .... [and so on through the 12 hours of the table and presumably the beginning of the night and the twelve hours of the remaining 23 tables].

"Nous avons donc ici une *table des influences*, analogue à celle qu'on avait gravée sur le fameux *cercle doré* du monument d'Osimandyas, et qui donnait, comme le dit Diodore de Sicile, les heures du lever des constellations *avec les influences de chacune d'elles.* Cela démontrera sans réplique, ..., que l'*astrologie* remonte, en Égypte, jusqu'aux temps les plus reculés; ....."

4. The key to Champollion's error is his repeated introduction of the word *"(influe)"* or its plural *"(influent)"* when there is no textual justification for such words. While the first scholar to point out his error was R. Lepsius, *Die Chronologie der Ägypter*, Part I (Berlin, 1849), p. 111, "Von Einflüssen, sieht man leicht, kann hier nicht die Rede sein," Lepsius' own proposal that the tables may be related to the seven "Climata oder Breitenzonen der Alten" is equally erroneous. It was Peter Le Page Renouf, "Calendar of Astronomical Observations Found in Royal Tombs of the XXth Dynasty," *Transactions of the Society of Biblical Archaeology*, Vol. 3 (1874), pp. 400-01 (full article, pp. 400-21), who first refuted Champollion's astrological solution, particularly in the extended form argued by Biot: "The most elaborate comment on this Calendar is to be found in a dissertation of the late eminent French astronomer, M. Biot, in the twenty-fourth volume of the *Mémoires de l'Académie des Sciences.* A French translation of the Calendar, by M. Emmanuel de Rougé, is appended to M. Biot's dissertation. The fundamental hypothesis of this dissertation is that the Calendar is a record, for astrological purposes, of the *risings* of stars and constellations. This hypothesis is entirely without foundation in the Egyptian text, which contains no allusion whatever either to astrology or to risings of stars. M. Biot's mistake was suggested by the old version of Champollion, 'la constellation d'Orion *(influe)* sur le bras gauche.' Had not the unfortunate word *influe* been introduced by the translator, M. Biot could hardly have missed the sense of 'Orion sur le bras gauche' or 'au milieu.' The document simply records the star's *position in the sky.*" Renouf then goes on to advance the almost certainly correct solution that the parts of the body listed in the tables stood for the transit points on lines of culminations and near culmination of the stars, the lines being those vertical lines of the accompanying grids which intersected the bodily parts of the target figure imposed on the grids next to the hour stars specified in the tables. As the reader will see in the long passage I have quoted from Renouf's account in my text over the next note, the positions of those stars were represented by star-symbols (asterisks) falling on one or another of the vertical lines of each chart or grid. One should note, however, that Renouf did not designate the Ramesside collection of tables as a "star clock" but simply as a "calendar of astronomical observations," i.e., a calendar of star transits hour by hour for each of the 24 semimonthly tables for a given year.

5. Renouf, *ibid.*, pp. 401-21.

6. L. Borchardt, "Ein altägyptisches astronomisches Instrument," *ZÄS*, Vol. 37 (1899), pp. 14-17 (but whole article, pp. 10-17). He explains how the sighting instrument is used in observing the meridian transit or transits before and after the meridian. As an example of the tables of the star clock he gives the hieroglyphic text for the first half of the second month of Akhet, and then describes the use of the exemplar table by two observers facing each other with their "horoscopes" consisting of a sighting stick and a shadow-clock type of a right-angled rule having a horizontal cross bar which supports a plumb line, the instruments of each observer being aligned on the meridian (represented by one observer looking north and the other, the target figure, looking south):

"(p.16) Die Benutzung solcher Tabellen hat man sich etwa so zu denken: Zwei Horoskopen — wir bleiben immer noch bei dem Beispeil vom 1. Paophi — wachen auf dem Dache des Tempels in den oben beschriebenen Positionen, d.h. beide im Meridian des Ortes sitzend, der nördliche nach Süden und der südliche nach Norden blickend. Nun wartet der nördliche den Moment ab, wann der 'Nacken des Riesen' genannte Stern culminirt, d.h. wann er ihn mit seinem *Merḫet* auf den Scheitel seines Gegenübers ablothen kann. In dem Momente verkündet er den Eintritt der Nacht. Ebenso meldet er nach einer geraumen Zeit bei Beobachtung der Culmination des Sternes *Bgs* des Riesen den Ablauf der 1. Stunde u.s.f. Bei Ablauf der 5. Stunde aber zeigt sich ihm eine kleine Schwierigkeit. Es culminirt nämlich zu dieser Zeit kein Stern von irgend nenneswerther Bedeutung, wohl aber ist der 'Kof des Vogels' gerade über die Culmination hinaus. Er beobacht also den Moment, in dem dieser Stern sich über dem linken Auge seines Collegen ablothen lässt, und bestimmt so den Ablauf des 5. Stunde. Die weiteren Variationen sind von selbst klar."

He notes finally that the use of the *Merkhet* with the sighting staff shown in Figs. III.20a and III.20b for the determining of the transits establishing the night hours, which the inscription on the sighting staff alludes to, explains to us not only why the sign for the rule with its dangling plumb line (see Fig. III.54) was employed in Ptolemaic [and Roman] times as the determinative for "hour" but also why the Greek term for such an instrument was ὡρολόγιον. The whole discussion is terminated by a patronizing and distasteful remark arising from the inaccuracy of the hour determinations of the ancient Egyptians: "Dass diese nicht allzu genau gehen würde, konnte man bei dem Charakter dieses Volkes, dessen Nachkommen heute noch nicht den Werth der Zeit kennen, schon von vorn herein erwarten."

That Borchardt knew of Schack-Schackenburg's views before publishing his own article of 1899 is clear from the fact that on p. 15, n. 2, of that article he cites the latter's *Studien* along with Renouf's article, and the further fact that Schack-Schackenburg adds at the end of his monograh (*op. cit.* in note 7 below, p. 128): "Beendet am 14 März 1894."

7. H. Schack-Schackenburg, *Ägyptologische Studien*, Vol. 1 (Leipzig, 1902), 2. Heft, pp. 61-64:

"Hierbei drängt sich uns die Erwägung auf, dass sich der Aufgang ohne, der Meridiandurchgang aber nur mit Hülfe eines Instruments beobachten lässt. Es will mir aber scheinen, dass die jeder Sterntafel beigegebene Abbildung ein Instrument darstellt, welches nicht nur diesen Zweck erfüllte, sondern auch in jeder Nacht gestattete die Stunden und Viertelstunden trotz ihrer wechselnden Länge direkt am Himmel abzulesen. Die Regeln der aegyptischen Perspektive gestatten nämlich sehr wohl die Abbildungen so perspektivisch wiederzugeben, wie aufnebenstehender Abbildung [= Fig. III.99c] geschehen ist: *ab* stellt einen in gleichen Abständen mit 7 parallelen Fäden bespannten Rahman dar, der senkrecht auf dem Brett *bc* steht. Dieses trägt ferner die Figur *de*, die senkrecht auf *bc* stehend dem Rahmen nach Belieben genähert werden kann. Beim Gebrauche musste der Apparat so aufgestellt werden, dass *bc* der Ebene des Aequators parallel war. Auf Seite 63 [= Fig III.99c] ist derselbe in der für Theben erforderlichen Stellung abgebildet. — Die wagerechten Linien in den überlieferten Abbildungen scheinen nur Fortsetzungen der Trennungslinien zwischen den einzelnen Stunden zu sein.

"Es ist nun zu beweisen: 1, dass mit dem geschilderten Apparate und mit Tafeln, die die Stellungen dazu geeigneter Sterne angaben, jederzeit die Sternnachtzeit am Himmel abgelesen werden konnte .... 2, dass die überlieferten Stundentafeln wirklich solche Angaben enthielten ..."

In his effort to prove these two points, the author, in the remainder of the monograph, presents a very clever argument involving many assumptions and numerous algebraic equations, that in my opinion do not tell us much of the knowledge and intent of the ancient Egyptian astronomers, though it does give us some sense of how such an apparatus might be used with the tables. Of course, as I have said, there is no archaeological evidence for such an apparatus. I prefer the graphic techniques of Neugebauer and Parker, *Egyptian Astronomical Texts*, Vol. 2, pp. 9-18 and 70-74, and the reader is urged to consult them and the hour length graphs at the bottom of each table (pp. 20-69). But in view of the insurmountable difficulties of star identifications and the inaccuracies of the determination of the hour lengths of the continually varying seasonal hours from table to table by the water clock probably used to calibrate them, neither analysis can give us complete confidence that we can reconstruct with surety the overall connection between the tables and the precise lengths of the hours they purport to measure. Also of interest is the analysis of seasonal hour lengths found in the pages from Bilfinger's monograph on the tables cited toward the beginning of this paragraph of my Introduction.

Returning to the monograph of Schack-Schackenburg, I cite further only his final results (p. 128):

"Die Ergebnisse dieser Studie können in den nachstehenden Sätzen ausgedrückt werden:

"1, Die Stundentafeln enthalten *nicht* Sternaufgänge, sondern Kulminationsbeobachtungen, (oder auch Beobachtungen des Durchganges durch einen dem Meridian nahe liegenden grössten Kreis am Himmel).

"2, Sternetzabscissen und die somatischem Relationen geben gleichmässig den (von der Beobachtungslinie aus gerechneten) Stundenwinkel des Sterns in der angeführten Stunde an.

"3, Hierbei wird die Länge der jeweiligen Viertel-sternnachtstunde als Masseinheit benutzt.

"4, Die den Stundentafeln beigefügten Abbildungen stellen den Apparat dar, an dem die Beobachtungen vorgenommen wurden."

There can be no objection to the basic idea of the first of these conclusions, i.e., that the tables contain not star risings but rather transits of the meridian culminations or near culmination of stars. But that the transit lines represent great circles of the sky (which makes clear the conclusion to the modern reader) would be entirely foreign to the ancient Egyptians. There is no objection to the second conclusion that both the star charts and their specified body locations equally state the measures of the star angles at the cited hour, measures that are calculated from hourly lines of observation. Or at least there is no objection so long as we realize that the measure of the angle is exclusively a time rather than angular degrees, since the latter were not employed by the Egyptians. That the unit of measurement was a quarter of an hour, as the third conclusion holds, is simply not proved. All we can know from the text is that the transit lines before and after the meridian are probably a short distance from each other. Nor is the last conclusion proved, namely, that the observations were made on the stringed siting instrument devised and described by the author at the beginning of the monograph.

8. L. Borchardt, *Die altägyptische Zeitmessung* (Berlin and Leipzig, 1920), pp. 55-58.

9. Neugebauer and Parker, *Egyptian Astronomical Texts*, Vol. 2, p. 5, n. 1.

10. I have mentioned Renouf's opinions regarding the identification of some of the stars in the notes to the document below. I tend to give somewhat more credence to Renouf's opinions than to those reported by Borchardt, *op. cit.* in note 6, p. 16, perhaps because the former had the help of the Astronomer Royal and his assistant.

11. *Ibid.*, pp. 1-3 briefly describe the four copies; pp. 22-69 contains the tabular lists (by hours) of the stars and their locations; Plates 1-28 are photographs of the remains of the clock in the four copies; and finally Plates 29-67 record, table by table, the hieroglyphic texts, copy by copy. The notes of NP are very informative about the state of the text of each table in each copy.

12. *Ibid.*, p. 2.

13. See the Plates mentioned in note 10 above, passim.

Document III.14

The Ramesside Star Clock

Table 1: [First half of Month] I of [Season] Akhet[1]

Beginning of the Night: The Two Feathers of the Giant *(šwty n nḫt).* [On the Left Ear] (i.e., on line "+2") —— |||||*|.

[End of] Hour 1: Head of <the Mace> of the Giant *(tp n <h̥ḏ> nt nḫt).* On the Right Ear (i.e., on line "-2") —— |*|||||.

[End of] Hour 2: His Neck *(nh̥bt.f).* Opposite the Heart (i.e., on line "0") —— |||*|||.

[End of] Hour 3: His Hip (?) *(bgs.f).* Opposite the Heart (i.e., on line "0") —— |||*|||.

[End of] Hour 4: His Shank *(sḏh̥.f).* On the Left Shoulder (i.e., on line "+3") —— ||||||*.

[End of] Hour 5: His Pedestal *(pt.f).* On the Left Eye (i.e., on line "+1") —— ||||*||.

[End of] Hour 6: Aryt *(ʿryt).* Opposite the Heart (i.e., on line "0") —— |||*|||.[2]

[End of] Hour 7: Head of <the Bird[3]> *(tp n <ꜣpd>).* Opposite the Heart (i.e., on line "0") —— |||*|||.

[End of] Hour 8: Its Rump *(kft.f).* On the Right Eye (i.e., on line "-1") —— ||*||||.

[End of] Hour 9: Star of the Thousands *(sbꜣ n ḫꜣw).*[4] Opposite the Heart (i.e., on line "0") —— |||*|||.

[End of] Hour 10: Star of Sar *(sbꜣ n sʿr).*[5] Opposite the Heart (i.e., on line "0") —— |||*|||.

[End of] Hour 11: Predecessor [of the Star] of Orion *(tpy-c s3ḥ).* Opposite the Heart (i.e., on line "0") —— |||✳|||.

[End of] Hour 12: The Star of Orion *(sb3 n s3ḥ).*[6] On the Right Shoulder (i.e., on line "-3") —— ✳||||||.

Table 2: Day 16 of the [Second] Half-Month of Month I of Season Akhet[7]

Beginning of the Night: Head of the Giant *(tp nḫt).* [On the Left Eye] (i.e., on line "+1") —— ||||✳||.[8]

[End of] Hour 1: His Neck *(nḥbt.f).* On the Left Eye (i.e., on line "+1") —— ||||✳||.

[End of] Hour 2: His Hip (?) (*Renouf*: His Back) *(bgs.f).* Opposite the Heart (i.e., on line "0") —— |||✳|||.

[End of] Hour 3: His Shank (*Renouf:* His Knee) *(sḏḥ.f).* On the Left Shoulder (i.e., on line "+3") —— ||||||✳.

[End of] Hour 4: His Pedestal *(pt.f).* Opposite the Heart (i.e., on line "0") —— |||✳|||.

[End of] Hour 5: Aryt *(cryt).* On the Left Eye (i.e., on line "+1") —— ||||✳||.

[End of] Hour 6: Head of the Bird *(tp 3pd).* Opposite the Heart (i.e., on line "0") —— |||✳|||.

[End of] Hour 7: Its Rump *(kft.f).* Opposite the Heart (i.e., on line "0") —— |||✳|||.

[End of] Hour 8: Star of the Thousands *(sb3 n ḫ3w).* Opposite the Heart (i.e., on line "0") —— |||✳|||.

[End of] Hour 9: Star of Sar *(sb3 n scr).* On the Left Eye (i.e., on line "+1") —— ||||✳||.

[End of] Hour 10: Predecessor of Orion *(tpy-c s3ḥ).* Opposite the Heart (i.e., on line "0") —— |||✳|||.

[End of] Hour 11: Star of Orion *(sb3 n s3ḥ).* Opposite

the Heart (i.e., on line "0") -- | | | * | | |.
[End of] Hour 12: Star of Sothis (Sirius) *(sbꜣ n spdt).* On the Right Shoulder (i.e., on line "-3") -- * | | | | | |.

Table 3: II Akhet 19

Beginning of the Night: Neck of the Giant *(nḫbt nḫt).* Opposite the Heart (i.e., on line "0") -- | | | * | | |.
[End of] Hour 1: His Hip (?) *(bgs.f).* On the Left Eye (i.e., line "+1") -- | | | | * | |.
[End of] Hour 2: His Shank *(sḏḫ.f).* Opposite the Heart (i.e., on line "0") -- | | | * | | |.
[End of] Hour 3: His Pedestal *(pt.f).* Opposite the Heart (i.e., on line "0") -- | | | * | | |.
[End of] Hour 4: Aryt *(ᶜryt),* On the Left Eye (i.e., on line "+1") -- | | | | * | |.
[End of] Hour 5: Head of the Bird *(tp n ꜣpd).* On the Left Eye (i.e., on line "+1") -- | | | | * | |.
[End of] Hour 6: Its Rump *(kft.f).* Opposite the Heart (i.e., on line "0") -- | | | * | | |.
[End of] Hour 7: Star of the Thousands *(sbꜣ n ḫꜣw).* Opposite the Heart (i.e., on line "0") -- | | | * | | |.
[End of] Hour 8: Star of Sar *(sbꜣ n sᶜr).* On the Right Eye (i.e., on line "-1") -- | | * | | | |.
[End of] Hour 9: Predecessor of Orion *(tpy-ᶜ sꜣḥ).* On the Right Eye (i.e., on line "-1") -- | | * | | | |.
[End of] Hour 10: Star of Orion *(sbꜣ n sꜣḥ).* On the Right Eye (i.e., on line "-1") -- | | * | | | |.
[End of] Hour 11: Star of Sothis (Sirius) *(sbꜣ n spdt).* On the Left Eye (i.e., on line "+1") -- | | | | * | |.
[End of] Hour 12: Predecessor of The Two Stars *(tpy-ᶜ sbꜣwy).* Opposite the Heart (i.e., on line "0") -- | | | * | | |.

Table 4: II Akhet 16[10]

Beginning of the Night: The Hip (?) of the Giant *(bgs nḫt).* Opposite the Heart (i.e., on line "0") —— | | | * | | |.

[End of] Hour 1: Knee of the Giant *(pd nḫt).* Opposite the Heart (i.e., on line "0") —— | | | * | | |.

[End of] Hour 2: His Pedestal *(pt.f).* Opposite the Heart (i.e., on line "0") —— | | | * | | |.

[End of] Hour 3: Aryt *(ᶜryt).* On the Left eye (i.e., on line "+1") —— | | | | * | |.

[End of] Hour 4: Beak of the Bird *(bᶜnt nt ʾpd).* On the Right Eye (i.e., on line "-1") —— | | * | | | |.

[End of] Hour 5: Its Rump *(kft.f).* Opposite the Heart (i.e., on line "0") —— | | | * | | |.

[End of] Hour 6: Star of the Thousands *(sbʾ n ḫʾw).* On the Left Eye (i.e., on line "+1") —— | | | | * | |.

[End of] Hour 7: Star of Sar *(sbʾ n sᶜr).* On the Left Eye (i.e., on line "+1") —— | | | | * | |.

[End of] Hour 8: Predecessor of Orion *(tpy-ᶜ sʾḥ).* <Opposite the Heart> (i.e., on line "0") —— | | | * | | |.

[End of] Hour 9: Star of Orion *(sbʾ n sʾḥ).* On the Left Shoulder (i.e., on line "+3") —— | | | | | | *.

[End of] Hour 10: The One Coming after the Star of Sothis *(ʾiy ḥr-sʾ sbʾ n spdt).* On the Right Shoulder (i.e., on line "-3") —— * | | | | | |.

[End of] Hour 11: Predecessor of The Two Stars *(tpy-ᶜ sbʾwy).* On the Right Shoulder (i.e., on line "-3") —— * | | | | | |.

[End of] Hour 12: The Stars of the Water *(sbʾw nw mw).* Opposite the Heart (i.e., on line "0") —— | | | * | | |.[11]

J=CTable 5: III Akhet 1[12]

Beginning of the Night: Knee of the Giant *(pd nḫt)*. Opposite the Heart (i.e., on line "0") -- | | | ✳ | | |.

[End of] Hour 1: His Pedestal *(pt.f)*. Opposite the Heart (i.e., on line "0") -- | | | ✳ | | |.

[End of] Hour 2: Aryt *(ᶜryt)*. On the Left Eye (i.e., on line "+1") --- | | | | ✳ | |.

[End of] Hour 3: Head of the Bird *(tp n ꜣpd)*. On the Left Eye (i.e., on line "+1") -- | | | | ✳ | |.

[End of] Hour 4: Its rump *(kft.f)*. Opposite the Heart (i.e., on line "0") -- | | | ✳ | | |.

[End of] Hour 5: Star of the Thousands *(sbꜣ n ḫꜣw)*. Opposite the Heart (i.e., on line "0") -- | | | ✳ | | |.

[End of] Hour 6: Star of <Sar> *(<sᶜr>)*. Opposite the Heart (i.e., on line "0") -- | | | ✳ | | |.

[End of] Hour 7: Predecessor of Orion *(tpy-ᶜ sꜣḥ)*. Opposite the Heart (i.e., on line "0") -- | | | ✳ | | |.

[End of] Hour 8: Star of Orion *(sbꜣ n sꜣḥ)*. On the Right Eye (i.e., on line "-1") -- | | ✳ | | | |.

[End of] Hour 9: The One Coming After Sothis *(ꜣy ḥr-sꜣ sbꜣ n spdt)*. On the Left Eye (i.e., on line "+1") -- | | | | ✳ | |.

[End of] Hour 10: Predecessor of The Two Stars *(tpy-ᶜ sbꜣwy)*. Opposite the Heart (i.e., on line "0") -- | | | ✳ | | |.

[End of] Hour 11: The Stars of the Water *(sbꜣw nw mw)*. Opposite the Heart (i.e., on line "0") -- | | | ✳ | | |.

[End of] Hour 12: Head of the Lion *(tp mꜣi)*. Opposite the Heart (i.e., on line "0") -- | | | ✳ | | | .[13]

Table 6: III Akhet 16[14]

Beginning of the Night: The Pedestal <of the Giant>

(pt ⟨nḫt⟩). Opposite the Heart (i.e., on line "0") -- | | | ∗ | | |.

[End of] Hour 1: Aryt *(ᶜryt).* On the Left Eye (i.e., on line "+1") -- | | | | ∗ | |.

[End of] Hour 2: Head of the Bird *(tp n ꜣpd).* Opposite the Heart (i.e., on line "0") -- | | | ∗ | | |.

[End of] Hour 3: Its Rump *(kft.f).* Opposite the Heart (i.e., on line "0") -- | | | ∗ | | |.

[End of] Hour 4: Star of the Thousands *(sbꜣ n ḫꜣw).* Opposite the Heart (i.e., on line "0") -- | | | ∗ | | |.

[End of] Hour 5: Star of Sar *(sbꜣ n sᶜr).* On the Right Eye (i.e., on line "-1") -- | | ∗ | | | |.

[End of] Hour 6: Predecessor of Orion *(tpy-ᶜ sꜣḥ).* Opposite the Heart (i.e., on line "0") -- | | | ∗ | | |.

[End of] Hour 7: Star of Orion *(sbꜣ n sꜣḥ).* On the Left Eye (i.e., on line "+1") -- | | | | ∗ | |.

[End of] Hour 8: The One Coming After Sothis *(ꞽy ḥr-sꜣ spdt).* On the Left Eye (i.e., on line "+1") -- | | | | ∗ | |.

[End of] Hour 9: Predecessor of The Two Stars *(typ-ᶜ sbꜣwy).* Opposite the Heart (i.e., on line "0") -- | | | ∗ | | |.

[End of] Hour 10: The Stars of the Water *(sbꜣ nw mw).* Opposite the Heart (i.e., on line "0") -- | | | ∗ | | |.

[End of] Hour 11: Head of the Lion *(tp n mꜣꞽ).* Opposite the Heart (i.e., on line "0") -- | | | ∗ | | |.

[End of] Hour 12: His Tail *(sd.f).* Opposite the Heart (i.e., on line "0") -- | | | ∗ | | |.[15]

Table 7: IIII Akhet 1[16]

Beginning of the Night: Star of the Back of the Pedestal *(sbꜣ n sꜣ pt) (or perhaps as an alternate, Aryt).*[17] On the Right ⟨Eye⟩ (i.e., on line "-1")

-- | | * | | | |.

[End of] Hour 1: Head of the Bird *(tp n 'pd).* On the Right Eye (i.e., on line "-1") -- | | * | | | |.

[End of] Hour 2: Its rump *(kft.f).* On the Left Eye (i.e., on line "+1") -- | | | | * | |.

[End of] Hour 3: Star of the Thousands *(sb' n ẖ'w).* On the Left Shoulder (i.e., on line "+3") -- | | | | | | *.

[End of] Hour 4: Star of Sar *(sb' n sʿr).* On Left Shoulder (i.e., on line "+3") -- | | | | | | *.

[End of] Hour 5: Star of Orion *(sb' n S'ḥ).* On the Right Eye (i.e., on line "-1") -- | | * | | | |.[18]

[End of] Hour 6: The One Coming After <the Star> of Sothis *(iy ḥr-s' <sb'> n spdt).* Opposite the Heart (i.e., on line "0") -- | | | * | | |.

[End of] Hour 7: Predecessor of The Two Stars *(tpy-ʿ sb'wy).* On the Right Shoulder (i.e., on line "-3") -- * | | | | | |.

[End of] Hour 8: The Two Stars *(sb'wy).* On the Left Eye (i.e., on line "+1") -- | | | | * | |.

[End of] Hour 9: The Stars of the Water *(sb'w nw mw).* On the Left Ear (i.e., on line "+2") -- | | | | | * |.

[End of] Hour 10: Head of the Lion *(tp n m'i).* On the Left Shoulder (i.e., on line "+3") -- | | | | | | *.

[End of] Hour 11: His Tail *(sd.f).* On the Left Shoulder (i.e., on line "+3") -- | | | | | | *.

[End of] Hour 12: The Many Stars *(sb'w ʿš'w).* On the Left Eye (i.e., on line "+1") -- | | | | * | |.[19]

Table 8: IIII Akhet 16[20]

Beginning of the Night: Head of the Bird *(tp n 'pd).* Opposite the Heart (i.e., on line "0") -- | | | * | | |.

[End of] Hour 1: Its Rump *(kft.f).* Opposite the Heart (i.e., on line "0") -- | | | * | | |.

[End of] Hour 2: Star of the Thousands *(sb' n ẖ'w).*

On the Left Ear (i.e., on line "+2") -- | | | | | ✱ |.

[End of] Hour 3: Star of Sar *(sbꜣ n sꜥr).* On the Left Ear (i.e., on line "+2") -- | | | | | ✱ |.

[End of] Hour 4: Star of Orion *(sbꜣ n sꜣḥ).* Opposite the Heart (i.e., on line "0") -- | | | ✱ | | |.

[End of] Hour 5: Star of Sothis *(sbꜣ n spdt).* On the Left Shoulder (i.e, on line "+3") -- | | | | | | ✱.

[End of] Hour 6: The Two Stars *(sbꜣwy).* On the <Right> Shoulder (i.e., on line "-3") -- ✱ | | | | | |.

[End of] Hour 7: The Stars of the Water *(sbꜣw nw mw).* On the Right Eye (i.e., on line "-1") -- | | ✱ | | | |.

[End of] Hour 8: Head of the Lion *(tp n mꜣi).* On the Right Eye (i.e., on line "-1") -- | | ✱ | | | |.

[End of] Hour 9: His Tail *(sd.f).* On the Right Eye (i.e., on line "-1") -- | | ✱ | | | |.

[End of] Hour 10: The Many Stars *(sbꜣw ꜥšꜣw).* Opposite the Heart (i.e., on line "0") -- | | | ✱ | | |.

[End of] Hour 11: Tja Nefer *(ṯꜣ nfr).*[21] Opposite the Heart (i.e., on line "0") -- | | | ✱ | | |.

[End of] Hour 12: Follower of the Front <of> the Mooring Post *(šmsw n ḥꜣt <n> mnit).*[22] On the Left Eye (i.e., on line "+1") -- | | | | ✱ | |.

Table 9: I Peret 1[23]

Beginning of the Night: Its (i.e., the Bird's) Rump *(kft.f).* Opposite the Heart (i.e., on line "0") -- | | | ✱ | | |.

[End of] Hour 1: Star of Sar *(sbꜣ n sꜥr).* On the Right Eye (i.e., on line "-1") -- | | ✱ | | | |.

[End of] Hour 2: Predecessor of the Star of Orion *(tpy-ꜥ sbꜣ n sꜣḥ).* Opposite the Heart (i.e., on line "0") -- | | | ✱ | | |.

[End of] Hour 3: Star of Orion *(sbꜣ n sꜣḥ).* On the

Left Eye (i.e., on line "+1") -- | | | |✱| |.

[End of] Hour 4: The One Coming After <the Star> of Sothis *(ỉy h̥r-sʾ <sbʾ> n spdt).* On the Right Eye (i.e., on line "-1") -- | |✱| | | |.

[End of] Hour 5: The Two Stars *(sbʾwy).* On the Right Ear (i.e., on line "-2") -- |✱| | | | |.

[End of] Hour 6: The Stars of the Water *(sbʾw nw mw).* On the Right Ear (i.e., on line "-2") -- |✱| | | | |.

[End of] Hour 7: Head of the Lion *(tp n mʾỉ).* On the Right Eye (i.e., on line "-1") -- | |✱| | | |.

[End of] Hour 8: His Tail *(sd.f).* Opposite the Heart (i.e., on line "0") -- | | |✱| | |.

[End of] Hour 9: Many Stars *(sbʾw ʿšʾw).* <On the Left Eye> (i.e., on line "+1") -- | | | |✱| |.

[End of] Hour 10: The Follower of the Front of the Mooring Post *(šmsw h̥ʾt n mnỉt).* On the Right <Shoulder> (i.e., on line "-3") --✱| | | | | |.

[End of] Hour 11: The Mooring Post *(mnỉt).* On the Right Eye (i.e., on line "-1") -- | |✱| | | |.

[End of] Hour 12: Follower of the Mooring Post *(šmsw n mnỉt).* On the Right Eye (i.e., on line "-1") -- | |✱| | | |.

Table 10: I Peret 16[24]

Beginning of the Night: Predecessor of Orion *(tpy-ʿ sʾh̥).* On the <Right> Shoulder (i.e., on line "-3") --✱| | | | | |.

[End of] Hour 1: Star of Orion *(sbʾ n sʾh̥).* On the Right Shoulder (i.e., on line "-3") --✱| | | | | |.

[End of] Hour 2: Star of Sothis *(sbʾ n spdt).* Opposite the Heart (i.e., on line "0") -- | | |✱| | |.

[End of] Hour 3: Predecessor of The Two Stars *(tpy-<ʿ> sbʾwy).* Opposite the Heart (i.e., on line "0")

-- | | | ✱ | | |.

[End of] Hour 4: The Two Stars *(sb'wy).* On the Right Ear (i.e., on line "-2") -- | ✱ | | | | |.

[End of] Hour 5: The Stars of the Water *(sb'w nw mw).* Opposite the Heart (i.e., on line "0") -- | | | ✱ | | |.

[End of] Hour 6: Head of the Lion *(tp n m'i).* Opposite the Heart (i.e., on line "0") -- | | | ✱ | | |.

[End of] Hour 7: His Tail *(sd.f).* On the Left Eye (i.e., on line "+1") -- | | | | ✱ | |.

[End of] Hour 8: Many Stars *(sb'w ᶜš'w).* Opposite the heart (i.e., on line "0") -- | | | ✱ | | |.

[End of] Hour 9: Follower of the Front of the Mooring Post *(šmsw ḫ't n mnit).* On the Right Shoulder (i.e., on line "-3") -- ✱ | | | | | |.

[End of] Hour 10: The Mooring Post *(mnit).* On the Right Eye (i.e., on line "-1") -- | | ✱ | | | |.

[End of] Hour 11: Follower of the Mooring Post *(šmsw mnit).* On the Right Shoulder (i.e., on line "-3") -- ✱ | | | | | |.

[End of] Hour 12: Knee of the Hippopotamus *(pd n rr<t>).*[25] On the Right Shoulder (i.e., on line "-3") -- ✱ | | | | | |.

Table 11: II Peret 1[26]

Beginning of the Night: Star of Orion *(sb' n s'ḫ).* On the Right Shoulder (i.e., on line "-3") -- ✱ | | | | | |.

[End of] Hour 1: Star of Sothis *(sb' n spdt).* On the Left Ear (i.e., on line "+2") -- | | | | | ✱ |.

[End of] Hour 2:<Predecessor> of The Two Stars *(⸢<tpy-ᶜ>⸣ sb'wy).* On the Left Ear (i.e., on line "+2") -- | | | | | ✱ |.

[End of] Hour 3: The Stars of the Water *(sb'w nw mw).* On the Right Eye (i.e., on line "-1") -- | | ✱ | | | |.

[End of] Hour 4: Head of the Lion *(tp n mꜣỉ)*. On the Right Ear (i.e., on line "-2") -- |*|||||.

[End of] Hour 5: His Tail *(sd.f)*. On the Right Ear (i.e., on line "-2") -- |*|||||.

[End of] Hour 6: Many Stars *(sbꜣw ʿšꜣw)*. <On the Right Ear> (i.e., on line "-2") -- |*|||||.

[End of] Hour 7: Tja Nefer *(ṯꜣ nfr)*. Opposite the Heart (i.e., on line "0") -- |||*|||.

[End of] Hour 8: Follower <of> the Front of the Mooring Post *(šmsw <n> ḫꜣt mnỉt)*. Opposite the Heart (i.e., on line "0") -- |||*|||.

[End of] Hour 9: The Mooring Post *(mnỉt)*. Opposite the Heart (i.e., on line "0") -- |||*|||.

[End of] Hour 10: Foot of the Hippopotamus *(rd n rrt)*. On the Right Eye (i.e., on line "-1") -- ||*||||.

[End of] Hour 11: Her Knee *(pd.s)*. On the Right Eye (i.e., on line "-1") -- ||*||||.

[End of] Hour 12: <Middle> of her Thighs *(<ḫry-ỉb> mnty.s)*. Opposite the Heart (i.e., on line "0") -- |||*|||.

Table 12: II Peret 16[27]

Beginning of the Night: Star of <Sothis> *(sbꜣ n <spdt>)*. Opposite the Heart (i.e., on line "0") -- |||*|||.[28]

[End of] Hour 1: The One Coming After Sothis *(ỉy <ḫr>-sꜣ spdt)*. On the <Left> Shoulder (i.e., on line "+3") -- ||||||*.

[End of] Hour 2: The Two Stars *(sbꜣwy)*. On the Left Eye (i.e., on line "+1") -- ||||*||.

[End of] Hour 3: Head of the Lion *(tp n mꜣ ỉ)*. On the Right Shoulder (i.e., on line "-3") -- *||||||.

[End of] Hour 4: His Tail *(sd.f)*. On the Right Eye

(i.e., on line "-1") – | | ✱ | | | |.

[End of] Hour 5: Many Stars *(sbꜣw ʿšꜣw).* Opposite the Heart (i.e., on line "0") —— | | | ✱ | | |.

[End of] Hour 6: Tja Nefer *(ṯꜣ nfr).* Opposite the Heart (i.e., on line "0") —— | | | ✱ | | |.

[End of] Hour 7: Follower <of> the Front of the Mooring Post *(šmsw <n> ḥꜣt n mnit).* Opposite the Heart (i.e., on line "0") —— | | | ✱ | | |.

[End of] Hour 8: Mooring Post *(mnit).* On the Right Ear (i.e., on line "-2") —— | ✱ | | | | |.

[End of] Hour 9: Follower of the Mooring Post *(šmsw n mnit).* Opposite the Heart (i.e., on line "0") —— | | | ✱ | | |.

[End of] Hour 10: <Knee> of the Hippopotamus *(<pd> n rrt).* On the Right Ear (i.e., on line "-2") —— | ✱ | | | | |.

[End of] Hour 11: Middle <of> her Thig<hs> *(ḥry-ib n mn<ty>.s).* Opposite the Heart (i.e., on line "0") —— | | | ✱ | | |.

[End of] Hour 12: Her Buttocks *(ḫpd.s).* Opposite the Heart (i.e., on line "0") —— | | | ✱ | | |.

Table 13: III Peret [129]

Beginning of the Night: [Predecess]or of the [Two] Star[s] *([tpy]-ʿ sbꜣ[wy]).* [On the Left Shoulder] (i.e., on line "+3") —— | | | | | | ✱.

[End of] Hour 1: The Star[s of the Water] *(sbꜣ[w nw mw]).* [On the Right Eye] (i.e., on line "-1") —— | | ✱ | | | |.

[End of] Hour 2: [Head of the Lion] *([tp n mꜣi]).* [Opposite the Heart] (i.e., on line "0") —— | | | ✱ | | |.

[End of] Hour 3: [His] T[ail] *(s[d.f]).* [On the Left Eye] (i.e., on line "+1") —— | | | | ✱ | |.

[End of] Hour 4: Many Stars *(sbʾw ᶜšʾw).* [Opposite the Heart] (i.e., on line "0") -- | | | ✱ | | |.

[End of] Hour 5: <Tja Nefer> *(<ṯʾ nfr>).* [Opposite the Heart] (i.e., on line "0") -- | | | ✱ | | |.

[End of] Hour 6: <Follower of the Front of the Mooring Post> *(<šmsw n ḥʾt mnỉt>).* On the [Right] Eye (i.e., on line "-1") -- | | ✱ | | | |.

[End of] Hour 7: Follower of the Mooring Post *(šmsw n mnỉt).* On the Right Eye (i.e., on line "-1") -- | | ✱ | | | |.

[End of] Hour 8: Foot of the Hippopotamus *(rd n rrt).* On the Right Eye (i.e., on line "-1") -- | | ✱ | | | |.

[End of] Hour 9: Her Knee *(pd.s).* Opposite the Heart (i.e., on line "0") -- | | | ✱ | | |.

[End of] Hour 10: Middle of Her Thighs *(ḥry-ỉb n mnty.s).* Opposite the Heart (i.e., on line "0") -- | | | ✱ | | |.

[End of] Hour 11: Her Buttocks *(ḫpd.s).* Opposite the Heart (i.e., on line "0") -- | | | ✱ | | |.

[End of] Hour 12: Her Breast *(mndt.s).* Opposite the Heart (i.e., on line "0") -- | | | ✱ | | |.

Table 14: III Peret 16[30]

Beginning of the Night: The Star<s> of the Water *(sbʾ<w> nw mw).* Opposite the Heart (i.e., on line "0") -- | | | ✱ | | |.

[End of] Hour 1: <Head of> the Lion *(<tp n> mʾỉ).* On the Left Eye (i.e., on line "+1") -- | | | | ✱ | |.

[End of] Hour 2: His Tail *(sd.f).* Opposite the Heart (i.e., on line "0") -- | | | ✱ | | |.

[End of] Hour 3: Many Stars *(sbʾw ᶜšʾw).* Opposite the Heart (i.e., on line "0") -- | | | ✱ | | |.

[End of] Hour 4: Tja Nefer *(ṯʾ nfr).* Opposite the

Heart (i.e., on line "0") -- | | |✱ | | |.

[End of] Hour 5: Follower of the Front of the Mooring Post *(šmsw ḥꜣt n mnỉt).* Opposite the Heart (i.e., on line "0") -- | | |✱ | | |.

[End of] Hour 6: Mooring Post *(mnỉt).* On the Left Shoulder (i.e., on line "+3") -- | | | | | |✱.

[End of] Hour 7: Feet of the Hippopotamus *(rdwy n rrt).* On the Left ⌜Eye⌝ (i.e., line "+1") -- | | | |✱ | |.

[End of] Hour 8: Her Knee *(pd.s).* On the Left Eye (i.e., on line "+1") -- | | | |✱ | |.

[End of] Hour 9: Middle of Her Thighs *(ḥry-ỉb mnty.s).* Opposite the Heart (i.e., on line "0") -- | | |✱ | | |.

[End of] Hour 10: Her Buttocks *(ḫpd.s).* On the Left Shoulder (i.e., on line "+3") -- | | | | | |✱.

[End of] Hour 11: Her <Breasts> *(<mndt>.s).* On the Left Shoulder (i.e., on line "+3") -- | | | | | |✱.

[End of] Hour 12: Her Two Feathers *(šwty.s).* Opposite the Heart (i.e., on line "0") -- | | |✱ | | |.

Table 15: IIII Peret 1[31]

Beginning of the Night: Head of the Lion *(tp n mꜣỉ).* Opposite the Heart (i.e., on line "0") -- | | |✱ | | |.

[End of] Hour 1: His Tail *(sd.f).* On the Right Eye (i.e., on line "-1") -- | |✱ | | | |.

[End of] Hour 2: Many Star<s> *(sbꜣ<w> ꜥšꜣw).* On the Right Eye (i.e., on line "-1") -- | |✱ | | | |.

[End of] Hour 3: Tja Nefer *(ṯꜣ nfr).* On the Left Eye (i.e., on line "+1") -- | | | |✱ | |.

[End of] Hour 4: Follower of the Front of the Mooring Post *(šmsw ḥꜣt mnỉt).* On the Left Eye (i.e., on line "+1") -- | | | |✱ | |.

[End of] Hour 5: Follower Which Comes After The Mooring Post *(šmsw ỉy <ḥr>-sꜣ mnỉt).* On the Right Eye (i.e., on line "-1") -- | |* | | | |.

[End of] Hour 6: Feet of the Hippopotamus *(rdwy n rrt)*. On the Right Eye (i.e., on line "-1") -- | |* | | | |.

[End of] Hour 7: Her Knee *(pd.s)*. Opposite the Heart (i.e., on line "0") -- | | |* | | |.

[End of] Hour 8: Middle of her Thighs *(ḥry-ib mnty.s)*. Opposite the Heart (i.e., on line "0") -- | | |* | | |.

[End of] Hour 9: Her Buttocks *(ḫpd.s)*. Opposite the Heart (i.e., on line "0") -- | | |* | | |.

[End of] Hour 10: Her Breast *(mndt.s)*. On the Left Ear (i.e., on line "+2") -- | | | | |* |.

[End of] Hour 11: Her Tongue *(ns.s)*. On the Left Shoulder (i.e., on line "+3") -- | | | | | |*.

[End of] Hour 12: <Predecessor> of the Two Feathers of the Giant *(<tpy-ʿ> šwty nt nḫt)*. On the Left Ear (i.e., on line "+2") -- | | | | |* |.

Table 16: IIII Peret 16[32]

Beginning of the Night: <Tail of> the Lion *(<sn n> mꜣi)*. On the <Left> Ear (i.e., on line "+2") -- | | | | |* |.

[End of] Hour 1: Many Stars *(sbꜣw ʿšꜣ<w>)*. On the Right Eye (i.e., on line "-1") -- | |* | | | |.

[End of] Hour 2: Tja Nefer *(ṯꜣ nfr)*. On the Left Eye (i.e., on line "+1") -- | | | |* | |.

[End of] Hour 3: Follower of the Front of the Mooring Post *(šmsw ḥꜣt mnit)*. On the Left Eye (i.e., on line "+1") -- | | | |* | |.

[End of] Hour 4: Follower Which Comes <After> the Mooring Post *(šmsw iy <ḥr-sꜣ> mnit)*. On the Right Eye (i.e., on line "-1") -- | |* | | | |.

[End of] Hour 5: Feet of the Hippopotamus *(rdwy n rrt)*. On the Right Eye (i.e., on line "-1") -- | |* | | | |.

[End of] Hour 6: Her Knee *(pd.s).* Opposite the Heart (i.e., on line "0") —— | | | * | | |.

[End of] Hour 7: Her Vulva *(bꜣḫ.s).* Opposite the Heart (i.e., on line "0") —— | | | * | | |.

[End of] Hour 8: Her Buttocks *(ḫpd.s).* On the Left Shoulder (i.e., on line "+3") —— | | | | | | *.

[End of] Hour 9: Her Breast *(mndt.s).* On the Left Ear (i.e., on line "+2") —— | | | | | * |.

[End of] Hour 10: Her Tongue *(ns.s).*[33] On the Left Shoulder (i.e., on line "+3") —— | | | | | | *.

[End of] Hour 11: Predecessor of the Two Feathers of the Giant *(tpy-ʿ šwty nt nḫt).* On the Left Ear (i.e., on line "+2") —— | | | | | * |.

[End of] Hour 12: The Two Feathers of the Giant *(šwty nt nḫt).* Opposite the Heart (on line "0") —— | | | * | | |.

Table 17: I Shemu 1[34]

Beginning of the Night: ⟨Predecessor⟩ of the Mooring Post *(⟨tpy-ʿ⟩ mnỉt).* On the Left Ear (i.e., on line "+2") —— | | | | | * |.

[End of] Hour 1: Follower ⟨of⟩ the Front of the Mooring Post *(šmsw ⟨n⟩ ḥꜣt n mnỉt).* On the Right Eye (i.e., on line "-1") —— | | * | | | | |.

[End of] Hour 2: The Mooring Post *(mnỉt).* On the Right Eye (i.e., on line "-1") —— | | * | | | | |.

[End of] Hour 3: Follower ⟨of⟩ the Mooring Post *(šmsw ⟨n⟩ mnỉt).* On the Right Eye (i.e., on line "-1") —— | | * | | | | |.

[End of] Hour 4: Feet of the Hippopotamus *(rdwy n rrt).* Opposite the Heart (i.e., on line "0") —— | | | * | | |.

[End of] Hour 5: Her Knee *(pd.s).* Opposite the Heart (i.e., on line "0") —— | | | * | | |.

[End of] Hour 6: Her Vulva *(bꜣh̥.s).* On the Left Ear (i.e., on line "+2") -- | | | | | * |.

[End of] Hour 7: Her Buttocks *(ḫpd.s).* On the Left Shoulder (i.e., on line "+3") -- | | | | | | *.

[End of] Hour 8: Her Breast *(mndt.s).* On the Left Ear (i.e., on line "+2") -- | | | | * |.

[End of] Hour 9: Her Two Feathers *(šwty.s).* On the Left Ear (i.e., on line "+2") -- | | | | | * |.

[End of] Hour 10: Predecessor of the Two Feathers of the Giant *(tpy-ꜥ šwty nt nḫt).* On the Left <Ear> (i.e., on line "+2") -- | | | | | * |.

[End of] Hour 11: The Two Feathers of the Giant *(šwty nt nḫt).* On the Right Ear (i.e., on line "-2") -- | * | | | | |.

[End of] Hour 12: The Head of the Mace of the Giant *(tp h̥ḏ nt nḫt).* On the Right Ear (i.e., on line "-2") -- | * | | | | |.

Table 18: I Shemu 16[35]

Beginning of the Night: Follower <of the Front of the Mooring Post> *(šmsw <n h̥ꜣt n mnit>).* On the Right Eye (i.e., on line "-1") -- | | * | | | |.

[End of] Hour 1: The Mooring Post *(mnit).* On the Left Eye (i.e., on line "+1") -- | | | | * | |.

[End of] Hour 2: Follower <of> the Mooring Post *(šmsw <n> mnit).* On the Right Eye (i.e., on line "-1") -- | | * | | | |.

[End of] Hour 3: Feet of the Hippopotamus *(rdwy n rrt).* Opposite the Heart (i.e., on line "0") -- | | | * | | |.

[End of] Hour 4: Her Knee *(pd.s).* Opposite the Heart (i.e., on line "0") -- | | | * | | |.

[End of] Hour 5: Her Vulva *(bꜣh̥.s).* On the Left Shoulder (i.e., on line "+3") -- | | | | | | *.

[End of] Hour 6: Her Buttocks *(h̲pd.s)*. On the Left Shoulder (i.e., on line "+3") — | | | | | | |*.

[End of] Hour 7: Her Breast *(mndt.s)*. On the <Left> Ear (i.e., on line "+2") — | | | | | |* |.

[End of] Hour 8: Her Two Feathers *(šwty.s)*. On the Left Ear (i.e., on line "+2") — | | | | | |* |.

[End of] Hour 9: Predecessor of the Two Feathers of the Giant *(tpy-^{c} šwty nt nh̲t)*. On the Left Ear (i.e., on line "+2") — | | | | | |* |.

[End of] Hour 10: The Two Feathers of the Giant *(šwty nt nh̲t)*. On the Right Ear (i.e., on line "-2") — |* | | | | | |.

[End of] Hour 11: Predecessor of his Mace *(tpy-^{c} h̥ḏ.f)*. Opposite the Heart (i.e., on line "0") — | | | |* | | | |.

[End of] Hour 12: Nape of his Neck *(h̲ꜣb.f)*.[36] On the Left Shoulder (i.e., on line "+3") — | | | | | | |*.

Table 19: II Shemu 1[37]

Beginning of the Night: <Follower> of the Mooring Post *(<šmsw> n mnit)*. Opposite the Heart (i.e., on line "0") — | | | |* | | | |.

[End of] Hour 1: Feet of the Hippopotamus *(rdwy n rrt)*. Opposite the Heart (i.e., on line "0") — | | | |* | | | |.

[End of] Hour 2: Her Knee *(pd.s)*. On the Right Eye (i.e., on line "-1") — | | |* | | | | |.

[End of] Hour 3: Her Vulva *(bꜣh̥.s)*. Opposite the Heart (i.e., on line "0") — | | | |* | | | |.

[End of] Hour 4: Her Buttocks *(h̲pd.s)*. Opposite the Heart (i.e., on line "0") — | | | |* | | | |.

[End of] Hour 5: Her Breast *(mndt.s)*. On the Right Eye (i.e., on line "-1") — | | |* | | | | |.

[End of] Hour 6: Her Tongue *(ns.s)*. On the Right Eye (i.e., on line "-1") — | | |* | | | | |.

[End of] Hour 7: Her Two Feathers *(šwty.s)*. On

the Left Ear (i.e., on line "+2") -- | | | | | * |.

[End of] Hour 8: Predecessor of the Two Feathers of the Giant *(tpy-ʿ šwty nt nḫt).* On the Left Ear (i.e., on line "+2") -- | | | | | * |.

[End of] Hour 9: The Two Feathers of the Giant *(šwty nt nḫt).* Opposite the Heart (i.e., on line "0") -- | | | * | | |.

[End of] Hour 10: The Nape of his Neck *(ẖꜣb.f).* Opposite the Heart (i.e., on line "0") -- | | | * | | |.

[End of] Hour 11: His Breast *(mndt.f).* Opposite the Heart (i.e., on line "0") -- | | | * | | |.

[End of] Hour 12: <His> Hip (?) *(bgs.<f>).* Opposite the Heart (i.e., on line "0") -- | | | * | | |.

Table 20: II Shemu 16[38]

Beginning of the Night: Feet of the Hippopotomus *(rdwy n rrt).* Opposite the Heart (i.e., on line "0") -- | | | * | | |.

[End of] Hour 1: Her Knee *(pd.s).* Opposite the Heart (i.e., on line "0") -- | | | * | | |.

[End of] Hour 2: Her Vulva *(bꜣḥ.s).* Opposite the Heart (i.e., on line "0") -- | | | * | | |.

[End of] Hour 3: Her Buttocks *(ḫpd.s).* Opposite the Heart (i.e., on line "0") -- | | | * | | |.

[End of] Hour 4: Her Breast *(mndt.s).* On the Left Eye (i.e., on line "+1") -- | | | | * | |.

[End of] Hour 5: Her Tongue *(ns.s).* On the Left Eye (i.e., on line "+1") -- | | | | * | |.

[End of] Hour 6: Her Two Feathers *(šwty.s).* On the Left Ear (i.e., on line "+2") -- | | | | | * |.

[End of] Hour 7: <The Two Feathers> of the Giant *(<šwty> nt nḫt).* On the Right Ear (i.e., on line "-2") -- | * | | | | |.

[End of] Hour 8: His Head *(tp.f)*. Opposite the Heart (i.e., on line "0") -- |||*|||.

[End of] Hour 9: The Nape of his Neck *(ḫʾb.f)*. Opposite the Heart (i.e., on line "0") -- |||*|||.

[End of] Hour 10: His Breast *(mndt.f)*. On the Left Shoulder (i.e., on line "+3") -- ||||||*.

[End of] Hour 11: His Hip (?) *(bgs.f)*. On the Left Ear (i.e., On line "+2") -- |||||*|.

[End of] Hour 12: His Knee *(pd.f)*. Opposite the Heart (i.e., on line "0") -- |||*|||.

Table 21: III Shemu 1[39]

Beginning of the night: The Vulva of the Hippopotamus *(bʾḫ n rrt)*. On the Left Eye (i.e., on line "+1") -- ||||*||.

[End of] Hour 1: Her Buttocks *(ḫpd.s)*. Opposite the Heart (i.e., on line "0") -- |||*|||.

[End of] Hour 2: Her Breast *(mndt.s)*. On the Right Eye (i.e., on line "-1") -- ||*||||.

[End of] Hour 3: Her Tongue *(ns.s)*. Opposite the Heart (i.e., on line "0") -- |||*|||.

[End of] Hour 4: Her Two Feathers *(šwty.s)*. Opposite the Heart (i.e., on line "0") -- |||*|||.

[End of] Hour 5: Predecessor of the Two Feathers of the Giant *(tpy-ᶜ šwty nt nḫt)*. Opposite the Heart (i.e., on line "0") -- |||*|||.

[End of] Hour 6: The Two Feathers of the Giant *(šwty nt nḫt)*. On the Left Eye (i.e., on line "+1") -- ||||*||.

[End of] Hour 7: His Neck *(nḫbt.f)*. Opposite the Heart (i.e., on line "0") -- |||*|||.

[End of] Hour 8: <His> Breast *(mndt.<f>)*. Opposite the Heart (i.e., on line "0") -- |||*|||.

[End of] Hour 9: His <Hip?> *(<bgs.>f)*. Opposite the

Heart (i.e., on line "0") -- | | | * | | |.

[End of] Hour 10: His Knee *(pd.f)*. On the Right Eye (i.e., on line "-1") -- | | * | | | |.

[End of] Hour 11: His Foot *(sbḳ.f)*. Opposite the Heart (i.e., on line "0") -- | | | * | | |.

[End of] Hour 12: His Pedestal *(pt.f)*. Opposite the Heart (i.e., on line "0") -- | | | * | | |.

Table 22: III Shemu 16[40]

Beginning of the Night: Bu⟨ttocks⟩ of the Hippopotamus *(ẖ⟨pd⟩ n rrt)*. Opposite the Heart (i.e., on line "0") -- | | | * | | |.

[End of] Hour 1: Her Breast *(mndt.s)*. Opposite the Heart (i.e., on line "0") -- | | | * | | |.

[End of] Hour 2: Her Tongue *(ns.s)*. Opposite the Heart (i.e., on line "0") -- | | | * | | |.

[End of] Hour 3: Her Two Feathers *(šwty.s)*. Opposite the Heart (i.e., on line "0") -- | | | * | | |.

[End of] Hour 4: Predecessor of the Two Feathers of the Giant *(tpy-ᶜ šwty nt nẖt)*. Opposite the Heart (i.e., on line "0") -- | | | * | | |.

[End of] Hour 5: The Two Feathers of the Giant *(šwty nt nẖt)*. On the Left Eye (i.e., on line "+1") -- | | | | * | |.

[End of] Hour 6: The Nape of his Neck *(ẖꜣb.f)*. Opposite the Heart (i.e., on line "0") -- | | | * | | |.

[End of] Hour 7: His Breast *(mndt.f)*. Opposite the Heart (i.e., on line "0") -- | | | * | | |.

[End of] Hour 8: His Hip (?) *(bg⟨s⟩.f)*. Opposite the Heart (i.e., on line "0") -- | | | * | | |.

[End of] Hour 9: His Knee *(pd.f)*. Opposite the Heart (i.e., on line "0") -- | | | * | | |.

[End of] Hour 10: His Foot *(sbḳ.f)*. On the Right Eye (i.e., on line "-1") -- | | * | | | |.

[End of] Hour 11: The One Coming After his Pedestal *(iy sꜣ pt.f)*. On the Right Eye (i.e., on line "-1") -- | | ✱ | | | | |.

[End of] Hour 12: Aryt *(ʿryt)*. Opposite the Heart (i.e., on line "0") -- | | | ✱ | | | |.

Table 23: IIII Shemu 1[41]

(See NP's Emended Table 23 following Table 24)

Beginning of the Night: Breast of the Hippopotamus *(mndt n rrt)*. <On the Lef>t <Ear> (i.e., on line "+2") -- | | | | | ✱ | |.

[End of] Hour 1: Predecessor of the Two Feathers of <the Giant> *(tpy-ʿ šwty nt <nḫt>)*. Opposite the Heart (i.e., on line "0") -- | | | ✱ | | | |.

[End of] Hour 2: Mace of the Giant *(ḥḏ nt nḫt)*. Opposite the Heart (i.e., on line "0") -- | | | ✱ | | | |.

[End of] Hour 3: Na<pe> of his Neck *(ẖꜣ<b>.f)*. Opposite the Heart (i.e., on line "0") -- | | | ✱ | | | |.

[End of] Hour 4: His Hip (?) *(bgs.f)*. On the Left Eye (i.e., on line "+1") -- | | | | ✱ | | |.

[End of] Hour 5: His Knee *(pd.f)*. Opposite the Heart (i.e., on line "0") -- | | | ✱ | | | |.

[End of] Hour 6: His Foot *(sbḳ.f)*. Opposite the Heart (i.e., on line "0") -- | | | ✱ | | | |.

[End of] hour 7: His Pedestal *(pt.f)*. On the Left Eye (i.e., on line "+1") -- | | | | ✱ | | |.

[End of] Hour 8: Aryt *(ʿryt)*. On the Left Shoulder (i.e., on line "+3") -- | | | | | | ✱.

[End of] Hour 9: Throat of the Bird *(ḥtyt nt ꜣpd)*. On the Left Shoulder (i.e., on line "+3") -- | | | | | | ✱.

[End of] Hour 10: Its Rump *(kft.f)*. On the Left Shoulder (i.e., on line "+3") -- | | | | | | ✱.

[End of] Hour 11: Star of the Thousands *(sbꜣ n ḫꜣw)*. On the Left Shoulder (i.e., on line "+3") -- | | | | | | ✱.

[End of] Hour 12: Star of Orion *(sbꜣ n sꜣḥ)*. Opposite the Heart (i.e., on line "0") -- | | | * | | |.

Table 24: IIII Shemu 16[42]

Beginning of the Night: The Two Feathers of the Hippopotamus *(šwty n rrt)*. <On the Left Ear> (i.e., on line "+2") -- | | | | | * |.

[End of] Hour 1: The Two Feathers <of the Giant> *(šwty <nt nḫt>)*. Opposite the Heart (i.e., on line "0") -- | | | * | | |.

[End of] Hour 2: .... [Opposite the Heart (i.e., on line "0")] -- | | | * | | |.

[End of] Hour 3: <His> Ne<ck> *(nḥ[bt.f])*. On the Left Eye (i.e., on line "+1") -- | | | | * | |.

[End of] Hour 4: <His> Brea<st> *(mn[dt.f])*. On the Right Eye (i.e., on line "-1") -- | | * | | | |.

[End of] Hour 5: [His] Hip (?) *(bgs.[f])*. On the Right Ear (i.e., on line "-2") -- | * | | | | |.

[End of] Hour 6: [His] F[oot] *(s[bḳ.f])*. On the Left Shoulder (i.e., on line "+3") -- | | | | | | *.

[End of] Hour 7: His <Pedestal> *(<pt>.f)*. On the Left Eye (i.e., on line "+1") -- | | | | * | |.

[End of] Hour 8: Aryt *(ꜥryt)*. Opposite the Heart (i.e., on line "0") -- | | | * | | |.

[End of] Hour 9: <Head of> the Bird *(<tp n> ꜣpd)*. On the Left <Ear> (i.e., on line "+2") -- | | | | | * |.

[End of] Hour 10: Its <Rump> *(<kft>.f)*. On the Left Eye (i.e., on line "+1") -- | | | | * | |.

[End of] Hour 11: Star of the Thousands *(sbꜣ n ḫꜣw)*. On the Right Eye (i.e., on line "-1") -- | | * | | | |.

[End of] Hour 12; Star of [Sar] *(sbꜣ n [sꜥr])*. Opposite the Heart (i.e., on line "0") -- | | | * | | |.

Table 23: IIII Shemu 1

(Emended Version by NP)[43]

Beginning of the Night: Breast of the Hippopotamus *(mndt n rrt).* <On the Lef>t <Ear> (i.e., on line "+2") -- | | | | | * |.

[End of] Hour 1: <Her Two Feathers> *(<šwty.s>).* ....

[End of] Hour 2: Predecessor of the Two Feathers of <the Giant> *(tpy-ꜥ šwty nt <nḫt>).* Opposite the Heart (i.e., on line "0") -- | | | * | | |.

[End of] Hour 3: <The Two Feathers of the Giant> *(<šwty nt nḫt>).* ....

[End of] Hour 4: Mace of the Giant *(ḥḏ nt nḫt).* Opposite the Heart (i.e., on line "0") -- | | | * | | |.

[End of] Hour 5: Na<pe> of his Neck *(ẖꜣ<b>.f).* Opposite the Heart (i.e., on line "0") -- | | | * | | |.

[End of] Hour 6: His Hip (?) *(bgs.f).* On the Left Eye (i.e., on line "+1") -- | | | | * | |.

[End of] Hour 7: His Knee *(pd.f).* Opposite the Heart (i.e., on line "0") -- | | | * | | |.

[End of] Hour 8: His Foot *(sbḳ.f).* Opposite the Heart (i.e., on line "0") -- | | | * | | |.

[End of] Hour 9: His Pedestal *(pt.f).* On the Left Eye (i.e., on line "+1") -- | | | | * | |.

[End of] Hour 10: Aryt *(ꜥryt).* On the Left Shoulder (i.e., on line "+3") -- | | | | | | *.

[End of] Hour 11: Throat of the Bird *(ḫtyt nt ꜣpd).* On the Left Shoulder (i.e., on line "+3") -- | | | | | | *.

[End of] Hour 12: Its Rump *(kft.f).* On the Left Shoulder (i.e., on line "+3") -- | | | | | | *.

Notes to Document III.14

1. NP give the following commentary to Table 1: "In 1501 B.C., the date about which our star clock was constructed, I *ꜣḫt* 1 in the

Egyptian Year was August 20 greg. June 21, the shortest night, was III *šmw* 6 or Table 21. December 21, the longest night, was I *prt* 4 or Table 9. From Table 21 to Table 9 the nights and the hours of the night were growing longer. The star of the 12th hour in Table 21 is *pt.f*...and there are seven stars (six hours) from *šwty nt nḫt* ... to ... [*pt.f*]. In Table 1 there are only six stars (five hours) from ... [*šwty nt nḫt*] ... to ... [*pt.f*]. Clearly the five seasonal hours of Table 1 cover essentially the same time span as the six seasonal hours of Table 21. And we may expect that the remaining hours of Table 1 are relatively longer than the hours in Table 21." Similar detailed comments on seasonal hour lengths and the distances between transit lines are given after each table and, since I shall not report those comments completely in the succeeding notes, they should be carefully studied by anyone interested in what may be gleaned concerning the given star transits in relationship to the varying hour lengths from one half-month to the next. However we can note here that I Akhet 16 in 1501 B.C. was identified by NP in their commentary to Table 2 as September 4 greg. and hence the nighttime is lengthening.

2. I have already mentioned that the star *ᶜryt* was probably not identical with the decanal star *ᶜrt*. As usual, we cannot surely identify the star (though Renouf identified it as β Adromedae in the long quotation from his article in my Introduction to this document), nor has the name of the star been successfully translated.

3. This star is deduced from its position in Table 2 as the marker of the end of the sixth hour. Renouf translated the name of this star as "Goose's head." In fact, the word *ỉpd* was used specifically for "duck" or on occasion perhaps for "goose," but is more generally rendered as "bird." In the celestial map of Petrie (see Fig. III.84a), the constellation is named "Goose" and so pictured. Renouf in the passage mentioned in the previous note, had "no hesitation in identifying" the star α Arietis with it.

4. As I remarked, this constellation also appears in the decanal tables. It was believed by Renouf to be the Pleïades.

5. I do not know how to translate the name of this star; however, its determinative consists of an intersection to two roads. Accordingly we could perhaps render it as "The Crossways" without knowing to what this refers. This star is identified by Renouf with α Tauri (Aldebaran).

6. This is probably the same star from the Constellation of Orion that appears as a decanal star. We do not know which specific star it was.

7. See NP, Vol. 2, pp. 7-8. In other words, what is almost surely meant is that these twelve stars of Table 2 will mark the hours during the night of day 16 and the rest of the nights of the half-month, i.e., during the last fifteen days of the designated month. Indeed, when the name "half-month" appears in the festival lists it seems always to have meant "the second half-month" (see the various festival lists given in Document III.1). Henceforth, for the remaining even-numbered tables, I shall merely write "Month No., Season Name, Day 16" (with the appropriate insertions of the month number and the name of the season), it being assumed that this shortened expression is always to be understood as comprising the whole second half of a 30-day month. Similarly, I shall use the shortened "Month No., Season Name, Day 1" for the odd-numbered tables, i.e., the tables for the first-halves of the months.

8. Notice that the first star of Table 1 has dropped out of Table 2 and in Table 2 all of the succeeding stars have become markers of the hour preceding those they marked in Table 1. This left room for the inclusion of Sirius as the marker of the end of the twelfth hour. Of course, this was the basic procedure assumed for the tables. But in fact, some stars were dropped before they would have disappeared from the next table.

9. II Akhet 1 in 1501 B.C. was September 19 greg. As NP, p. 25, note, this was almost full equinox, the first eleven hours have lengthened slightly.

10. Akhet 16 in 1501 B.C. was October 4 greg. with the length of the night still increasing.

11. This does not appear to be identical with a star named in other sources as "The Star of the Water" (see NP, p. 6, n. 1).

12. III Akhet 1 in 1501 was October 19 greg., with the length of the nights still growing.

13. Renouf (p. 407) believed that the stars of the constellation of the Lion were a part of our constellation of the same name.

14. In the year 1501 B.C. III Akhet 16 fell on November 3 greg. Obviously the lengths were still increasing.

15. See note 13.

16. In 1501 B.C. IIII Akhet 1 fell on November 18 greg. The length of the night continues to increase.

17. See NP, p. 6 (E 16).

18. NP (p. 33, Comm.) notes the effect of dropping from Table 7 the star called "Predecessor of Orion" (which marked the end of Hour 6 in Table 6) and that of adding to Table 7 "Two Stars" to mark the end of Hour 8: "In Table 7 we encounter the first instance of the dropping of one hour star and the introduction of another hour star at some distance from the former with considerable positional shifting of hour boundaries in compensatory adjustment. This is best seen in the diagram [p. 32]. K1 [NP's symbols designating the "Predecessor of Orion"] is the star which drops out. The three before it shift increasingly toward + positions while the three which follow it and which jump an hour all shift in the opposite direction. The result is that when ... [the "Predecessor of Orion"] drops from between ... [the "Star of Sar"] and ... [the "Star of Orion"] the hour boundaries shift closer together by six positions. Similarly when ... ["Two Stars"] is introduced as a new hour star after ... [the "Predecessor of Two Stars"], the three following hour boundaries all shift in a + direction and those for ... [the "Predecessor of Two Stars"] and ... [the "Stars of the Water"] have moved farther apart by five positions."

19. "The Many Stars" was thought by Renouf (p. 407) to be a part of the Coma Berenices.

20. IIII Akhet 16 falls on December 3 greg. and the night continues to lengthen.

21. The translation is uncertain. Renouf (p. 407) calls it "The Lute Bearer" and identifies it with α Virginis. NP (p.7) say that it may be outside of the constellation *mnit*.

22. Renouf (p. 407) believes that α Boötis and α Scorpionis are part of the constellation Menat *(mnit)*. In the table he calls this entry "scouts of Menat."

23. I Peret 1 falls on December 18 greg. in 1501 B.C. Hence the longest night should be embraced by Table 9. NP (p. 37) believe that their comparison of Table 8 with Table 9 and Table 9 with Table 10 confirm this. Again I leave the details of their argument to the interested reader.

24. I Peret 16 falls on January 2 greg. in 1500 B.C. NP (p. 39) remarks: "The night ought, accordingly, to begin to shorten and there is clear indication that Table 10 is, indeed, shorter than its predecessor." The authors go on to analyze the shifts from Table 9 that they believe confirm this. They conclude: "This turn toward a shorter night confirms in the clearest possible way the accuracy of the date ascribed to the construction of the star clock as the early

Eighteenth Dynasty."

25. In his reconstruction of the celestial map (see Fig. 84b), Petrie identifies this with the standing hippopotamus which is a part of the conventional collection of "northern" constellations. He draws it as being of enormous size. But NP (p. 7) indicate that the Hippo of the northern constellations is never called *rrt* but usually Ipet, "*never* is shown wearing two feathers as a headdress [as she is said to have in Hour 9 of Table 17], and very frequently has a crocodile on its back." But they report at R on the same page concerning the Hippopotamus: "As the sequence of hour stars will show, the animal was conceived as standing upright on its hind legs, like the goddess Ipet."

26. In the year 1500 II Peret 1 falls on January 17 greg. The night is decreasing in length. Concerning our table NP (p. 41) remark: "The [first] eleven hours ... have shortened by two positions and though this may have been more than compensated for by the new [last] hour ... it is more likely that the whole twelve hours are some minutes shorter than those of Table 10."

27. In the year 1500 II Peret 16 fell on Feb. 1 greg. NP (p. 43) believe that no shortening of the night can be detected in Table 12, though the shortening should be happening.

28. In the introduction to this document (n. 1), I have included NP's use of the fact that the meridian culmination of Sothis (Sirius) occurred at the beginning of the night on II Peret 16 and the further Egyptian belief that its rising would occur 160 days afterwards to deduce a date of 1470 B.C. for the origin of the Ramesside Star Clock. Of course it should be pointed out that the name "Sothis" is not on the only copy (**A**) to have any names of stars for Table 12. However the assumption of Sothis as the name is surely correct if we note that Sothis was the star marking the end of Hour 1 in Table 11 and that "The One Coming After Sothis" marked the end of the first hour in Table 12.

29. III Peret 1 fell on February 16 greg. in the year 1500, with a slight shortening of the night evident from Table 13.

30. III Peret 16 fell on March 3 greg. in 1500 B.C. NP (p. 47) say: "There is no direct evidence in the table that the night is shortening. The eleven-hour span from ... [the first star to the eleventh] actually lengthens by two positions, but again it is probable that the hour dropped at the beginning of the night was some minutes longer than the one added at the end ...."

31. IIII Peret I fell on March 18 greg. in year 1500 B.C. This

table then included the vernal equinox. NP (p. 49): "It is likely that the shortening night has been taken into account in [hours] 11 and 12."

32. In the year 1500 B.C. IIII Peret 16 fell on April 2 greg. The nighttime shortens and apparently the Table confirms this.

33. In this table we can see the enormous size of the Constellation of the Hippopotamus, since its parts culminate or nearly culminate over a period of five hours, and indeed these hours are approximately equal to equinoctial hours.

34. I Shemu 1 in the year 1500 B.C. fell on April 17 greg. NP (p. 55) say: "The shortening of the night is evidenced to some extent by the shifts at the beginning and the end of Table 17."

35. I Shemu 16 in the year 1500 B.C. fell on May 2 greg. NP (p. 55) say: "It is not possible to detect any shortening of the night from Table 17 to Table 18."

36. This is apparently equivalent to the star called "His Neck."

37. II Shemu 1 in the year 1500 B.C. fell on May 17 greg. NP (p. 57): "There can be little doubt that Table 19 evidences a delay in beginning the night, ..."

38. II Shemu 16 in the year 1500 B.C. fell on June 1 greg. The shortening of the night proceeds further. NP (p. 59) believe this is shown by comparing the diagrams they have added to Tables 19 and 20.

39. III Shemu 1 fell in the year 1500 B.C. on June 16 greg. NP (p. 61): "Table 21 then should take in the shortest night of the year. It is immediately clear that the beginning of the night is later than in Table 20 since two hours preceded .... [the Vulva of the Hippopotamus] there and a shift to the left of one position contributes to a delay in beginning the night."

40. III Shemu 16 fell in the year 1500 on July 1 greg. NP (p. 63): "The night should begin to lengthen again but so little that there can be no evidence of this in the transition from Table 21 to Table 22."

41. IIII Shemu 1 in the year 1500 B.C. fell on July 16 greg. NP (p. 65): "The night is lengthening but the obvious corruption of Table 23 prevents any attestation of this. One has only to look at the diagram [included below the table] and to recall that *sbꜣ n sꜣḫ* ... was the 12th hour star in Table 1 to see the difficulties which confront an attempt to reconstitute Table 23. This must be done but our task requires that Table 24 be taken into account as well and it will be better to postpone all discussion until we have

examined that text." I have included NP's emended version of Table 23 following Table 24.

42. IIII Shemu 16 in the year 1500 fell on July 31 greg. There is a long commentary by NP which explains how Table 23 ought to be emended (with the emended table itself following Table 24). It also assumes that the epagomenal days that follow the end of the year are included in Table 24. A series of diagrams is included to enlighten the account. As I have said before, I find this graphic analysis of interest but in view of its speculative character I shall not include the full account here but only the first paragraph and the beginning of the second (NP, p. 67): "A comparison of Table 23 with Table 24 reveals unmistakably that one or both are corrupt and we have omitted a diagram [below Table 24]. The first step must therefore be to compare Table 24 with Table 1 as a check on the accuracy of the former in its emended form. This is done in Fig. 15 and the result is such as to inspire confidence. We have to recall that the epagomenal days separate the two tables. The fact that Table 24 covers not fifteen but twenty days, and the necessary lengthening of some interim hours (the *nḫt*-constellation spans one hour less in Table 1) are quite sufficient to account for the changes and to support Table 24 as being very probably correct.

"The next step should be to compare Tables 22 and 24 to see what might best fall between them ...."

Then follows the explanation of the new diagrams added with the conclusions to draw from them, the whole epitomized in the emended Table 23, which includes in abbreviated form the data of Tables 22, 24, and 1 together with the data of Table 23 inserted between the columns for 22 and 24 so that all can be easily compared.

43. See the preceding note.

Document III.15: Introduction

Amenemhet's Water Clock

In February 1885, some fellahin discovered a ruined tomb of an 18th-dynasty dignitary named Amenemhet near the top of the hill of Sheikh Abd el-Gurnah in Western Thebes and Ernesto Schiaparelli was able to take advantage of this discovery.[1] An unusual bas-relief shows the deceased attending a bull fight, and beneath it is an inscription of 16 columns which is now so worn that only about 1/3 of it could be accurately copied by Schiaparelli and his friend Golénischeff. Although the copy they made was good so far as it went, the editor did not understand it properly so that he missed the essential fact that it includes the description of Amenemhet's discoveries which led him to construct a water clock in honor of the king, Amenhotep I (ca. 1525-1504). It was only later, after Golénischeff sent a better copy of the inscription to Sethe, that the latter discovered the importance of the inscription for the history of the water clock and so he communicated the improved text with his own German translation to Borchardt. It was this text and Sethe's translation (altered somewhat by Borchardt) that was published by Borchardt as a Nachtrag to his *Die altägyptische Zeitmessung*, pp. 60-63 and Tafel 18 (see my Fig. III.25).

We learn from the text of the inscription that Amenemhet lived during the reigns of the first three kings of Dynasty 18: Ahmose (the last 10 years of his reign, i.e., ca. 1535-1525 B.C.), Amenhotep I (all 21 years of his reign), and Tuthmosis I (some unknown number of years from 1504 B.C.). Amenemhet tells us that he discovered the 14:12 ratio of the lengths of the (longest) night in the winter to the (shortest) night in the summer (see note 4 to line 8 below). Recall that this was the ratio of the longest to shortest monthly hour scales in the Karnak clock. He also speaks of the increase and then decrease of the lengths from month to month. Whether he was the first to "discover" the common 14:12 ratio and the varying lengths of the scales, or whether he found these facts in his reading of the earlier literature (see line 7) we cannot really know. At any rate, the inscription, in line 14, informs us that he constructed a *mrḫyt*-instrument for telling time (an instrument which in line 15 he calls a *dbḥ(t)*; this latter word has as its determinative the small pot-shaped vessel used in measuring grain and liquids). This was surely a water clock that embraced his discoveries concerning the hour-scales and their divisions (see the notes to lines 15 and 16). He tells us that he made it in honor of Amenhotep I. This seems to accord with the suggestion made in Chapter Three, that the Ebers Calendar was prepared in year 9 of Amenhotep's reign for use with a water clock. Finally the very last sentence of the inscription to the effect that the water runs out through a single exit ensures us that the water clock was of the outflow variety.

Note to the Introduction of Document III.15

1. E. Schiaparelli, "Di un'iscrizione inedita del regno di Amenofi I," *Actes du 8me Congrès International des Orientalistes tenu en*

1889 à Stockholm et à Christiana, 4me partie (Leiden, 1892), pp. 203-208.

Document III.15

Amenemhet's Water Clock:

1....the land which is called Mitanni; the enemy
2....Thebes; the announcement (?)[1] of his majesty in this land; he made it as retribution for the unjust
3....[great] in his office, elevated in dignity, a prince before whom
4 [the people]....[Amenemhet. He said: Listen to what I say to you,][2] you who are upon the earth: I lived ten years under King Nebpehtire [Ahmose].
5....[I lived] 21 years under the Horus who subjected the lands, the Lord of the Two Lands Djeserkare [Amenhotep I].
6....The first time that I was honored was under the majesty of the King of Upper and Lower Egypt, Djeserkare [Amenhotep I].
7....while reading in all of the books of the divine word[3]
8....[I found the (longest) night of wintertime to be] 14 [hours long] when the [shortest] night of the summertime is 12 hours [long][4]
9....[I found an increase in the night's length from] month to month [and then] a decrease from month to month.
10....[I found them represented on the interior of the water clock][5] and the movements of Re [and the moon-god?][6] with the utterances of both, and an offering (*ʾwt-ʿ*)[7]
11....[one?] like [the other?] with the utterance before him. The sign of Life as well as that of Good Fortune

is in their hands
12....Re gives it to Nekhbet, who approaches Re
13....[She holds the sign of Life,] which is in her hand, to the nose of his majesty who goes down before her; meanwhile they
14....[he, the king, is] happy that he sees these goddesses as they go forth and descend in front of him. I made a *mrḫyt*-instrument (i.e., an instrument for telling time), reckoned upon the year, for the Good Fortune of the King of Upper and Lower Egypt, Djeserkare [Amenhotep I], now deceased.[8]
15....season...every.[9] Never was made the like of it since the beginning of time. I made this remarkable instrument *(dbḥ(t))* in honor of the King of Upper and Lower Egypt, Djeserkare [Amenhotep I], deceased; dividing it [i.e., each scale] in half
16.... [and then the halves into halves, and finally the four thirds into thirds].[10] It was correct (?) for the beginning of Akhet (?), for Peret, [for Shemu],[11] for embracing (?) the moon at its times,[12] and for every hour at its times. The water runs out only through a single exit.

Notes to Document III.15

1. Or "exalting." In the word (*ṯst*), the problem is that two different determinatives are here included, namely that for "exalting" (=) and that for "announcement" ( ).

2. This addition in brackets was suggested by Sethe and Borchardt.

3. Sethe and Borchardt suggest that this means "the whole of Egyptian literature."

4. What I believe is being measured here is the scale lengths for the longest and shortest nights. This fits in with the analysis

of the Karnak clock where the scales are respectively 14 and 12 fingers long. Hence, more exactly the meaning of this statement probably is "I found the scale of the longest night of wintertime to be 14 fingers long if the scale of the shortest night of the summertime is 12 fingers, with each finger marking one hour." Needless to say, we are not to believe that there was a universally accepted standard hour length, for the hours were continually variable in length as the total nighttime varied in length from season to season. However, the fact that he compared the total length of winter hours to that of the summer hours, might suggest that although the concept of variable seasonal hours was everywhere accepted in the development of Egyptian clocks and in the commonplace telling of time, the use of the summer hour as a unit for the comparison of the varying seasonal hours hints of a later trend to divide the whole period of a day and a night into 24 hours each equal to a summer hour.

5. The bracketed material was suggested by Sethe and Borchardt. Like the Karnak clock later, Amenemhet's clock seems to have had, on the exterior surface, reliefs depicting gods who were connected with time.

6. This was also suggested by Sethe and Borchardt. But what the Moon-god's movements have to do with the water clock is not clear.

7. Nor is this passage clear. Presumably there is an offering scene that involves the king and Re and Thoth.

8. Here the *merkhyt* is not the simple sighting instrument used with shadow clocks but rather seems to be the water clock that Amenemhet constructed. For the sighting instrument that was used with the shadow clock, see Figs. III.20a and III.20b.

9. Sethe and Borchardt would expand this to read "It was correct for that season."

10. Sethe-Borchardt simply add "[and thirds]" where I have added the longer statement. Borchardt suggests that the "r" sign (𓂋) that can be seen near the beginning of line 16 may be the "r" sign used for fractions and thus would presumably have had three vertical strokes underneath it to indicate "thirds." He explains that to arrive at the 12 individual hours not only is division by 2 necessary but also by 3. Another possibility is that the use of division into thirds, if it was actually in the inscription, may have been a reference to the fact that, as we have seen in both the Karnak and Edfu clocks, the successive monthly scale lines decrease

or increase in length by 1/3 of a finger's breadth.

11. I have added this bracketed phrase with the idea in mind that perhaps here the author simply meant that his water clock was good for all three seasons, that is, it was of the kind that the Ebers Calendar was perhaps to be used with. This makes good sense, for recall that the Ebers Calendar was composed for the ninth year of Amenhotep I's reign. Amenemhet might have made his water clock when Amenhotep was still alive, and then dedicated it to him upon his death, to be included among the things available for Amenhotep I in his afterlife.

12. This is surely not clear, since the lunar calendar has nothing to do with such water clocks. Perhaps it merely is a poetic way of saying that each scale is good for the whole month in which it is used.

Document III.16: Introduction

The Shadow Clock described in the Cenotaph of Seti I

As I stressed in Chapter Three, this is the only text from Ancient Egypt to describe a shadow clock. I also indicated that the clock so described was probably earlier than the period of Seti I (ca. 1306-1290 B.C.) since only four hours are marked out on its base to determine the four hours before and the four hours after noon, leaving two hours of daylight before and two hours after the eight hours measured by the shadow clock, the first two hours of light being perhaps divided by sunrise and the last two by sunset. But the earliest extant shadow clock, which goes back to the time of Tuthmosis III (ca. 1479-1425 B.C.), was a clock with five hours on the board to determine five hours before and five hours after noon, thus leaving unmeasured on the board one hour before and one hour after the 10 measured hours. Hence this clock probably embraced a more advanced concept of hour division starting with sunrise and ending with sunset.

The substance of the account in the cenotaph of Seti I has already been presented in Chapter Three and need not be repeated here. It is of some interest that the text in large part confirmed the analysis of shadow clocks made by Ludwig Borchardt on the basis of actual shadow clocks which had been added to the Berlin Museum (see the section on Shadow Clocks included in

Chapter Three above). His analysis focused on the clock shown in Fig. III.41, Berlin Museum 19743.

The cenotaph text with an English translation was first published by Henri Frankfort in his *The Cenotaph of Seti I at Abydos* (London, 1933), Vol. I, pp. 77-78; Vol. 2, Plate 83 (reproduced as my Fig. III.38). The latter plate is the transcription of the text located on the west side of the roof of the Sarcophagus Chamber. In Fig. III.38 we notice the presence in the first seven vertical lines of lacunas that represent open spaces where no signs have been cut. According to Frankfort (Vol. 1, p. 77, n. 1) this suggests "that our copyist worked from an original that was already defective."

Frankfort's English translation was slightly revised by Neugebauer and Parker, *Egyptian Astronomical Texts*, Vol. I, pp. 116-18, who also included a photograph of the text on the left side of their plate 32. As I have mentioned in the course of Chapter Three, E.M. Bruins, "The Egyptian Shadow Clock," *Janus*, Vol. 52 (1965), pp. 127-37, devised a clever and coherent interpretation of the text differing in crucial places from that of the previous authors. His account concludes that the shadow clock was adjusted for the seasons by adding a strip with a thickness of 1 or 2 fingers (depending on the season) to the top of the crossbar in order to increase its height and thus compensate for the changing declination of the sun during the year. Hence he alters the translation of the text to accommodate his theory, as I have indicated in note 2 to the translation below. No extant clocks include such a strip or strips, which is perhaps not surprising since no clock includes the crossbar itself.

In the translation below I have followed Frankfort, adding some of the bracketed phrases of Neugebauer and Parker and a few of my own. With Neugebauer and Parker I have replaced Frankfort's "spans" with "palms." Note that 5 palms (as marked on the rule or base) = 20 fingerbreadths = ca. 15 inches. I have followed Bruins in abandoning in line 8 the early translations of *hp* as a unit of linear measurement in favor of "rule" (see note 6 to line 8 below). The line numbers given in my translation are those of the vertical columns in Fig. III.38. I have started a new line in my printed text with each line of the inscription's text so that my translation may be more readily compared with the prior translations.

Document III.16

The Shadow Clock described in the Cenotaph of Seti I

Knowing the Hours of the Day and the Night: An Example of Fixing Noon[1]

\1\ The hours of the day, beginning from fixing the location. [Knowing] the hours:[2]
\2\ the hour after the first landing (*or* mooring post);[3]
\3\ the hour after the second landing;
\4\ the hour after the third landing.
\5\ Knowing the ho[urs by means of a shadow clock *(sṯ't)* whose base *(mrtwt)* is][4] 5 palms in its length,
\6\ height [.... with a crossbar *(mrḫyt)*][5] of two fingers in its height [placed]
\7\ upon the head *(tp)* of the shadow clock *(sṯ't)*. [You shall divide up] these 5 palms into 4 parts

\8\ [which are branded] on this shadow clock. According to the [accepted] rule (*n hp*)[6], you shall put 12 [units] therefrom [i.e., from the 5 palms] for [marking] the first hour; you shall put 9 therefrom for [marking] the second hour; you shall put 6 therefrom for [marking]
\9\ the third hour; you shall put 3 therefrom for the fourth hour.

When you have adjusted this shadow clock on a level with the sun, its head being toward the east, [the head] on which
\10\ is this crossbar, the shadow of the sun will be correct on this shadow clock.

Now after the fourth hour has ended [and] after the sun has stood in the opening (*wpt*) [i.e., crown] of the crossbar,[7] you shall turn around
\11\ this shadow clock, its base [now] being toward the east, Moreover, you shall reckon these
\12\ hours until the sun passes the four hours according to the previous rule.

It sums at [only] eight hours, for two hours
\13\ have passed in the morning before the sun shines [on the shadow clock] and another two hours [will] pass after [which] the sun enters [the Duat][8] in order to fix the location of the hours of night.

Notes to Document III.16

1. This superscribed title is not wholly pertinent to the text involving the shadow clock (and hence its omission by Neugebauer and Parker). Not only is the reference to the hours of the night superfluous (unless the title is also meant to refer to the text on night hours which is to the right of the text on the shadow clock, which, however has its own superscribed title: "Knowing the Hours of the Night") but the text on the shadow clock is not really an

example of "fixing noon" except in the vague sense that when the shadow disappears from the horizontal board it is assumed to be noon and the clock is reoriented with its head to the west. Bruins believed that from the last part of the title (*tp n ir.t mtr.t*) a prepositional expression equivalent to "till and from" had been dropped. And so adding that phrase he would translate the last part of the title as follows: "The method for operating till and from noon" (E. M. Bruins, "The Egyptian Shadow Clock," *Janus*, Vol. 52 [1965], pp. 136-37). However, I remind the reader that it is not always prudent to correct the text to fit the reader's fancy. For example, Bruins notes that he believes that there is enough room between the eye glyph for *(ir)* and the loaf glyph for *t* so that a preposition like 𓂋 *(r)* could have been in the original text. But in fact it seems certain that nothing has dropped out between these two signs since together they properly constitute the necessary infinitive, *ir.t*, which as a verbal noun with a feminine ending means "making" or "fixing" and clearly belongs intact in the title. And, furthermore, the spacing in the title is not particularly unusual, though to be sure there are, as I have said, open gaps in the text on which signs have not been cut.

2. The translation of the whole text by Bruins, "The Egyptian Shadow Clock," pp. 135-36, is worth recording. It is the most coherent of the translations but is the one which has been most widely altered from what can be read in the text. Still, his is the first translation and interpretation that accounts rationally for the first four lines of the text. (Note that he has given in italics the changes he has made in the Neugebauer-Parker translation.) They are made in order to support his view that the shadow clock used additional strips placed on top of the crossbar to account for the seasonal variation in the declination of the sun.

> 1 The hour of the day, beginning with fixing the place: [*Determining*] the hours:
>
> 2 the hour, *corresponding to* the first *domain*;
>
> 3 the hour, *corresponding to* the second *domain*;
>
> 4 the hour, *corresponding to* the third *domain*.
>
> 5 *Determining* the hours by means of a shadow clock, whose scale-bar is 5 palms in its length,
>
> 6 the height [*2 1/2 palms*] with a *top-bar-strip* of 2 fingers in its height
>
> 7 on top of the shadow clock. [You shall divi]de these 5 palms into 4 parts....

> 8 branded on this shadow clock. You shall put 12 therefrom, *according to a rule*, for the first hour; you shall put 9 therefrom for the 2nd hour; you shall put 6 therefrom for
>
> 9 the 3rd hour; you shall put 3 therefrom for the 4th hour. When you have adjusted this shadow clock *in accordance with* the sun, its *end* being to the east, (that is to say:) the *end* on which the cross-bar is (mounted)
>
> 10 the shadow of the sun will be correct on this shadow clock. Now after the 4th hour has ended you shall turn around
>
> 11 this shadow clock, its (other) *end* being to the east, after the sun has stood on the crown of this top-bar. Moreover you shall reckon these
>
> 12 hours until the sun enters into the four hours, according to the former rule. It is totaling *(only)* to 8 hours, for 2 hours
>
> 13 have passed in the morning before the sun shines *(on the clock)* and another two pass after *(which)* the sun enters for determining the hours of the night.

The keys to Bruins' reinterpretation of the clock are (1) his substitution in lines 2-5 of "corresponding to" and "domain" for "after" and "landing" (neither of which seems fully appropriate) and (2) his reconstruction of line 6 where he adds in brackets the full height of 2 1/2 palms for the crossbar and his mention without brackets of the hypothesized top-bar-strip, all of which is not present in the lacuna of the text. The doubtful substitutions and unsupported reconstruction, one gathers, would roughly satisfy his conception of the clock, which I now quote (*ibid.*, p. 135): "For measuring the time by means of a shadow clock at [latitude] $p = 30^{\circ}$ one has to divide the regions in which the sun moves in the sky into *three* domains, one about the winter solstice, one about the equinoxes and one about the summer solstice. When the sun moves to a following domain, one has to put [or remove, depending on the season] a strip on top of the top-bar." Then later (p. 137) he specifies the use of a strip as follows: "The method for operating till and from noon, which the shadow clock in fact permits, with a high accuracy, measuring seasonal hours and adjusting the height of the top-bar by means of a thin strip for the compensation of the change in declination of the sun during the year. This adjustment, depending on the season only, can easily be effected: Nov., Dec., Jan. 0 [i.e., no strip is added]; Feb., March, Apr. 1 [i.e., a strip with thickness of 1 finger is added]; May, June, July 2 [i.e., a strip with thickness of 2 fingers is added]; Aug., Sept., Oct. 1 [i.e., a 1-finger strip is added]; Nov., Dec., Jan. 0 [i.e., as before no strip is added]."

3. This language assumes the imagery of the sun-bark sailing across the sky from one landing place to another in the course of the day, or, if you believe Bruins, from one domain to another in the course of the year.

4. The material added by Neugebauer and Parker in brackets is "intended to suggest the thought rather than exact wording." As I remarked in note 2 there is a lacuna in the text at this point.

5. Again I give Neugebauer's and Parker's bracketed material, except that I use *mrḫyt* instead of their *mrḫt*, since the former is the form found in lines 10-11, and of course the word in question has been added by them here in line 6 in an actually existing lacuna in the text so that we cannot tell which of the forms appeared there. The reader should also note, as I have said before in Chapter Three when describing the Seti text, we cannot be sure that the term *merkhyt* refers to a crossbar rather than to the vertical block on the end of the base on which a crossbar has been presumed to rest. For the term is so used for the block when it contains a groove for mounting a plummet line. However "two fingers" seems rather a modest figure for the block since the next line appears to tell us that whatever is being measured is being mounted on the "head" of the shadow clock, which itself might be the vertical block. It is possible, however, that the "head" is merely the "end" of the clock on which the vertical block stands.

6. It seems probable that the clock maker initially marked off the position where the end of the crossbar's shadow first fell on the baseboard; this length is assumed by rule as 30 units from the upright (this being by definition the beginning of the first of four measured hours before noon). Presumably the period of daylight before the shadow-line fell on the board was divided into two uneven hours by sunrise. Then, from that end point at a distance of 12 units away a mark is made for the end of the first measured hour. The succeeding marks delineate the succeeding three hours before noon, the marks having been placed successively 9, 6, and 3 units closer to the upright. These distances of 12, 9, 6, and 3 units are those specified by the text ("[You shall divide up] these 5 palms into 4 parts"). By a curious error the illustration of the clock

above the text shows a division into *five* parts (see Fig. III.38). Perhaps, in a more convenient fashion, the clock maker (having assumed from experience with the sun's rising at the location where the clock was to be used a total length of 30 units from the upright for the measured time of the four hours before noon) simply started from the upright and placed the marks at intervals of 3, 6, 9, 12 units. As I have already said, the shadow clock was turned around at noon and the markings showed the four hours after noon. Note that in the case of these length-markings Frankfort, but with some doubts expressed, and Neugebauer and Parker thought that *n hp* referred to units of length called *hp* or *nhp*. But I think that here Bruins' translation "according to a rule" is the correct one; this is bulwarked by the clear use of *hp* with that meaning in line 12 below.

7. The "opening of the crossbar" means here the vertical plane of the plumbline, or as Neugebauer and Parker say "in transit of this crossbar." Since the word translated as "crossbar" is again *merkhyt*, we must once more stress that the key element of a *merkhyt* is the sighting instrument with its plumbline.

8. I have added "[the Duat]" even though this is missing in the text, since this certainly is the sense of the passage, the nighttime hours being determined from the beginning of total darkness. I have also added (though it is not in the text) the preceding "[which]" at Bruins' suggestion to escape the incorrect sense of the literal text that the sun enters the Duat immediately after no longer shining on the shadow clock.

Document III.17

The Rectangular Zodiac from the Temple of Khnum (Esna A) and the Round Zodiac from the Temple of Hathor (Dendera B)

The form of this document differs somewhat from that of other documents in this volume. The "document" consists of Figs. III.75a and III.76a. The "introduction" to the "document" (which immediately follows) is a discursive description and analysis of the zodiacs pictured in the figures. The change in form has been dictated by my reluctance merely to repeat in brackets all of the names of the astronomical elements that the deities represent. In a sense, there is not much to translate but for the most part only something to describe and evaluate. Hence, I shall merely identify, locate and characterize the figures that represent the deities making up the zodiacs: the zodiacal signs, the constellations, the planets, and the families of decans appearing on the two zodiacs that I give as models of the hybrid Egyptian zodiacs. Needless to say, I shall be primarily discussing in this "document" the figures of the various Egyptian elements associated with the zodiacal signs. It will be evident from my citations that I base my discussion of this document almost entirely on the treatment found in Neugebauer and Parker's *Egyptian Astronomical Texts*, which, in my opinion, exceeds in detail and profundity any previous (or in fact succeeding) discussion of the Egyptian zodiacs.

This class of Egyptian astronomical monuments, i.e., the zodiacs, arose and flourished exclusively in the Hellenistic and Roman periods. It consists of a small number of rectangular and/or near-round zodiacs found in temples, tombs, and coffin lids.[1] These zodiacs are combinations of older Egyptian and newer Hellenistic themes, and, as I have said, I wish to describe two such zodiacs as models, namely the earliest examples of each type of Egyptian zodiac. They are ultimately of Babylonian origin, at least so far as the pictorial representation of the zodiacal figures is concerned.[2] Woven above, below, between or around the zodiacal signs themselves are the Egyptian elements: the Egyptian decanal deities (with only a few names), the planets, as for example their positions when they are in exaltations (i.e., their positions in relation to the zodiacal signs where they assert the most astrological influence), the so-called Northern Constellations, and other Egyptian mythological depictions. In short, these Egyptian elements (regardless of where they occur in relation to the zodiacal signs) are those found in various versions of the celestial diagram that appear in Documents III.3-III.4 and III.11-III.14 and are pictorially represented in the drawings and plates to which my renderings of the documents refer.

The research of the last hundred years or so has rather clearly shown that the Egyptian zodiacs are not so astronomically accurate as the students of Egyptian astronomy in the early nineteenth century thought.[3] In this regard the remarks of Neugebauer and Parker are especially pertinent:[4]

> Almost all zodiacs known to us occupy a position either on the ceiling of a temple or a tomb or on the lid of a coffin. Obviously they are understood as belonging to the sky that stretches out above us as does the ceiling of a room. But this qualitative similarity is of little use for accurate astronomical representations. The terrestrial observer never sees more than six of the twelve signs of the zodiac above the horizon at any one time. If one desires, nevertheless, to represent the complete zodiac on a ceiling the question may be asked whether there exists any "natural" order of arrangement for such a representation. One might perhaps argue that an observer facing south sees the constellations rise, culminate, and set in a clockwise sense of rotation but finds that the order of the zodiacal signs is opposite to that rotation. Hence, an observer looking up to a ceiling might expect a counterclockwise sequence of the twelve signs.

The authors then note that the actual monuments do not uniformly confirm that expectation, since nine of the extant zodiacs show a counterclockwise sequence and sixteen of them a clockwise sequence. They go on to conclude:

> It is, of course, meaningless to ascribe to such a small number of cases any statistical significance. But it is clear that both modes of orientation occur in all types of our monuments, temples, private tombs, and coffins, with the latter having mixed orientation as well.... Hence no astronomical

> principle is responsible for the orientation of the zodiacs.
>
> Equally without significance is the way of dividing a zodiac into two halves, e.g., to either side of Nut on a coffin lid or on parallel strips of a ceiling. One might expect some uniformity but what the monuments show is again quite different ....
>
> All that one can conclude from the.... [extant zodiacs] is that there is a certain tendency to divide the zodiac more or less along the solstices; only Athribis A [see Fig. III.104, A] divides nearer to the equinoxes—right beside a different division in the other half of the ceiling (B) [*ibid.*, B]. It also follows from our list, which is chronologically arranged, that the distribution of signs is not chronologically determined. Attempts to date zodiacal representations astronomically according to the arrangement of the signs disregard the accumulated evidence of the available monuments.

In my treatment of Egyptian zodiacs, I shall not attempt any discussion of the iconography of the zodiacal signs (other than listing them in their positions relative to the decans or other Egyptian elements).

The Esna A Zodiac

Now I am prepared to discuss the earliest of the rectangular zodiacs, that of Esna A (dating from about 200 B.C.). As is evident from my brief comments on the Esna A zodiac at the end of Chapter Three (see "Egyptian Zodiacs"), it was part of a now destroyed and dismantled Temple of Khnum near Esna whose blocks have disappeared (for its preservation in the *Description de l'Égypte*, see Fig. III.75a):[5]

> The temple faces east and the columned hall has five ceiling strips, the decorations of which run west-east on the southern two and east-west on the northern two. The northernmost and southernmost strips [respectively at the bottom and the top of Fig. III.75a] have each three registers. The zodiac with constellations and planets in exaltations forms the middle register, with six signs to each strip, beginning with Pisces on the south.

As I describe the Esna A zodiac in more detail, the reader should also consult Figs. III.75b and III.75c. In Fig. III.75b I have included, from Neugebauer and Parker, (1) the common symbols used to identify the zodiacal figures, (2) the common planetary symbols for the deities representing the planets that accompany the signs of the zodiac, and (3) decanal numbers under the gods representing two differing sets of decans. In Fig. III.75c I have supplemented the preceding figure with the names of the signs and of the principal Egyptian elements of the zodiac.

The top registers of both strips of the zodiac together include the figures of decanal gods associated with the decans that are collectively designated as Esna A_1. This collection derives (with some divergencies) from the family of decans named Seti I B (see Fig. III.102).[6] Most of these figures are of standing deities, though a few show seated gods. The bottom registers of the strips present a different collection of decanal gods (A_2) that come from the Tanis family of decans (see Fig. III.102 again).[7] On comparison of the two sets, Neugebauer and Parker conclude: "that there are only twelve decans which occupy the same position in the zodiac [i.e., relative to the zodiac], that there are eight decans in Esna A_1 not found in Esna A_2 and the same number in Esna A_2 not appearing in Esna A_1."[8]

It is from these two families (Seti I B and Tanis) that the earliest complete list of Greek decans in the zodiac of Hephaestion of Thebes (4th century A.D.) comes (see Fig. III.103 and also NP, Vol. 3, pp. 170-71). I have stressed elsewhere that it was only with the Greeks and Romans that the Egyptian decanal names became simply designations for 10-degree divisions of the zodiacal belt rather than the actual stars themselves. The reason for this shift is obvious: by the time the names were absorbed the lists of the decans had become corrupted, out-of-date, and virtually useless. And though they were to be retained for funerary purposes, they apparently had little significance as the basis of nighttime astronomical horology. And furthermore since the placement of the Egyptian celestial reliefs on monuments was never accurately determined by the use of degrees, either in right ascension or declination, it was no doubt difficult to place them exactly in relationship to the newly encountered zodiacal constellations. But I need not go into these matters once more.

The first half of the decanal figures of A_1, i.e., those from the northern strip, come from a group of four expanded decanal lists, which may contain as many as fifty-nine decans. According to Neugebauer and Parker:[9]

> These no longer constitute a decan list primarily but represent the deities of the dual year, the combined lunar-civil year .... The fifty-nine deities are divided into forty-eight

> and eleven. The forty-eight consist of the thirty-six decans expanded by the addition of one new name among every three true decans. The remaining eleven deities are built up about the epagomenal days and represent the days between the lunar year of 354 days and the civil year of 365, the so-called "epact."

The numbers of the added deities in the truncated northern strip are 13a, 16a, 19a, and 22a. The next deity (the first in the southern strip) should have been 25a, but no "a"-decans are included in the southern strip. Furthermore the deities to the left of the god of decan 34, do not appear to be the normal deities associated with decans 35 and 36, and those for decans 5-13, which ought to be at the beginning of the northern strip, are absent because of the destruction of the right side of that strip. Presumably the decans 7a and 10a were added to the regular deities in the missing section of the northern strip. Therefore, because of these divergencies from a normal list of 36 decans, the total number of deities found in A_1 was not that given in the almost canonical list of 36 decans. This is not surprising in view of the probable correctness of the doubts expressed above by Neugebauer and Parker that the decans were still used in this late period to tell time at night as they were formerly. Furthermore, the astronomical exactness attributed to the round zodiac by early authors (Biot and others) and based on the ill-founded belief that the Dendera B ceiling represented a careful geometric projection of the so-called Egyptian celestial sphere cannot be established either by comparing with each other the many representations of the celestial diagram or by considering the mathematical prowess of the Egyptians.

The quite different group of deities in A_2, when it was intact, apparently consisted of a complete collection of 36 deities, without any of the "a" or added deities appearing in A_1, each register when intact consisting of 18 decans, i.e., three to a sign (see Fig. III.102, right column). On the northern strip before the god of decan no. 14 is found the figure of Hippopotamus (see bottom, Figs. III.75a and III.75b), the northern constellation, with a crocodile on her back and holding on to the Mooring Post, a constellation that played a significant role in the Ramesside star clock. There is also a bit of the Foreleg or Thigh representing the constellation of the Big Dipper. These are the only vestiges of the so-called northern constellations in the Esna A zodiac. But note that between decans 31 and 32 in the southern strip appears the constellation of Sothis (here written *stt*) with a cow and the goddess Satis in a bark (see *ibid.,* top). Finally notice that the names of only five decans of A_2 have been added to the figures of their gods, all but one having been written badly.

The central register in each strip consisted of the figures of six zodiacal signs, Pisces to Leo on the southern strip and, when intact, Virgo to Scorpio on the northern strip. Traces of the figures of Virgo, Libra, and Scorpio, now missing, were reported by de Villiers in his *Journal* of the French expedition to Egypt in 1798-1801.[10] As I have said before, with the zodiacal signs are the deities of the planets. They are located in the signs when they were thought to be in exaltations (see III.75c): Venus in Pisces, the sun in Aries, the moon in Taurus, Jupiter in Cancer, Mercury in Virgo, Saturn in Libra, and Mars in Capricorn. Actually it is because of the fact that the planets are in exaltations that we are able to affirm that Mercury and Saturn are placed as I have indicated, since they were no doubt in the now missing part of the northern strip. We should also notice that Orion in his bark is correctly placed between Taurus and Gemini. I have noted on Fig. III.75c the location of the figures of the winds and other unidentifiable mythological figures. Though the artists preparing these reliefs were trying in some fashion to reproduce the Egyptian celestial diagram and have the zodiac conform to it so that horoscopes could be made, the Egyptian elements of the Esna A zodiac, represented largely by the figures of deities, were apparently more reverential and decorative than they were precisely placed elements that would be astronomically useful to the deceased in his life in the Otherworld. For the most part, the protective deities of the decans in both lists were shown standing with a scepter in hand, usually on a platform or apparently a horizontal plane but occasionally on a bark.

The Round Dendera B Zodiac

Passing to our second model zodiac, namely Dendera B (Figs. III.76a and III.76b), we first note that it was originally located in the Eastern Chapel of Osiris on the roof of the Temple of Hathor at Dendera – more specifically, on the western half of the ceiling of the central (i.e., first enclosed) chamber. It was, however, removed from Dendera and transferred to Paris in 1828 where it remains at the Musée du Louvre. It was replaced in the Osiris chapel by a cast made from the original. The date is "Late Ptolemaic, before 30 B.C."[11]

Looking at Fig. III.76b, the reader will see that the zodiac is supported by four human-headed standing goddesses of the cardinal points of the compass, correctly oriented, and by four pairs of kneeling deities with falcon-heads. Similar deities are found among the supporters of the Sky-goddess Nut in earlier times (e.g., see Volume One, Fig. II.2a).

As for the disk itself, we see that the outer row contains the figures of 36 decans from the Tanis family (see Fig. III.76a where the numbers of the decans have been added on the rim; and see note 7). Then inside the decan-ring are located the signs of the zodiac. Interspersed among them are the planets in exaltations and some constellations, as they were in the Esna A zodiac, but in a far more detailed fashion. In the center are the northern constellations. Before we list the constellations, a somewhat more detailed description by Neugebauer and Parker of the plan of the ceiling is worth quoting:[12]

> Study of the ceiling makes it clear that its organization is far from haphazard and that it represents an attempt to picture the larger relationships in the heavens with some approach to fidelity. In *EAT* I, pp. 97-100, we found that the decanal stars were located in a band roughly parallel to and south of the ecliptic [see Fig. III.15]. The two northern constellations in the center of the ceiling [i.e., Hippopotamus and the Foreleg or Big Dipper, letters **A** and **C** in Fig. III.76a] place the pole star there as well. The decans are at the perimeter of the circular sky, and between them and the pole is the circle of the zodiac, askew as we should expect and not centered at the pole. Between the zodiac and the pole are various figures of constellations (**A** to **M** ...) which must be considered as north of the ecliptic. *Hippo* and *Mes[khetiu]* have been

> selected from the usual northern group as being, perhaps, most representative. The other eleven fill the remaining space, even crowding in between *Hippo* and *Mes* where they may be suspected of being quite out of place.
>
> Between the zodiac and the decanal band, because of the skewness of the former, there is a crescent-shaped area filled with constellations **N** to **Y**. Since the two that are identifiable of these, Orion (**P**) and Sothis (**S**), are decanal constellations, it is a safe conclusion that the others are either in that band as well or perhaps somewhat south of it. ....
>
> In this connection it should be remarked that the figures **T** and **U** are hardly constellations in themselves but are present because of association with Sothis. **T** is the goddess Sothis ... and **U** is Anukis ..., companion goddess of Elephantine with Satis, who herself has become identified with Sothis. This assemblage of cow in bark and two goddesses surely relates only to the constellation of Sirius. Other figures as well among those south of the zodiac may conceivably be aspects of decanal constellations instead of independent ones.

Now I shall give a list of the constellations referred to by the bold-face letters in the previous passage.[13] My list is based on that in NP, Vol. 3, pp. 200-03, but in a considerably truncated form (consult Figs. III.76a-c; the last of these gives the best view of some of the constellations despite its incorrect orientation and reversed image). There is an occasional reference to

Esna A and the rectangular zodiac Dendera E (NP, Vol. 3, Plate 40.) By "above" I mean toward the pole, by "below" or "under" I mean away from the pole, i.e., toward the rim of the disk.

Constellations North of the Ecliptic

A = Hippopotamus, one of the very familiar northern constellations.

B = Jackal on a hoe.

C = The Big Dipper as the Foreleg or Thigh of a Bull *(Meskhetiu)*.

D = Lion (?) lying on The Big Dipper.

E = Human-headed god with a two-feathers crown, near Gemini.

F = Small seated, human-headed god with a white crown, above Leo.

G = Group of a falcon-headed god with disk on head, a seated god above him, and a walking jackal below the latter, near Libra and Scorpio.

H = Group of human-headed god with a mace and ceremonial tail plus a goose below, with the latter between Sagittarius and Capricorn.

J = A headless body in an animal position, i.e., a four-legged stance. Perhaps a human body should have been there (see NP). The marking letter **J** is missing from Fig. III.76a, but the figure is easily found directly above Aquarius.

K = Group of a human-headed god with ceremonial tail and holding an animal by the horns. The latter is so badly drawn in Dendera B that the animal cannot be identified. In Dendera E the animal appears to be an oryx. The group is above **J** and between Aquarius and Pisces.

L = An oryx back to back with a baboon, above

Aries.

M = A disk with a wadjat-eye between Pisces and Aries. A full moon? If so, what does it mean? In Esna A a sundisk is above Aries.

Constellations South of the Ecliptic

N = A human-headed goddess holding a pig by the hind foot, both within a disk. Below Pisces.

O = A lion-headed goddess and a human-headed goddess, each holding a was-scepter. Below Aries.

P = The constellation of Orion as Osiris with white crown, ceremonial tail, and was-scepter. Below and between Taurus and Gemini.

Q = A crested bird behind and close to Orion's leg. Probably in Gemini.

R = A papyrus column surmounted by a falcon with a double crown. Under Gemini.

S = Sothis (Sirius) as a recumbent cow in a bark with a star between the horns, under Cancer. Compare Esna A, where she is accompanied by the goddess Satis. See fuller account in NP.

T = The goddess Satis, with bow and arrow, probably as an associate of Sothis. Under the front legs of Leo.

U = The goddess Anukis with a headdress of feathers and holding up two water vases, immediately behind Satis.

V = A seated woman, balancing a child on one hand and keeping it upright with the other, near and below the hindquarters of Leo.

W = A bull-headed god holding a hoe, under Virgo. Perhaps it is the Greek constellation Boötes; but the latter lies north of the ecliptic. It could be that it was placed here under Virgo for the spatial convenience of the artist. So reason NP.

X = A lion with forefeet in water, near Virgo and Libra. Perhaps the constellation of the Lion in the Ramesside clock (see Document III.14 and NP).

Y = A deity with a mixed body, upper part human, lower part hippopotamus. Near Libra and Scorpio.

Constellations either North or South of the Ecliptic [not in Dendera B]

Z = Human-headed goddess in a disk, with white crown. Only in Dendera E, where it is in Libra.

Aa = A serpent with four coils, inside a rectangle. In Leo in Dendera E. In Esna A it is above Leo.

A Bibliographical Conclusion

Finally, I add a short, bibliographical section. The controversy between J.B. Biot and A.J. Letronne in which Letronne demolished the older view that the Egyptian zodiacs were to be dated to the Pharaonic era some several centuries prior to the Ptolemaic and Roman periods appeared in one volume as follows: Biot, "Mémoire sur le zodiac circulaire de Denderah," *Mémoires de l'Institut Royal de France, Académie des Inscriptions et Belles-Lettres*, Tome 16,2 (Paris, 1846), pp. 1-92, and Letronne, "Analyse critique des représentations zodiacales de Dendéra et d'Esné," *ibid.*, pp. 102-210, with plates. The planispheric projection of the ceiling of Dendera B, which was drawn by Gau, based on Biot's calculations, and was one of the objects of Letronne's criticism, is Plate I of that volume (=Fig. III.76c in my volume).

Franz Boll ninety years ago discussed in a measured and useful way the relationship of the Dendera B constellations to those of the Greek constellations.[14]

The crucial parts treating Egyptian zodiacs in the third volume of Neugebauer and Parker, *Egyptian Astronomical Texts,* are as follows: Esna A zodiac (pp. 62-64), Seti I B family of decans (pp. 133-40), Tanis family of decans (pp. 140-49), the decans of the zodiac (pp. 168-74), constellations in zodiacs (pp. 199-202), the zodiacs (pp. 203-12).

Notes to Document III.17

1. For a list of Egyptian zodiacs, see Neugebauer and Parker, *Egyptian Astronomical Texts*, Vol. 3, pp. 204-05. To these add the interesting examples described by O. Neugebauer, R.A. Parker, and D. Pingree (with color notes by J. Osing), "The Zodiac Ceilings of Petosiris and Petubastis," in J. Osing et al., *Denkmäler der Oase Dachla aus dem Nachlass von Ahmed Fakhry,* (Mainz am Rhein, 1982), pp. 96-101, and Tafeln 36-44. Two of the zodiacs (one almost round and one clearly elliptical; see Fig. III.100a and III.100b) are in the tomb of Petosiris (dated between A.D. 54 and 84) and one (round; see Fig. III.101) is in that of Petubastis. Both tombs are on the Qaret el-Muzawwaqa toward the western extremes of the Oasis of Dachla in an area having Roman burials. As the authors note (on p. 96), these zodiacs were an important find "since nothing to compare with the Petosiris ceilings had yet been found in the valley of the Nile and the Mithraic elements were quite unexpected." In the general comment concerning the Petosiris ceilings we read (p. 100): "Both ceilings seem to be artistic expressions of the escape of the soul from the material to the spiritual world, based on a conflation of Egyptian, Greek, and, probably, Mithraic symbols. In this syncretism the artist displays a tendency parallel to that of the magical papyri and amulets produced in Hellenistic and Roman Egypt; only a Jewish element is missing from the ceilings to make the parallelism exact." The Petubastis ceiling is simpler and not so distinctive; its elements can be found in other zodiacs of the Nile valley. In the cases of all three ceilings comparison to other Egyptian zodiacs is made by the editors. We can also point to a tomb of the Roman period with two horoscopes that contain zodiacs of Egyptian brothers during the Roman period, namely Athribis A and Athribis B (see Fig.

III.104). There the arrangements of the zodiacs are loosely rectangular tending toward square.

The refutation of J.B. Biot's opinion of the much earlier origin of the Egyptian zodiacs and as well as that of Biot's conclusion that the round zodiac of Dendera B was an accurate planisphere was announced in the subtitle of the *Mémoire* of Letronne of 1846 mentioned in the concluding section of my treatment of Document III.17: "où l'on établit, 1° que ces représentations ne sont point astronomiques, 2° que les figures, autres que celles des signes du zodiaque, ne sont pas des constellations; 3° que le zodiaque circulaire de Dendéra n'est point un planisphère soumis à une projection quelconque." Letronne's second point that the figures (save those of zodiacal signs) were themselves not constellations is surely wrong, as is clear from my listing of the constellations on Dendera B.

2. Neugebauer and Parker, *Egyptian Astronomical Texts*, Vol. 3, p. 203. The Babylonian origin is particularly evident in the figures of Capricorn (the goat-fish), the double-headed archer Sagittarius on a winged horse with a scorpion-tail, and the spike of grain held by Virgo.

3. *Ibid.*

4. *Ibid.*, p. 205.

5. *Ibid.*, p. 62.

6. For an excruciatingly detailed treatment of the Seti I B family of decans, see Neugebauer and Parker, *Egyptian Astronomical Texts*, Vol. 3, pp. 133-40, and their previous discussion of the family in Vol. 1, pp. 83-86, 113-15. The main point to realize is that this family differs from all other families and reflects the change from decanal risings to decanal transits that is apparent in Document III.12. The figures associated with the decans, which is almost all we have from the decans in Esna A_1 (though there are a few names of decans poorly written and added near the figures), are quite different from those linked to any other family.

7. The Tanis family of decans is given a detailed examination in Neugebauer and Parker, *Egyptian Astronomical Texts*, Vol. 3, pp. 140-49.

8. *Ibid.*, p. 169.

9. *Ibid.*, p. 133.

10. E. de Villiers du Terrage, *Journal et souvenirs sur l'expédition d'Égypte (1798-1801)* [Paris, 1899], p. 161. Speaking of

the portico he says:

"La portique a huit colonnes. Dans plusieurs endroits du temple on remarque des murs, qui se sont enfoncés verticalement.

"Les sculptures peu soignées ne sont même pas d'un dessin correct.

"Ce qu'il y a de plus remarquable dans cette construction est un zodiaque disposé comme celui du portique du grand temple d"Esné, seulement les signes en separés par le portique.

"A gauche, en entrant, se trouvent le lion, le cancer, les gémeaux, le taureau, le bélier et les poissons; de l'autre côté sont sculptés le capricorn, le verseau et la moitié du sagittaire.

"Nous avons retrouvé par terre le morceau de la pierre sur lequel se trouvait le reste de ce signe.

"Le scorpion, la balance et la vierge se trouvent certainement sur les pierres tombées en monceau à l'entrée du temple, car, â travers les jours que le hasard a laissés dans cet amas, avec Jollois nous avons pu reconnaitre une portion de la queue du scorpion, un plateau de la balance et l'épi de la vierge."

11. Neugebauer and Parker, *Egyptian Astronomical Texts*, Vol. 3, p. 72.

12. *Ibid.*, p. 73.

13. *Ibid.*, pp. 200-02.

14. *Sphaera. Neue griechische Texte und Untersuchungen zur Geschichte der Sternbilder* (Leipzig, 1903), pp. 232-44. He analyzes and discusses the Dendera B and the Dendera E zodiacs and follows Letronne's conclusion concerning the lateness of their dates, and indeed of all Egyptian zodiacs. His comments on the mix of Greek, Babylonian, and Egyptian elements of these zodiacs is worth quoting (p. 235):

"Das Rätzel, was diese Bilder bezwecken, hat damit, wie ich meine, in einer sehr einfachen Weise seine endgültige Lösung gefunden. Aber von dem Charakter des in ihnen dargestellten Himmelsbildes ist noch Einiges zu sagen. Dass sie unter griechischem Einfluss stehen, hat man bisher stets aus dem Zodiacus geschlossen. Aber der Schütze ist, wie wir oben sahen, nicht griechisch, sondern echt altbabylonisch; man mass daraus mindestens soviel entnehmen, das die altägyptischen Priester auch von dort her direkte Einflüsse durch die Astrologen erfahren haben. Den Schluss, dass der ganze ägyptische Tierkreis unmittelbar babylonischer Herkunft ist, will ich noch nicht ziehen. Freilich weichen auch verschiedene andere Bilder des Zodiacus im

Einzelnen von den griechischen Darstellungen ab; die Zwillinge, um noch einmal das Wichtigste kurz zusammenzustellen, sind ein Paar verschiedenem Geschlechts, auf dem rechteckigen Bild mit dem Kopfschmuck ägyptischer Gottheiten ausgestattet; der Stier (als ganze Figur dargestellt) trägt ebenda auf dem Rücken die Mondscheibe; der Krebs ist hier in einer 'mehr ägyptischen Gestalt', um mit Brugsch zu reden, als Käfer dargestellt. Der Löwe steht in beiden Bildern unmittelbar auf der Schlange; die Haltung der flügellosen Jungfrau mit der Ähre ist ägyptisch stilisiert; zur Wage gehört eine Scheibe mit einer sitzenden Figur darin, die den einen Arm erhebt. Der Wassermann 'erscheint als Nilgott, mannweiblich mit hängenden Brüsten, in beiden Händen ausströmende Libationsvasen, auf dem Kopfe das obere Pschent oder auch Nilblumen'; die Fische sind durch eine Darstellung des Wassers von einander getrennt. Dass der Steinbock genau die Gestalt des babylonischen Ziegenfisches hat, macht den meisten griechischen Darstellungen gegenüber keinen Unterschied, mag aber hier doch erwähnt werden. Ob ausser dem Schützen nicht noch einige dieser Züge auf unmittelbaren babylonischen Einfluss zurückzuführen sind, können wir heute noch nicht beurteilen; ....

"Haben aber die ägyptischen Priester ... in dieser späten Zeit den fremden Einflüssen soweit nachgegeben, um die astrologischen Lehren und mit ihnen die Sternbilder des Tierkreises aufzunehmen, so muss die weitere Frage gewagt werden, ob nicht auch darüber hinaus andere Bestandteile der griechischen oder der babylonischen Sphäre in die ägyptische übergegangen sind. Vergleichen wir die Gestalten des Rundbildes von Dendera mit denen der griechischen Sphäre, so ist von den Bildern um den Nordpol, in der Mitte des Ganzen, die Nilpferdgöttin mit dem Schiffspflock und der Stierschenkel unzweifelhaft altpharaonisch, nicht griechisch auch das Übrige, was hier zu sehen ist, Pflug mit Schakal und das Tier auf dem Stierschenkel." In the remaining description of the various figures of Dendera B, Boll has added in brackets efforts at their approximate identification with modern constellations. I have already mentioned more than once the difficulty of such identifications, but the reader may want to consult them. I do not include Boll's descriptions of the constellations since I have already given in brief the later descriptions of Neugebauer and Parker.

Document III.18: Introduction

Inscriptions on the Statue of the Astronomer Harkhebi

In 1906 Ahmed Kamal reported the finding of a statue on a farm near Tell Faraoun by its owner and he immediately published it.[1] The statue is of the astronomer, snake charmer, and controller of scorpions Harkhebi. Made of basalt, it contains a very interesting inscription on its back pillar, where it appears in three vertical lines and describes Harkhebi's abilities and duties. A further short inscription is found on the left side in two vertical columns. George Daressy in 1916 republished the inscriptions with some corrections and accompanied the text with a French translation.[2] Finally in 1969 Neugebauer and Parker rendered an English translation of the parts of the inscriptions pertinent to astronomy, making use of Daressy's texts and translations and some improvements furnished by De Meulenaere from his own collation.[3] My translation of the inscriptions employs all three of the previous publications, adopting more often the suggestions of Neugebauer and Parker than those of Daressy.

The statue which is the object of our study in this document dates from the third century B.C. Hence we should also mention, from the Ptolemaic period, the statute of another stargazer or astronomer with similar but less elaborately described duties, one named Senty, the son of Pen-Sobek, justified. This astronomer was a priest in the Temple of Sobek in Khenty.[4] He tells us:

> [I have been designated among the chiefs] of men, the guides of the country chosen by the king. One will not find anyone more favored than I, [since] telling the hour conforms to the desire of the god so that he [i.e., the king] may give the order to erect constructions [such as temples, at the right time]. [My duties include] announcing to man his future, telling him about his youth and his death; telling the years, the months, the days, and the hours, the course of every star by the observation of its path, [I] Senty, son of Pen-Sobek, justified. He has said, "Oh, Master of the gods, in whose retinue you have caused me to reach a ripe old age, [and have given me] a beautiful tomb in the temple of.... I having been an astronomer (*wnwnw*) in the temple of his lord. [I] Senty, son of the same Pen-Sobek, justified."

It is evident from the inscriptions on both of the statues that the stargazers or astronomers were in fact priests of temple organizations, and it is tempting to follow Günther Roeder in believing that the chief priest of Heliopolis called "The Great Seer" (known from at least the third dynasty) was responsible for the hour watches or rather for astronomical observations in the temple of Re.[5] In fact, Roeder would equate the title with "supreme observer (namely of the sky), and the panther's skin he wears as a vestment is trimmed with stars." This makes some sense, for Heliopolis seems to have been the chief religious center of ancient Egypt where the civil calendar was set, and possibly where the observation of the heliacal rising of Sirius was first linked to that calendar when it was noticed that its appearance coincided with the sudden rising of the Nile.

The earliest extant inscriptions that describe his duties were placed on two astronomical instruments by an astronomer (*imy wnwt*) of the sixth century B.C. named Hor. These two inscriptions were transcribed by Borchardt.[6] The first appears on the underside of a rule (shaped like a shadow clock; see Figs. III.20a and III.20b, Berl. Mus. Nr. 14085). It tells us that the stargazer "knew *(rḫ)* the movements of the two disks (i.e., the sun and the moon) and every star to its abode (*demy*)." The second inscription on a notched palm rib (Figs. III.20a and III.28b, Berl. Mus. Nr. 14084) tells us that it was for "attending to the guiding (or introduction) of festivals and giving all people their hours."

Such activities are reflected in the list of books in the library room of the temple of Edfu (built by Ptolemy VIII Euergetes II, 170-63, 145-116 B.C.). This catalogue, which we reported on in Volume One (pp. 45-46), had books on the Knowledge of the Periodic Returns of the Two Celestial Spirits: the Sun and the Moon, and on The Governing of the Periodic returns of the Stars. As Otto Neugebauer has shown, Clement of Alexandria (2nd. century A.D.) appears to have read that list as he describes the four Hermetic books on astronomy studied by the Egyptian Horoscopist in order that he might know them by heart: books on the arrangement of the fixed stars, on the position of the sun and the moon and the five planets, on the syzygies and phases of the sun and the moon, and on the risings.[7]

Our astronomer Harkhebi's duties are described here in Document III.18 in more detail than is evident in the inscriptions of the stargazers noted above. We see that Harkhebi observed everything observable on heaven and earth, that he predicted risings and culminations of stars, that he was particularly concerned with predicting, presumably at the beginning of the civil year, when the heliacal rising of Sirius was to take place during that year. Harkhebi actually seems to have had the duty of checking his prediction with the actual rising. It was just such a prediction and later confirmation that was evident in entries for 1872 B.C. from a temple register at Illahun (see Document III.10). It is clear that Harkhebi also kept track of day and night hours, and the risings and settings of the sun.

Finally we should note at the end of our document that Harkhebi was also a priest of Selket (=Serqet) as well as a stargazer, that is, a medical priest of the sort we described in Volume One (p. 19) who knew the charms to pacify scorpions and specialized in the treatment of snake bites and stings. Furthermore, it is evident that his duties as adviser to the god (i.e., the king) were extensive and were such as to advise the king on the time to travel and to protect him in the course of his journeys.

For the hieroglyphic text of the document as published by Daressy, see Fig. III.105. I have indicated in my translation where the line breaks occur by arabic numerals inserted between slant lines. On my Fig. III.105 they occur where the blocks of slant hatch lines appear.

Notes to the Introduction of Document III.18

1. A. Kamal, "Rapport sur quelques localités de la Basse-Égypte," *ASAE*, Vol. 7 (1906), pp. 239-40 (full article, pp. 232-40).

2. G. Daressy, "La statue d'un astronome," *ASAE*, Vol. 16 (1916), pp. 1-5.

3. Neugebauer and Parker, *Egyptian Astronomical Texts*, Vol. 3, pp. 213-16.

4. G. Daressy, "Antiquités trouvées à Fostat," *ASAE*, Vol. 18 (1919), pp. 275-78.

5. See my Volume One, pp. 141 and 380, and G. Roeder, "Die Himmelsbeobachtung der alten Ägypter," *Rundschau der gesamten Sternforschung für Himmelskunde und Fachastronomen: "Sirius,"* 1917 Heft. 1/2, p. 4 (full article, pp. 1-11; at least these are pages of a Sonder Abdruck I possess).

6. L. Borchardt. "Ein altägyptisches astronomisches Instrument," *ZÄS*, Vol. 37 (1899), p. 11 (full article, pp. 10-17). Also see above, Chapter Three, nn. 70-71, and the remarks in the chapter to which they refer.

7. Neugebauer, "Egyptian Planetary Texts," *Transactions of the American Philosophical Society*, N.S., Vol. 32, Part II (Jan. 1942), pp. 237-39 (full article, pp. 209-50). Note that when Clement speaks of the four works as "Hermetic," he uses vocabulary that in his time referred to Egyptian sacred literature. The reader will also find useful the citation to other classical authors concerning the Egyptian knowledge of astronomy given by R. Lepsius, *Die Chronologie der Ägypter*, Einleitung etc. (Berlin, 1849), pp. 55-56, 58-60.

Document III.18

Inscriptions on the Statue of the Astronomer Harkhebi

[A. Inscription on the Back Pillar:]

/1/ Hereditary prince, count, sole friend [of the king], skilled in sacred writings, one who observes everything observable in the heaven and on earth, skilled in observing the stars with no erring, one who announces rising(s) and setting(s) at their times, with the gods who arrange (*or* foretell) the future, for which [activity] he purified himself on their days when [the decan] Akh rose [heliacally] beside Bennu (Venus) from earth, and he made the lands content by his predictions; one who observes the culmination of every star in the sky, knowing the [heliacal] rising of /2/ every....in a good year; one who announces (*or* foretells) the [helical] rising of Sothis at the beginning of the year[1] and [then] observes her on her first festival day [i.e., when she actually rises heliacally], calculating her course at the designated times, observing what she does every day; everything she has ordered (*or* foretold) is in his charge; one knowing the northing and southing of the sun disk, announcing all of its wonders [i.e., the special phenomena of the disk] and appointing for them [i.e., establishing] their times; he declares when they have occurred, coming at their times; one who divides the hours of the two times (i.e., day and night) without erring at night..../3/ one knowledgeable in everything which is seen in the sky, [i.e., everything][2] for which he has waited [or expected], one who is skilled with respect to their conjunction(s) and regular movement(s),[3] who does not disclose [anything] at all concerning his report [to the king] after [his] judgment [has been made], discrete with everything he has seen; no master can refute one of his counsels to the Lord of the Two Lands; [he is] one who pacifies scorpions, understands the removal of serpents [by] indicating their places and drawing the serpents to them, closing the mouths of their inhabitants, their serpents....

[B. Inscription on the Left Side:]

/4/ Initiated in his [i.e., the king's or god's] mysteries, favoring [i.e., bringing favor to] his voyages and protecting his route, overcoming [opponents] of his expedition..../5/ [the king or god is] pleased with his counsel, the god loving him as the controller of the scorpion, [he is] Harkhebi, son of the one honored before Wadjet.

Notes to Document III.18

1. Neugebauer and Parker, *Egyptian Astronomical Texts*, Vol. 3, p. 215, believe this to be the beginning of the original lunar year, which they (or at least Parker) believed to be controlled by the heliacal rising of Sirius. But I have expressed my doubts about this aspect of Parker's description of the original lunar calendar, and hence I believe this reference to be to the beginning of any ad hoc Sothic year.

2. *Ibid.*

3. *Ibid.*

Postscript

A Petroglyph Discovered at Nekhen with Possible Astronomical Significance

While I was serving on the Advisory Board of the Egyptian Studies Association of the University of South Carolina, James O. Mills of the Hieraconpolis excavation team communicated to me a paper presented to the Society for African Archaeologists at its meeting of March 22-25, 1990. The paper was entitled "Predynastic Astronomy at Hieraconpolis" and contained the description of a petroglyph (see Figs. III.106a amd III.106b) discovered at the ancient site of Hieraconpolis (i.e., Nekhen) by the author and Ahmed Irawy Radwan, which in all likelihood dates from predynastic times and may have served as the recording of solar risings and settings during the year from solstice to solstice. Though the investigation of the glyph is only in a preliminary form and the author presents a somewhat negative conclusion concerning its applicability to solar risings and settings, he has recently told me that he wishes to investigate further the rock which bears the glyph and which is positioned on a steep slope to see whether it has shifted from a previous position; for if a shift of the rock of 10-degrees has somehow taken place since its original recording, then it could well be a record of actual annual solar risings and settings from solstice to solstice. If such is the case, and if the glyph is of Predynastic date (as are the many other petroglyphs of the site), then it would surely provide evidence for the knowledge of a 365-day solar year, which, when expressed as the sum of 12 schematized lunar months of 30 days each and five epagomenal days, could have produced the antecedent to the Egyptian civil year that remained in force throughout Pharaonic history.

I shall now, with Mr. Mills' permission, quote much of his paper with the hope that he will be able to solve the puzzles presented by the glyph in the very near future (I have omitted the section on the civil calendar already discussed at length in my volume). Note that throughout Mills has adopted the alternate spelling "Hierakonpolis."

> Astronomy at Hierakonpolis
> James O. Mills
> Paper presented at The 1990 Society for Africanist Archaeologists. Biennial Conference, March 22-25, 1990, University of Florida, Gainesville, Florida.
>
> The site of Hierakonpolis, located approximately half-way between modern Aswan and Luxor in Upper Egypt, is best known for its extensive and well preserved Predynastic settlement and cemeteries (4th mil. to early 3rd mil. B.C.), and for the Archaic Period temple and walled compound site of Nekhen in the modern alluvium.
>
> Since the work of Quibell and Green (Quibell 1900; Quibell and Green 1902) at the site at the turn of the century, a number of expeditions have conducted excavations at

Hierakonpolis (see Hoffman 1982:3, and Adams 1974). The present expedition, initially under the direction of Walter Fairservis, is now under the direction of Michael Hoffman [but at the time of writing this book (1993) he is deceased and Fairservis is once more directing excavations at Nekhen].

The regional focus and multi-disciplinary aspect of our research has generated a diversity of research questions and projects subsumed under the broader goal of understanding the rise of the Egyptian state and the development of Ancient Egyptian culture.

An ancillary project to our ongoing investigations at numerous sites throughout the concession has been the building of a corpus of graffiti and other art forms found on pot sherds, and other objects as well as on rock surfaces. Together with such pieces as the Painted Tomb and the Narmer Palette found at Hierakonpolis by Quibell and Green at the end of the 19th century (Quibell 1900; Quibell and Green 1902), the corpus serves as an initial step in understanding the emergence of symbolic and textual iconography in its nascent stages.

Isolate examples of graffiti occur throughout the Hierakonpolis region but are concentrated in several localities. One such site, graffiti hill, is a prominent sandstone inselberg, at the juncture between the Great Wadi (the primary relict drainage system for the desert region) and the Pleistocene floodplain, which today forms the low desert. At the base of this formation is a complex of Predynastic sites including a cemetery, pottery kilns, several clusters of domestic structures, and well preserved trash middens over a meter in depth (Hoffman 1970, 1974, 1982; and Harlan 1985). These sites range in date from ca. 3900-3300 B.C.

In succeeding field seasons, surveys of the adjacent hill continue to reveal unrecorded graffiti (Berger 1982; Hoffman 1987). In 1980, in a cluster of boulders at the hill's base, and adjacent to the Gerzean period cemetery, were found four finely executed incurved boats, analogous in form and detail to boats depicted on Gerzean vessels (Berger 1982). Other petroglyphs on the hill include depictions of undulates (likely gazelles or ibexes), elephants, a giraffe, and most recently a large cat, probably a lion (see Berger 1982; and Hoffman 1987:234-37).

Only one human figure has been found at the site. It is of a water carrier and is located near the base of the hill (Hoffman 1987:235-36). We have also found the usual abundance of isolate crosses and single incisions which remain to be explained. It has been postulated that they might be vantage and even surveying points. Alternatively, they may be sharpening grooves as are found on modern masonry throughout the Egyptian country-side (personal observations).

While surveying the hill in 1986, Ahmed Irawy Radwan of the Egyptian Antiquities

Service and I discovered a complex geometric design etched and pecked into the bedrock, near the crest of the hill at about 40 meters above the low desert (see figures in Hoffman 1987:237 [see my Figs. III.106a and III.106b]). The glyph lies level with the slope of the hill, which faces the flood plain. 12 or 13 pecked divots form a bulging "V" shape which points eastward. 2 and possibly 3 additional divots form a horizontal row within the "V". The divots are superimposed over an earlier series of short parallel incised marks along the left arm of the "V" and across its center. In addition, a series of incisions above the "V" are arranged in an arch. This petroglyph is a totally unique representation.

The glyph's symmetry and repetition of form suggest that it may have been used for Pneumonic or orientation purposes. Its bilateral symmetry along an east-west axis presented the question of its being a record or device for astronomical observations.

Of primary interest are the incisions arranged in an arch above the "V", as they might mark the cyclic change in rise-location on the horizon of celestial bodies within the solar system.

To test this hypothesis the magnetic direction of each incision was recorded using a Brunton pocket transit: and with the same compass, readings were taken of the setting sun on two arbitrary dates, Jan. 26, and Feb. 18, 1988. The difference in azimuth readings over this 23 day period provided us with a rough field measure of the rate of azimuth change per day. We then extrapolated back to the Winter solstice, Dec. 22, to determine where on the horizon the sun should set on that date (or alternatively the suns rising at the summer solstice at the inverse angle). The calculated figure was 215 degrees; the glyph's left-most hash mark, also measured 215 degrees.

These calculations could only serve as tentative estimations however, since the rate of change for azimuth risings and settings is non-linear, decreasing its rate with proximity to the solstices (Aveni 1981:15).

After returning from the field we consulted "The Rising and Setting Azimuth Tables of Principal Astronomical Bodies" (see Aveni 1972), a computer generated almanac which retrodicts for each whole degree of latitude, the solstice azimuths over 500-yr. increments as far back as 1500 BC. Over the millennia the solstice rise set azimuths at the site have varied to within a degree, well within the resolution of our rather crude petroglyph. The difference between winter and summer solstice azimuths, over the past 3,000 years, has been between 52 and 53 degrees (see Aveni 1972), which coincides well with the 54 degrees measured difference between the incisions at either extreme of the glyph's arch; additional markings may have been present, however, and subsequently lost on the right side of the arch.

A final and crucial check remained:

rectifying the compass readings and the almanac data, listed relative to true north. Local maps present a declination of 22 1/2 degrees while the World Magnetism Charts from the U.S. Department of Commerce's Coast and Geodetic Survey (Nelson *et al.* 1962:12) show the 0 degree declination isobar passing through the site. Such extreme variation between readings taken in the past 50 years is unlikely to be explained by wandering of the magnetic poles or local anomalies. Regardless, using the compass readings of sun set and comparing with the U.S. Naval Almanac for those dates a declination of 17 +/- 1 degrees east for our compass was determined. Adjusting the glyph measurements accordingly there is a 10 decree variance with the rise/set tables - in effect the glyph is cocked 10 degrees clockwise from effectively marking the solar cycle. (The rock into which the glyph was cut may have shifted). At 25 degrees of latitude the extreme rising and setting points of the moon fall 6 degrees outside of the solstice points which likewise is not reflected by the arch of incisions.

Thus the specific hypothesis that the glyph marked the solar or lunar horizon cycle looks doubtful. Various other astronomical events will have to be checked against the glyph before the larger astronomical hypothesis can be laid to rest....

The enigmatic petroglyph at Hierakonpolis provides a potentially important clue to prehistoric methods of timekeeping and the

> genesis of Ancient Egyptian calendric system(s). We can date the glyph only indirectly by association with Predynastic sites (ca. 3900-3300 B.C.) and over 70 other pieces of rock art on graffiti hill; none of these are suggestive of a post Dynasty Zero date.
>
> We have begun our inquiry by testing the most obvious celestial phenomena, and only on the upper part of the glyph. Further research will inquire into the potential significance of the "V" shaped portion of the glyph and its relationship to other graffiti and features in the region.

I now append the expanded bibliographical references given by Mills at the end of his paper, omitting only those which refer to the passages not included in this postscript.

REFERENCES CITED

Adams, Barbara A.

1974: *Ancient Hierakonpolis (and Supplement)*, Aris & Phillips Ltd., Warminster, England.

Aveni, Anthony F.

1981: "The Astronomical and Technical Background for Archaeoastronomical Studies," *Advances in Archaeological Method and Theory,* 4, ed. by M.B. Schiffer, pp. 1-77, Academic Press, New York.

1972: "Astronomical tables intended for use in astro-archaeological studies," *Amer. Antiq.*, 37:531-40.

Berger, Michael

1982: "The Petroglyphs at Locality 61," *The*

Predynastic of Hierakonpolis. An Interim Report, edited by Michael A. Hoffman, pp. 61-65. Egypt. Stud. Assoc. Publ., No. 1, Alden Press, Oxford.

Harlan, John Frederick
1985: *Predynastic Settlement Patterns: A view from Hierakonpolis*, Ph.D. dissertation, Dept. of Anthropology, Washington University, St. Louis, Missouri.

Hoffman, Michael Allen
1970: *Culture, History and Cultural Ecology at Hierakonpolis from Palaeolithic Times to the Old Kingd om*, Ph.D. dissertation, Dept. of Anthropology, University of Wisconsin, Madison, Wisconsin.
1974: "The Social Context of Trash Disposal in an Early Dynastic Egyptian Town," *Amer. Antiq.*, 39, 35-50.
1982: *The Predynastic of Hierakonpolis: Interim Report* (Michael A. Hoffman, editor). Egyptian Studies Association, Publication No. 1, Alden Press, Oxford.
1987: *Final Report to The National Endowment for the Humanities on Predynastic Research at Hierakonpolis, 1985-86*. (N.E.H. Grant No. RO-20805-85). Earth Sciences and Resources Institute, Univ. of South Carolina, Columbia, South Carolina.

Nelson, James H., Louis Hurwitz, and David G. Knapp
1962: *Magnetism of the Earth*, U.S. Department of Commerce and Coast and Geodetic Survey, Publication 40-1, U. S. Government Printing Office: Washington D.C.

Quibell, J.E.
1900: *Hierakonpolis I*, Egyptian Research Account, 4, London.

Quibell, J.E. and F.W. Green
1902: *Hierakonpolis II*, Egyptian Research Account, 5, London.

Part Three

Bibliography and Indexes

Abbreviations Used in Text and Bibliography

ASAE = *Annales du Service des Antiquités de l'Égypte.*

BIFAO = *Bulletin de l'Institut Français d'Archéologie Orientale.*

JEA = *The Journal of Egyptian Archaeology.*

JNES = *Journal of Near Eastern Studies.*

LN = Lange, H.O., and O. Neugebauer, *Papyrus Carlsberg No. I. Ein hieratisch-demotischer kosmologischer Text (Det Kongelike Danske Videnskabernes Selskab. Historisk-filologiske Skrifter, Band I, Nr. 2)* (Copenhagen, 1940).

MDAIK = *Mitteilungen des Deutschen Archäologischen Instituts, Abteilung Kairo.*

NP = Neugebauer, O. and Parker, R.A., *Egyptian Astronomical Texts*, 3 vols. (Providence, Rhode Island and London, 1960-1969).

PSBA = *Proceedings of the Society of Biblical Archaeology.*

Urkunden = G. Steindorff, ed., *Urkunden des ägyptischen Altertums* (For *Urkunden* I, II, IV, and VII, see the Bibliography, Sethe, K., *Urkunden des Alten Reichs*, *Hieroglyphische Urkunden der griechisch-römischen Zeit, Urkunden der 18. Dynastie*, and *Historisch-biographische Urkunden des Mittleren Reiches.*).

Wb, see the Bibliography, Erman, A., and H. Grapow, *Wörterbuch.*

ZÄS = *Zeitschrift für ägyptische Sprache und Altertumskunde.*

Bibliography

(In multiple entries under an author's name, articles are listed before books.)

Bakir, Abd el-Moshen, *The Cairo Calendar No. 86637* (Cairo, 1966).

Barta, W., "Die ägyptischen Mondaten und der 25-Jahr-Zyklus des Papyrus Carlsberg 9," *ZÄS*, Vol. 106 (1979), pp. 1-10.

Barta, W., "Die Chronologie der 12. Dynastie nach den Angaben des Turiner Königspapyrus," *Studien zur Altägyptischen Kultur*, Vol. 7 (1979), pp. 1-9.

Bayoumi, A., and O. Guéraud, "Un nouvel exemplaire du Décret de Canope," *Annales du Service des Antiquités de l'Égypte*, Vol. 46 (1947), pp. 373-82, plus Plate LXXXI.

Berlin Museum, *Ägyptische Inschriften aus den Königlichen Museen zu Berlin*, Vol. 1 (Leipzig, 1913). Edited by the Generalverwaltung, i.e., by H. Schäfer and others, q.v.

Bernand, A., *De Thèbes à Synène* (Paris, 1989), pp. 48-49. It concerns a fragment of the Greek text of the Decree of Canopus.

Biegel, R.A., *Zur Astrognosie der alten Ägypter* (Zurich, 1921).

Bilfinger, G., *Die Sterntafeln in den ägyptischen Königsgräbern von Bibân el Molûk* (Stuttgart, 1891).

Biot, J.B., "Mémoire sur le zodiac circulaire de Denderah," *Mémoires de l'Institut Royale de France, Académie des Inscriptions et Belles-Lettres*, Tome 16,2 (Paris, 1846), pp. 1-92.

Boas, G., *The Hieroglyphics of Horapollo* (New York, 1950); and see Horapollo.

Böker, R., "Über Namen und Identifizierung der ägyptischen Dekane," *Centaurus*, Vol. 27 (1984), pp. 189-217.

Boll, F., *Sphaera. Neue griechische Texte un Untersuchungen zur Geschichte der Sternbilder* (Leipzig, 1903), pp. 231-44.

Borchardt, L., "Altägyptische Sonnenuhren," *ZÄS*, Vol. 48 (1910), pp. 9-17.

Borchardt, L., "Der zweite Papyrusfund von Kahun und die zeitliche Festlegung des mittleren Reiches der ägyptischen Geschichte," *ZÄS*, Vol. 37 (1899), pp. 89-102.

Borchardt, L., "Ein altägyptisches astronomisches Instrument," *ZÄS*, Vol. 37 (1899), pp. 10-17.

Borchardt, L., "Eine Reisesonnenuhr aus Ägypten," *ZÄS*, Vol. 49 (1911), pp. 66-68.

Borchardt, L., *Die altägyptische Zeitmessung* (Berlin and Leipzig, 1920) (E. von Bassermann-Jordan, ed., *Die Geschichte der Zeitmessung und der Uhren*, Vol. 1).

Borchardt, L., *Quellen und Forschungen zur Zeitbestimmung der ägyptischen Geschichte*, Vol. 1: *Die Annalen und die zeitliche Festlegung des alten Reiches der ägyptischen Geschichte*, (Berlin, 1917), Vol. 2: *Die Mittel zur zeitlichen Festlegung von Punkten der ägyptischen Geschichte und ihre Anwendung* (Cairo, 1935).

BIBLIOGRAPHY

Borchardt, L., see Neugebauer, P.V., and.

Breasted, J.H., *Ancient Records of Egypt*, Vol. I (1908), pp. 101-03.

Breasted, J.H., *The Edwin Smith Papyrus*, Vol. 1 (Chicago, 1930).

Brugsch, H., "Ein neues Sothis-Datum," *ZÄS*, Vol. 8 (1870), pp. 109-10.

Brugsch, H., *Die Ägyptologie*, (Leipzig, 1891).

Brugsch, H., *Matériaux pour servir la construction du calendrier des Égyptiens* (Leipzig, 1864).

Brugsch, H., *Recueil de monuments égyptiens*, I (Leipzig, 1862).

Brugsch, H., *Thesaurus inscriptionum aegyptiacarum*, (Graz, 1968, unaltered repr. of Leipzig ed. of 1883-91).

Bruins, E.M., "Egyptian Astronomy," *Janus*, Vol. 52 (1965), pp. 161-80.

Bruins, E.M., "The Egyptian Shadow Clock," *Janus*, Vol. 52 (1965), pp. 127-37.

Budge, E.A. Wallis, *The Decree of Canopus* (London, 1904). This is Volume III of his *The Decrees of Memphis and Canopus in Three Volumes*.

Budge, E.A. Wallis, *An Egyptian Hieroglyphic Dictionary*, 2 Vols. (Dover reprint, New York, 1978).

Bull, L.S., "An Ancient Egyptian Astronomical Ceiling-decoration," *Bulletin of the Metropolitan Museum of Art*, Vol. 18 (1923), pp. 283-86.

Caminos, R.A., *The New-Kingdom Temples of Buhen*, Vol. 2 (London, 1974).

Černý, J., "The Origin of the Name of the Month Tybi," *ASAE*, Vol. 43 (1943), pp. 173-81.

Champollion, J.F., *Lettres écrits d'Égypte et de Nubie en 1828 et 1829*, new ed. (Paris, 1868).

Champollion, J.F., *Monuments de l'Égypte et de la Nubie*, 4 Vols. (Paris, 1835-45); I have used the photographic reduction of the original edition produced in 1970-71 by the Centre de Documentation Oriental in Geneva.

Champollion, J.F., *Monuments de l'Égypte et de la Nubie. Notices descriptives conformes aux manuscrits autographes rédigés sur les lieux*, 2 vols. (Paris, 1844-89). The second volume, i.e., the volume of 1889, was written in the hand of G. Maspero.

Chatley, H., "Egyptian Astronomy," *The Journal of Egyptian Archaeology*, Vol. 26 (1940), pp. 120-26.

Chatley, H., "The Egyptian Celestial Diagram," *The Observatory*, Vol. 63 (1940), p. 69.

Clère, J.J., "Un texte astronomique de Tanis," *Kêmi*, Vol. 10 (1949), pp. 3-27.

Daressy, G., "Antiquités trouvées à Fostat," *ASAE*, Vol. 18 (1918), pp. 275-78.

Daressy, G., "Deux clepsydres antiques," *Bulletin de L'Institut Égyptien*, 5me Série, Vol. 9 (l'Année 1915), pp. 5-16.

Daressy, G., "Grand vase en pierre avec graduations," *ASAE*, Vol. 3 (1902), pp. 236-39.

Daressy, G., "La statue d'un astronome," *ASAE*, Vol. 16 (1916), pp. 1-4.

Daressy, G., "Une ancienne liste des décans égyptiens," *ASAE*, Vol. 1 (1900), pp. 79-90.

Daumas, F., *Les moyens d'expression du Grec et de l'Égyptien comparé dans les Décrets de Canope et de Memphis*, in *Supplements annales du Service des Antiquités de l'Égypte*, Cahier No. 16 (Cairo, 1952).

de Buck, A., "The Dramatic Text," in H. Frankfort, *The Cenotaph of Seti I at Abydos* (London, 1933), Vol. I, Chapter IX, Vol. 2, Plates 84-85.

de Buck, A., *The Egyptian Coffin Texts*, II (Chicago, 1938).

Description de l'Égypte, ou, Recueil des observations et des recherches qui ont été faites en Égypte pendant l'expédition de l'armée française, Plates, 5 vols. (Antiquity), 2nd ed. (Paris, 1809, et seq.). Cf. Gillispie, C.

Devauchelle, D., "Wasseruhr," *Lexikon der Ägyptologie*, Vol. 6 (Wiesbaden, 1986), cc. 1156-57.

Dittenberger, W., *Orientis graeci inscriptiones selectae*, Vol. 1 Leipzig (1903), pp. 91-110 (Greek text of the Decree of Canopus).

Dümichen, J., "Namen und Eintheilung der Stunden bei den alten Ägyptern," *ZÄS*, Vol. 3 (1865), pp. 1-4.

Ebers, G.M., *Papyros Ebers, das hermetische Buch über die Arzeneimittel der alten Ägypter in hieratischer Schrift.... Mit hieroglyphisch-lateinischem Glossar von Ludwig Stern*, Vol. 1 (Leipzig, 1875).

Edgar, C.C., *Sculptors' Studies and Unfinished Works: Catalogue général des antiquités égyptiennes du Musée du Caire, Nos. 33301-33506* (Cairo, 1906).

Edgerton, W.F., "Chronology of the Twelfth Dynasty," *JNES*, Vol. 1 (1942), pp. 307-14.

BIBLIOGRAPHY

Edgerton, W.F., "Chronology of the Early Eighteenth Dynasty (Amenhotep I to Thutmose III)," *The American Journal of Semitic Languages and Literature*, Vol. 53 (1936), pp. 188-97.

Eggebrecht, A., *Suche nach Unsterblichkeit: Totenkult und Jenseitsglaube im Alten Ägypten* (Hildesheim, 1990).

Eisenlohr, A., "Der doppelte Kalender des Herrn Smith," *ZÄS*, Vol. 8 (1870), pp. 165-67.

Epigraphic Survey of the University of Chicago, The, see Nelson, H.H., et al.

Erman, A., "Monatsnamen aus dem neuen Reich," *ZÄS*, Vol. 39 (1901), pp. 128-30.

Erman, A., and H. Grapow, *Wörterbuch der ägyptischen Sprache*, 7 vols. (Leipzig, 1926-53), *Die Belegstellen*, 5 vols. (Leipzig, 1953). The whole work was reprinted in Berlin in 1971. It is abbreviated as *Wb.*

Faulkner, R.O., *The Ancient Egyptian Coffin Texts*, Vol. 1 (Westminster, England, 1973), pp. 134-35.

Frankfort, H., *The Cenotaph of Seti I at Abydos* (London, 1933), Vol. I: Texts; Vol. II: Plates.

Gardiner, A.H., "Mesore as first month of the Egyptian Year," *ZÄS*, Vol. 43 (1906), pp. 136-44.

Gardiner, A., *Egyptian Grammar*, 3rd ed. rev. (London, 1973).

Gardiner, A., *The Tomb of Amenemhēt (No. 82)* (London, 1915).

Gayet, A., *Le Temple de Louxor* (Paris, 1894), Plate LXVIII (Pl. LXXIV, Fig. 212).

Gensler, F., *Die thebanischen Tafeln stündlicher Sternaufgänge* (Leipzig, 1872).

Gibbs, S.L., *Greek and Roman Sundials* (New Haven and London, 1976).

Gillispie, C., *Monuments of Egypt. The Napoleonic Edition.* (Princeton, N.J., 1987).

Godron, G., "Études sur l'Époque Archaique," *BIFAO*, Vol. 57 (1958), pp. 143-55.

Goodwin, C.W., "Notes on the calendar in Mr. Smith's papyrus," *ZÄS*, Vol. 11 (1873), pp. 107-09.

Grapow, H., see Erman, A., and.

Grenfell, B.P., and A.S. Hunt, *The Oxyrhynchus Papyri*, Part III (London, 1903), No. 470, pp. 141-46.

Griffith, F.Ll., "Notes on Egyptian Weights and Measures," *Proceedings of the Society of Biblical Archaeology*, Vol. 14 (1892), pp. 403-50, Vol. 15 (1893), pp. 301-16.

Griffith, F.Ll., *Hieratic Papyri from Kahun and Gurob*, 2 vols. (London, 1908).

Griffith, F.Ll., and W.M.F. Petrie, *Two Hieroglyphic Papyri from Tanis* (London, 1889).

Groff, W.N., *Les deux versions démotiques du décret du Canope* (Paris 1888).

Gundel, W., *Dekane und Dekansternbilder: Ein Beitrag zur Geschichte der Sternbilder der Kulturvölker* (Glückstadt und Hamburg, 1936).

Gundel, H.G., *Weltbild und Astrologie in den griechischen Zauberpapyri* (Munich, 1968), *Münchener Beiträge zur Papyrusforschung und antiken Rechtsgeschichte*, 53. Heft (1968).

Gunn, B., "The Coffins of Ḥeny," *ASAE*, pp. 166-71.

Hayes, W.C., "Chapter VI. Chronology," *The Cambridge Ancient History,* Vol. 1, Part 1 (Cambridge, 1970), p. 182.

Helck, W., *Urkunden des ägyptischen Altertums: IV. Abt. Urkunden der 18. Dynastie*, Heft 19 (Berlin, 1957), 21 (Berlin, 1958).

Helck, W., and W. Westendorf, eds., *Lexikon der Ägyptologie*, Vol. 6 (Wiesbaden, 1986), cc. 1156-57 ("Wasseruhr" by D. Devauchelle).

Horapollo, *Hieroglyphica*, ed. of C. Leemans (Amsterdam, 1835).

Hornung, E., "Chronologie in Bewegung," *Festschrift Elmar Edel 12 März 1970. Unter Mitwirkung von Agnes Wuckelt und Karl-Joachim Seyfried herausgegeben von Manfred Görg und Edgar Pusch* (Bamberg, 1979), pp. 247-52.

Hornung, E., *The Tomb of Pharaoh Seti I...photographed by Harry Burton* (Zurich and Munich, 1991).

Hornung, E., *Untersuchungen zur Chronologie und Geschichte des Neuen Reiches* (Wiesbaden, 1964).

Hornung, E., (mit Beiträgen von: S. Bickel, E. Staehelin, D. Warburton), *Zwei ramessidische Königsgräber: Ramses IV. und Ramses VII.* (Mainz am Rhein, 1990).

Hughes, G.H., "The Sixth Day of the Lunar Month and the Demotic Word for 'Cult Guild'," *MDAIK*, Vol. 16, II. Teil (1958), pp. 147-160.

James, T.G.H. and W.V. Davies, *Egyptian Sculpture* (London, 1983).

Kamal, A., "Rapport sur quelques localités de la Basse-Égypte," *ASAE*, Vol. 7 (1906), pp. 232-40.

Kamal, Ahmed Bey, *Stèles Ptolémaiques et romaines*, 2 vols. (Cairo, 1904-05) 183 (=*Catalogue général des antiquités du Musée du Caire. Nos. 22007-22208)*.

Krall, J., "Der Kalender des Papyrus Ebers," *Recueil de travaux relatifs à la philologie et à l'archéologie égyptiennes et assyriennes*, Vol. 6 (1885), pp. 57-63.

Krauss, R.K., "Sothis, Elephantine und die altägyptische Chronologie," *Göttinger Miszellen*, Heft 50 (1981), pp. 71-80.

Krauss, R.K., *Probleme des altägyptischen Kalenders und der Chronologie des Mittleren und Neuen Reiches in Ägypten* (Diss. Berlin, 1981).

Lacau, P., *Catalogue général des antiquités égyptiennes du Musée du Caire: Nos. 28087-28126: Sarcophages antérieurs au novel empire*, Vol. 2 (Cairo, 1906), pp. 101-28, and Pl. IX.

Lange, H.O., and O. Neugebauer, *Papyrus Carlsberg No. I. Ein hieratisch-demotischer kosmologischer Text (Det Kongelike Danske Videnskabernes Selskab. Historisk-filologiske Skrifter, Band I, Nr. 2)* (Copenhagen, 1940).

Leitz, C., *Studien zur äyptischen Astronomie* (W. Helck, ed., *Ägyptologische Abhandlungen*, Bd. 49) (Wiesbaden, 1989).

Lepsius, R., "Entdeckung eines bilinguen Dekretes durch Lepsius," *ZÄS*, Vol. 4 (1866), pp. 29-34.

Lepsius, R., "Das Dekret von Kanopus," *ZÄS*, Vol. 4 (1866), pp. 49-52.

Lepsius, R., "Über den Kalendar des Papyrus Ebers und die Geschichtlichkeit der ältesten Nachrichten," *ZÄS*, Vol. 13 (1875), pp. 145-57.

Lepsius, R., *Das bilingue Dekret von Kanopus in der Originalgrösse mit Übersetzung und Erklärung beider Texte* (Berlin, 1866).

Lepsius, R., *Die Chronologie der Aegypter*, Einleitung und erster Theil (Berlin, 1849).

Lepsius, R., *Denkmäler aus Ägypten und Äthiopen*, Abt. I-VI (Berlin, 1849-58; photographic reprint, Geneva, 1972); Text, 5 vols., ed. Naville et al. (Leipzig, 1897-1913, phot. repr., Geneva, 1975).

Letronne, A.J., "Analyse critique des représentations zodiacales de Dendéra et d'Esneé," *Mémoires de l'Institut Royale de France, Académie des Inscriptions et Belles-Lettres*, Tome 16,2 (Paris, 1846), pp. 102-210, with plates; and see Biot.

Macalister, R.A.S., *The Excavation of Gezer 1902-1905 and*

1907-1909, 2 Vols. (London, 1912).

Mahler, E., *Études sur le calendrier égyptien (Annales du Musée Guimet, Bibliothèque d'études*, Vol. 24), Paris, 1907.

Mariette, A.E., *Denderah. Description générale du grand temple de cette ville*, Vol. 4 (Paris, 1878).

Mariette, A.E., *Les Mastabas de l'Ancien Empire. Fragment du dernier ouvrage de Auguste Edouard Mariette. Publié d'après le manuscrit de l'auteur par Gaston Maspero* (Hildesheim and New York, 1976).

Meyer, E., *Ägyptische Chronologie*, (Berlin, 1904).

Meyer, E., *Nachträge zur Ägyptischen Chronologie* (Berlin, 1908).

Miller, E., "Découverte d'un nouvel exemplaire du Décret de Canope," *Journal des savants*, Avril, 1883, pp. 214-29, and plate opp. p. 240.

Mills, J.O., "Astronomy at Hierakonpolis," an unpublished paper presented at the 1990 *Society for Africanist Archaeologists*, Biennial Conference, March 22-25, 1990, University of Florida. See my Postscript.

Möller, G., ed., *Hieratische Lesestücke für den akademischen Gebrauch*, 1. Heft (Leipzig, 1909).

De Morgan, J., et al., *Catalogue des monuments et inscriptions de l'Égypte antique. Première Série: Haute Égypte*, Tome 1: *De la frontière de Nubie à Kom Ombos* (Vienne, 1894).

Murray, M.A., *Saqqara Mastabas*, Parts 1-2 (London, 1905-37).

Nelson, H.H. (and edited by W.J. Murnane), *The Great Hypostyle Hall at Karnak*, Vol. 1, Part 1: *The Wall Reliefs* (Chicago, 1981), (*The University of Chicago Oriental Institute Publications Volume 106*).

Nelson, H.H., et al. (i.e., The Epigraphic Survey), *The University of Chicago Oriental Institute Publications, Volume XXII: Medinet Habu; Volume III, Plates 131-192; The Calendar, the "Slaughterhouse," and Minor Records of Ramses III* (Chicago, 1934), Vol. VI: *Plates 363-482: The Temple Proper* (Chicago, 1963).

Nelson, H.H., and U. Hölscher, *The Oriental Institute of the University of Chicago, Oriental Institute Communications No. 18, Work in Western Thebes, 1931-33* (Chicago, 1934?).

Neugebauer, O., "Die Bedeutungslosigkeit der Sothisperiode für die älteste ägyptische Chronologie," *Acta orientalia*, Vol. 17 (1938), pp. 169-95.

Neugebauer, O., "The Egyptian 'Decans'," *Vistas in Astromomy*, Vol. 1 (1955), pp. 47-51.

Neugebauer, O., "Egyptian Planetary Texts," *Transactions of the American Philosophical Society*, N.S., Vol. 32, Part II (Jan. 1942), pp. 209-50.

Neugebauer, O., "The Origin of the Egyptian Calendar," *Journal of Near Eastern Studies*, Vol. 1 (1942), pp. 396-403. This and the preceding article were reprinted in O. Neugebauer, *Astronomy and History, Selected Essays* (New York / Berlin / Heidelberg / Tokyo, 1983).

Neugebauer, O., *A History of Ancient Mathematical Astronomy*, Part 2 (Berlin, Heidelberg, and New York, 1975).

Neugebauer, O., R.A. Parker, and D. Pingree, (with notes on the colors by R. Stadelmann and J. Osing), "The Zodiac Ceilings of Petosiris and Petubastis," in J. Osing et al., *Denkmäler der Oase Dachla aus dem Nachlass von Ahmed Fakhry* (Mainz am Rhein, 1982), pp. 96-101, and Tafeln 36-44.

Neugebauer, O. and R.A. Parker, *Egyptian Astronomical Texts*, 3 vols. (Providence, Rhode Island, and London, 1960-1969).

Neugebauer, O. and A. Volten, "Untersuchungen zur antiken Astronomie IV," *Quellen und Studien zur Geschichte der Mathematik, Astronomie, und Physik,* Abt. B, Vol. 4 (1938), pp. 401-02.

Neugebauer, P.V., *Astronomische Chronologie*, 2 vols. (Berlin, 1929).

Neugebauer, P.V., and Borchardt, L., "Beobachtungen des Frühaufgangs des Sirius in Ägypten im Jahre 1926," *Orientalistische Literaturzeitung,* Vol. 29 (1926), cc. 309-16, and Vol. 30 (1927), cc. 441-49.

Oppolzer, T. v., "Über die Länge des Siriusjahres und der Sothisperiode," *Sitzungsberichte der Kaiserlichen Akademie der Wissenschaften. Mathematisch- naturwissenschaftliche Classe*, Vol. 90, II. Abt. (1884; publ. Wien, 1885), pp. 557-84.

Osing, J. et al., *Denkmäler der Oase Dachla aus dem Nachlass von Ahmed Fakhry* (Mainz am Rhein, 1982). See Neugbauer, O.. R.A. Parker, and D. Pingree.

Parker, R.A., "Ancient Egyptian Astronomy," *Philosophical Transactions of the Royal Society of London*, A.276 (1974), pp. 51-65.

Parker, R.A., "The Beginning of the Lunar Month in Ancient Egypt," *Journal of Near Eastern Studies*, Vol. 29 (1970), pp.

217-20.

Parker, R.A., "The Calendars and Chronology," *The Legacy of Egypt*, 2nd ed., edited by J.R. Harris (Oxford, 1971), pp. 13-26.

Parker, R.A., "Egyptian Astronomy, Astrology, and Calendrical Reckoning," in C.C. Gillispie, ed., *Dictionary of Scientific Biography*, Vol. 15 (New York, 1978), pp. 706-27.

Parker, R.A., "The Problem of the Month-Names. A Reply," *Revue d'Égyptologie*, Vol. 11 (1957), pp. 85-107.

Parker, R.A., "Sothic Dates and Calendar 'Adjustment'," *Revue d'Égyptologie*, Vol. 9 (1952), pp. 101-08.

Parker, R.A., *The Calendars of Ancient Egypt* (Chicago, 1950).

Petrie, W.M.F., *Ancient Weights and Measures* (British School of Archaeology in Egypt, 1926; reprinted, Warmouth, Wiltshire, and Encino, Calif., 1974).

Petrie, W.M.F., *Athribis* (London, 1908).

Petrie, W.M.F., *Dendereh 1898* (London, 1900).

Petrie, W.M.F., *The Royal Tombs of the Earliest Dynasties*, Part II (London, 1901).

Petrie, W.M.F., *Wisdom of the Egyptians* (London, 1940).

Petrie, W.M.F., see Griffith. F.Ll., and.

Pierret, P., *Le Décret trilingue de Canope* (Paris 1881).

Pogo, A., "The Astronomical Ceiling-decoration in the Tomb of Senmut (XVIIIth Dynasty)," *Isis*, Vol. 14 (1930), pp. 301-25.

Pogo, A., "The Astronomical Inscriptions on the Coffins of Ḥeny (XIth dynasty?)," *Isis*, Vol. 18 (1932), pp. 7-13.

Pogo, A., "Calendars on Coffin lids from Asyut (Second half of the third millennium)," *Isis*, Vol. 17 (1932), pp. 6-24.

Pogo, A., "Egyptian Water Clocks," *Isis*, Vol. 25 (1936), pp. 403-425.

Pogo, A., "Zum Problem der Identifikation der nördlichen Sternbilder der alten Ägypter," *Isis*, Vol. 16 (1931), pp. 102-14.

Porter B., and R.L.B. Moss, *Topographical Bibliography of Ancient Egyptian Texts, Reliefs and Paintings*, Vol. III2, revised by J. Mālek (Oxford, 1978-79, 1981).

Reinisch, S.L., and E.R. Roesler, *Die zweisprachige Inschrift von Tanis* (Vienna, 1866).

Renouf, P. Le Page, "Calendar of Astronomical Observations Found in Royal Tombs of the XXth Dynasty," *Transactions of the Society of Biblical Archaeology*, Vol. 3 (1874), pp. 400-21.

Révillout, É., *Chrestomathie démotique. Études égyptienne,*

Livraisons 13-16 (Paris 1880).

Roeder, G., "Die Himmelsbeobachtung der alten Ägypter," *Rundschau der gesamten Sternforschung für Himmelskunde und Fachastronomen: "Sirius,"* 1917, 1./2. Heft, pp. 1-11.

Roeder, G., "Eine neue Darstellung des gestirnten Himmels in Ägypten aus der Zeit um 1500 v. Chr.," *Das Weltall*, 28. Jahrgang (1928), Heft 1, pp. 1-5.

Roesler, E.R., see Reinisch, S.L. and.

Roullet, A., *The Egyptian and Egyptianizing Monuments of Imperial Rome* (Leiden, 1972), pp. 145-46, item nos. 326-328, 330, and Figs. 334-36, 337-38, 339-42, 344.

Sambin-Nivet, C., "L'offrande de la pretendue clepsydre et la phrase specifique," *Akten des Vierten Internationalen Ägyptologen Kongresses München 1985*, Vol. 3, pp. 368-378.

Sambin-Nivet, C., "L'offrande de la soi-disant 'clepsydre,', le symbole sbt\wnsb\wtt," *Studia Aegyptiaca*, Vol. XI (1988). This is an excellent monograph, which I read after i had completed my volume and thus too late for the inclusion of its results. The same is true for Sambin-Nivet's article in the preceding entry.

Samuel, A.E., *Ptolemaic Chronology (Münchener Beiträge zur Papyrusforschung und antiken Rechtsgeschichte)*, 43. Heft (Munich, 1962).

Schack-Schakenburg, H., *Ägyptologische Studien*, Vol. 1, 2. Heft (Leipzig, 1902): *Die Sternnetzabscissen und die somatischen Relationen der Thebanischen Stundentafeln.*

Schäfer, H., et al., *Ägyptische Inschriften aus den Königlichen Museen zu Berlin*, Vol. 1 (Leipzig, 1913).

Scharff, A.E., "Die Bedeutungslosigkeit des sogennanten ältesten Datums der Weltgeschichte und einige sich daraus ergebende Folgerungen für die ägyptische Geschichte und Archäologie," *Historische Zeitschrift*, Vol. 161 (1939), pp. 3-32.

Schiaparelli, E., "Di un'iscrizione inedita del regno di Amenofi I," *Actes du 8ème Congrès International des Orientalistes tenu en 1889 à Stockholm et à Christiana,* 4me partie (Leiden, 1892), pp. 201-208.

Schott, S., "Die altägyptischen Dekane," in W. Gundel, *Dekane und Dekansternbilder: Ein Beitrag zur Geschichte der Sternbilder der Kulturvölker* (Glückstadt and Hamburg, 1936), pp. 1-21.

Schott, S., *Altägyptische Festdaten*, in *Akademie der Wissenschaften und der Literatur in Mainz, Abhandlungen der Geistes- und Sozialwissenschaftlichen Klasse*, Jahrgang 1950. Nr.

10. Published as a separate volume by the Messers. Scheel on 27 Oct., 1950.

Sethe, K., "Die Zeitrechnung der alten Ägypter im Verhältnis zu der der andern Völker," *Nachrichten von der Königlichen Gesellschaft der Wissenschaften zu Göttingen,* Philologisch-historische Klasse aus dem Jahre 1919 und aus dem Jahre 1920 (Berlin, 1920), 1. Heft, Part I, "Das Jahr," pp. 287-320; 2. Heft, Part II, "Jahr und Sonnenlauf," pp. 28-55; Part III, "Einteilung des Tages- und des Himmelskreises," pp. 97-141.

Sethe, K., *Ägyptische Lesestücke: Texte des Mittleren Reiches*, reprint (Hildesheim / Zurich / New York, 1983).

Sethe, K., *Die altägyptischen Pyramidentexte*, 4 vols. (Leipzig, 1908-22).

Sethe, K., *Hieroglyphische Urkunden der griechisch-römischen Zeit (=Urkunden II, Heft II)* (Leipzig, 1904), Vol. 2.

Sethe, K., *Historisch-biographische Urkunden des Mittleren Reiches (=Urkunden VII)*, Vol. 1 (Leipzig, 1935).

Sethe, K., *Übersetzung und Kommentar zu den altägyptischen Pyramidentexten* (Hamburg, 1962), Vols. 1 and 2.

Sethe, K., *Urkunden des Alten Reichs (=Urkunden I)*, Vol. 1, rev. (Leipzig, 1933).

Sethe, K., *Urkunden des 18. Dynastie (=Urkunden IV)*, Vol. 3 (Leipzig, 1907).

Sloley, R.W., "Ancient Clepsydrae," *Ancient Egypt*, Vol. ix (1924), pp. 43-50.

Sloley, R.W., "Primitive Methods of Measuring Time with Special Reference to Egypt," *The Journal of Egyptian Archaeology*, Vol. 17 (1931), pp. 166-78.

Stricker, B.H., "Spreuken tot Beveiliging gedurende de Shrikkeldagen naar Pap. I 346," *Oudheidkundige Mededeelingen uit het Rijksmuseum van Oudheden te Leiden,* Vol. 29 (1948), pp. 55-70.

Toomer, G.J., "Mathematics and Astronomy," *The Legacy of Egypt*, 2nd ed., edited by J.R. Harris (Oxford, 1971), pp. 27-54.

Villiers du Terrage, E. de, *Journal et souvenirs sur l'expédition d'Égypte (1798-1801)* [Paris, 1899].

Wainwright, G.A., "A Pair of Constellations," *Studies Presented to F.Ll. Griffith* (London / Oxford, 1932), pp. 373-82.

Wainwright, G.A., "A Subsidiary Burial in Ḥap-Zefi's Tomb at Assiut," *ASAE*, Vol. 26 (1926), pp. 160-66.

Weill, R., *Bases, méthodes et résultats de la chronologie*

égyptienne (Paris, 1926).

Weill, R., *Bases, méthodes et résultats de la chronologie égyptienne. Compléments* (Paris, 1928). This *Compléments* was bound with the preceding work in the copy I used.

Wells, R.A., "Some Astronomical Reflections on Parker's Contributions to Egyptian Chronology," in L.H. Lesko, ed., *Egyptological Studies in Honor of Richard A. Parker* (Hanover and London, 1986), pp. 165-71.

Wiedemann, A., "Bronze Circles and Purification Vessels in Egyptian Temples," *PSBA*, Vol. 23 (June, 1901), pp. 263-74.

Winlock, H.E., "The Egyptian Expedition 1925-1927," *Section II of the Bulletin of the Metropolitan Museum of Art* (February, 1928), pp. 3-58.

Winter, E., "Zur frühesten Nennung der Epagomenentage und deren Stellung am Anfang des Jahres," *Wiener Zeitschrift für die Kunde des Morgenlandes,* 56. Band (1960), pp. 262-266.

Index of Egyptian Words and Phrases

Note: I have added a +-sign to the page number when there is more than one reference to the word or phrase being indexed and the letter "g" when the glyph of the word also appears on that page.

Most often I have not capitalized the phonetic letters representing glyphs, since there is no capitalization in hieroglyphic writing. But occasionally capitals are used by the modern authors I quote or discuss and hence they sometimes appear in this index. When both capitals and lower case letters have been used for the same word in this volume, I have almost always simply given the latter to represent all cases.

The distinction between "s" (—) and "ś" ( ), which current practice seldom maintains (instead using "s" to represent both letters), is only partially followed in this index. I too would have abandoned the distinction were it not for the fact that many earlier authors who are discussed and quoted here do make the distinction. Hence the reader is advised to consult, for any word using letter "ś", entries under "s" as well.

The reader is also encouraged to consult the Index of Subjects for instances of the many Egyptian words rendered pronounceable by the insertion of arbitrary vowels and used in this form in my translations, as, for example, Akhet, Peret, Shemu, Sepdet or Sopdet, etc. Those words are here represented only by their consonants, i.e., by the conventional letters and signs used by scholars to replace the consonantal hieroglyphs.

The reader will also recognize that here words are arranged according to the alphabetic order adopted by modern Egyptologists for those letters and signs.

Index of Proper Names and Subjects

Note that in this Index of Proper Names and Subjects I often refer the reader to the preceding Index of Egyptian Words and Phrases since the latter is in fact a very important supplement. This is particularly true for references to the names of feasts, seasons, planets, decans, and other stars which have frequently been presented in this volume in consonantal, phonetic form only. Even when readers do not know the ancient Egyptian language, they will learn, almost by osmosis, the Egyptian words for such items. In the following index I have often given only the page numbers for the lists of names of the various items noted above, without repeating every page number for each entry of the lists. Such is especially the case when I have given the item here in the index but with only the page numbers for the first couple of references to it and then followed it by a reference to the Index of Egyptian Words where all instances of the item are noted. I have adopted this procedure to keep the current index from being redundant and too detailed.

Further note that I have here abandoned the practice followed in the shorter Index of Egyptian Words of adding a plus sign to those numbers of pages on which more than one reference to the indexed item appears. The reason is that most of the items indexed here are found more than once on the cited pages.

I must also point out that the subject indexing of the texts is more complete than that of the notes, since in most cases the texts include references to the notes, and hence the reader would be led to more information by the normal process of consulting the notes specified in the texts. On the other hand, proper names are fully indexed in both the texts and notes, except as noted above in the first paragraph. Incidentally, the contemporary locations of the modern stars and constellations given in the index can be found on the star maps of Figures III.70a and III.70b, while the locations of many of them (i.e., the northern stars) in 3000 B.C. and 2000 B.C. may be seen in Figs. III.64a and III.64b. Petrie's effort to reconstruct the ancient Egyptian star map (i.e., to give the positions of their stars and constellations) is embraced by Figs. III.84a and III.84b.

In searching for specific feasts, the reader should look in one or more of three places: the proper name of the god of the feast (e.g. "Min, Feast of"), or under "Feast of...." (e.g., Feast of Millions of Years), or "Going Forth of...." (e.g., "Going Forth of Sem" or "Going Forth or Procession of Min").

INDEX OF PROPER NAMES

INDEX OF PROPER NAMES

INDEX OF PROPER NAMES

INDEX OF PROPER NAMES

INDEX OF PROPER NAMES

Part Four

Illustrations

A List of Illustrations

(Sources and Details given with the Illustrations which follow this List)

the Ebers Calendar with the gratuitous (and, I believe, erroneous) additions that each entry represented the first day of successive lunar months. Given by Borchardt.

Fig. III.13 Schematic version of a star clock.

Fig. III.14 Hour decans from the lid of the coffin of *Idy*, Dynasty ?, from Asyut.

Fig. III.15 The decanal band south of, and roughly parallel to, the ecliptic, with Orion and Sirius shown as hour stars by dots near the right margin.

Fig. III.16 Hour decans from the lid of the coffin of *It-ib*, Dynasty IX or X, from Asyut.

Fig. III.17 Detail from the lid of the Coffin of *It-ib*, called by Pogo "Tefabi."

Fig. III.18 A Concordance of Decans from Coffin Lids.

Fig. III.19a A copy of the tables of the Ramesside star clock appearing in Hall K of the tomb of Ramesses VI in the Valley of the Kings. Only bits of the first table remain, and table 24 is missing. The tables are successively for the first and sixteenth day of each month and are used for the nighttime hours for these days plus the fourteen days succeeding each of them.

Fig. III. 19b Another copy of the hour tables of the Ramesside star clock appearing north and south of the ceiling of Corridor B of the tomb of Ramesses IX.

Fig. III.19c Table 4 from the Ramesside star clock as given by J. F. Champollion.

Fig. III.20a Two Egyptian astronomical instruments that belong together: a sighting instrument (Inv. Nr. 14084) and a right angled shadow board (Inv. Nr. 14085). Both were bought together in Cairo and were probably found in Abydos. They are in (West) Berlin in the Ägyptische Museum.

Fig. III.20b A Drawing of the instruments photographed in Fig. III.20a.

Fig. III.21a Four views of a water clock from the reign of Amenhotep III. Found at Karnak and now in the Cairo Museum.

Fig. III.21b Exterior, interior, and edge views of a fragment of a water clock from the British Museum (cat. no. 938). Time of Philip Arrhidaeus (323-16 B.C.).

Fig. III.21c Drawing of the interior surface of the water clock pictured in Fig. III.21b.

Fig. III.21d Exterior and interior views of a fragment of a second water clock in the British Museum (cat. no. 933). Also of

Fig. III.36 A votive offering in the form of a clepsydra? Time of Amenhotep III (ca. 1391-1343 B.C.).

Fig. III.37 Table taken from Pogo, "Egyptian Water Clocks."

Fig. III.38 Transcription of the description of a shadow clock on the west side of the roof of the sarcophagus chamber of the cenotaph of Seti I.

Fig. III.39 Drawing of a shadow clock of the type given in the cenotaph of Seti I.

Fig. III.40 A graphical representation of four-hour shadow clocks.

Fig. III.41 Shadow clocks in the Ägyptische Museum (West Berlin).

Fig. III.42 Names of the Goddesses of the first six hours.

Fig. III.43 Borchardt's reconstruction of a complete shadow clock, with the base ruler pointed toward the east and the crossbar toward the north.

Fig. III.44 Papyrus fragments (ca. 100 A.D.) depicting a shadow clock with its hour markings.

Fig. III.45 Borchardt's investigation of the shadow clock of Berlin (No. 19743) for the latitude 29° north.

Fig. III.46 The use of a five-edged rule on the Berlin shadow clock (No. 19743) to obtain equal hours.

Fig. III.47 A model combining three kinds of shadow clocks: (1) one having a flat shadow-receiving surface, (2) the second having a flight of steps as the shadow-receiving surface, and (3) the last having an inclined plane as the shadow-receiving surface.

Fig. III.48 A drawing of the Cairo model in use.

Fig. III.49 Determination of three sets of hour scales for different times of the year, i.e., for the summer and winter solstices and for the equinoxes; to be mounted on an inclined-plane shadow clock like that included in the Cairo model.

Fig. III.50 Remains of some sun-pointing shadow clocks.

Fig. III.51 A sun-pointing shadow clock in use.

Fig. III.52 Several views of a sun-pointing shadow clock with an inclined plane as the shadow-receiving surface.

Fig. III.53a Investigation of the sun-pointing shadow clock from Qantara.

Fig. III.53b Investigation of the sun-pointing shadow clock from Paris.

Fig. III.54 Shadow-clock glyphs used as determinatives for the word "hour" or as ideograms.

predicting the rising of Sothis in the 7th year of the reign of Sesostris III.

Fig. III.93 A notice of the rising of Sothis in an unknown year of Tuthmosis III.

Fig. III.94a The hieroglyphic version of the Decree of Canopus, published according to a paper impression from the Stela of Tanis.

Fig. III.94b The text of the hieroglyphic version of the calendric passages of the Decree of Canopus edited by Sethe.

Fig. III.94c The Decree of Canopus inscribed on the Stela of Kom-el-Hisn (Egyptian Museum, Cairo, no. 22186).

Fig. III.95a West side of the roof of the Sarcophagus Chamber in the Cenotaph of Seti I at Abydos.

Fig. III.95b Chart locating the texts in the preceding illustration.

Fig. III.95c Concordance of Texts of Seti I (**S**), Ramesses IV (**R**), P. Carlsberg I (**P**) and Ia (**Pa**).

Fig. III.96a Chart locating the texts of the Cosmology of Seti I and Ramesses IV on the ceiling of the Sarcophagus Chamber in the Tomb of Ramesses IV in the Valley of the Kings (Western Thebes).

Fig. III.96b Brugsch's drawing of the Figure of Nut with accompanying texts from the Sarcophagus Chamber in the Tomb of Ramesses IV.

Fig. III.96c An earlier drawing of the same figure and text as in Fig. III.96b, taken from Champollion.

Fig. III.97 Figure of the sky godess Nut in the vaulted position, with the earth god Geb below, lying on his back in a contorted position. In the Western Chapel of Osiris (Chamber no. 2) on the roof of the Temple of Hathor at Dendera.

Fig. III.98a A drawing of the first 18 lines of the Dramatic Text near the figure of Nut in the west side of the roof of the Sarcophagus Chamber of the Cenotaph of Seti I.

Fig. III.98b A hand copy of the first part of the Dramatic Text prepared by Neugebauer and Parker.

Fig. III.99a An English translation of the tables of the Ramesside star clock by P. Le Page Renouf.

Fig. III.99b A diagram to illustrate the phrases used to indicate positions of stars transiting in the Ramesside star clock.

Fig. III.99c Stringed apparatus for sighting star transits, suggested for use with the Ramesside star clock by H.

Part Four

Illustrations

Fig. III.1 Inscription to Isis-Sothis at Aswan from the reign of Ptolemy IV. Taken from Morgan et al., *Catalogue des monuments et inscriptions de l'Égypte antique: Première Série: Haute Égypte*, Tome 1, p. 55. See my Document III.10.

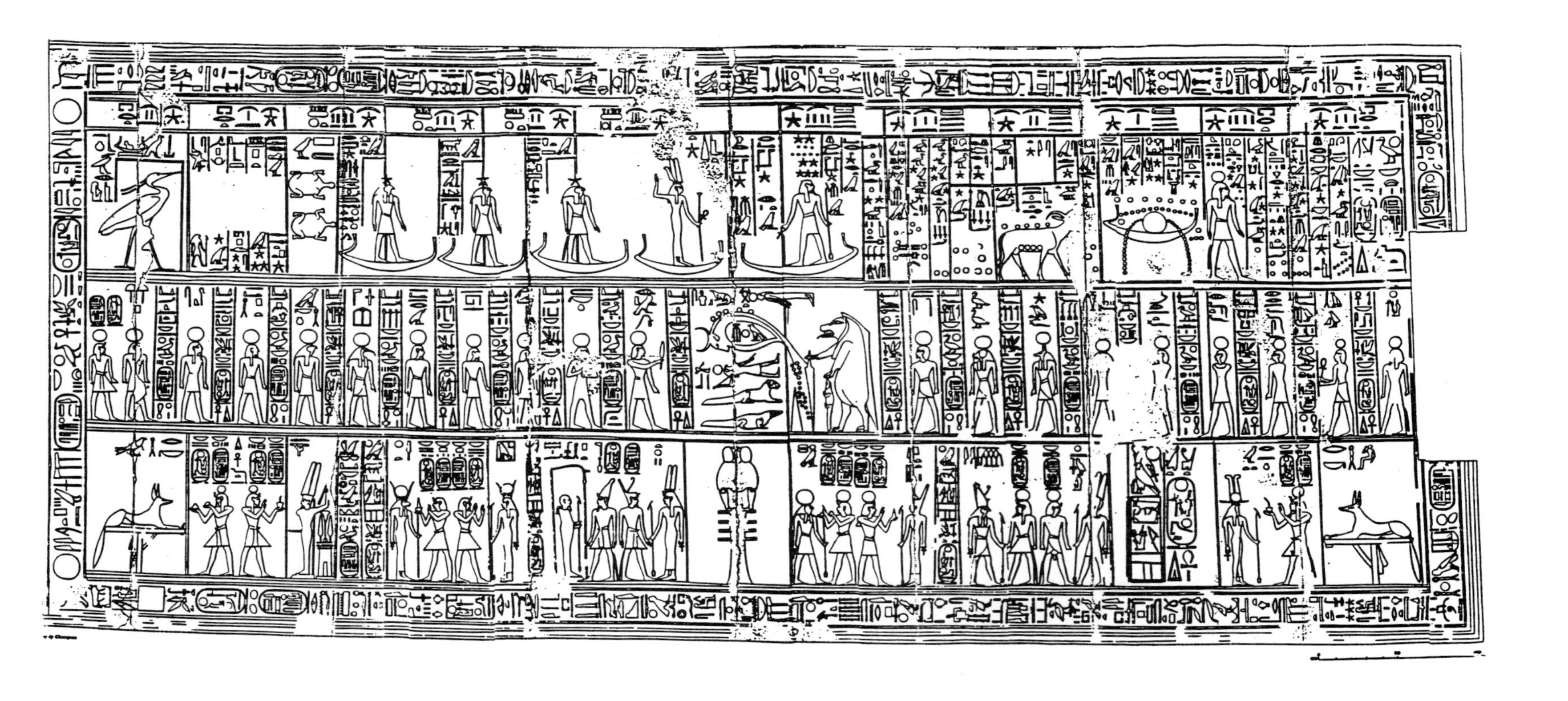

Fig. III.2 The Astronomical Ceiling of the Ramesseum, Second Hypostyle Hall, Ceiling of Axis Aisle. Taken from H. H. Nelson et al., *Medinet Habu*, Vol. 6, Plate 478.

Fig. III.3a Tablet from the Reign of Djer, Dynasty I, probably representing, on the right side, the goddess Sekhet-Hor as a reclining cow with a feather between her horns. University of Pennsylvania Museum, E. 9403. I must thank the University Museum for supplying the photograph (Neg. no. S4-54899) and granting permission for its publication.

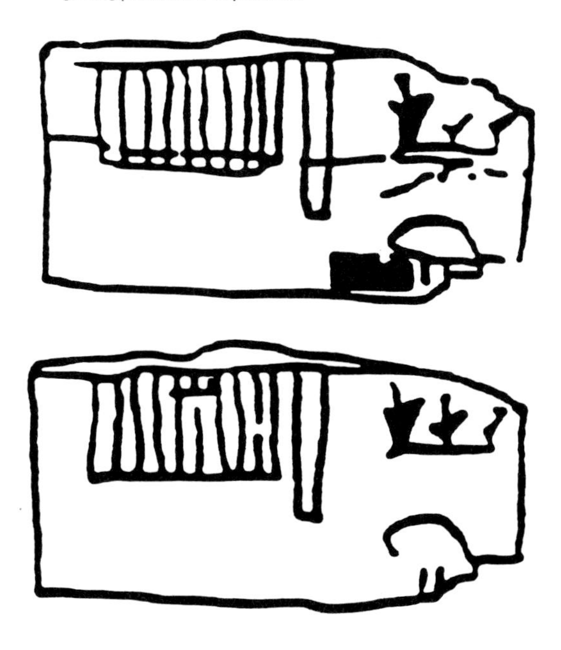

Fig. III.3b Above is a fragmentary doublet presenting the same theme as that given in Fig. III.3a. The bottom copy is a facsimile with the drawing corrected by Borchardt. Both drawings are from Godron, " Études," p. 145, Fig. 2.

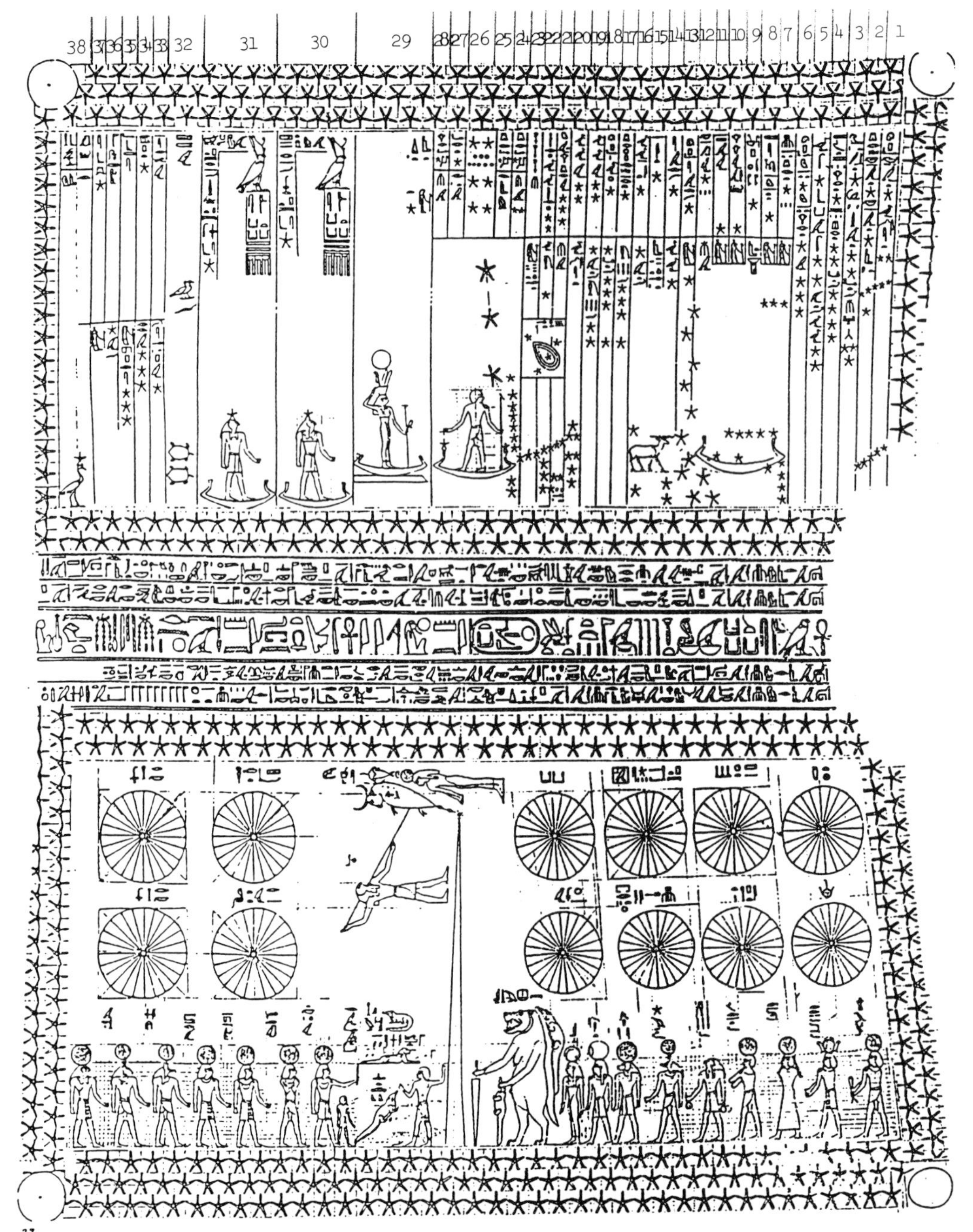

27

Fig. III.4 Astronomical Ceiling from the Secret Tomb of Senmut near the Temple of Hatshepsut at Dar el-Bahri, Western Thebes. The south-half of the ceiling is above; the north-half below. Cf. Pogo, "The Astronomical Ceiling-decoration in the Tomb of Senmut (XVIIIth Dynasty)," *Isis*, Vol. 14 (1930), Plates A-K, for photographs. The column numbers at the top of the figure have been added by the author.

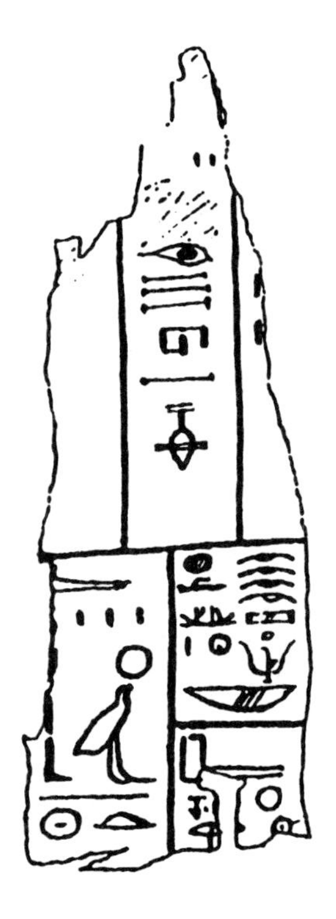

Fig. III.5 A geographical fragment from Tanis naming (in the second register, on the right) the fourth month of Shemu as *Wp rnpt*. Taken from Griffith and Petrie, *Two Hieroglyphic Papyri from Tanis*, Plate IX,2.

Documents	Fêtes et leurs dates calendériques.	Mois homonymes
Papyrus de Kahoun Amenemhat III, au 35.	*Navigation de Hathor*, 4e mois, 1er jour. *Fête de Neheb-Kaou* (= *Ka-her-ka*) [1], 5e mois, 1er jour.	*Hathor-Athyr* = 3e mois. *Kaherka-Khoiak* = 4. mois.
Fragment de calendrier de Karnak Thoutmès III (Brugsch, *Thes.*, 362).	*Fête de Neheb-Kaou* (v. ci-dessus), 5e mois, 1er jour.	*id.*
Tombeaux de Khamhat et de Nofirhotep. 18e dynastie (Brugsch, *Thes.*, 303-304, Prisse, *Mon.*, pl. 42).	*Fête de Renenouti*, 9e mois, 1er jour.	*Renenouti-Pharmouti* = 8e mois.
Tombeau de Zosirkaresenb 18 dynastie (Scheil, *Mém. mission française*, v. p. 577-578).	*Fête de Renenouti*, 8e mois (jour perdu).	*id.*
Calendrier de Medinet-Habou. Ramsès III (Brugsch, *Thes.*, 364).	*Fête de Hathor*, 4e mois, 1er jour, *Fête de Neheb-Kaou* (voir ci-dessus), 5e mois, 1er jour.	*Hathor-Athyr* = 3e mois. *Kaherka-Khoiak* = 4. *mois*.
Papyrus *Boulaq 19* 20e dynastie.	*Fête d'Epep*, 12e mois, 15e jour (donnée douteuse).	*Epet-Epiphi* = 11e mois.
Papyrus Gardiner Ramsès IX, an 13.	*Naissance de Re-Hor-Khouti*, 1er mois, 1er jour.	*Re-Hor-Khouti — Mesore* = 12e *mois*.
Papyrus Gardiner Ramsès XI, an 3.	*Fête d'Epep*, 12e mois, 1er-2e jours.	*Epet-Epiphi* = 11e mois.
Calendrier d'Edfou 100 av. J.-C. (Brugsch, *Thes.*, 368-373).	*Naissance de Re*, 1er mois, 1er jour.	*Re-Hor-Khouti — Mesore* = 12e mois.
	Fête de Tekhou du 1er mois, 20e jour au 2e mois, 4e jour.	*Tekhi-Thot* = 1er mois.
»	*Procession de Hathor* du 3e mois, 29e jour au 4e mois, 1er jour.	*Hathor-Athyr* = 3e mois
»	*Fête de Neheb-Kaou* commence le 4e mois, 28e jour. (voir ci-dessus).	*Kaherka-Khoiak* = 4e mois.
»	*Fête de Shefbedet* commence le 5e mois, 20e jour.	*Shefbedet-Tybi* = 5e mois.
»	*Fête de Renenouti* 9e mois, 1er jour.	*Renenouti-Pharmouti* = 8e mois.
»	*Procession de Khonsou* commence le 9e mois, 19e jour.	*Khonsou-Pakhon* = 9e mois.
Autre texte d'Edfou (Brugsch, *Drei Festkalender*, pl. II, 11).	*Fête de Mekhi-ir*, ([hieroglyphs]) 6e mois, 21e jour.	......*-Mekhir* = 6e. mois.
Calendrier d'Esneh (Brugsch, *Thes.*, 304, 380-383).	*Fête de Kaherka*, 4e mois (*sic* ?), 1er jour. *Fête de Renenouti*, 9e mois, 1er jour.	*Kaherka-Khoiak* = 4e mois. *Renenouti-Pharmouti* = 8e mois.
Rituel de Denderah (*Rec. de trav.*, IV, p. 24).	*Fête de Shefbedet*, 5e mois, 20e jour.	*Shefbedet-Tybi* = 5e mois.

Fig. III.6a A List of Egyptian Feasts and their eponymous months. Taken from Weill, *Bases, méthodes et résultats de la chronologie égyptienne*, pp. 121-22.

I. Alte Ordnung.			II. Neue Ordnung.				
1. Pap. Ebers.	2. Ältere Monatsfeste.	3. Hierat. inscr. 28.	4. Ramesseum.	5. Edfu.	6. Spätere Varianten.	7. Aramäisch.	8. Griechische und koptische Formen.
(ḥeb) wepet ronpet I	Me-su(t)reʿ I, Gard. Thoutfest am 19. I. zu allen Zeiten, Techufest vom 20. I. bis 5. II.	»das Gehen des Horus« I	Reʿ Ḥor-achuti XII	Reʿ Ḥor-achuti XII	ḥeb wepet ronpet = Mesore XII (Edfu), s. S. 3	מסורי XII	Μεσορη, theban. Μεσορει, Μεσωρηι. Μεσωρη, Μεσουρη XII, k. Μεσωρη
Techi II	— Fest des Amon von Opet vom 19. II. bis 12. III., Kal. von Med. Habu	Thout II	Techi I (verbunden mit Isis-Sothis)	Techi I		תחות I	Θωυθ, Θωυτ, theban. alt Θαυτ I, k. Θωουτ : Θοουτ, Θαυτ
Menchet III	—	pen-Opet »der von Opet (Karnak)« III	Ptaḥ von Memphis II	Menchet (Gott Ptaḥ) II	Ḥeb Opi u. ä.	פאפי II, פאופי (Berl. 5, 1)	Φαωφι, theban. Παωπι II, k. Παοπι : Πααπε, Ποοπε
Ḥathor IV	»Ausfahrt der Ḥathor« IV, Kahun (12. Dyn.). Ḥathorfest am 1. IV., Kal. von Med. Habu	Ḥathor IV	Ḥathor III	Ḥathor III		התור III	Αθυρ, Αθαυρ III, k. Αθωρ : Ϩατωρ
Kaḥirka V	Neḥebkau 1. V., Kahun (12. Dyn.)	Kaḥirka V	Sochmet IV	Kaḥirka IV		כיחך IV	Χοιακ, Χοιαχ IV, k. Χοιακ : Κιαϩκ. Χοιαϩκ
Šefbedet VI	—	»die Fahrt der Mut« VI	Min V	Šefbedet V	Šefbedet	—	Τυβι, Τυβε V, k. Τωβι : Τωβε
Rekeḥ VII	—	pen pa Mechiru »der des Mechir[festes]« VII	Rekeḥ uer (Anubis) VI	Rekeḥ uer (Nilpferd) VI	Mechirfest am 21. VI.	מחיר VI	Μεχιρ VI, k. Μεχιρ : Μϣιρ
Rekeḥ VIII	—	pen Amenḥotep »der des Königs Amenophis« VIII	Rekeḥ nezes (Anubis) VII	Rekeḥ nezes (Nilpferd) VII		פמנחתפ VII (Berl. 14, 4, 2)	Φαμενωθ, theban. Φαμενωτ VII, k. Φαμενωθ : Παρμϩατ, Παρεμϩατπ
Renenutet IX	Renenutet 1. IX., 18. Dyn.	—	Renenutet VIII	Renenutet VIII		—	Φαρμουθι, theban. Φαρμουτι VIII, k. Φαρμουθι : Παρμουτε
Chonsu X	—	—	Chonsu IX	Ḥeb Chonsu IX		פחנס IX	Παχων, theban. alt Παχωνς IX, k. Παχων : Παϣονς
Chentechtai XI	—	—	Chenti (Horus) X	Ḥor Chentechtai X	ḥeb en Onet »Thalfest« 9. X.	פאני X	Παϋνι, Παυνη, Παονι X, k. Παωνι : Παωνε, Πααωνι
Epet XII	jpjp (Epiphi) XII. Gard.	—	Epet XI	Ḥeb Epet (Nilpferdgöttin) XI		אפיפ XI, אפף (Berl. 9, 1)	Επιφ, theban. Εφιπ, Εφειπ, Εφειφ XI, k. Επηπ : Επεπ

Fig. III.6b The so-called Old and New Orders of Egyptian Months and Feasts. Taken from Meyer, *Nachträge zur Ägyptischen Chronologie*, Table for p. 16.

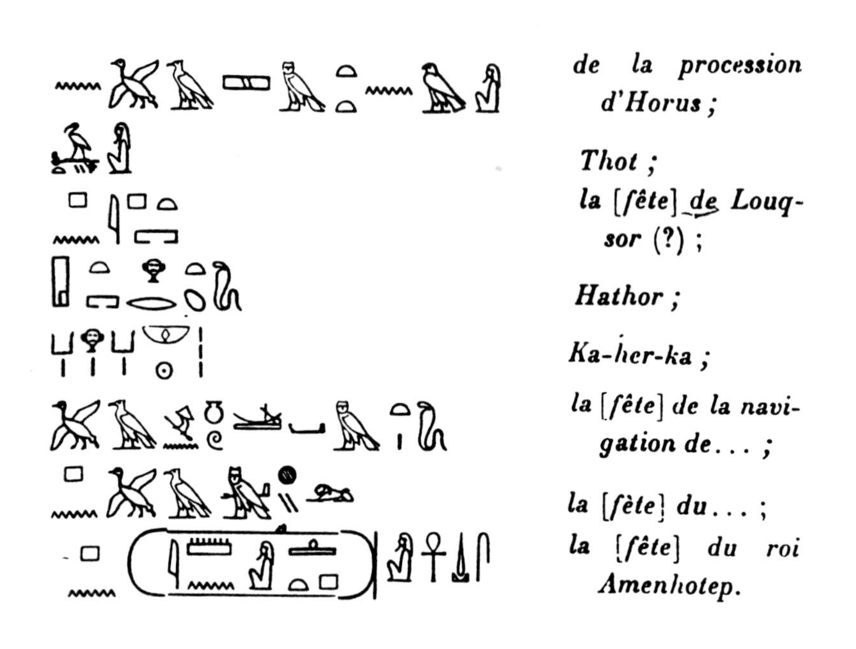

de la procession d'Horus ;
Thot ;
la [fête] de Louqsor (?) ;
Hathor ;
Ka-ḥer-ka ;
la [fête] de la navigation de. . . ;
la [fête] du. . . ;
la [fête] du roi Amenhotep.

Fig. III.7 A list of month-names discovered by A. Erman in an ostracon of the Late Kingdom, "Monatsnamen aus dem neuen Reich," *ZÄS*, Vol. 39 (1901), pp. 128-30. (See col. 3 in Fig. III.6b.) The list as given here with French translations of the names is taken from Weill, *Bases, méthodes et résultats de la chronologie égyptienne*, pp. 124-25.

Monate:	I_1 II_1 III_1 IV_1	I_2 II_2 III_2 IV_2	I_3 II_3 III_3 IV_3
Jahr 1	× 1 × 30	× 29 × 2[8]	× 27 × 26
2	× 20 × 19	× 18 × 17	× 16 × 15
3	× 9 × 8	× 7 × 6	× 5 × 4
4	× 28 × 2[7]	× 26 × 25	× 24 × 23
5	× 18 × 17	× 16 × 15	× 14 × 13
6	× 7 × 6	× 5 × 4	× 3 × 2
7	× 26 × 25	× 24 × 23	× 22 × 21
8	× 15 × 14	× 13 × 12	× 11 × 10
9	× 4 × 3	× 2 × 1	× 30 × 29
10	× 24 [×] 23	× 22 × 21	× 20 × 19
11	× 1[3 ×] 12	× 11 × 10	× 9 × 8
12	[× 2 × 1]	× 30 × 29	× 28 × 27
13	[× 21 × 20]	× 19 × 18	× 17 × 16
14	× [10 × 9]	× 8 × 7	× 6 × 5
15	× [30 × 29]	× 28 × 27	× 26 × 25
16	× [19 × 18]	× 17 × 16	× 15 × 14
17	× [8 × 7]	× 6 × 5	× 4 × 3
[18]	[× 27 × 26]	× 25 × 24	× 23 × 22
[19]	[× 16 × 15]	× 14 × 13	× 12 × 11
[20]	[× 6 × 5]	× 4 × 3	× 2 × 1
[21]	× [25 × 24]	× 23 × 22	× 21 × 20
[22]	× [14 × 13]	× 12 × 11	× 10 × 9
[23]	× [3 × 2]	× 1 × 30	× 29 × 28
[24]	× [22 × 21]	× 20 × 19	× 18 × 17
[25]	× 12 [× 11]	× 10 × 9	× 8 × 7

Fig. III.8a The even-numbered months of the Egyptian 25-year lunar cycle as tabulated by Neugebauer and Volten, "Untersuchungen zur antiken Astronomie IV," *Quellen und Studien zur Geschichte der Mathematik, Astronomie, und Physik* Abt. B, Vol. 4 (1938), p. 395.

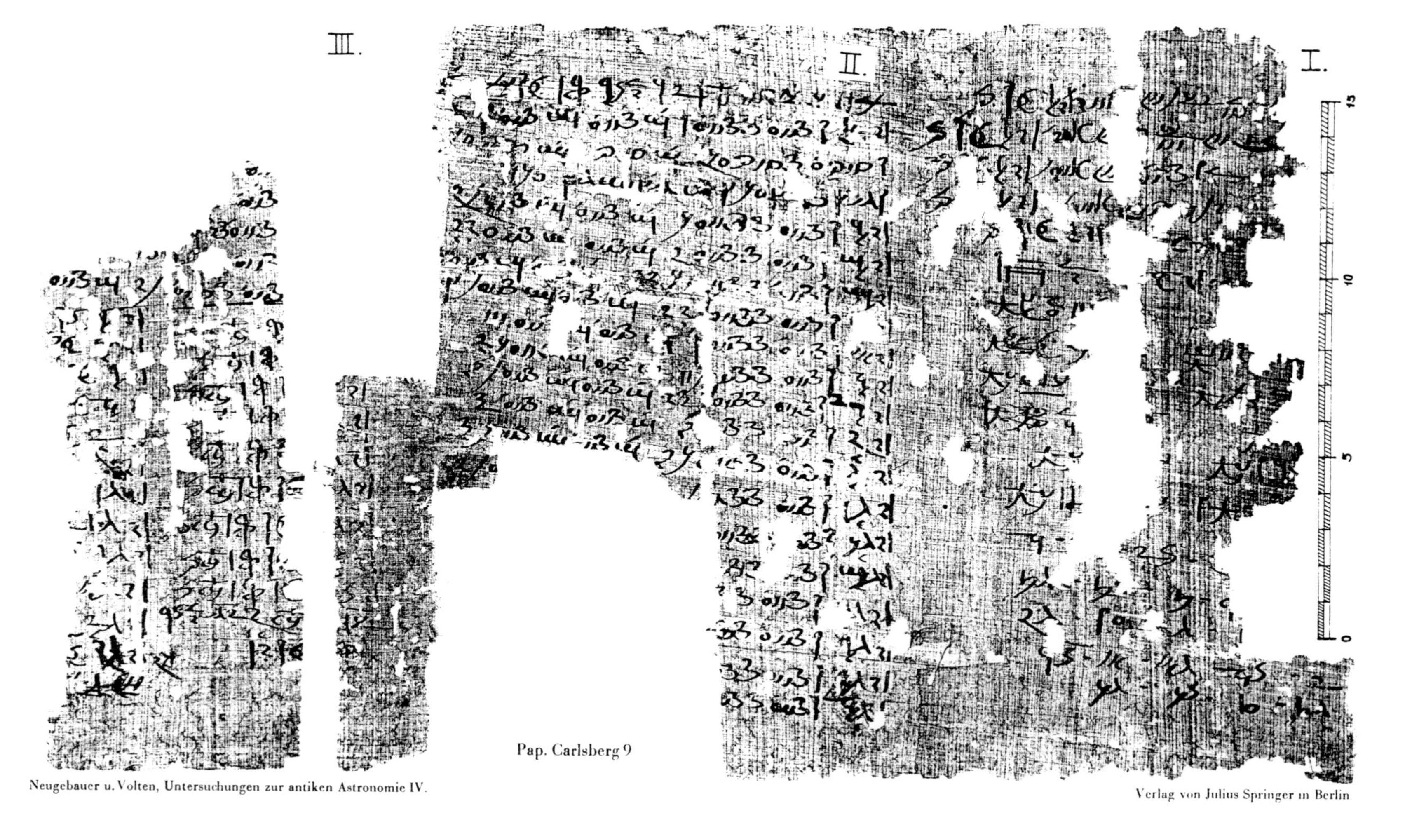

Fig. III.8b The demotic text of the 25-year lunar cycle in Pap. Carlsberg 9. Taken from the plate added to the transcription and translation of the demotic text prepared by Neugebauer and Volten, *ibid.*, pp. 383-406.

TABLE 5

THE COMPLETED 25-YEAR CYCLE

	ꜣḪT				PRT				ŠMW				EPAG.
Months:	I	II	III	IIII	I	II	III	IIII	I	II	III	IIII	
Year 1	1	1	1-30	30	29	29	29	28	27	27	27	26	
2	20	20	19	19	18	18	18	17	16	16	16	15	
3	9	9	8	8	7	7	7	6	5	5	5	4	4
4	28	28	27	27	26	26	26	25	24	24	24	23	
5	18	18	17	17	16	16	16	15	14	14	14	13	
6	7	7	6	6	5	5	5	4	3	3	3	2	2
7	26	26	25	25	24	24	24	23	22	22	22	21	
8	15	15	14	14	13	13	13	12	11	11	11	10	
9	4	4	3	3	2	2	2	1	1-30	30	30	29	
10	24	24	23	23	22	22	22	21	20	20	20	19	
11	13	13	12	12	11	11	11	10	9	9	9	8	
12	2	2	1	1	1-30	30	30	29	28	28	28	27	
13	21	21	20	20	19	19	19	18	17	17	17	16	
14	10	10	9	9	8	8	8	7	6	6	6	5	5
15	30	30	29	29	28	28	28	27	26	26	26	25	
16	19	19	18	18	17	17	17	16	15	15	15	14	
17	8	8	7	7	6	6	6	5	4	4	4	3	3
18	27	27	26	26	25	25	25	24	23	23	23	22	
19	16	16	15	15	14	14	14	13	12	12	12	11	
20	6	6	5	5	4	4	4	3	2	2	2	1	1
21	25	25	24	24	23	23	23	22	21	21	21	20	
22	14	14	13	13	12	12	12	11	10	10	10	9	
23	3	3	2	2	1	1	1-30	30	29	29	29	28	
24	22	22	21	21	20	20	20	19	18	18	18	17	
25	12	12	11	11	10	10	10	9	8	8	8	7	

Fig. III.9 The completed 25-year cycle as deduced by Parker, *The Calendars of Ancient Egypt*, p. 25.

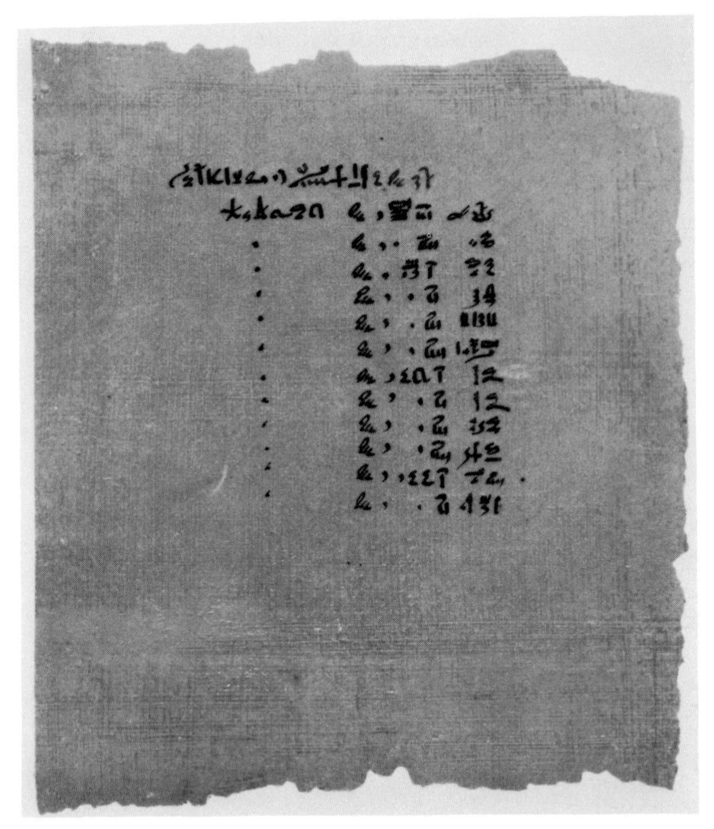

Ruckseite der Tafel I des Papyros Ebers.

Fig. III.10 Hieratic text of the Ebers Calendar from the Papyrus Ebers in the University of Leipzig Library. Year 9 of the reign of Amenophis I. Taken from G.M. Ebers, *Papyros Ebers,* Verso side of Tafel I.

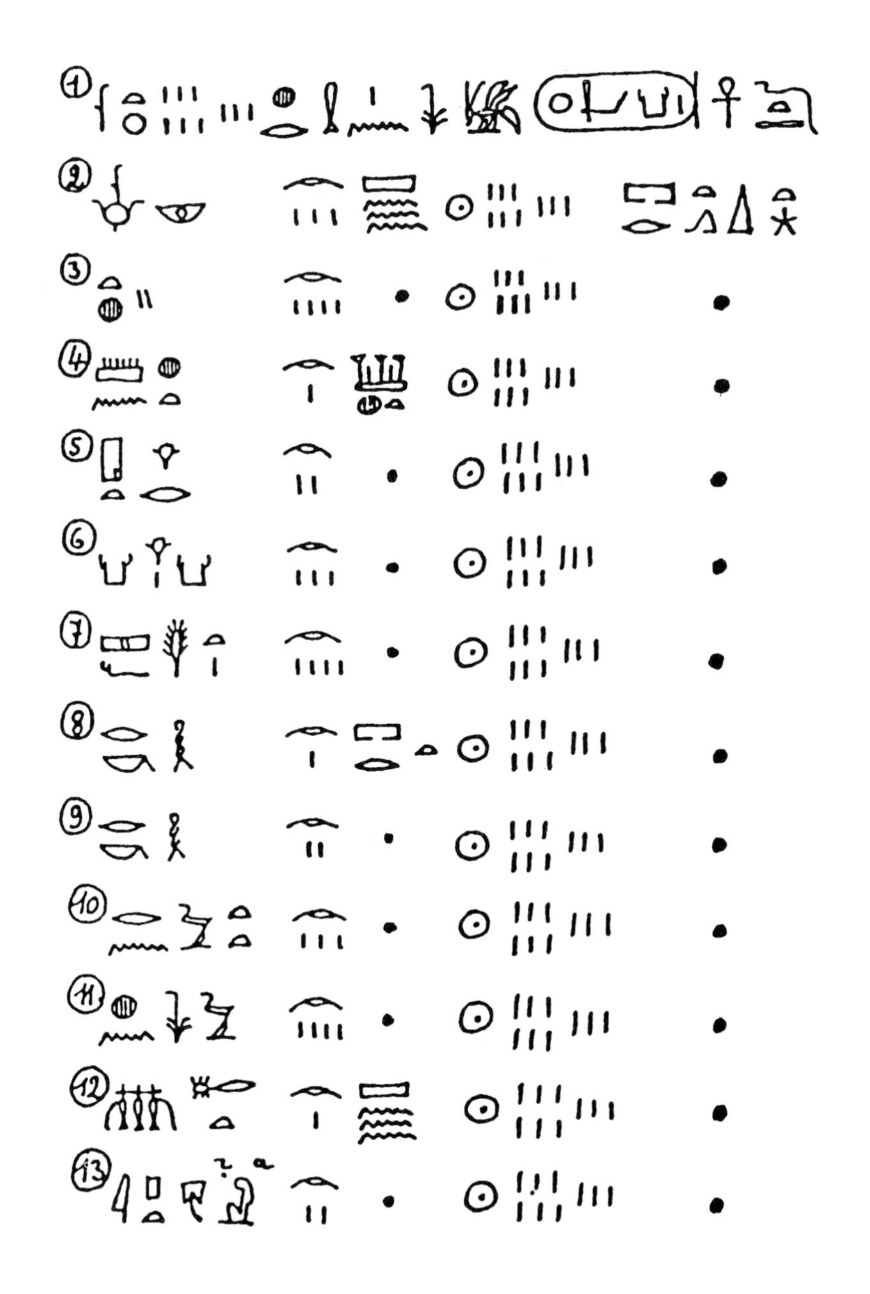

Fig. III.11 Hieroglyphic transcription of the Ebers Calendar. Taken from K. Sethe, *Urkunden der 18. Dynastie*, Vol. 1 (=*Urkunden IV*), p. 44.

(Amenophis I.)

Jahr 9 unter S. M. König *Ḏśr-kꜣ-rꜥ*, möge er ewig leben!

ḥb-wp·t-rnp·t	Monat 3	Erntejahreszeit	Neumondtag	Hundssternfrühaufgang
tḫj	Monat 4	Erntejahreszeit	Neumondtag	—
mnḫ·t	Monat 1	Überschwemmungsjahreszeit	Neumondtag	—
ḥꜣ·t-ḥr	Monat 2	Überschwemmungsjahreszeit	Neumondtag	—
kꜣ-ḥr-kꜣ	Monat 3	Überschwemmungsjahreszeit	Neumondtag	—
šf-bd·t	Monat 4	Überschwemmungsjahreszeit	Neumondtag	—
rkḥ	Monat 1	Winterjahreszeit	Neumondtag	—
rkḥ	Monat 2	Winterjahreszeit	Neumondtag	—
rnwt·t	Monat 3	Winterjahreszeit	Neumondtag	—
ḫnśw	Monat 4	Winterjahreszeit	Neumondtag	—
ḫnt-ḫtj	Monat 1	Erntejahreszeit	Neumondtag	—
ip·t	Monat 2	Erntejahreszeit	Neumondtag	—
(Mondmonate)	(Kalendermonate)		(Tage)	

Fig. III.12 Phonetic transcription and German translation of the Ebers Calendar with the gratuitous (and, I believe, erroneous) additions that each entry represented the first day of successive lunar months. Given by Borchardt, *Die Mittel zur zeitlichen Festlegung von Punkten der ägyptischen Geschichte und ihre Anwendung*, Blatt 1 opposite p. 20.

	40	39	38	37	36	35	34	33	32	31	30	29	28	27	26	25	24	23	22	21	20	19	V	18	17	16	15	14	13	12	11	10	9	8	7	6	5	4	3	2	1	
T	Epag.				36 ← 19 Decades																			18 ← 1 Decades																		T
1	A	25	13	1	↖	↖	↖	↖	↖	↖	↖	↖	↖	↖	↖	↖	↖	↖	↖	↖	↖	↖		↖	↖	↖	↖	↖	↖	↖	↖	↖	↖	↖	↖	↖	↖	↖	↖	↖	1	1
2	B	26	14	2	↖	↖	↖	↖	↖	↖	↖	↖	↖	↖	↖	↖	↖	↖	↖	↖	↖	↖		↖	↖	↖	↖	↖	↖	↖	↖	↖	↖	↖	↖	↖	↖	↖	↖	↖	2	2
3	C	27	15	3	↖	↖	↖	↖	↖	↖	↖	↖	↖	↖	↖	↖	↖	↖	↖	↖	↖	↖		↖	↖	↖	↖	↖	↖	↖	↖	↖	↖	↖	↖	↖	↖	↖	↖	↖	3	3
4	D	28	16	4	↖	↖	↖	↖	↖	↖	↖	↖	↖	↖	↖	↖	↖	↖	↖	↖	↖	↖		↖	↖	↖	↖	↖	↖	↖	↖	↖	↖	↖	↖	↖	↖	↖	↖	↖	4	4
5	E	29	17	5	↖	↖	↖	↖	↖	↖	↖	↖	↖	↖	↖	↖	↖	↖	↖	↖	↖	↖		↖	↖	↖	↖	↖	↖	↖	↖	↖	↖	↖	↖	↖	↖	↖	↖	↖	5	5
6	F	30	18	6	↖	↖	↖	↖	↖	↖	↖	↖	↖	↖	↖	↖	↖	↖	↖	↖	↖	↖		↖	↖	↖	↖	↖	↖	↖	↖	↖	↖	↖	↖	↖	↖	↖	↖	↖	6	6
R																																										R
7	G	31	19	7	↖	↖	↖	↖	↖	↖	↖	↖	↖	↖	↖	↖	↖	↖	↖	↖	↖	↖		↖	↖	↖	↖	↖	↖	↖	↖	↖	↖	↖	↖	↖	↖	↖	↖	↖	7	7
8	H	32	20	8	↖	↖	↖	↖	↖	↖	↖	↖	↖	↖	↖	↖	↖	↖	↖	↖	↖	↖		↖	↖	↖	↖	↖	↖	↖	↖	↖	↖	↖	↖	↖	↖	↖	↖	↖	8	8
9	J	33	21	9	↖	↖	↖	↖	↖	↖	↖	↖	↖	↖	↖	↖	↖	↖	↖	↖	↖	↖		↖	↖	↖	↖	↖	↖	↖	↖	↖	↖	↖	↖	↖	↖	↖	↖	↖	9	9
10	K	34	22	10	↖	↖	↖	↖	↖	↖	↖	↖	↖	↖	↖	↖	↖	↖	↖	↖	↖	↖		↖	↖	↖	↖	↖	↖	↖	↖	↖	↖	↖	↖	↖	↖	↖	↖	↖	10	10
11	L	35	23	11	↖	↖	↖	↖	↖	↖	↖	↖	↖	↖	↖	↖	↖	↖	↖	↖	↖	↖		↖	↖	↖	↖	↖	↖	↖	↖	↖	↖	↖	↖	↖	↖	↖	↖	↖	11	11
12	M	36	24	12	L	K	J	H	G	F	E	D	C	B	A	36	35	34	33	32	31	30		29	28	27	26	25	24	23	22	21	20	19	18	17	16	15	14	13	12	12

Fig. III.13 Schematic version of a star clock, slightly emended from the diagram in Neugebauer and Parker, *Egyptian Astronomical Texts*, Vol. 1, p. 1, Fig. 1. The diagonal lines representing the decans have been converted to arrows to indicate their upward movements from box to box.

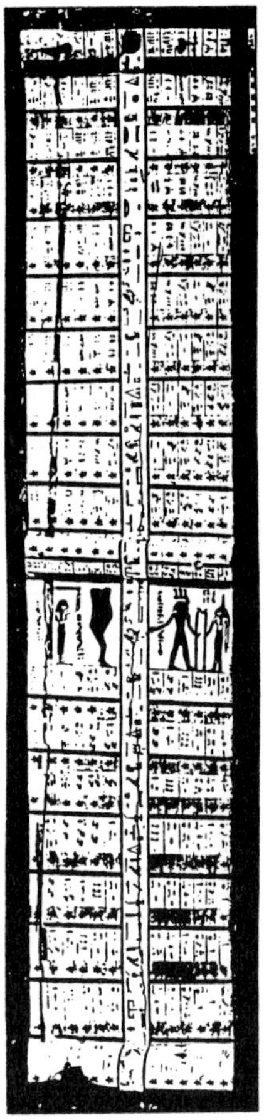

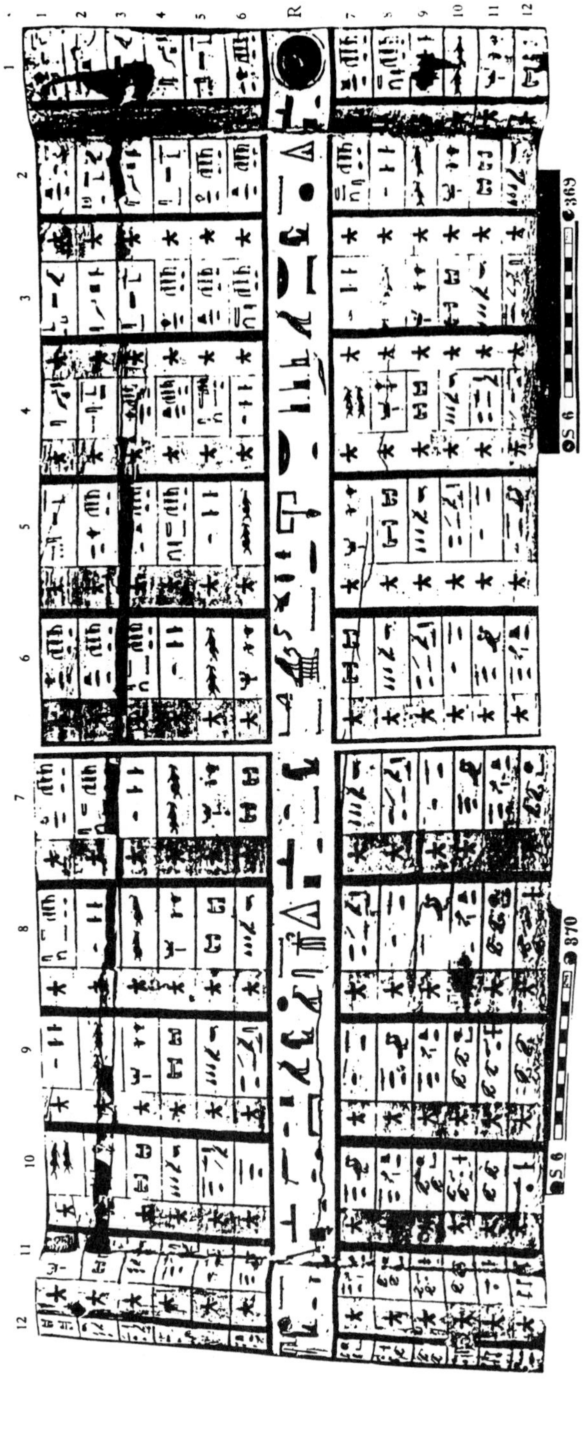

COFFIN 3

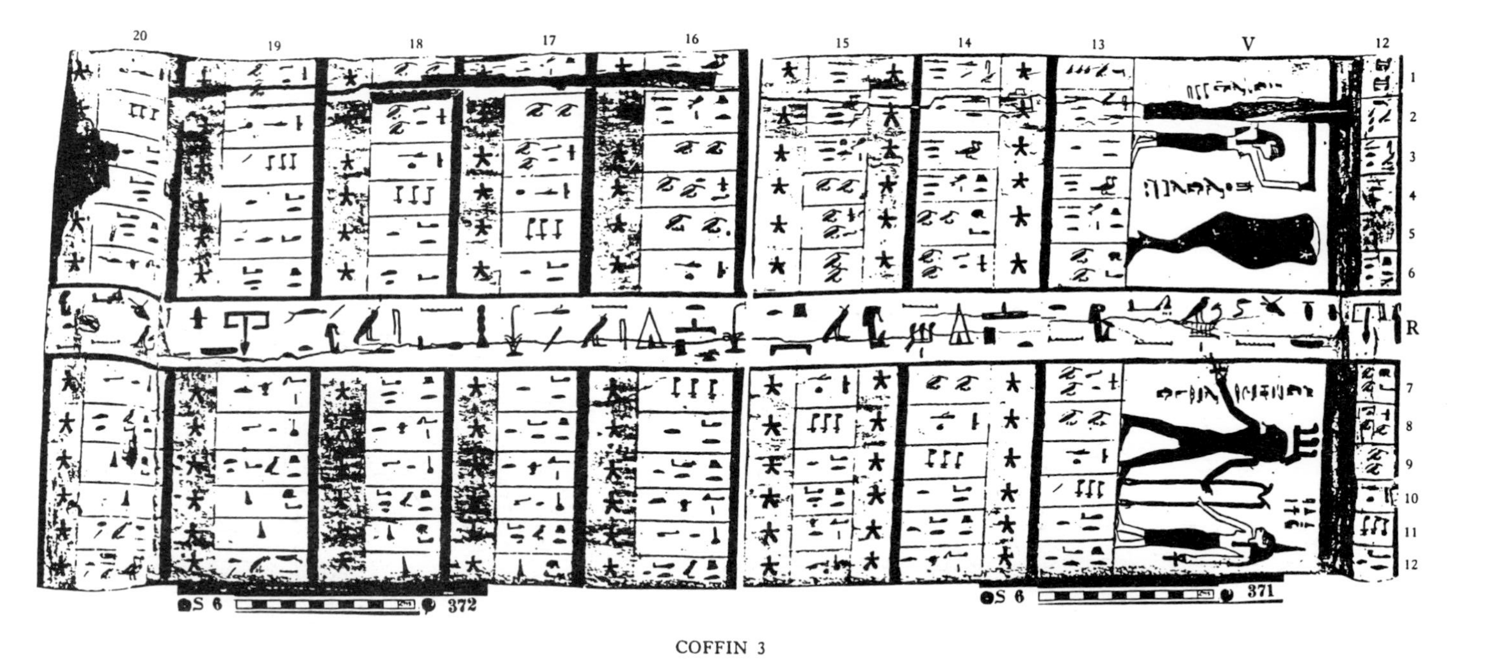

COFFIN 3

Fig. III.14 Hour decans from the lid of the coffin of *Idy*, Dynasty ?, from Asyut. Taken from Neugebauer and Parker, *Egyptian Astronomical Texts*, Vol. 1, Plates 5 and 6.

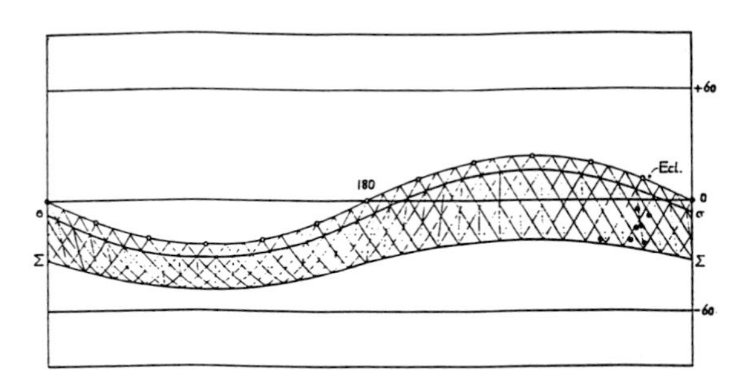

Fig. III.15 The decanal band south of, and roughly parallel to, the ecliptic, with Orion and Sirius shown as hour stars by dots near the right margin. Taken from Neugebauer and Parker, *Egyptian Astronomical Texts*, Vol. I, p. 100, Fig. 27.

PLATE 3

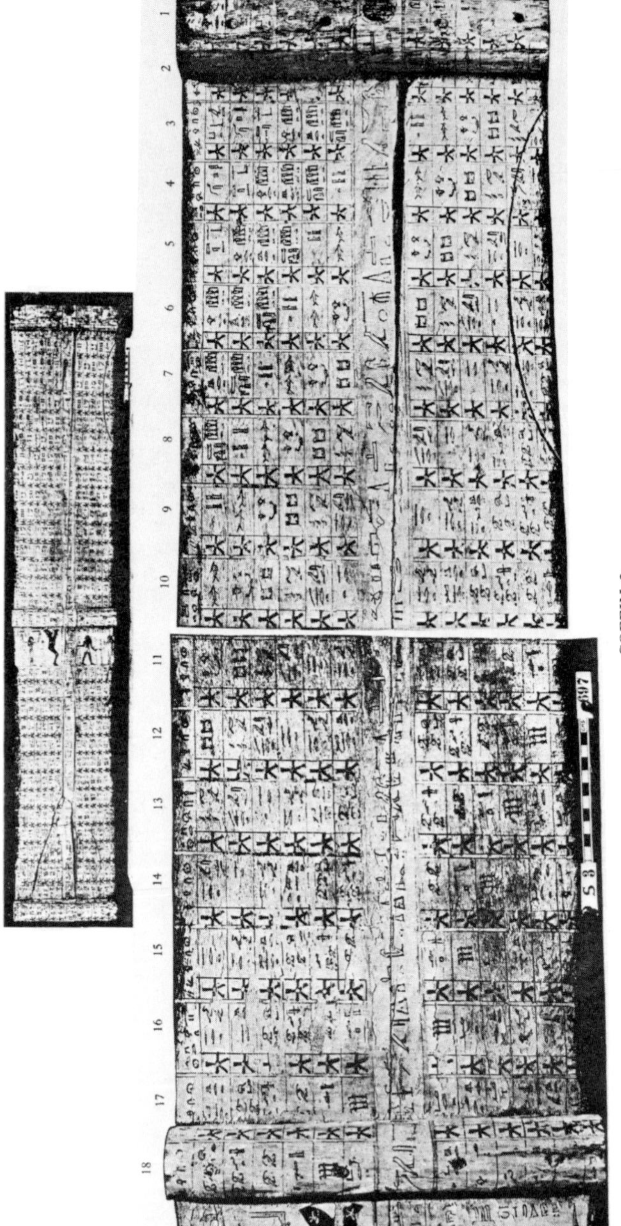

COFFIN 2

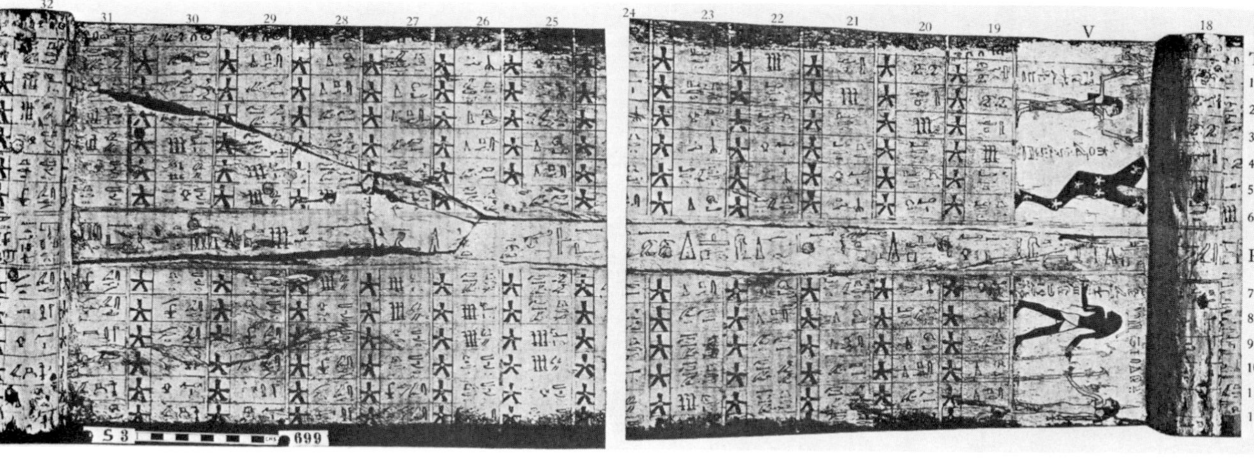

COFFIN 2

Fig. III.16 Hour decans from the lid of the coffin of *It-ib*, Dynasty IX or X, from Asyut. Taken from Neugebauer and Parker, *Egyptian Astronomical Texts*, Vol. 1, Plates 3 and 4.

Fig. III.17 Detail from the lid of the Coffin of *It-ib*, called by Pogo "Tefabi." A transversal picture given by A. Pogo, "Calendars on the Coffin Lids from Asyut," *Isis*, Vol. 17, Plate I.

Concordance

1. *ṯmꜣt ḥrt*			
2. *ṯmꜣt ẖrt*			
3. *wšt bkꜣt*	3a. *wšꜣti*	3b. *bkꜣti*	
4. *ỉpḏś*	4a. *śšpt*		
5. *śbššn*	5a. *tpy-ʿ ḫntt*		
6. *ḫntt ḥrt*			
7. *ḫntt ẖrt*			
8. *ṯmś n ḫntt*			
9. *ḳdty*	9a. *śpty*	9b. *śpty ḫnwy*	
10. ⌜*ḫnwy*⌝	10a. *ḫnwy*		
11. *ḥry-ỉb wỉꜣ*			
12. "crew"(?)	12a. *śšmw*		
13. *knm*	13a. *tpy-ʿ śmd*		
14. *śmd srt*	14a. *śmd*		
15. *srt*			
16. *sꜣwy srt*			
17. *ẖry ḫpd srt*			
18. *tpy-ʿ ꜣḫwy*			
19. *ꜣḫwy*			
20. *ỉmy-ḫt ꜣḫwy*			
21. *bꜣwy*	21a. *ḫntw ḥrw*	21b. *ḫntw ẖrw*	
22. *ḳd*	22a. *sꜣwy ḳd*		
23. *ḫꜣw*			
24. *ʿryt*			
25. *ẖry ʿryt*			
26. *rmn ḥry*	26a. *rmn ḥry Śꜣḥ*	26b. *ṯs ʿrḳ*	
27. *rmn ẖry*	27a. *rmn ẖry Śꜣḥ*	27b. *rmn Śꜣḥ*	27c. *Śꜣḥ*
28. *ʿbwt*			
29. *ḫrt wʿrt*			
30. *tpy-ʿ śpd*			
31. *śpd*	31a. *tpy-ʿ knmt*	31b. *štwy*	
32. *knmt*			
33. *sꜣwy knmt*			
34. *ẖry ḫpd n knmt*			
35. *ḥꜣt ḫꜣw*	35a. *ḥꜣt ḏꜣt*		
36. *pḥwy ḫꜣw*	36a. *pḥwy ḏꜣt*		

A *śmd rśy*	E *ḫꜣw* (II)	J *ḫꜣw* (=23)	W *Nwt*
B *śmd mḥty*	F *tpy-ʿ śpd* (=30)	K *nṯr ḏꜣ pt*	X *Mśḫtyw*
C *nṯr ḏꜣ pt*	G *ỉmy-ḫt śpd*	L *śꜣbw*	Y *Śꜣḥ*
D *rmn ẖry* (=27)	H *ꜣḫwy* (=19)	M *pḥwy śꜣbw*	Z *Śpdt*

Variants insofar as the arrangement of the hieroglyphs is concerned occur frequently in the texts; we pay no attention to these in the critical apparatus.

Fig. III.18 A Concordance of Decans from Coffin Lids. Taken from Neugebauer and Parker, *Egyptian Astronomical Texts*, Vol. I, pp. 2-3.

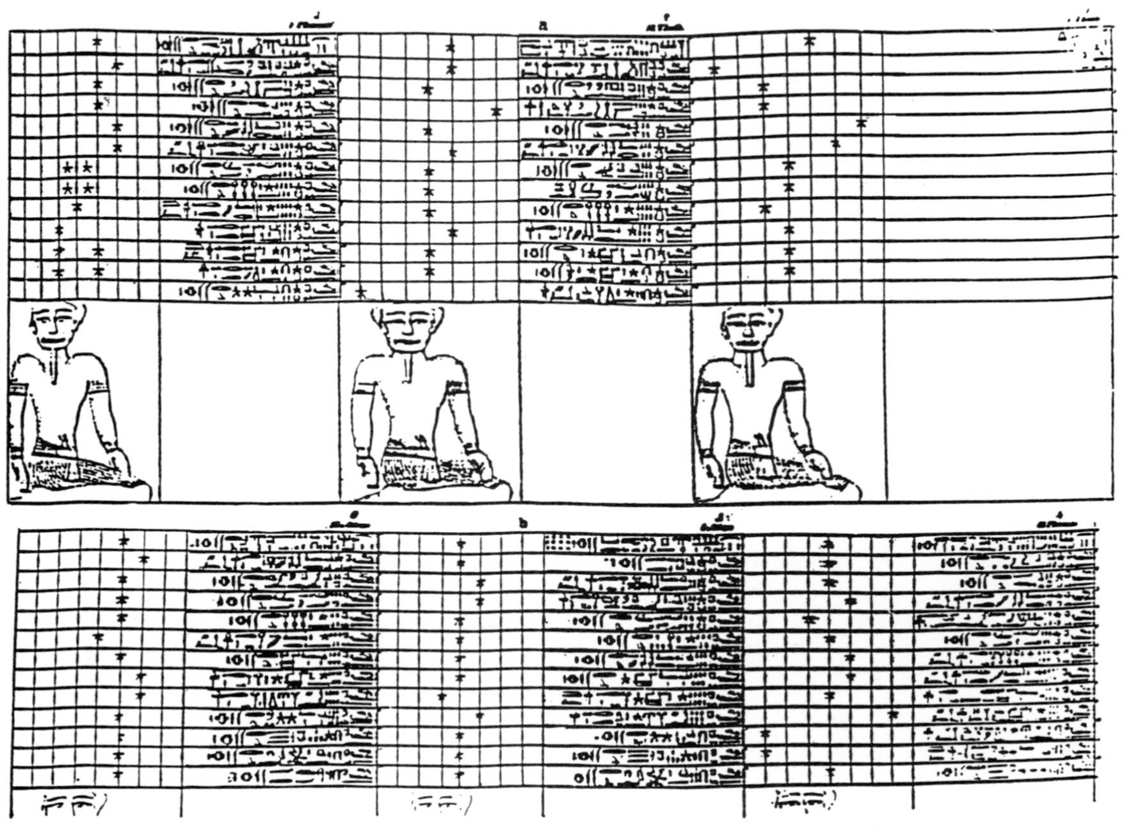

Fig. III.19a A copy of the tables of the Ramesside star clock appearing in Hall K of the tomb of Ramesses VI in the Valley of the Kings. Only bits of the first table remain, and table 24 is missing. The tables are successively for the first and sixteenth day of each month and are used for the nighttime hours for these days plus the fourteen days succeeding each of them. Taken from Lepsius, *Denkmäler aus Ägypten und Äthiopen*, Abt. III, Plates 227-28. Cf. the photographs of all extant copies of the twenty-four tables in Neugebauer and Parker, *Egyptian Astronomical Texts*, Vol. 2, Plates 9-28. Above all see the hand copy of the tables in Plates 29-67 of the same work.

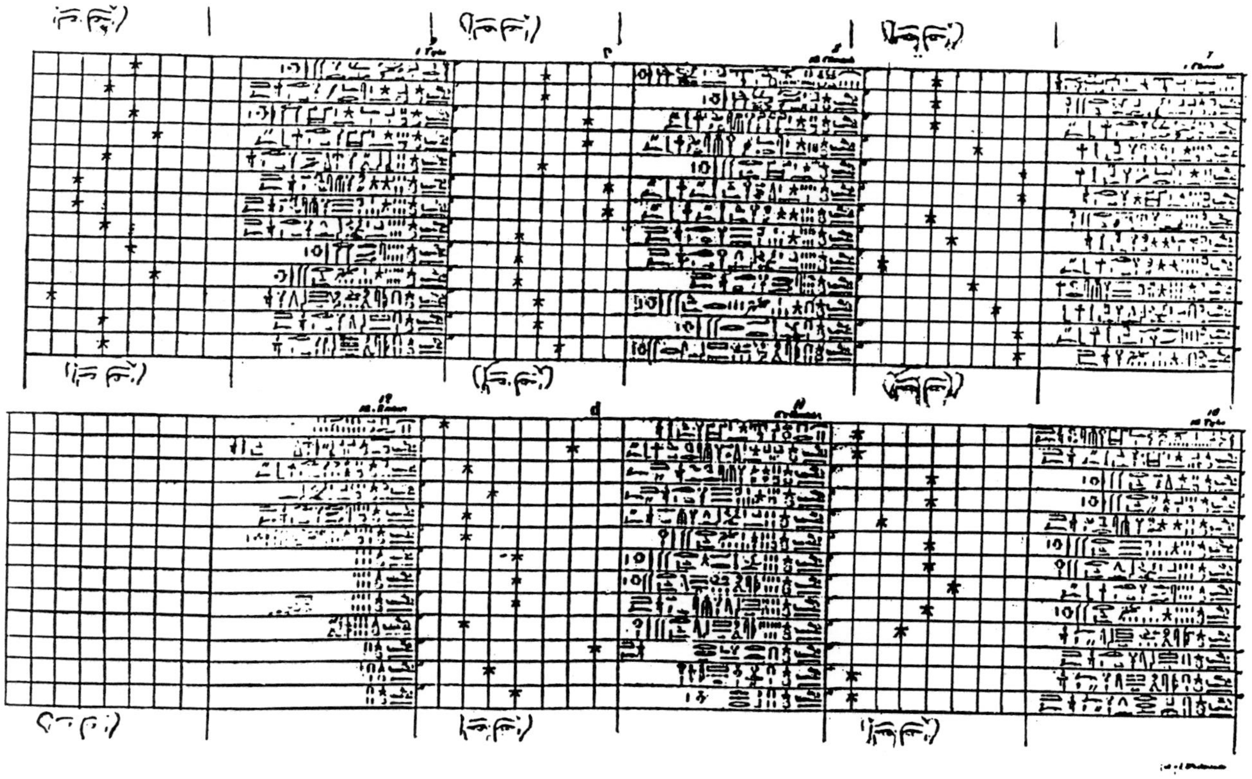

Fig. III.19a (Continued)

Fig. III.19a (Continued)

Fig. III.19a (concluded)

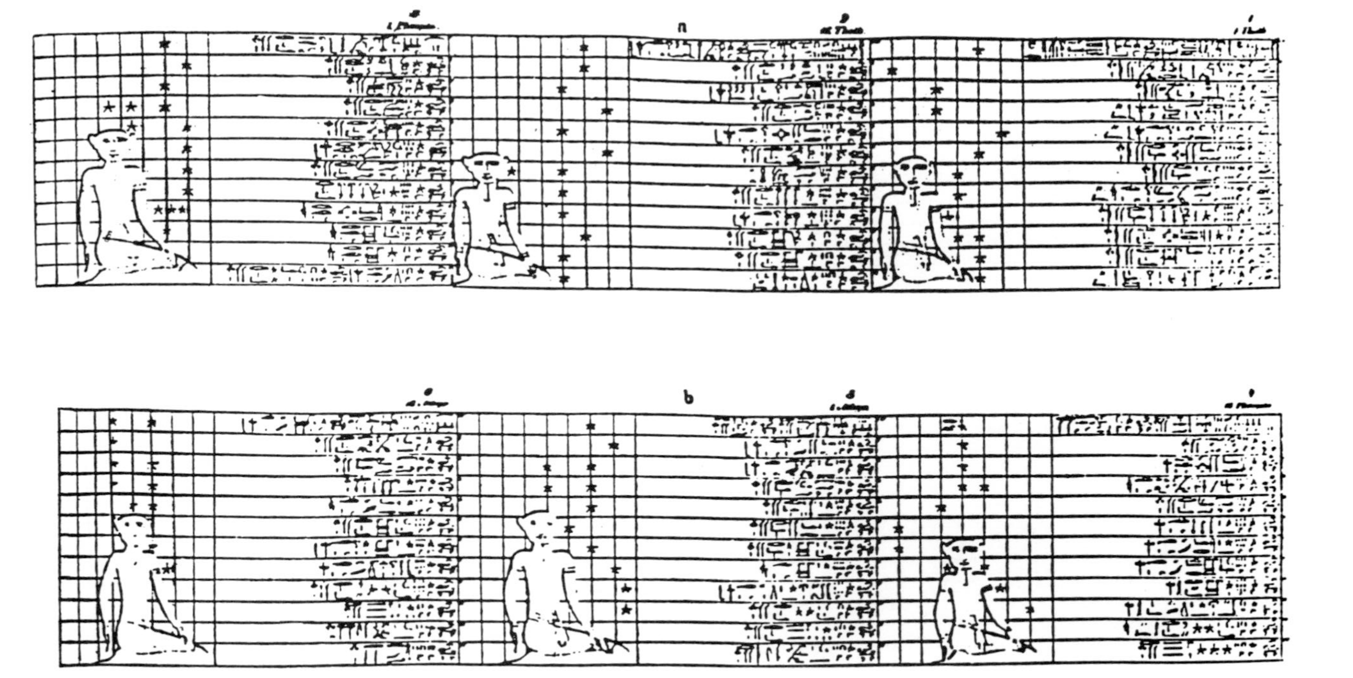

Fig. III. 19b Another copy of the hour tables of the Ramesside star clock appearing north and south of the ceiling of Corridor B of the tomb of Ramesses IX. This is also taken from Lepsius, *Denkmäler*, Abt. III, Plate 228bis.

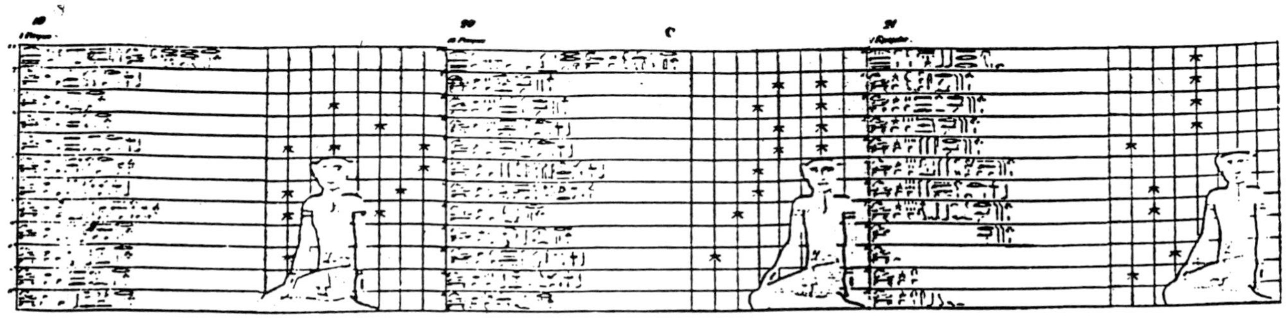

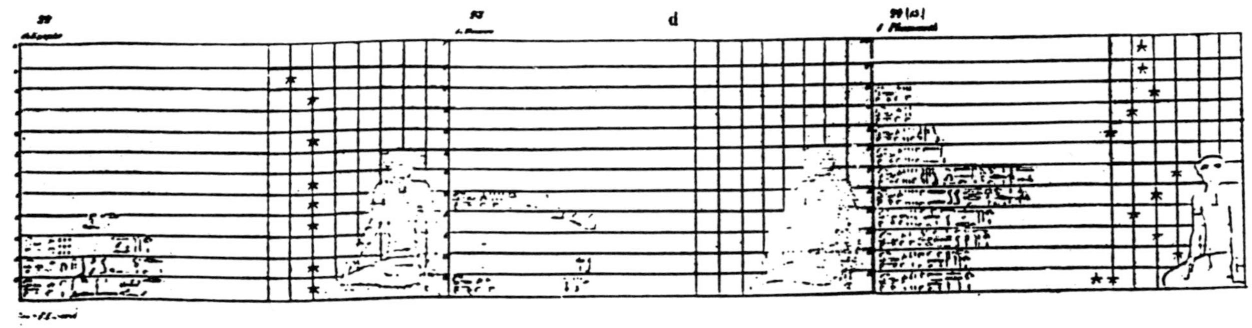

c

d

Fig. III.19b (Concluded)

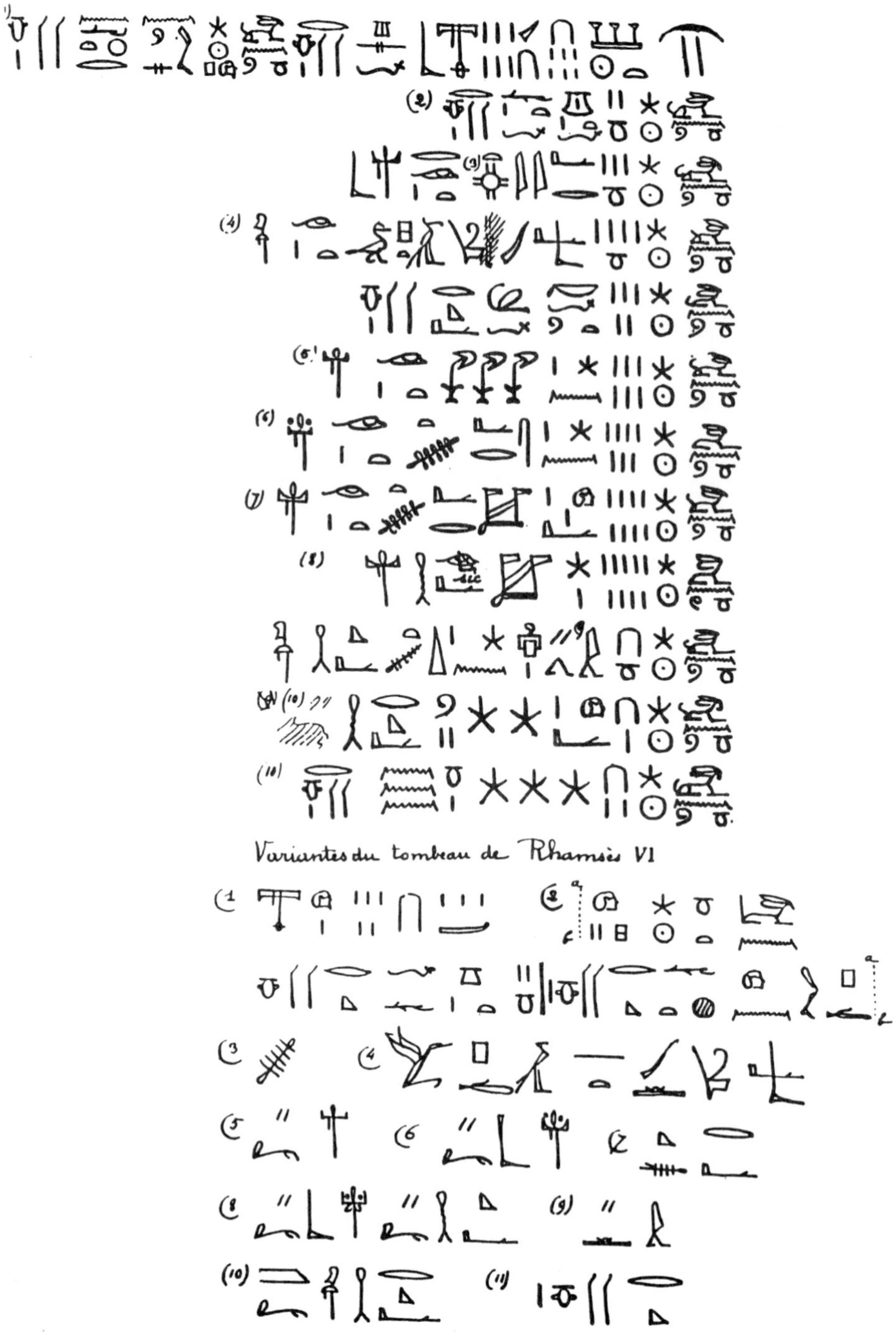

Variantes du tombeau de Rhamsès VI

Fig. III.19c Table 4 from the Ramesside star clock as given in J. F. Champollion, *Monuments de l'Egypte et de la Nubie. Notices descriptives conformes aux manuscripts autographes rédigés sur les lieux*, Vol. 2 (Paris, 1889), pp. 549-50. This volume was written in the hand of G. Maspero.

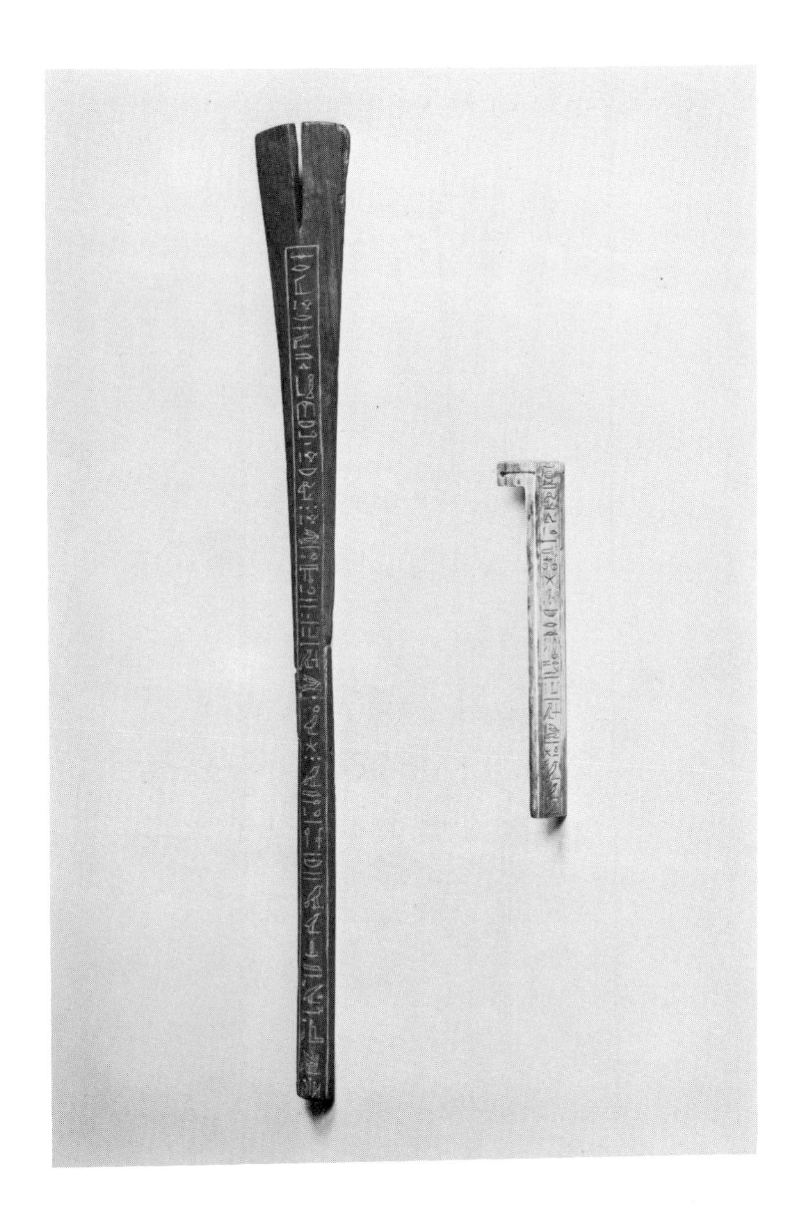

Fig. III.20a Two Egyptian astronomical instruments that belong together: a sighting instrument (Inv. Nr. 14084) and a right angled shadow board (Inv. Nr. 14085). Both were bought together in Cairo and were probably found in Abydos. They are in (West) Berlin in the Ägyptische Museum. The photograph was kindly provided by the Museum and was made by Margarete Büsing.

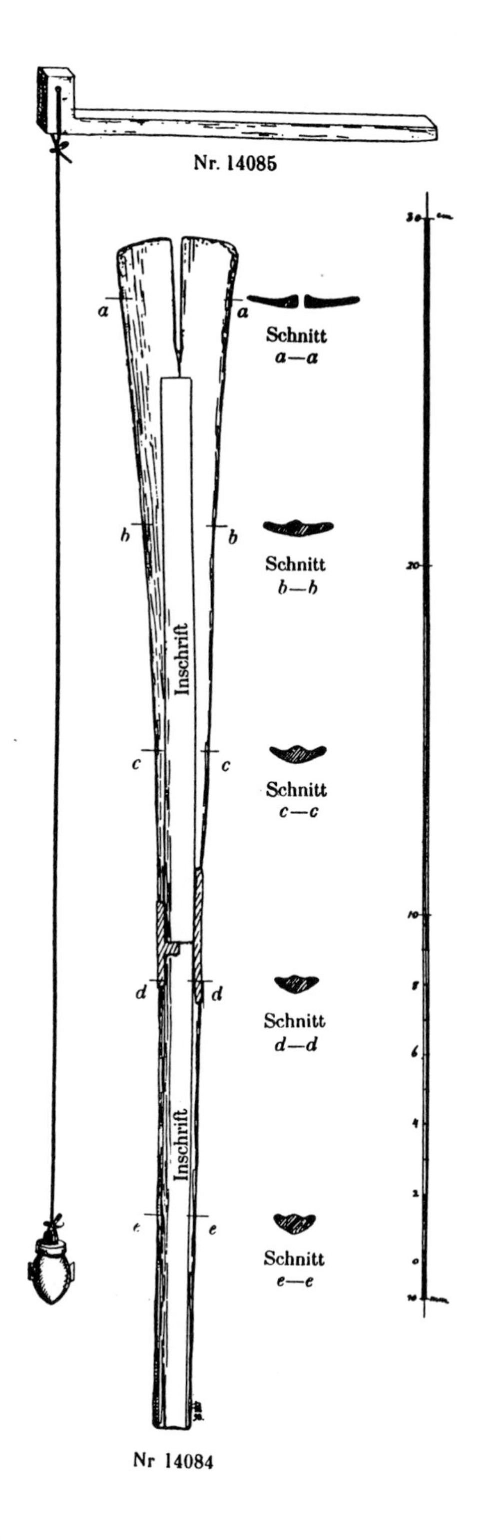

Fig. III.20b A Drawing of the instruments photographed in Fig. III.20a. Taken from Borchardt, "Ein altägyptisches astronomisches Instrument," *ZÄS*, Vol. 37 (1899), p. 10.

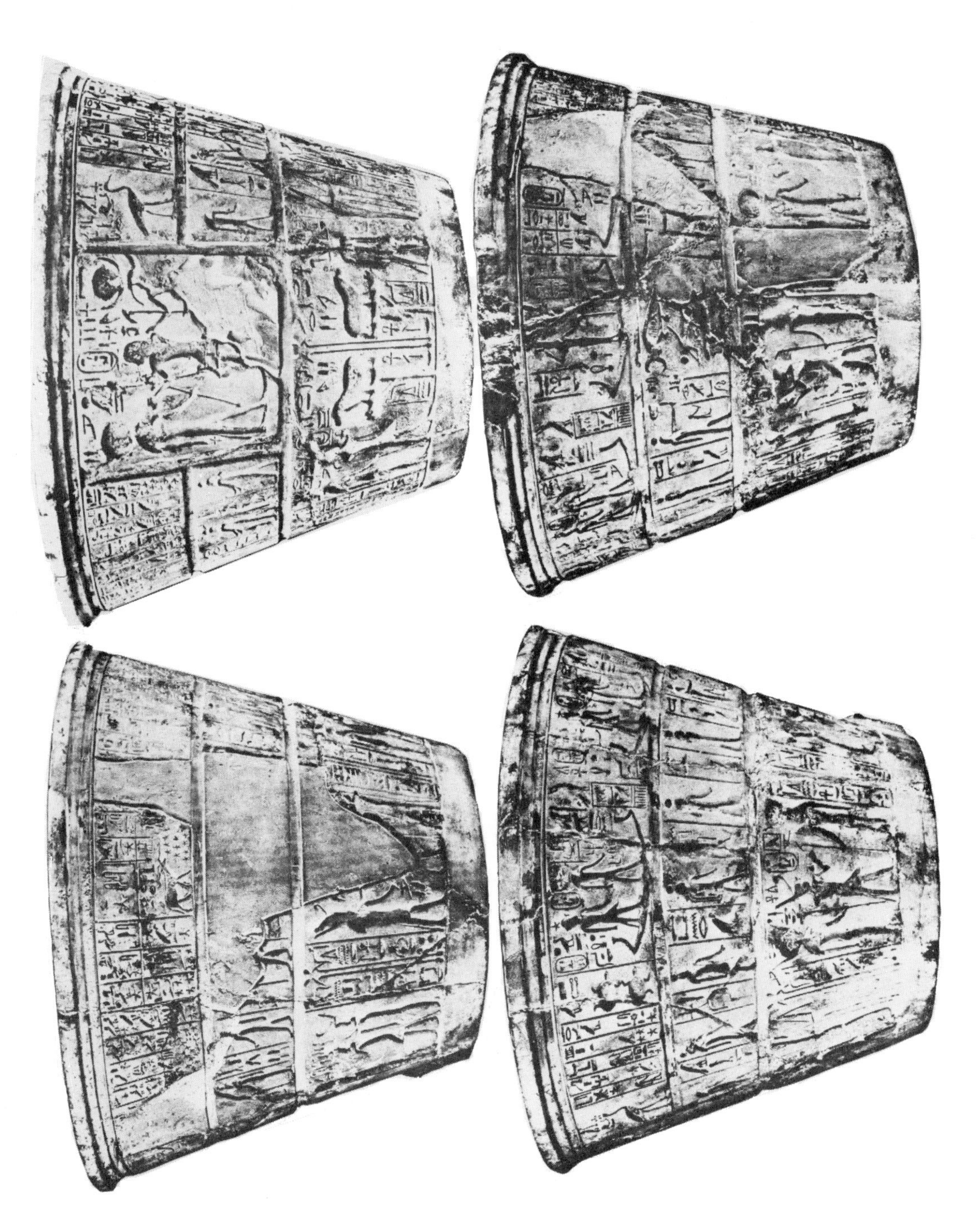

Fig. III.21a Four views of a water clock from the reign of Amenhotep III. Found at Karnak and now in the Cairo Museum.

Fig. III.21b Exterior, interior, and edge views of a fragment of a water clock from the British Museum (cat. no. 938). Time of Philip Arrhidaeus (323-16 B.C.). I thank the Trustees of the British Museum for permission to publish these photographs made by the Museum.

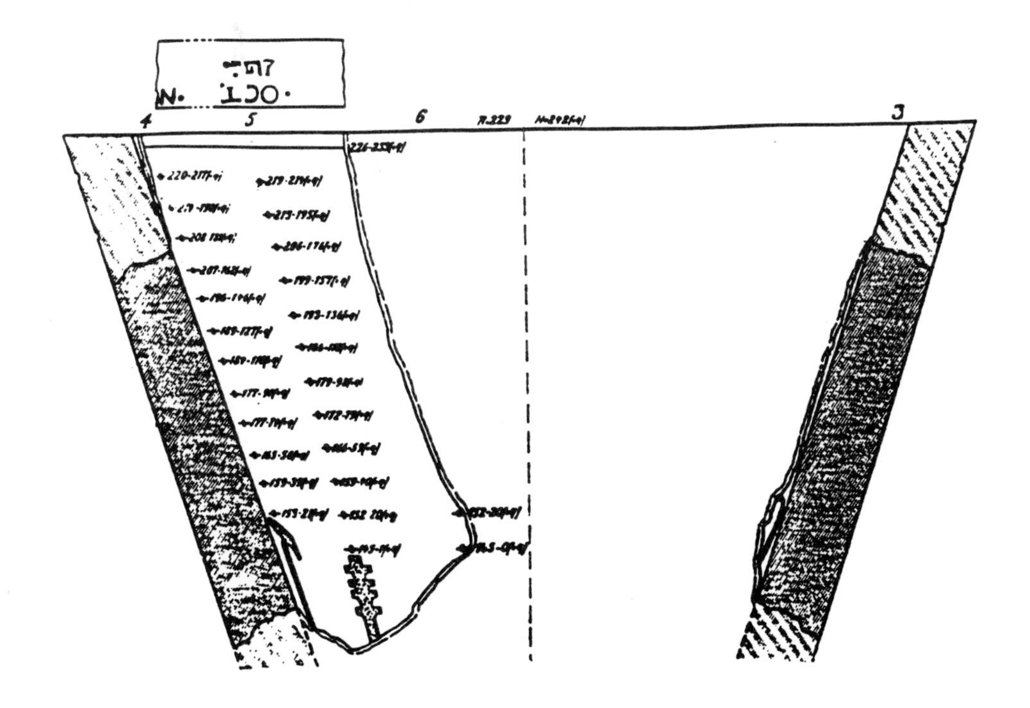

Fig. III.21c Drawing of the interior surface of the water clock pictured in Fig. III.21b. Given by Borchardt, *Die altägyptische Zeitmessung*, Tafel 6.

Fig. III.21d Exterior and interior views of a fragment of a second water clock in the British Museum (cat. no. 933). Also of the time of Philip Arrhidaeus. As in the case of Fig. III.21b, I thank the Trustees of the Museum for permission to publish these photographs supplied by the Museum.

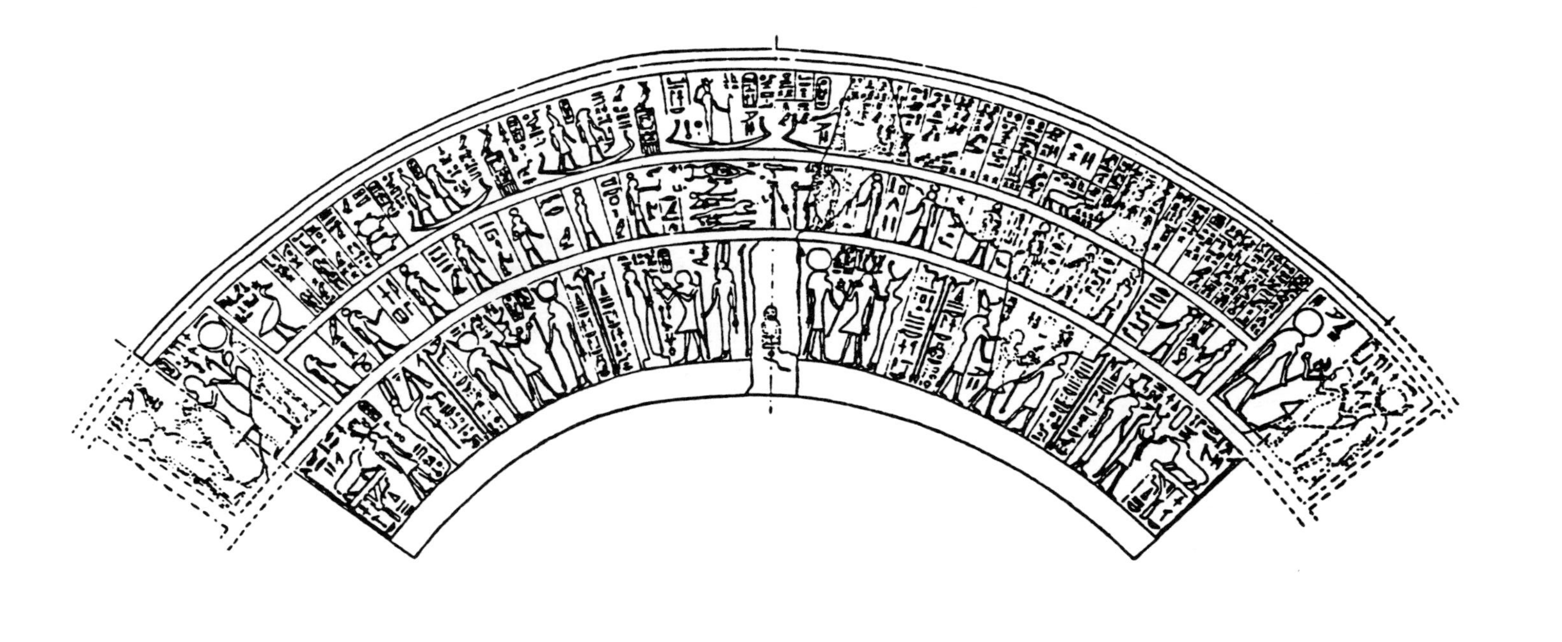

Fig. III.22 Drawing of the exterior decoration of the Karnak water clock. There are some errors of detail (see Neugebauer and Parker, *Egyptian Astronomical Texts*, Vol. 3. p. 13), but it shows the over-all structure of the celestial diagram neatly. Taken from Chatley, "The Egyptian Celestial Diagram," *The Observatory*, Vol. 63 (1940), p. 69.

Vorderansicht.

Fig. III.23 Front view of the Edfu Inflow Clock in the Cairo Museum. Taken from Borchardt, *Die altägyptische Zeitmessung*, Tafel 9.

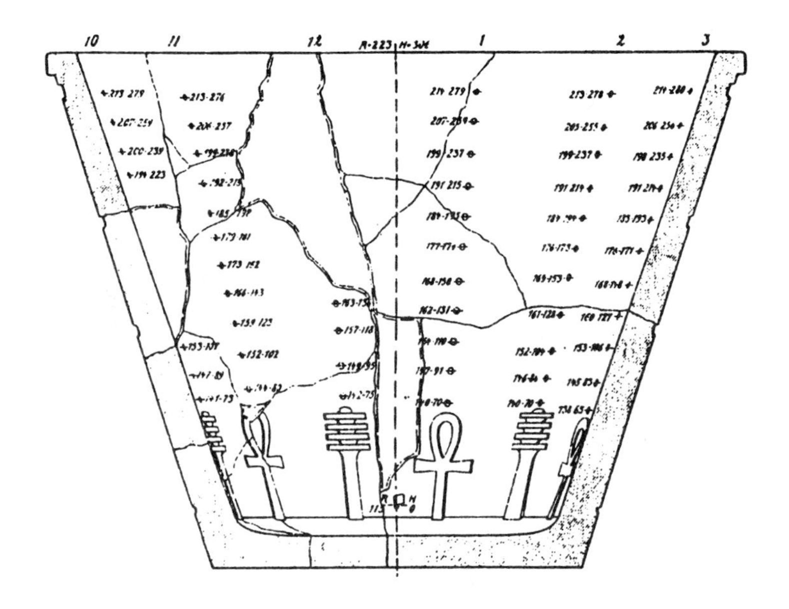

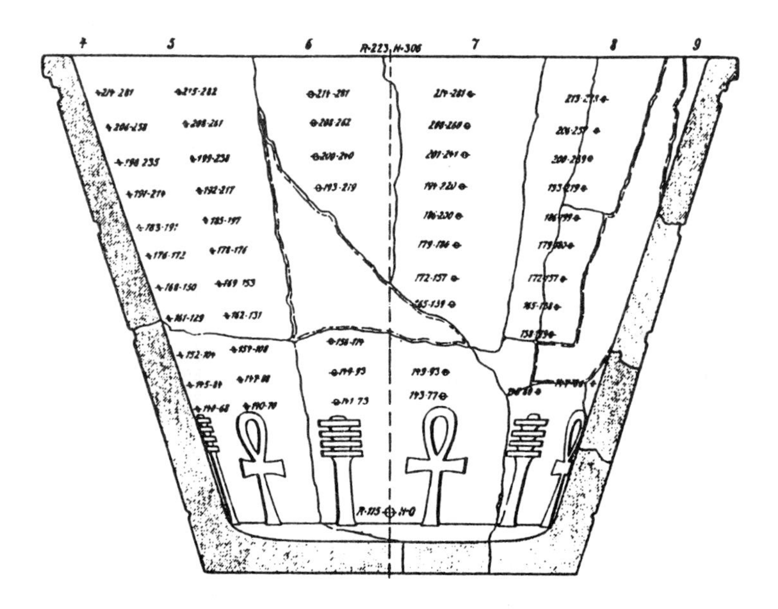

Fig. III.24a Sections of the interior of the Karnak water clock illustrating the positions of its monthly scales of hours, the first section at the top seen towards the front and the second one below it seen towards the rear. The first number at each point of the scales gives in millimeters the radius (R), the second the height (H) above the plane of outflow. These are Borchardt's measurements and the diagram is taken from his *Die altägyptische Zeitmessung*, Tafel 2.

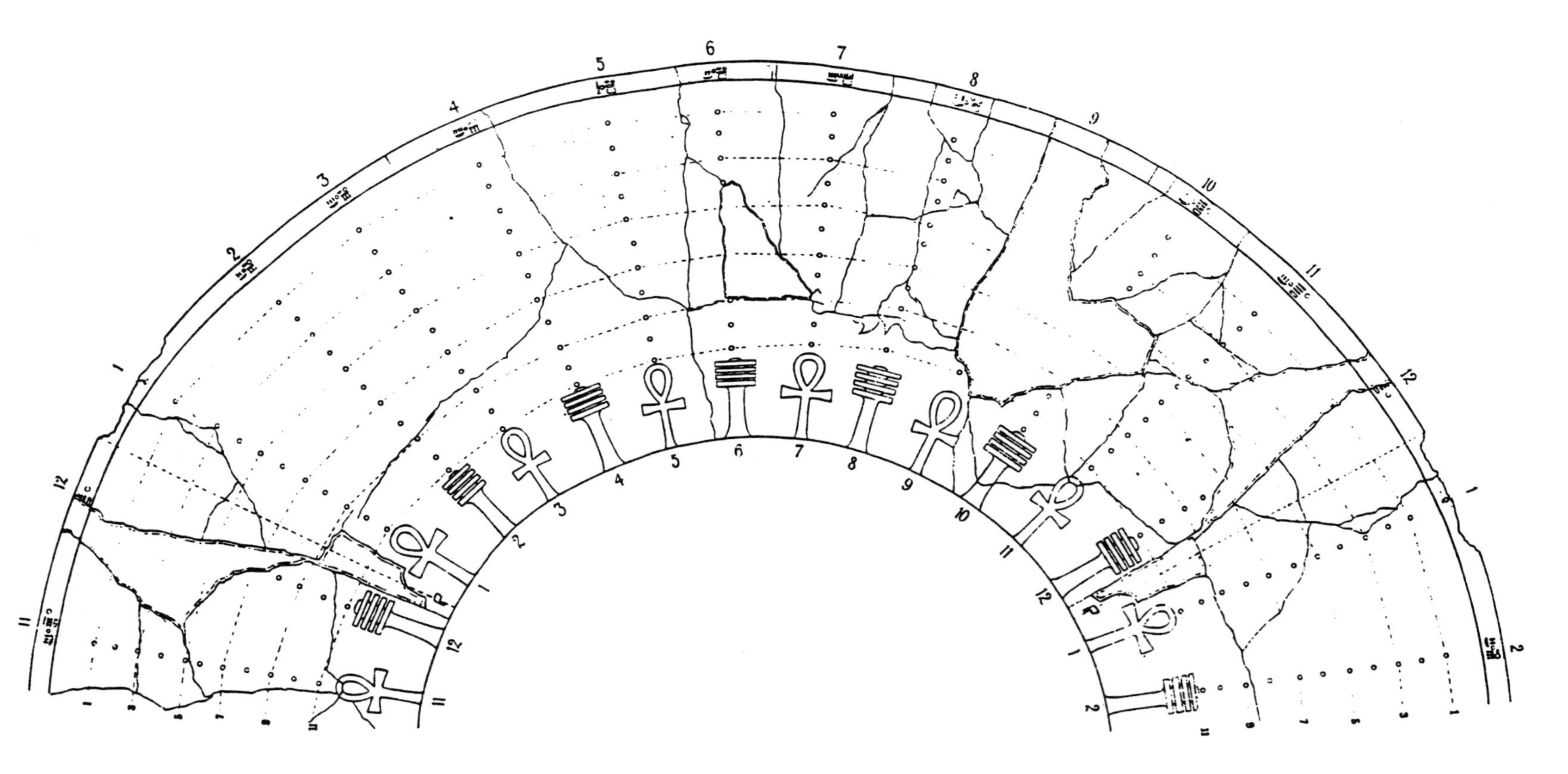

Fig. III.24b A drawing representing the conical interior surface of the Karnak water clock as a segment of a plane circular band. The broken lines were not present on the surface of the clock, but were added by Borchardt. The broken line between the 12th and first monthly scales constitutes the joining line of reference for unfolding the conical surface on the plane. The broken lines at the odd-numbered points on each of the scales were added to make comparison of the scales easier. The drawing was taken from Borchardt's *Die altägyptische Zeitmessung*, Tafel 3.

1 2 3 4 5 6 7 8 9 10 11 12 13 14 15 16

Berl. Mus. Nr. 14476.

Fig. III.25 Description by its inventor of an outflow water clock in the tomb of Amenemhet living under the reign of Amenhotep I. Taken from Borchardt's *Die altägyptische Zeitmessung*, Tafel 18.

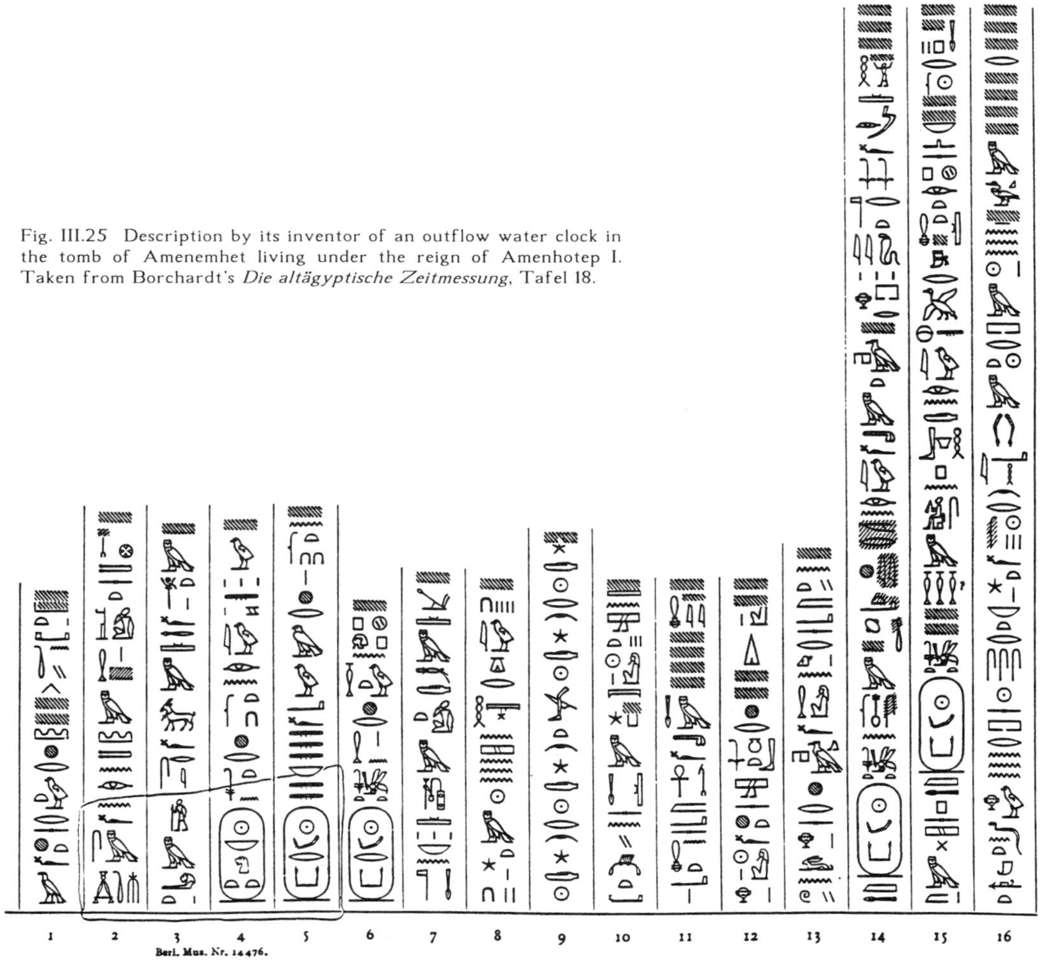

	2	3	4	
2. Monat . .	— Finger	— Finger	— Finger	
3. " . .	— "	— "	— "	. . 1. Monat
4. " . .	11 "	— "	11 "	. . 12. "
5. " . .	12 "	12 "	12 "	. . 11. "
6. " . .	13 "	13 "	— "	. . 10. "
7. " . .	— "	13 2/3 "	— "	. . 9. "
	— "	— "	— "	. . 8. "

Fig. III.26 The lengths in fingerbreadths of the remains of the monthly scales on the clocks designated by Borchardt as outflow clocks nos. 2-4. Taken from Borchardt's *Die altägyptische Zeitmessung*, p. 13.

	9	
3. Monat . .	? Finger	
4. " . .	11 "	. . 2. Monat
5. " . .	11 1/2 "	. . 1. "
6. " . .	12 "	. . 12. "
7. " . .	13 "	. . 11. "
8. " . .	13 2/3 "	. . 10. "
	14 "	. . 9. "

Fig. III.27 The lengths in fingerbreadths of the remains of the monthly scales of outflow clock no. 9. *Ibid.*.

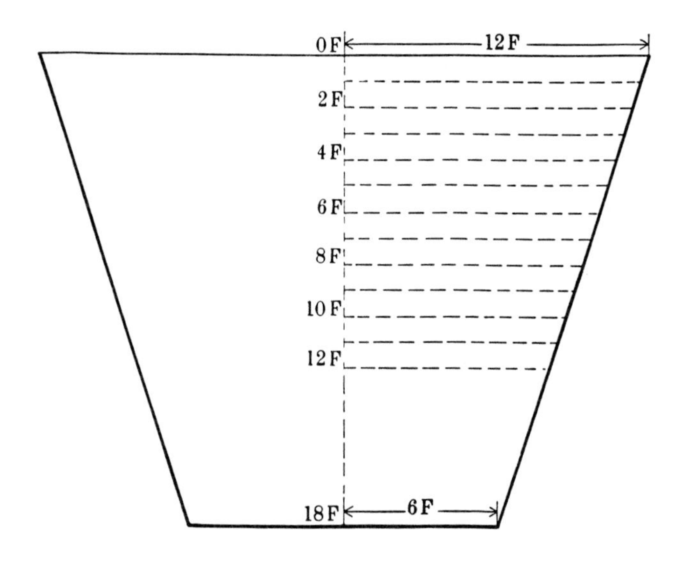

Fig. III.28 A vertical section of the outflow clock of Oxyrhynchus Papyrus No. 470. *Ibid.*, p. 11.

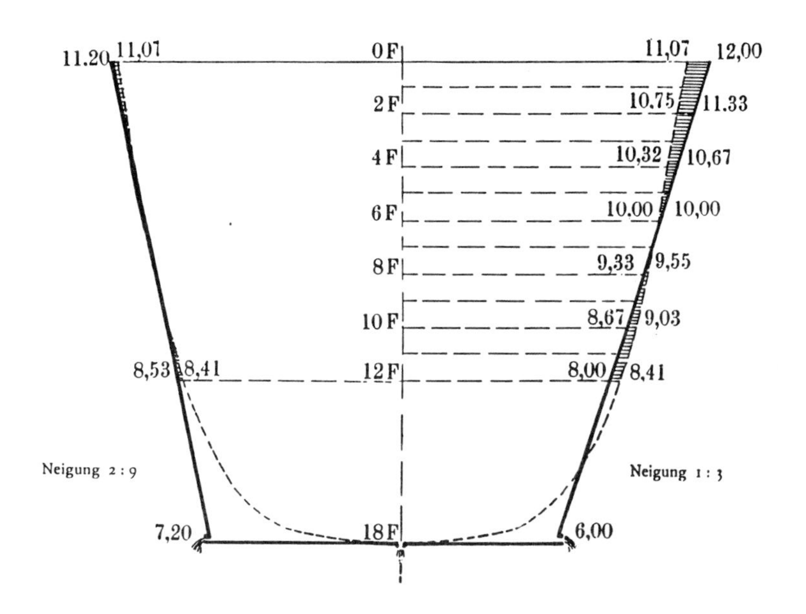

Fig. III.29 Comparison of the parabolic section (drawn in broken lines) of a paraboloidal outflow clock which in theory produces equal flow with the corresponding vertical section of an old Egyptian flowerpot clock with a slope of 1:3 (on the right) and with that of an ideal flowerpot clock with a slope of 2:9 (on the left). As can be seen, the clock with the latter slope would produce a closer approximation to the equal flow of the paraboloidal clock than would the actual Egyptian clock of 1:3 slope. Taken from Borchardt's *Die altägyptische Zeitmessung*, p. 16.

	Verhältniszahl der Auslaufdauer		Abweichung vom Mittel	
	bei 18 Finger Gefäßhöhe	bei 16 Finger Gefäßhöhe	bei 18 Finger Gefäßhöhe	bei 16 Finger Gefäßhöhe
1. Stunde	74,3	74,7	+ 20,5 %	+ 24 %
2. "	71,8	72,1	+ 16,5 %	+ 19,5 %
3. "	69,1	71,1	+ 12 %	+ 18 %
4. "	66,7	66,6	+ 8 %	+ 10,5 %
5. "	65,8	64,7	+ 6 %	+ 7,5 %
6. "	61,9	61,1	± 0 %	+ 1 %
7. "	59,8	58,8	− 2,5 %	− 2 %
8. "	58,1	56,8	− 5,5 %	− 5,5 %
9. "	55,6	53,6	− 10 %	− 11 %
10. "	54,1	51,5	− 12 %	− 14,5 %
11. "	52,8	49,6	− 14,5 %	− 18 %
12. "	51,0	43,5	− 17 %	− 27,5 %
Mittel:	61,7	60,3		

Fig. III.30 Table comparing Egyptian "hour"-durations in a vessel 18 fingerbreadths high and one 16 fingerbreadths high and their deviations from the mean. Borchardt, *Die altägyptische Zeitmessung*, p. 16.

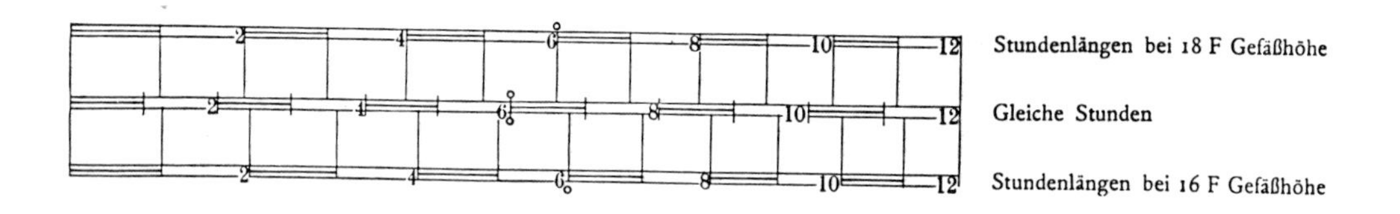

Fig. III.31 Linear comparisons of "hour"-durations in the water clocks of 18 and 16 fingerbreadths height with equal hours. "O" is the position of midnight. Borchardt, *Ibid.*.

Fig. III.32 Models of inflow water clocks and a votive offering, as given by Pogo, "Egyptian Water Clocks," *Isis*, Vol. 25 (1936), Fig. 4 (p. 416) and Plate 4 (oppos. p. 16).

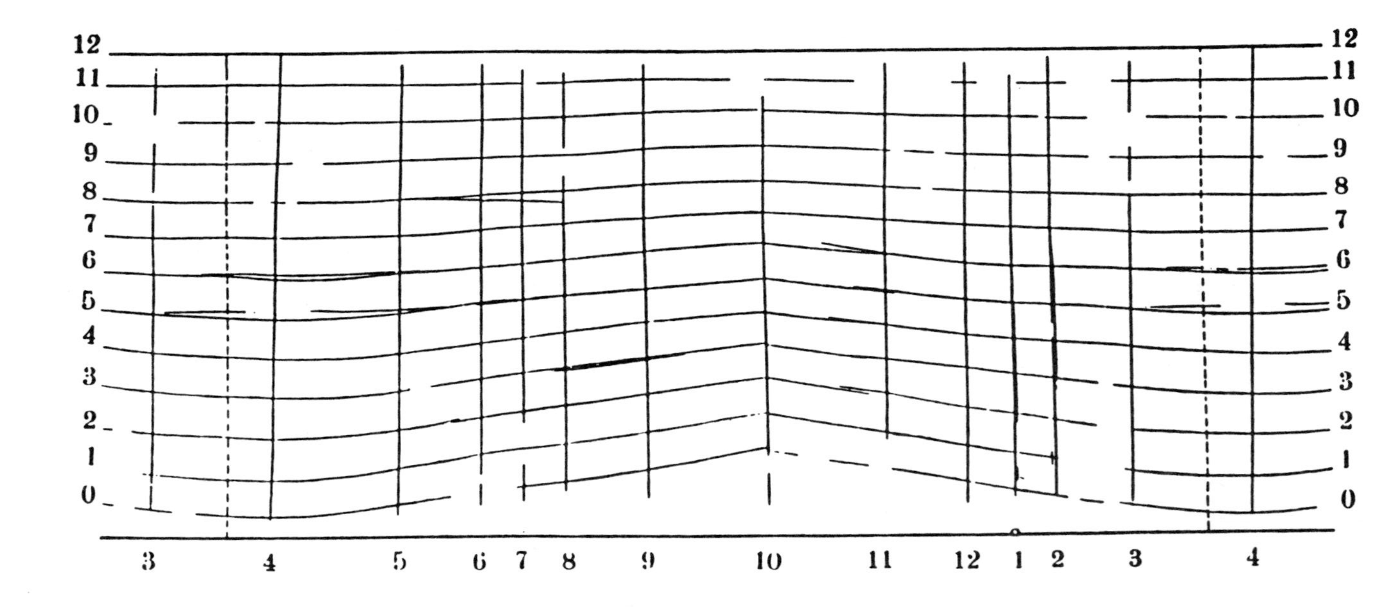

Fig. III.33 The grid of the Edfu inflow clock as reproduced from a paper impression of the interior surface taken by Borchardt, *Die altägyptische Zeitmessung*, Tafel 9.

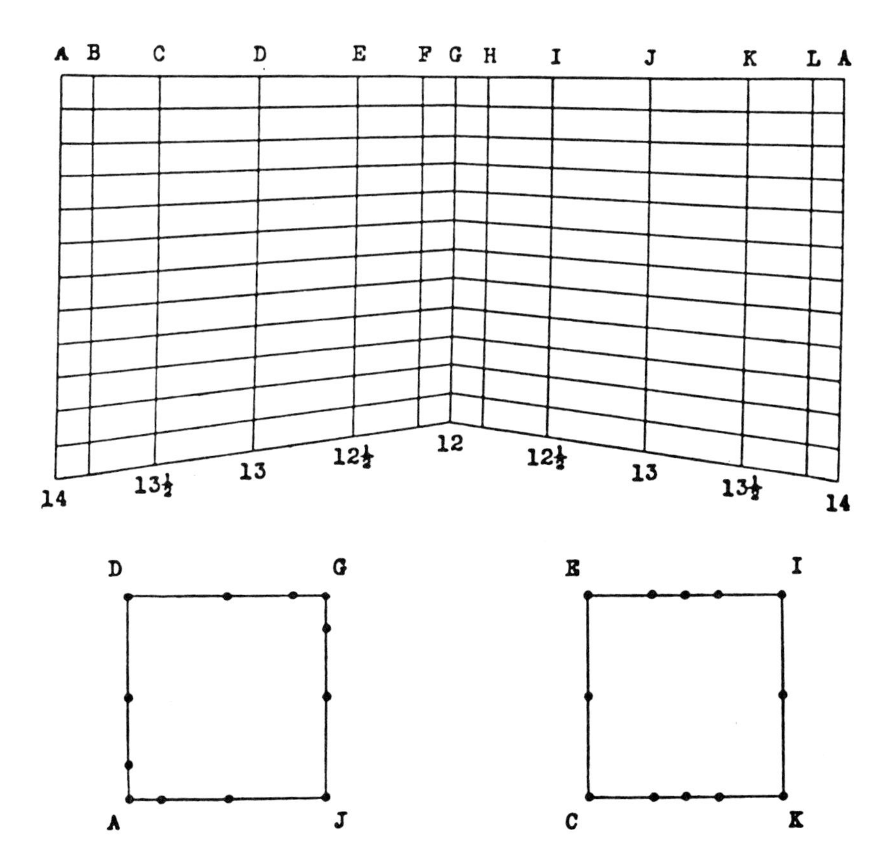

Fig. III.34 Graduation of the 1:2:3 type of grid for a square prismatic inflow clock. Taken from Pogo, "Egyptian Water Clocks," p. 408.

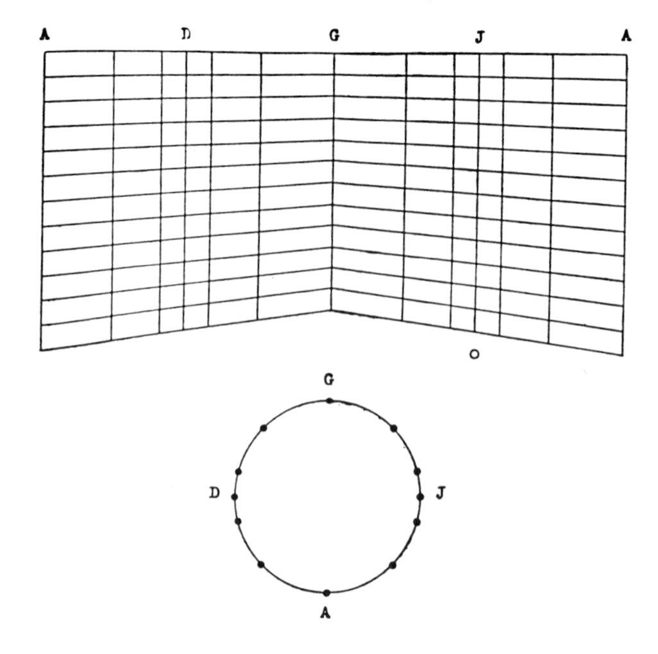

Fig. III.35 Graduation scheme of the cylindrical inflow clock from Edfu. This eliminates the careless imperfections of the original grid reproduced in Fig. III.33. Taken from Pogo, "Egyptian Water Clocks," p. 409.

Fig. III.36 A votive offering in the form of a clepsydra? Time of Amenhotep III (ca. 1391-1343 B.C.). Taken from Gayet, *Le Temple de Louxor* (Paris, 1894), Plate LXVIII (Pl. LXXIV, Fig. 212).

TABLE II

The civil year and the seasons

Year B.C.	Civil New Year's Day	Vernal equinox	Vernal equinox	Summer solstice	Autumnal equinox	Winter solstice
1617	Oct. 1	Apr. 4				
			VII	X	I	IV
1497	Sep. 1	Apr. 3				
			VIII	XI	II	V
1373	Aug. 1	Apr. 2				
			IX	XII	III	VI
1249	July 1	Apr. 1				
			X	I	IV	VII
1129	June 1	Mar. 31				
			XI	II	V	VIII
1105	May 1	Mar. 30				
			XII	III	VI	IX
885	Apr. 1	Mar. 29				
			I	IV	VII	X
761	Mar. 1	Mar. 28				
			II	V	VIII	XI
648	Feb. 1	Mar. 28				
			III	VI	IX	XII
524	Jan. 1	Mar. 27				
			IV	VII	X	I
401	Dec. 1	Mar. 25				
			V	VIII	XI	II
281	Nov. 1	Mar. 25				
			VI	IX	XII	III
157	Oct. 1	Mar. 24				
			VII	X	I	IV
37	Sep. 1	Mar. 23				

Note.—Roman numerals refer to Egyptian months.

The first two columns show the rapid recession, through the Julian calendar, of the civil New Year's Day, i.e., of the 1st day of the 1st month of the *ꜣḫt* season, or, in the post-Persian period, of the 1st day of the month Thōuth.

The first and third columns show the slow recession, through the Julian calendar, of the date of the vernal equinox.

The last four columns indicate, roughly, the Egyptian months of the equinoxes and solstices. Thus, during the fifteenth century, the civil year began in August, and the winter solstice fell on dates advancing through the 1st month of the *prt* season; in the fourth century, the civil year began in November, and the winter solstice progressed through the month Phaōphi.

Fig. III.37 Table taken from Pogo, "Egyptian Water Clocks," p. 410.

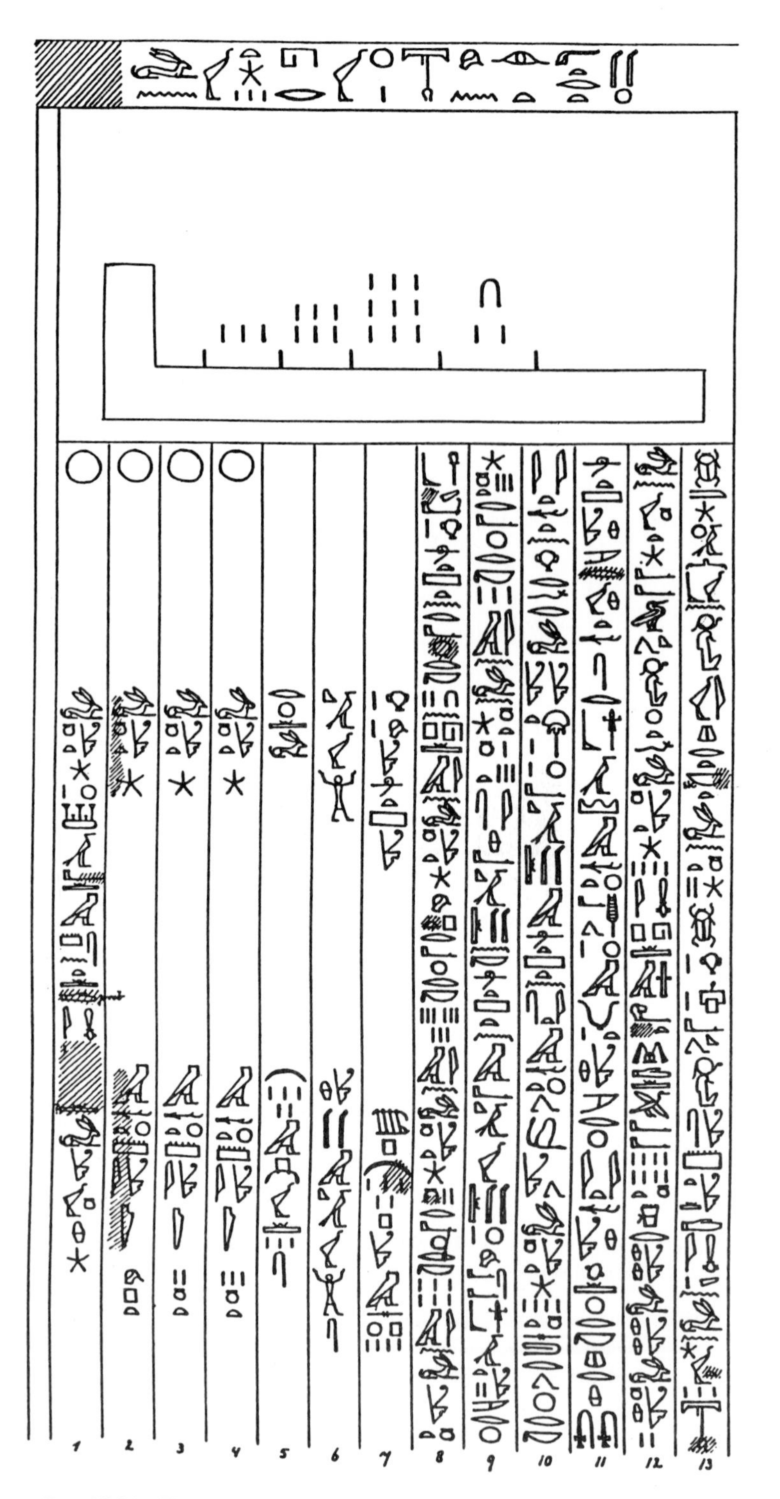

Fig. III.38 Transcription of the description of a shadow clock on the west side of the roof of the sarcophagus chamber of the cenotaph of Seti I. Taken from Frankfort, *The Cenotaph of Seti I at Abydos*, Vol. 2, Plate 83. For a photograph, see Neugebauer and Parker, *Egyptian Astronomical Texts*, Vol. 1, Plate 32.

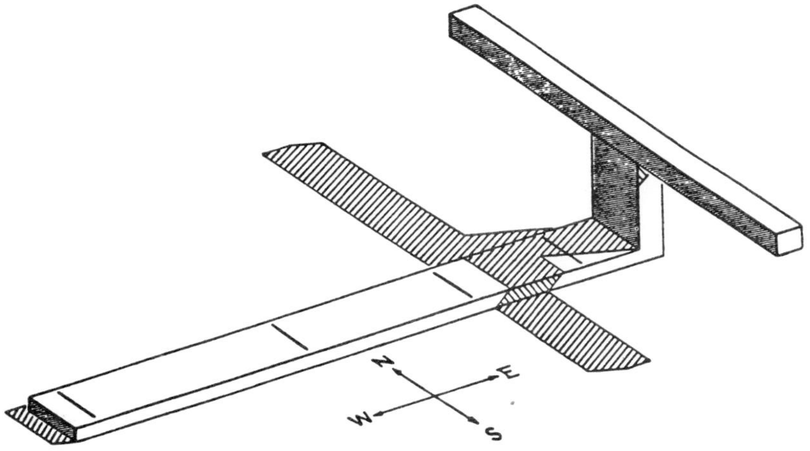

Fig. III.39 Drawing of a shadow clock of the type given in the cenotaph of Seti I. Given by Neugebauer and Parker, *Egyptian Astronomical Texts*, Vol. 1, p. 117, Fig. 36.

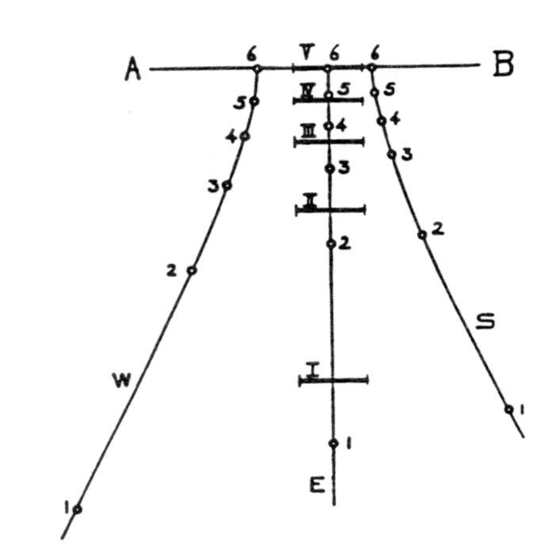

Fig. III.40 A graphical representation of four-hour shadow clocks. The hyperbolic curves at each side contain the theoretical locations of the hours at the winter and summer solstices. Given by Neugebauer and Parker, *Egyptian Astronomical Texts*, Vol. 1, p. 117, Fig. 37.

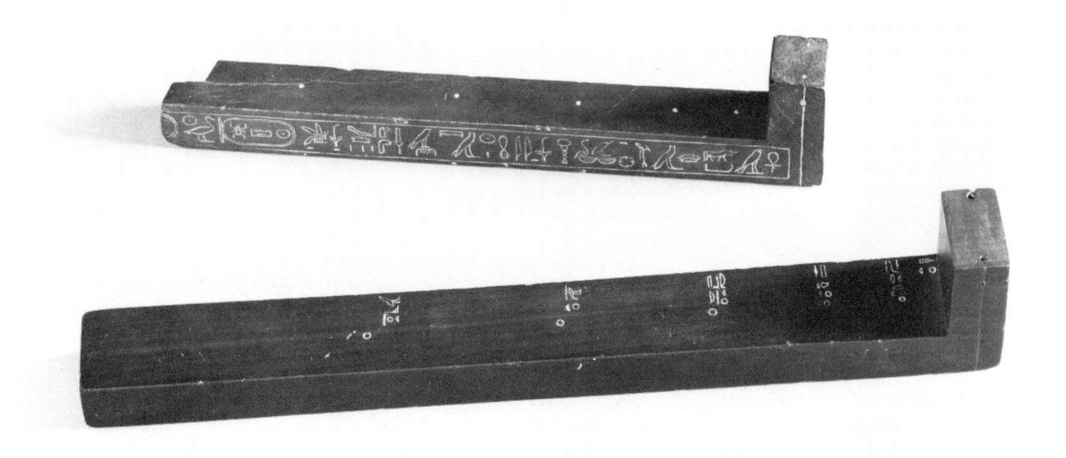

Fig. III.41 Shadow clocks in the Ägyptische Museum (West Berlin): at the top, a separate endpiece from the reign of Nebmaatre, i.e., Amenhotep III, No. 14573 (photograph by Elsa Postel); in the middle, a shadow clock from the reign of Menkheperre, i.e., Tuthmosis III, No. 19744; and, at the bottom, a shadow clock from Sais some 500 years later, No. 19743; the photograph of both shadow clocks is by Margarete Büsing. The photographs have been very kindly supplied by the Ägyptische Museum.

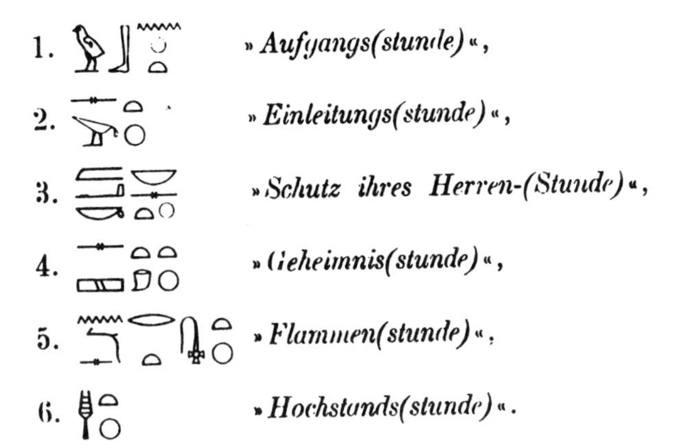

1. »Aufgangs(stunde)«,
2. »Einleitungs(stunde)«,
3. »Schutz ihres Herren-(Stunde)«,
4. »Geheimnis(stunde)«,
5. »Flammen(stunde)«,
6. »Hochstands(stunde)«.

Fig. III.42 Names of the Goddesses of the first six hours. Taken from Borchardt, "Altägyptische Sonnenuhren," *ZÄS*, Vol. 48 (1910), p. 10.

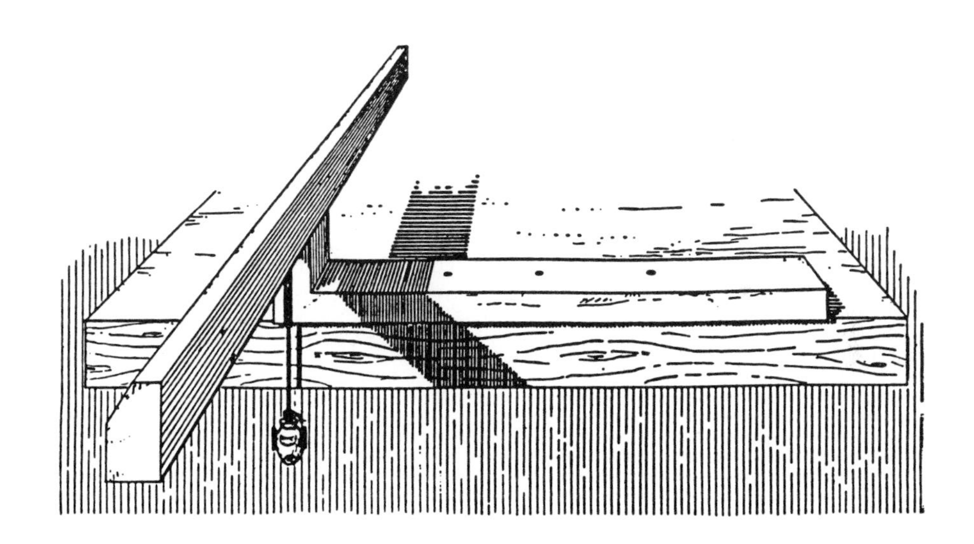

Fig. III.43 Borchardt's reconstruction of a complete shadow clock, with the base ruler pointed toward the east and the crossbar toward the north. The imagined crossbar was supposed by Borchardt to have had beveled edges allowing for changes in the height of the crossbar to approximate the measurement of equal hours. Taken from Borchardt, *Die altägyptische Zeitmessung*, p. 33.

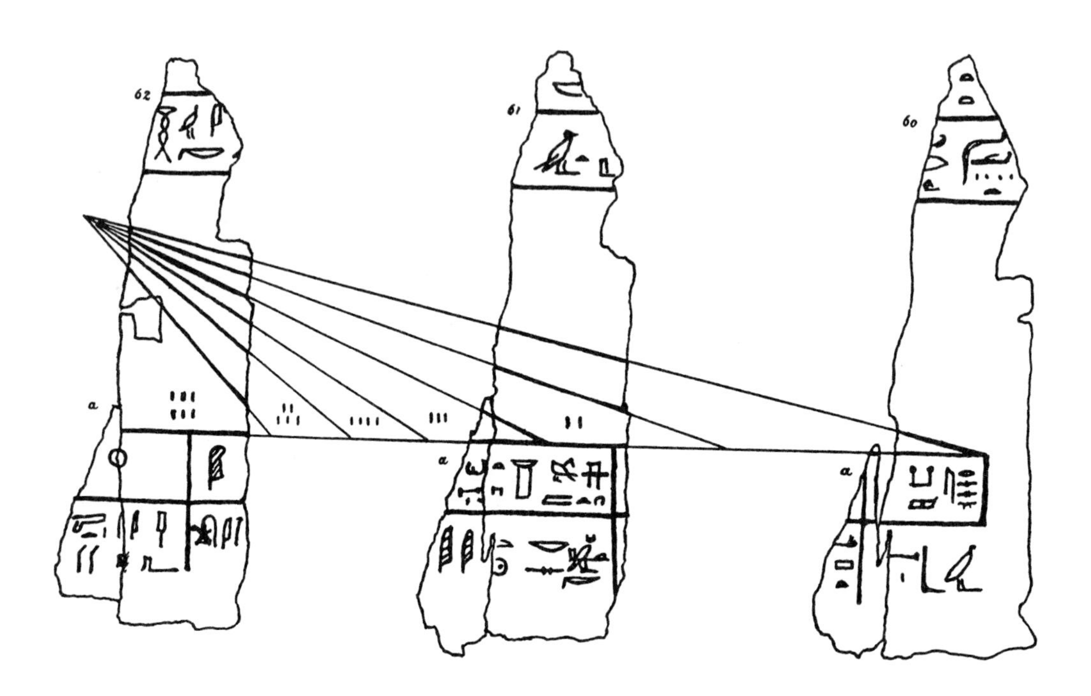

Fig. III.44 Papyrus fragments (ca. 100 A.D.) depicting a shadow clock with its hour markings. Taken from Griffith and Petrie, *Two Hieroglyphic Papyri from Tanis*, Plate 15.

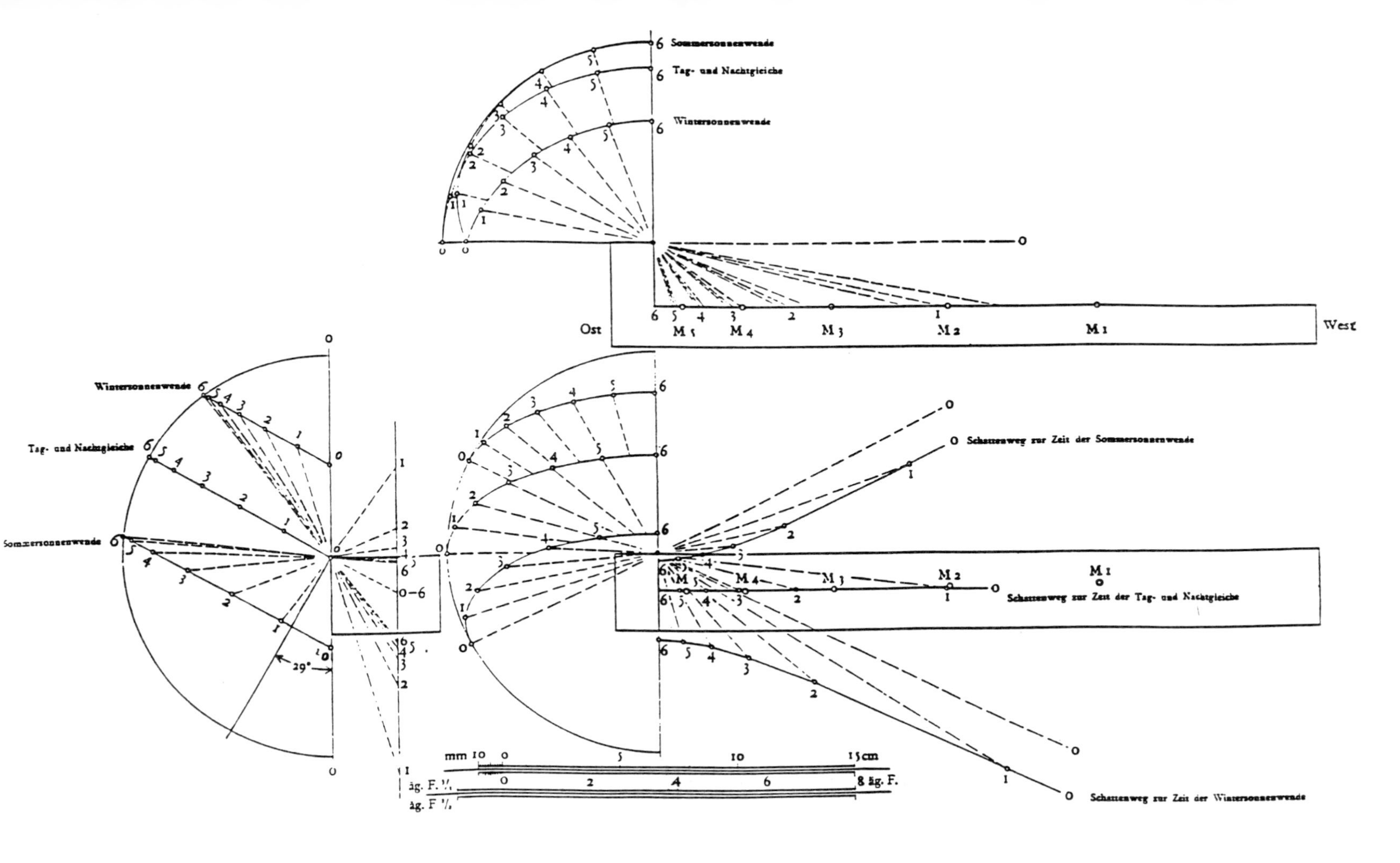

Fig. III.45 Borchardt's investigation of the shadow clock of Berlin (No. 19743) for the latitude 29° north. It shows the approximate deviations of the clock's markings from the actual shadow positions of equal hours at the times of the equinoxes and of the solstices. Taken from Borchardt, *Die altägyptische Zeitmessung*, Tafel 13.

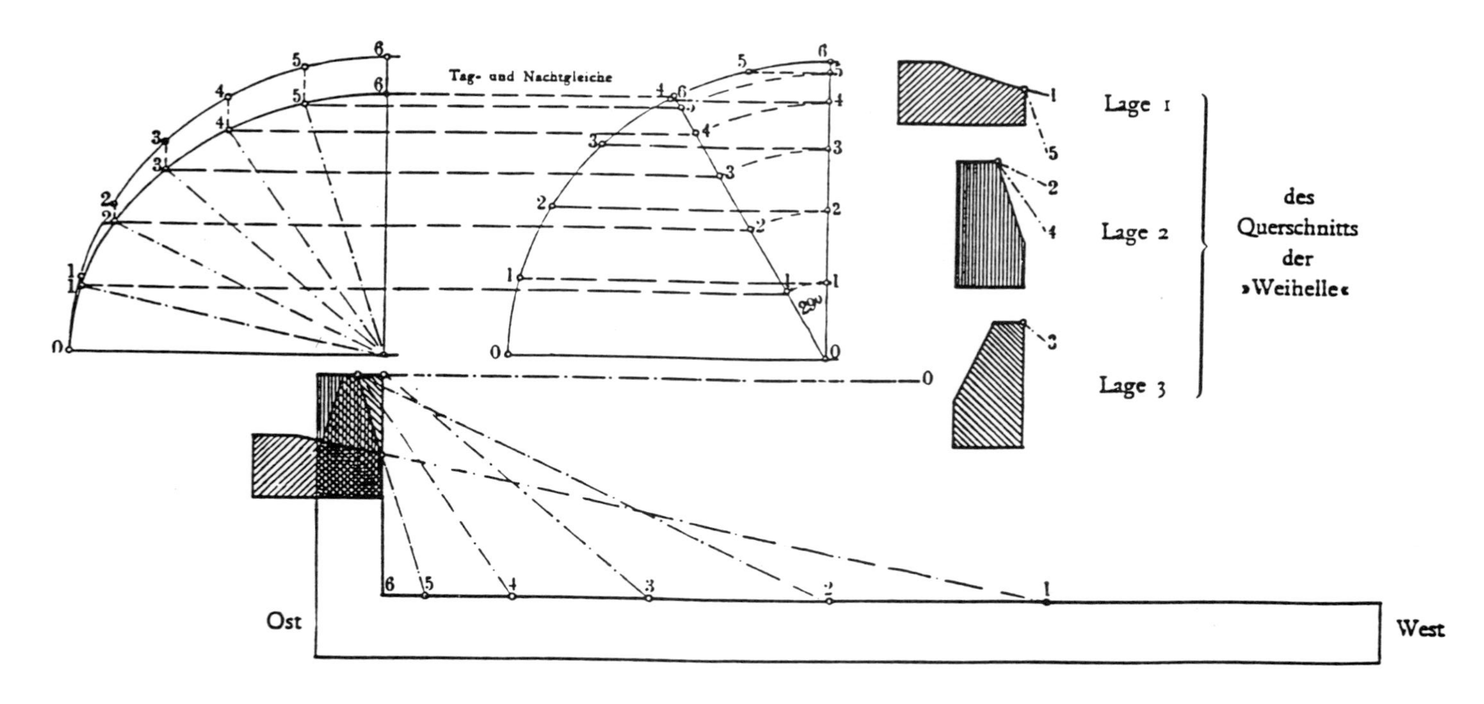

Fig. III.46 The use of a five-edged rule on the Berlin shadow clock (No. 19743) to obtain equal hours. Three positions of the rule are shown in vertical section, with the hours to be determined by each position indicated by broken lines. Taken from Borchardt, *Die altägyptische Zeitmessung*, p. 36.

Fig. III.47 A model combining three kinds of shadow clocks: (1) one having a flat shadow-receiving surface, (2) the second having a flight of steps as the shadow-receiving surface, and (3) the last having an inclined plane as the shadow-receiving surface. Now in the Cairo Museum. Dated later than the Berlin shadow clock No. 19743. Taken from Borchardt, *Die altägyptische Zeitmessung*, p. 37. Cf. Edgar, *Sculptors' Studies and Unfinished Works*, Plate XXI.

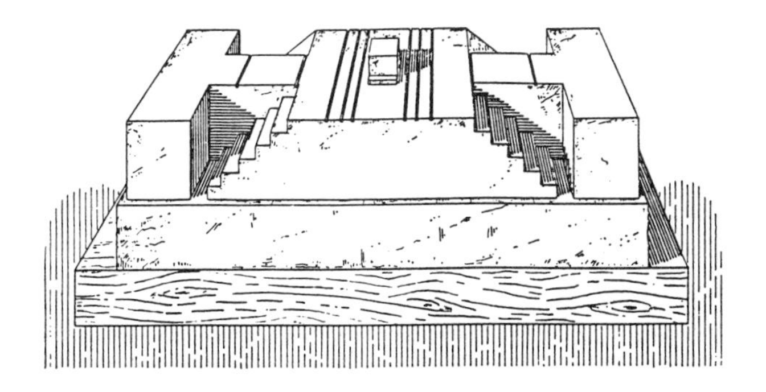

Fig. III.48 A drawing of the Cairo model in use. The front side is oriented in the east-west direction. Taken from Borchardt, *Die altägyptische Zeitmessung*, p. 37.

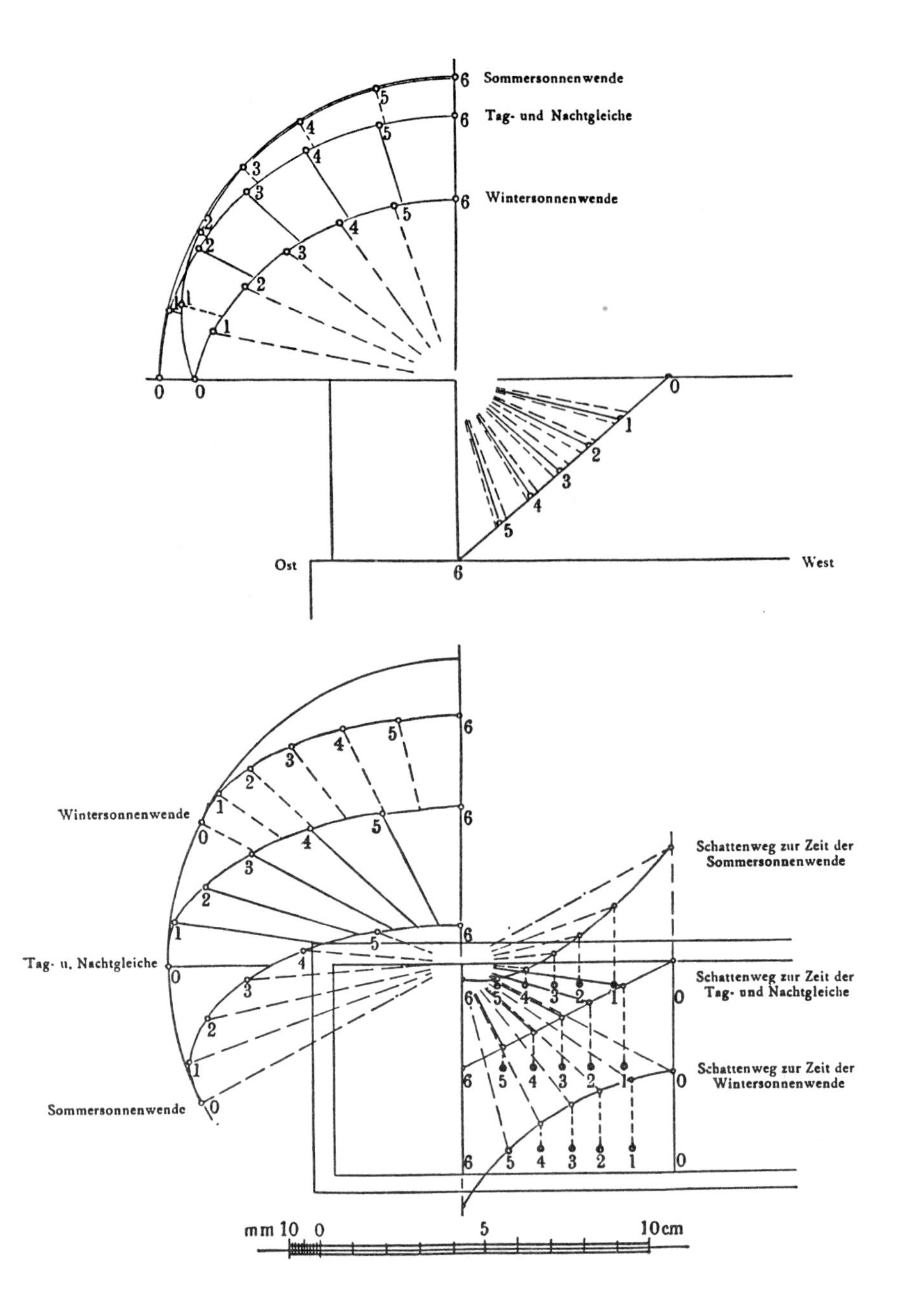

Fig. III.49 Determination of three sets of hour scales for different times of the year, i.e., for the summer and winter solstices and for the equinoxes; to be mounted on an inclined-plane shadow clock like that included in the Cairo model. Taken from Borchardt, *Die altägyptische Zeitmessung*, p. 42.

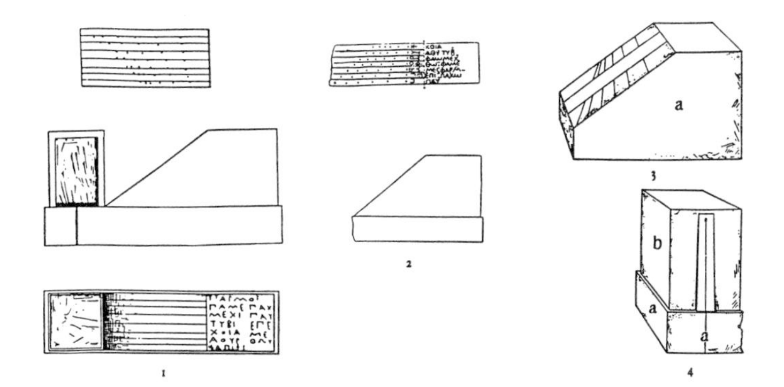

Fig. III.50 Remains of some sun-pointing shadow clocks: (1) one almost complete from Qantara (ca. 320 B.C.); (2) fragments of another from the Hofmann collection in Paris (Roman period); (3) one containing an inclined-plane surface in the Petrie Museum at University College, London (see Fig. III.52); (4) a fragment showing the vertical block with the ray-producing edge, from Turin. (1) and (2) bear the Greek names of Egyptian months. (3) and (4) have hieroglyphic inscriptions and appear to be older than (1) and (2). The group is taken from Borchardt, *Die altägyptische Zeitmessung*, p. 44.

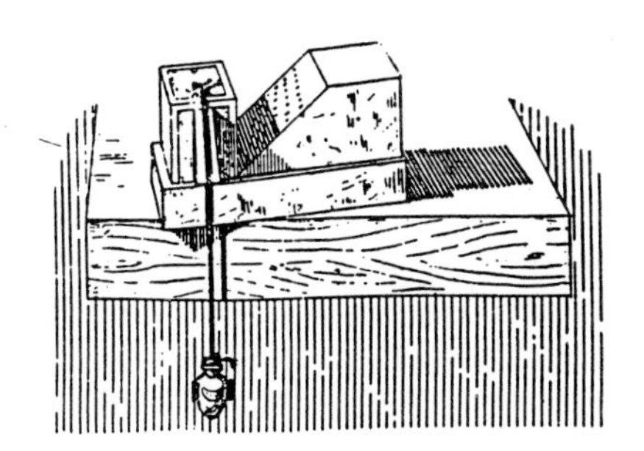

Fig. III.51 A sun-pointing shadow clock in use. Taken from Borchardt, *Die altägyptische Zeitmessung*, p. 45.

Viewed from the left side, front

Viewed from the right side, front

Fig. III.52 Several views of a sun-pointing shadow clock having an inclined plane as the shadow-receiving surface (cf. Fig. III.50, no. 3). The shadow-producing endpiece is missing. Numbered 16376 in the Petrie Museum of University College, London. I thank the Museum for providing me with this series of photographs.

Viewed from the left side, direct

Viewed from the right side, direct

Fig. III.52 (continued)

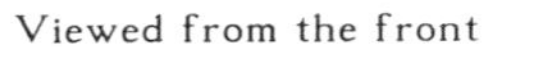

Viewed from the front

Viewed from the back

Fig. III.52 (concluded)

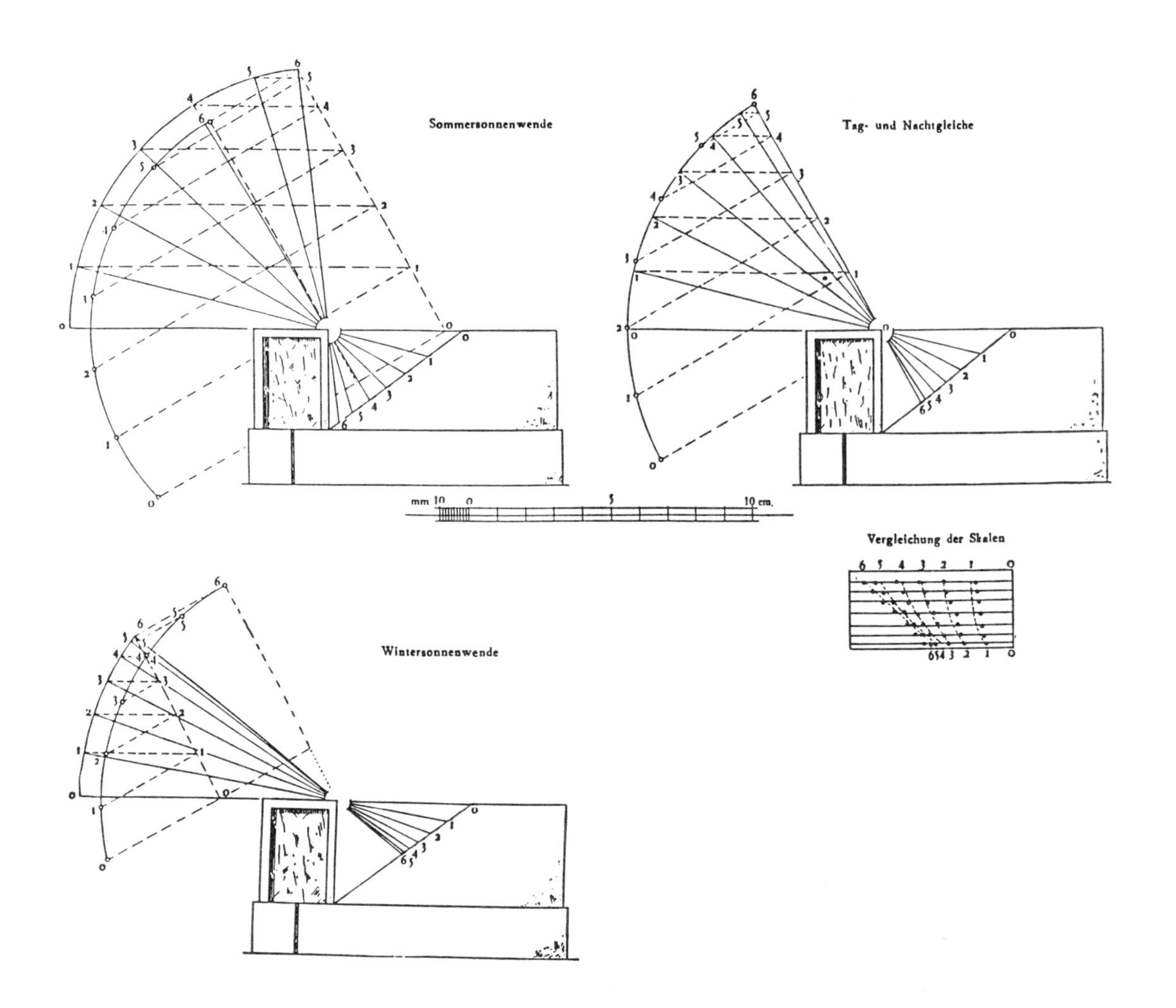

Fig. III.53a Investigation of the sun-pointing shadow clock from Qantara (Fig. III.50, no. 1) for the latitude 31° north. In the small diagram (right, below) comparing the scales, the small circles give the existing scales and the broken lines give the theoretically correct positions. Taken from Borchardt, *Die altägyptische Zeitmessung*, p. 46.

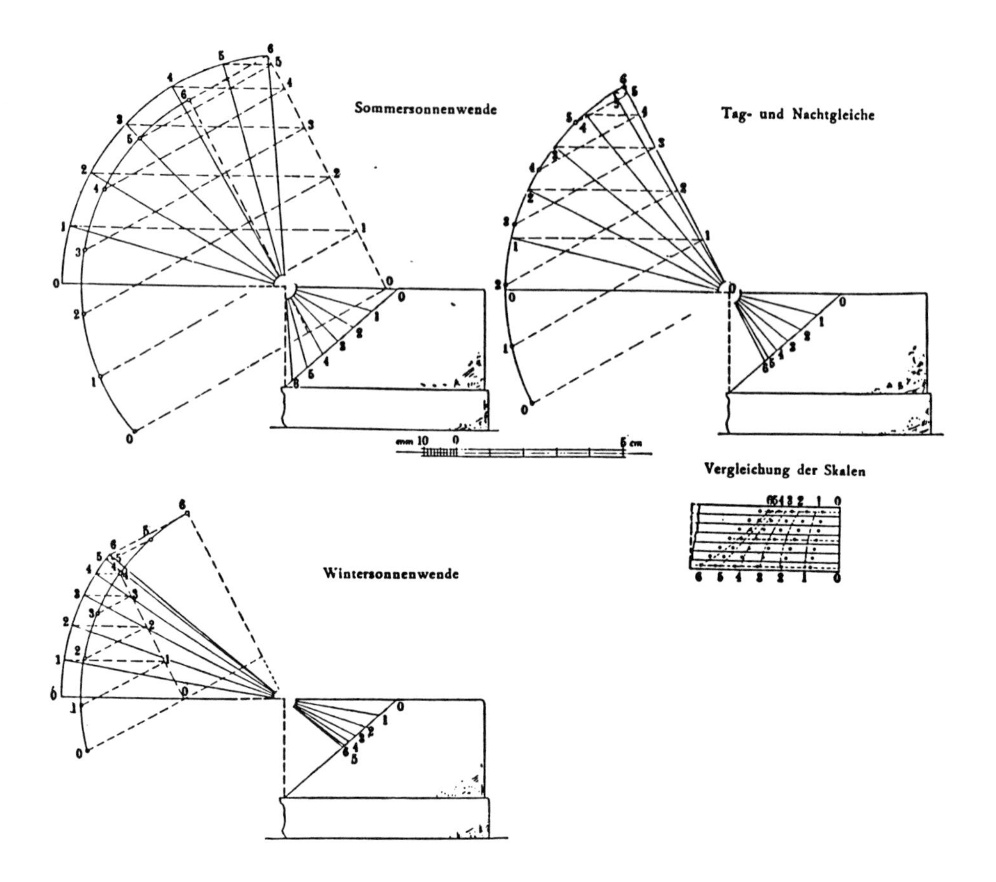

Fig. III.53b Investigation of the sun-pointing shadow clock from Paris (Fig. III.50, no. 2) for the latitude 29° north. In the small diagram (right, below) comparing the scales, the small circles give the existing scales and the broken lines give the theoretically correct positions. Taken from Borchardt, *Die altägyptische Zeitmessung*, p. 47.

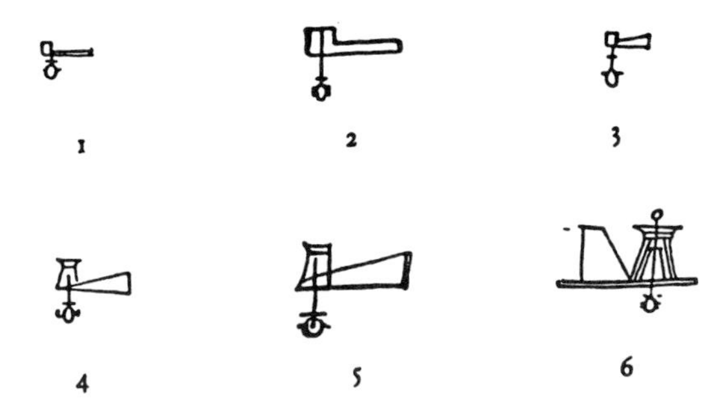

Fig. III.54 Shadow-clock glyphs used as determinatives for the word "hour" or as ideograms. All are from the Ptolemaic or Roman periods. Taken from Borchardt, *Die altägyptische Zeitmessung*, p. 52, where the source of each glyph is given.

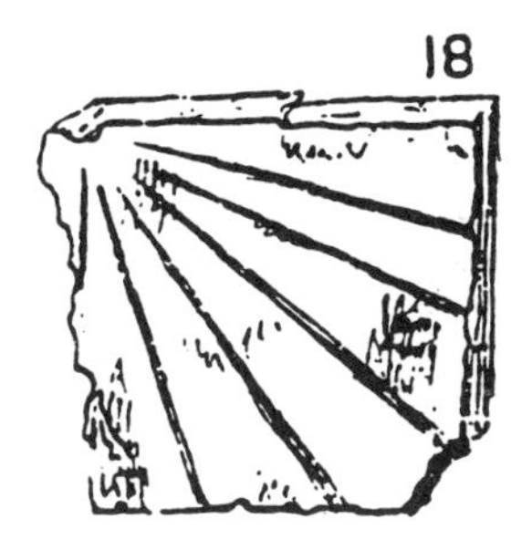

Fig. III.55a Fragment of a flat sundial from Dendera. Undated. Taken from Petrie, *Dendereh 1898* (London, 1900), Plate XIX, 18.

Fig. III.55b A flat sundial (to be hung vertically). Found in Geser (Palestine). Time of Merenptah (ca. 1224-1214 B.C.). Taken from Macalister, *The Excavation of Gezer 1902-1905 and 1907-1909*, Vol. 2, p. 331, Fig. 456.

Fig. III.56 Front and back views of a flat sundial (West Berlin, Ägyptische Museum, no. 20322). Found at Luxor. Dates from Greco-Roman times. The photograhs, by Jürgen Liepe, were kindly supplied by the Museum.

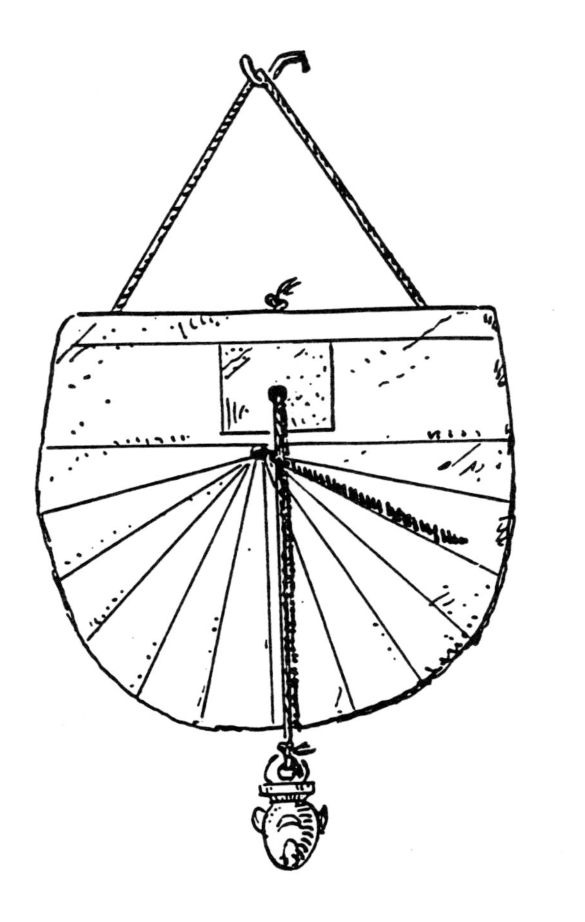

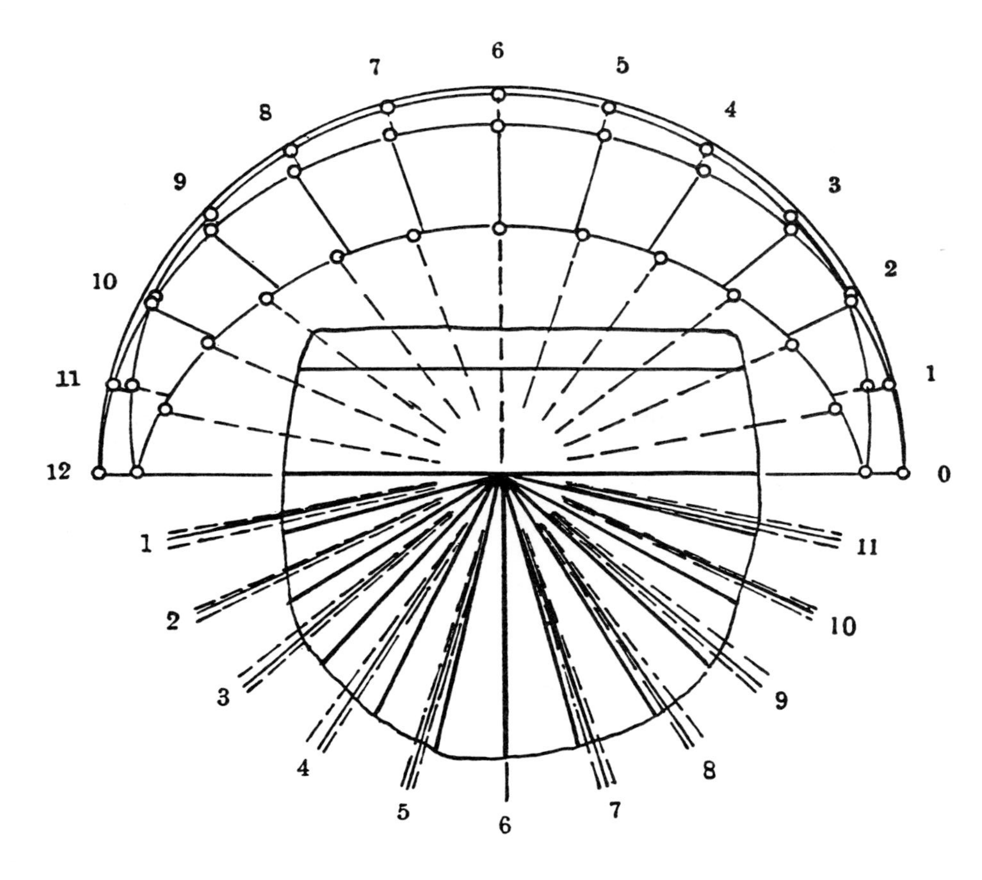

Fig. III.57 Drawing of the Berlin sundial given in Fig. III.56 and analyzed for the latitude of Thebes (25.5° north). Taken from Borchardt, *Die altägyptische Zeitmessung*, Tafel 15.

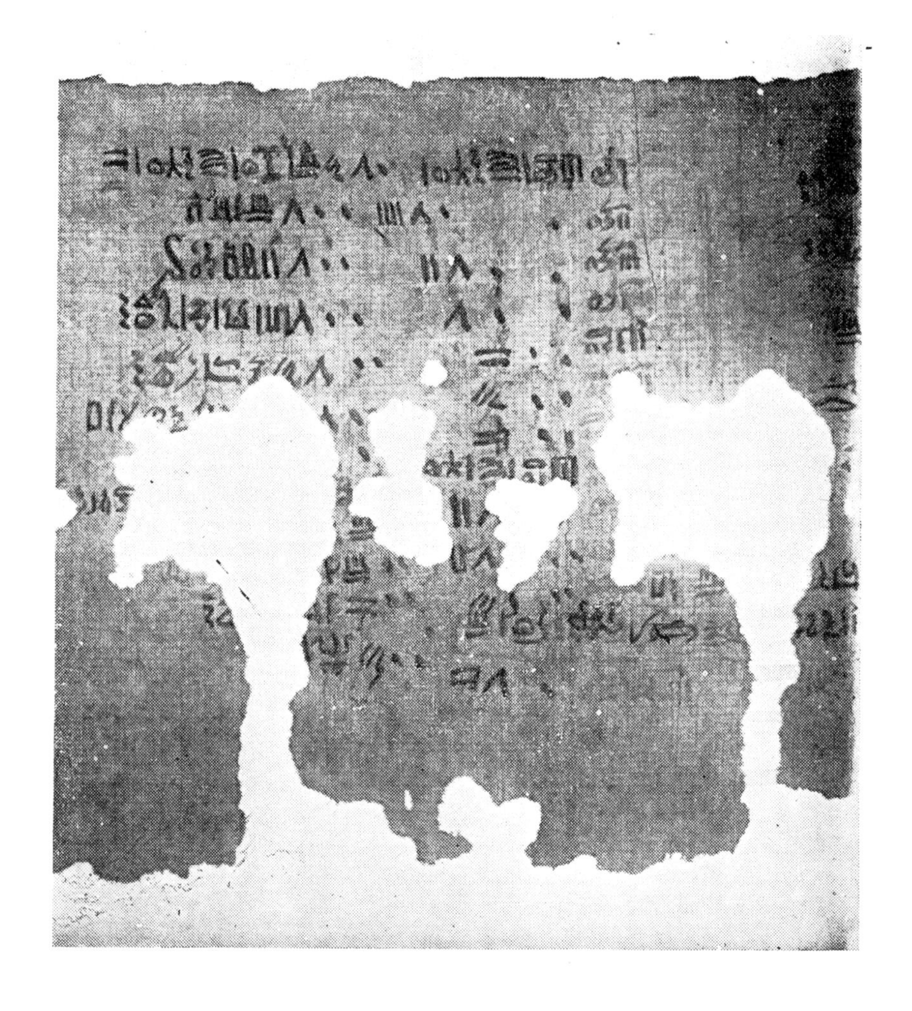

Fig. III.58a Hieratic table of the monthly 24-hour daylight and nighttime lengths from Cairo Papyrus No. 86637, folio XIV, verso. Taken from Bakir, *The Cairo Calendar No. 86637*, Plate XLV.

1.

2.

3.

sic

4.

5.

6.

7.

8.

9.

sic

10.

11.

12.

Verso XIV

Fig. III.58b Hieroglyphic transcription of the table given in Fig. III.58a. Taken from Bakir, *The Cairo Calendar No. 86637*, Plate XLIVA.

Fig. III.59 A table of semimonthly 24-hour daylight and nighttime lengths from Tanis. Dynasty 26 (?). Taken from Clère, "Un texte astronomique de Tanis," *Kêmi*, Vol. 10 (1949), p. 8, Fig. 2.

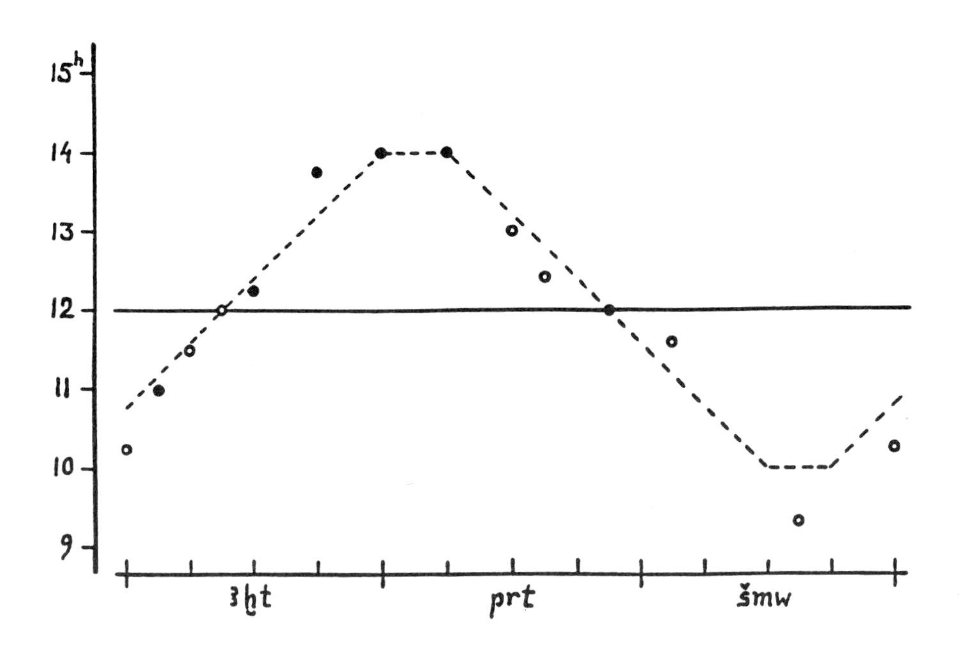

Fig. III.60 A graphing of the lengths of daylight and nighttime hours given in the table of Fig. III.59. Taken from Neugebauer and Parker, *Egyptian Astronomical Texts*, Vol. 3 (Text), p. 46.

TEXTE DE TANIS				DURÉE RÉELLE DU JOUR À			
DATE			DURÉE DU JOUR	TANIS	MEMPHIS	THÈBES	ASSOUAN
ꜣḫt	I	1	10h 15m	10h 40m	10h 45m	11h 0m	11h 5m
		15	11 0	11 5	11 10	11 20	11 20
	II	1	11 30	11 35	11 35	11 40	11 45
		15	12 0	12 0	12 0	12 0	12 0
	III	1	12 15	12 30	12 30	12 25	12 25
	IV	1	13 45	13 25	13 20	13 10	13 5
prt	I	1	14 0	14 0	13 55	13 35	13 30
	II	1	14 0	14 5	14 0	13 40	13 40
	III	1	13 0	13 35	13 30	13 20	13 10
		15	12 25	13 15	13 10	13 0	12 55
	IV	15	12 0	12 20	12 20	12 20	12 15
šmw	I	15	11 35	11 25	11 30	11 35	11 35
	III	15	9 20	10 5	10 10	10 30	10 40

Fig. III.61 A comparison of the values given in the table of Fig. III.59 with the actual values of the duration of daylight and nighttime at Tanis, Memphis, Thebes, and Aswan. The data lost from the Tanis table have been omitted. Taken from Clère, "Un texte astronomique de Tanis," p. 18, Fig. 4.

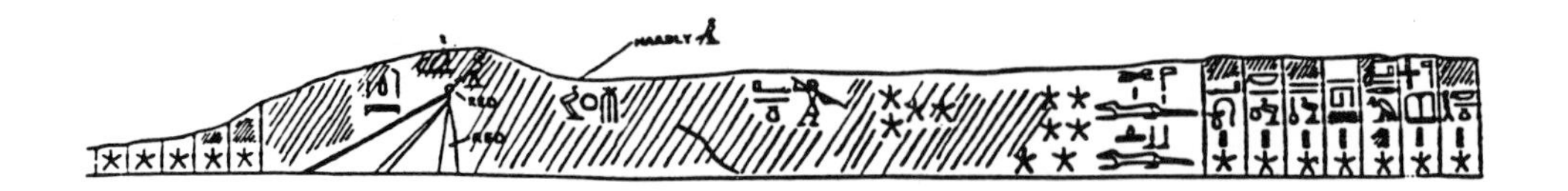

Fig. III.62a Board no. 1 from the coffin of Ḥeny (ca. Dynasty XI). Taken from Gunn, "The Coffins of Ḥeny," *ASAE*, p. 171.

4. Part of the left side of a coffin; outside, incised and painted blue (←) : [hieroglyphs] . . . . Inside, traces of three registers divided up by vertical lines : in the uppermost, names of stars or constellations, of which (←) [hieroglyphs]*, [hieroglyphs]*, [hieroglyphs]*, [hieroglyphs], [hieroglyphs]*, [hieroglyphs]*, [hieroglyphs]* were visible; in the second register, illegible signs in red; in the lowest, varying numbers of stars in blue.

Fig. III.62b Board no. 4 with its reference to Sothis ([hieroglyphs]). Taken from Gunn, "The Coffins of Ḥeny," p. 169, paragraph 4.

Fig. III.63 Sketch of Senmut, with his title "Overseer of the Estate of Amun" and his name, on the wall of the stairway leading to the first chamber of his secret tomb near Hatshepsut's temple at Dar el-Bahri. Taken from Winlock, "The Egyptian Expedition 1925-1927," p. 36.

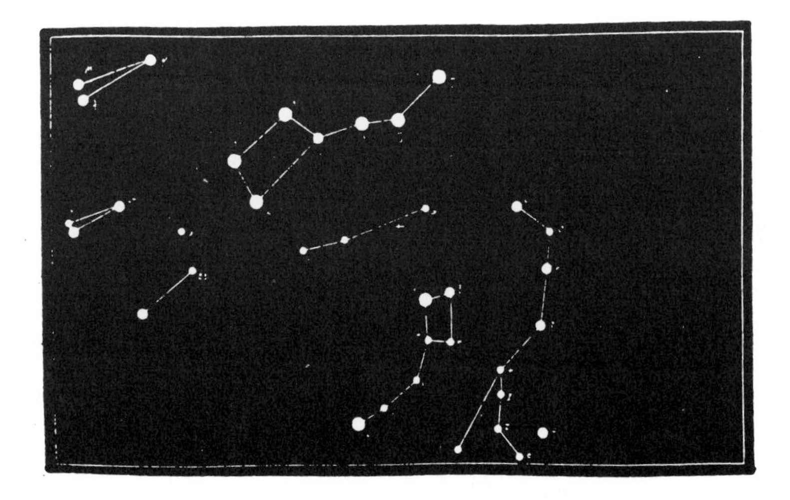

Fig. III.64a Sketch of the northern circumpolar constellations. Horizon of Thebes. About 2000 B.C. The approximate position of the pole is indicated by a cross. Taken from Pogo, "The Astronomical Ceiling-decoration in the Tomb of Senmut (XVIIIth Dynasty)," *Isis*, Vol. 14 (1930), p. 309, Fig. 4.

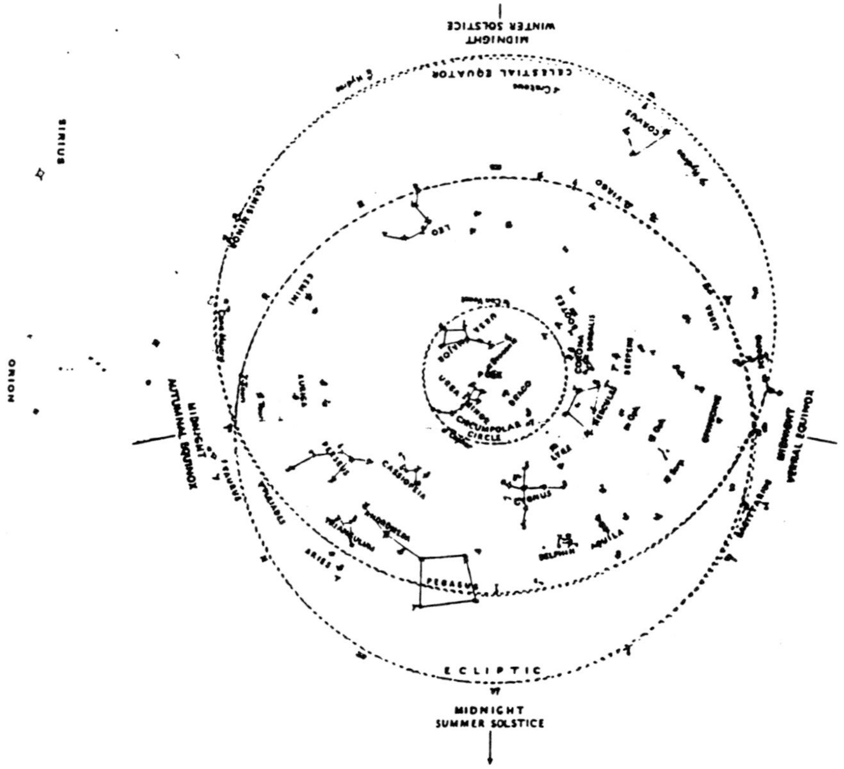

The Northern Stars at Heliopolis for 3500 B.C.

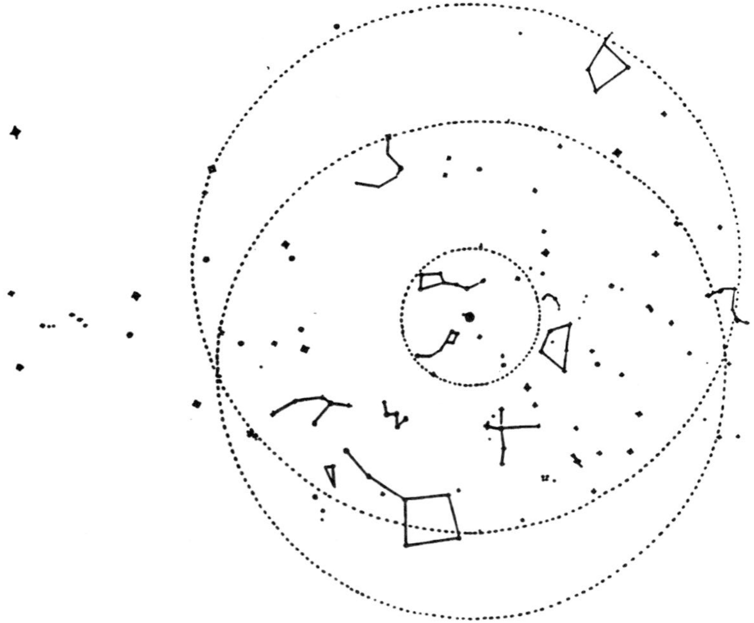

The Northern Stars at Heliopolis for 3500 B.C.

Fig. III.64b (continued on next page)

Fig. III.64b The northern stars at Heliopolis for 3500 B.C. Drawn by E. Williamson and included in G. A. Wainwright, "A Pair of Constellations," *Studies Presented to F. LL. Griffith* (London/Oxford, 1932), Plate 58. Williamson notes on p. 383: "The map is drawn on the stereographic projection, for North Latitude 30° the point of projection being the north pole of the ecliptic. It shows the brighter stars of the northern hemisphere. It is corrected for Precession and shows the North Celestial Pole and the equator as they were in 3500 B.C. The changes of position of the stars relatively to each other caused by their Proper Motions were investigated. If the motion in 5,400 years (computed from Boss's catalogue) did not excceed 20^m of Right Ascension and 20' of declination they were neglected. The most noticeable changes of positions are those of β Cassiopeiae (which alters the shape of the constellation considerably), α Boötis (Arcturus), and α Canis Minoris (Procyon). The observer's horizon is not shown on the map, but the circumpolar circle touches it in the North point. Thus the appearance of the northern sky at any season can be found with the aid of the meridians drawn on the transparent sheet [not included here but in the original article]. These meridians give the direction of North and South for the observer at midnight at the equinoxes and solstices, and intermediate times can be estimated. The map only goes down to the equator, so that the observer's southern horizon cannot be shown."

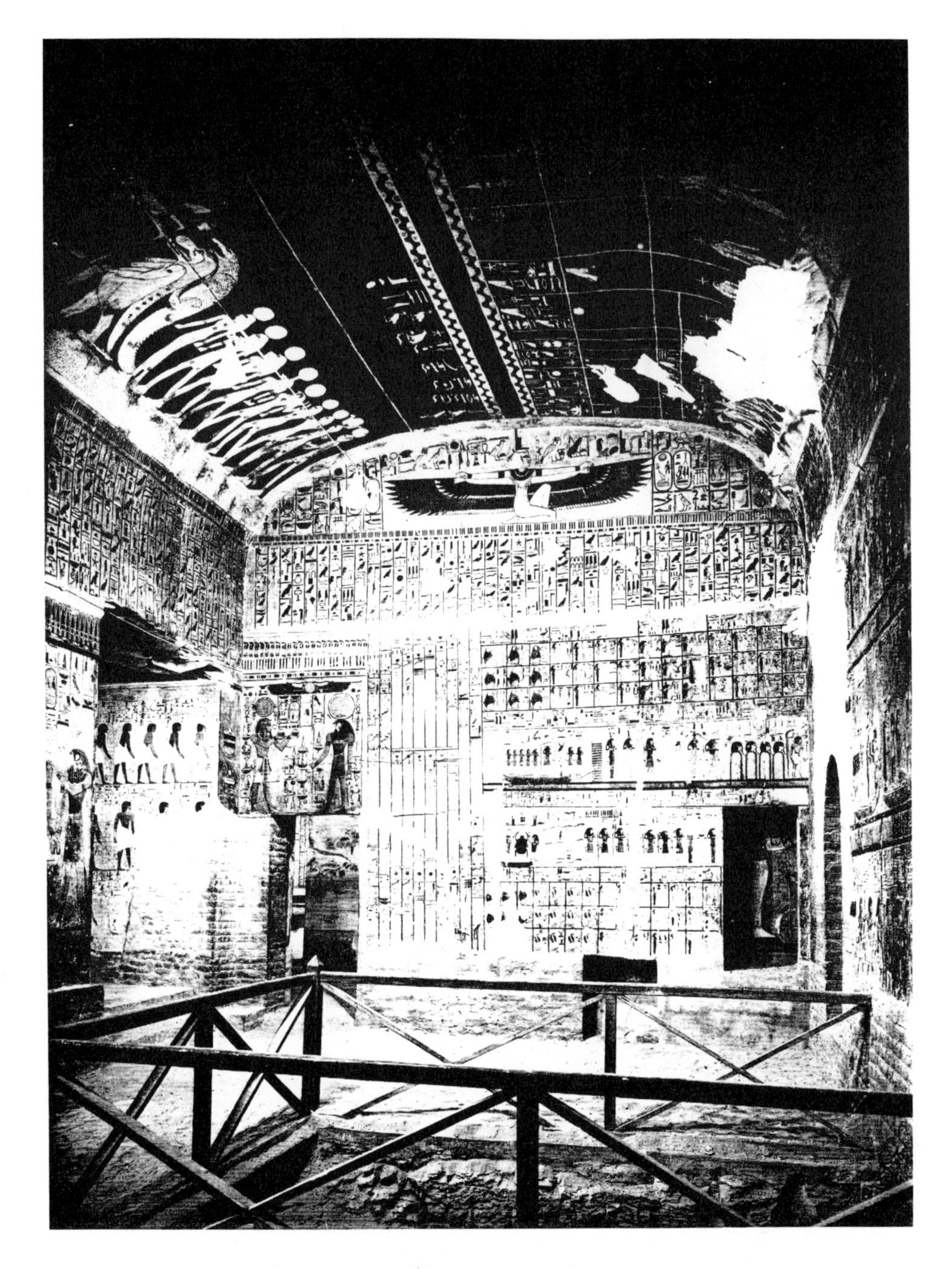

Fig. III.65a The sepulchral chamber of the tomb of Seti I displaying its astronomical ceiling. Taken from a photograph by Harry Burton in Hornung, *The Tomb of Pharaoh Seti I...photographed by Harry Burton*, p. 224.

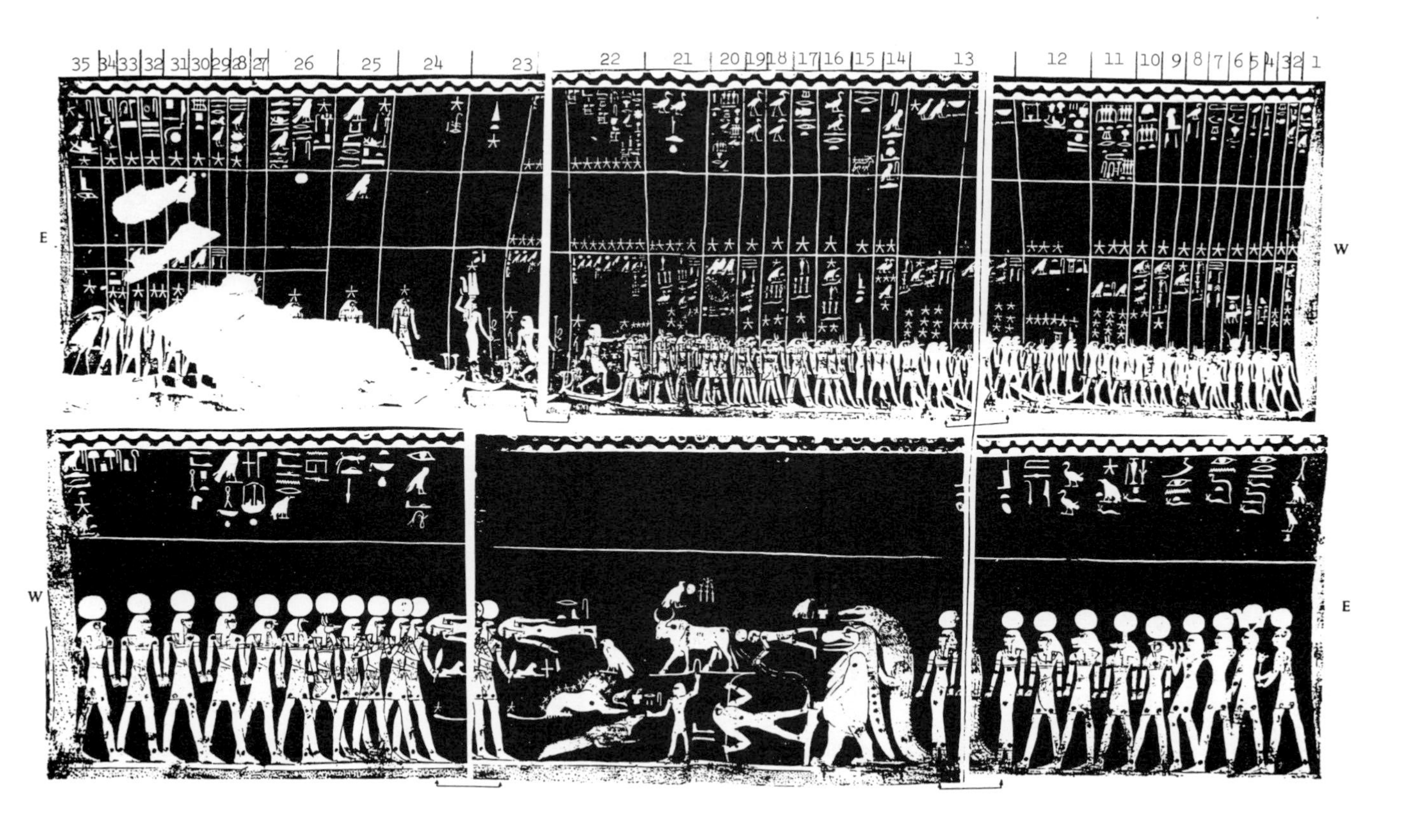

Fig. III.65b Photocopy of a photograph of the astronomical ceiling of Seti I noted in Fig. III.65a. Reproduced from Neugebauer and Parker, *Egyptian Astronomical Texts*, Vol. 3 (Plates), Plate 3.

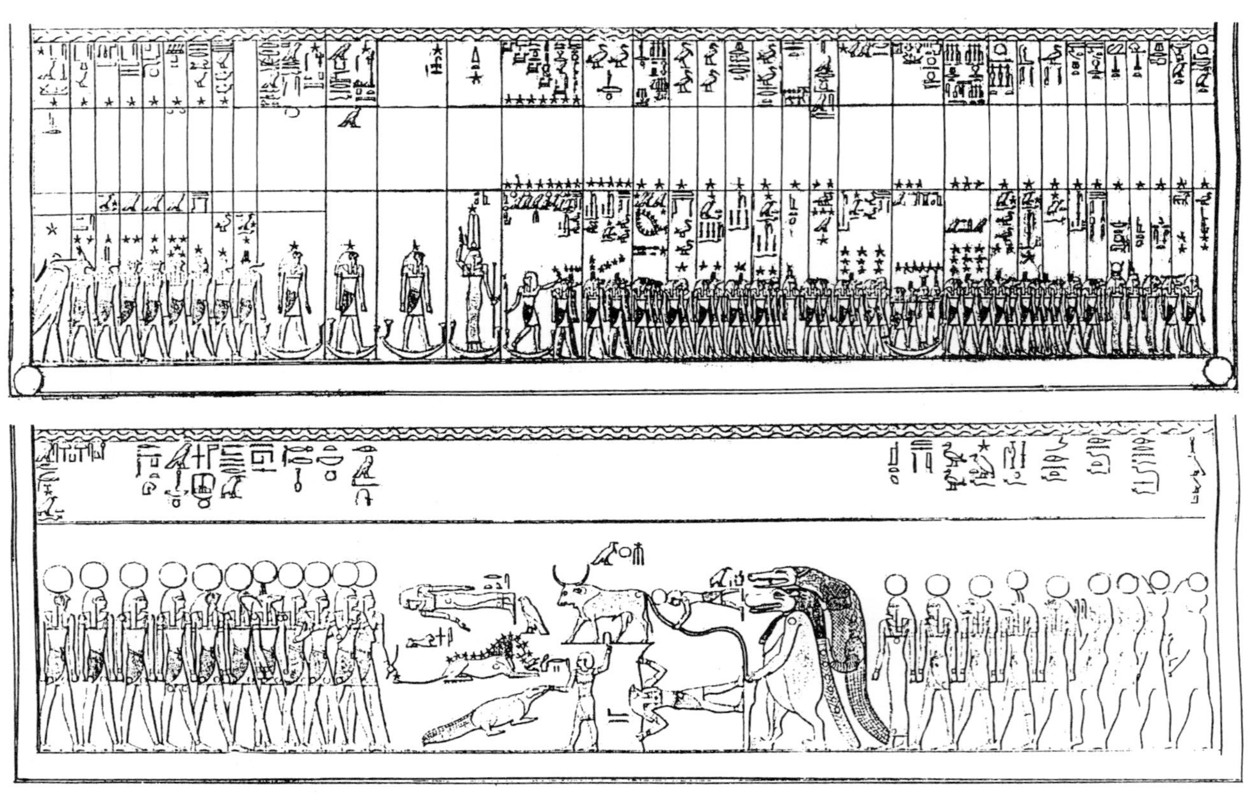

Fig. III.65c A copy of the previously noted astronomical ceiling of Seti I before parts of the ceiling had fallen. Given by Lepsius, *Denkmäler aus Ägypten und Äthiopen*, Abt. III, Blatt 137.

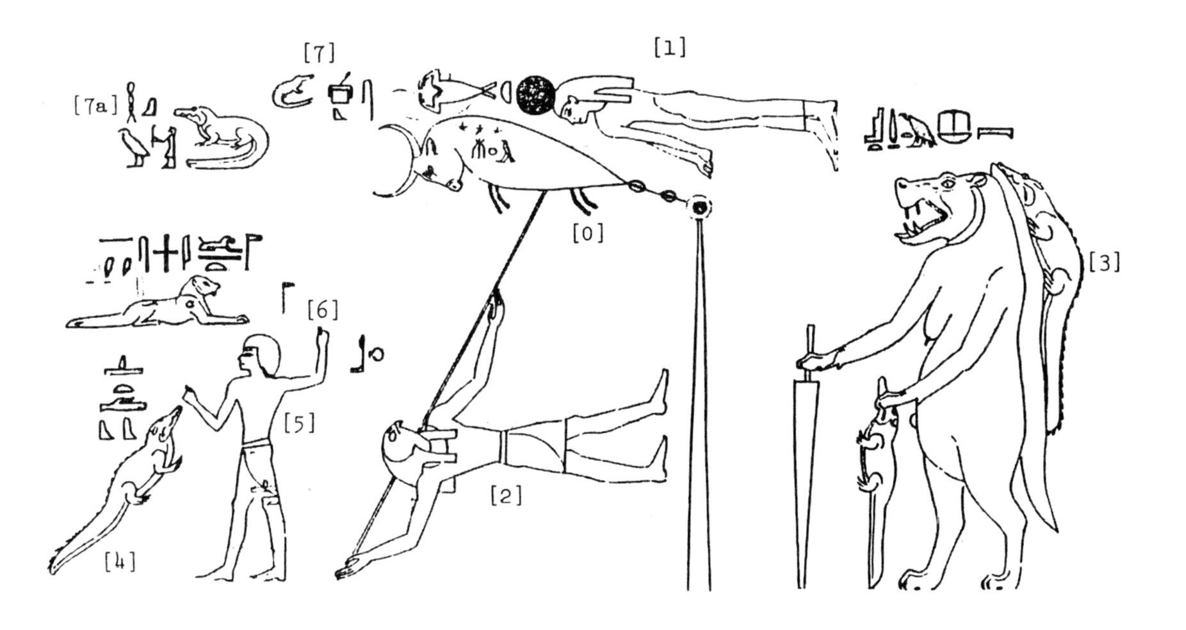

Fig. III.66 A reconstruction of the arrangement of the northern constellations found in the Senmut ceiling and its families. The reconstruction is from Neugebauer and Parker, *Egyptian Astronomical Texts*, Vol. 3 (Texts), p. 184, Fig. 27, with bracketed numbers added.

Fig. III.67 Arrangement of the northern constellations on the astronomical ceiling of the Ramesseum (cf. Fig. III.2). Taken from Neugebauer and Parker, *Egyptian Astronomical Texts*, Vol. 3 (Texts), p. 185, Fig. 28.

Fig. III.68 Arrangement of the northern constellations on the astronomical ceiling of corridor B of the tomb of Ramesses VI. Taken from Neugebauer and Parker, *Egyptian Astronomical Texts*, Vol. 3 (Texts), p. 187, Fig. 30.

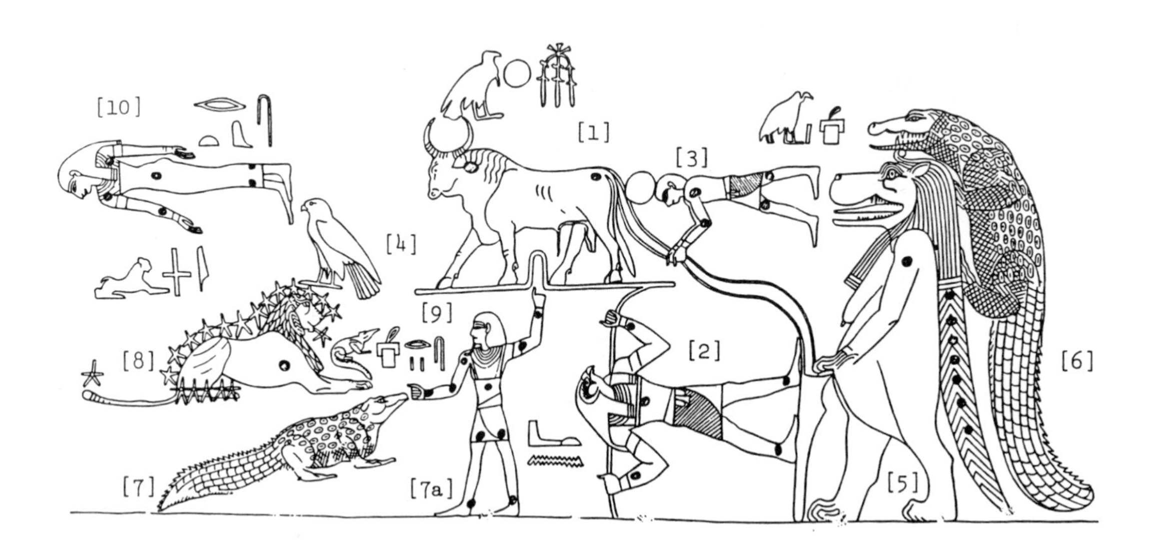

Fig. III.69 Arrangement of the northern constellations on the astronomical ceiling of Hall K of the tomb of Seti I (cf. Figs. III.65b and III.65c). Taken from Neugebauer and Parker, *Egyptian Astronomical Texts*, Vol. 3 (Texts), p. 188, Fig. 31, but with bracketed numbers added.

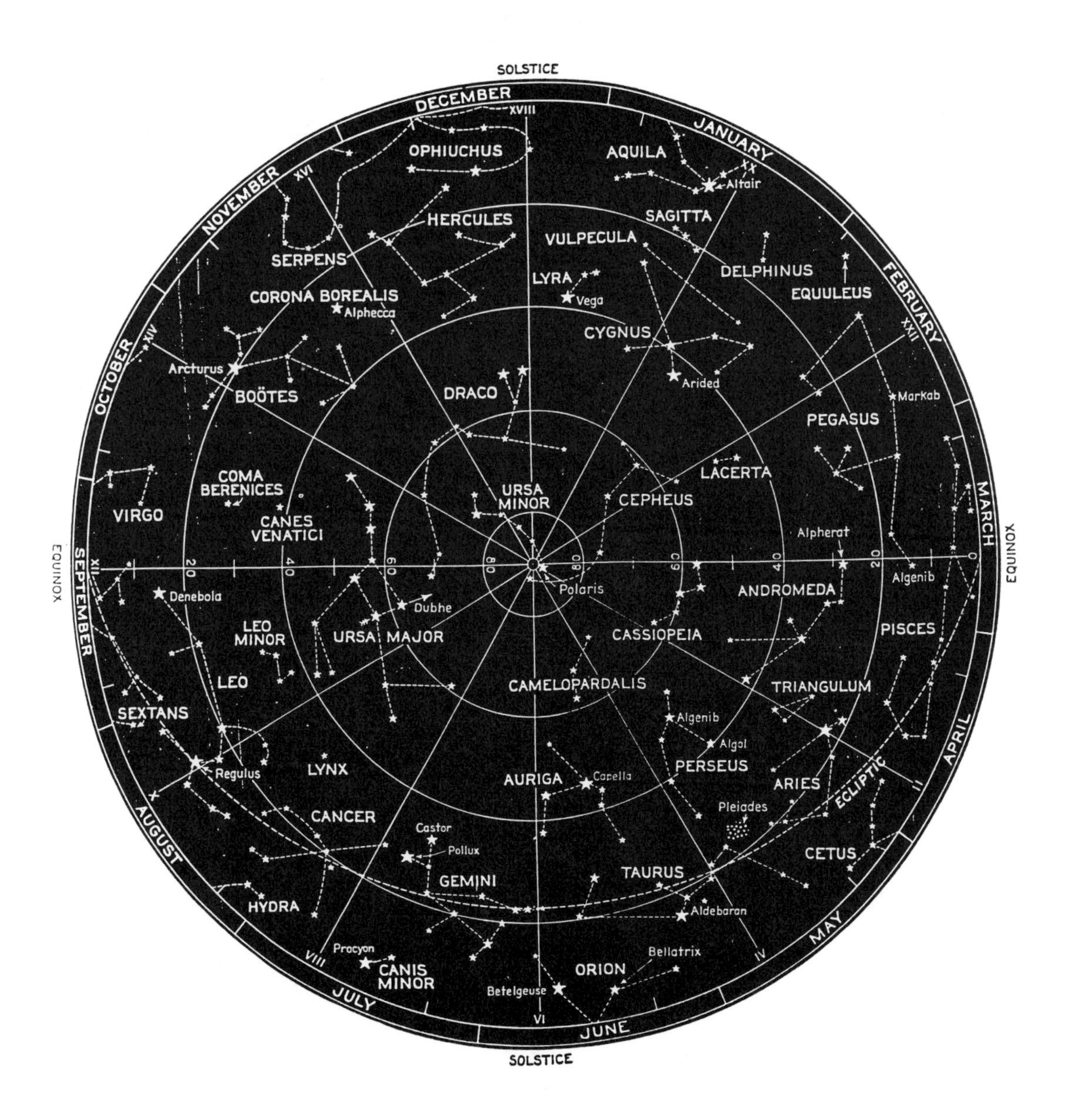

Fig. III.70a A contemporary view of the constellations and stars of the Northern Hemisphere.

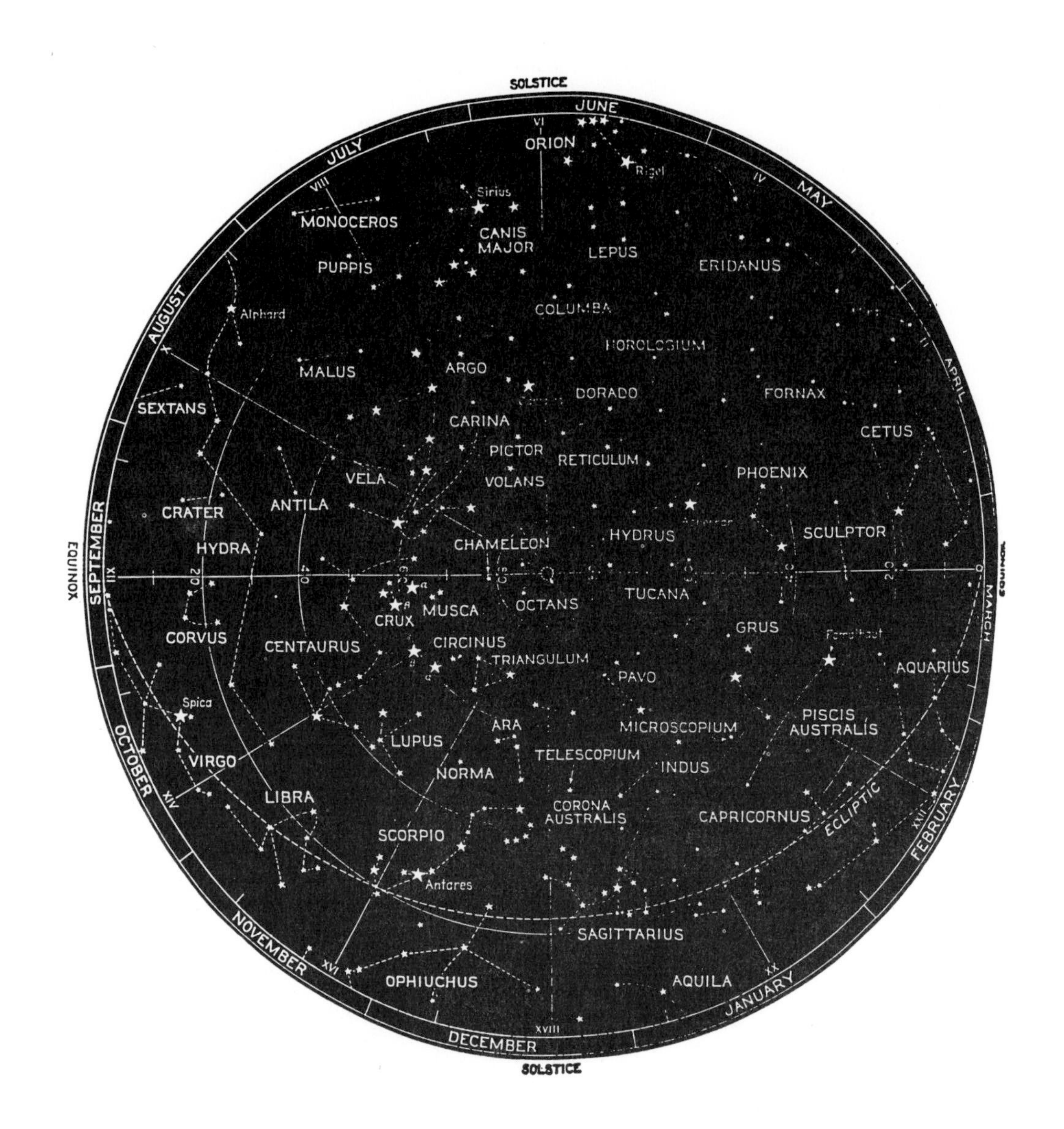

Fig. III.70b A contemporary view of the constellations and stars of the Southern Hemisphere.

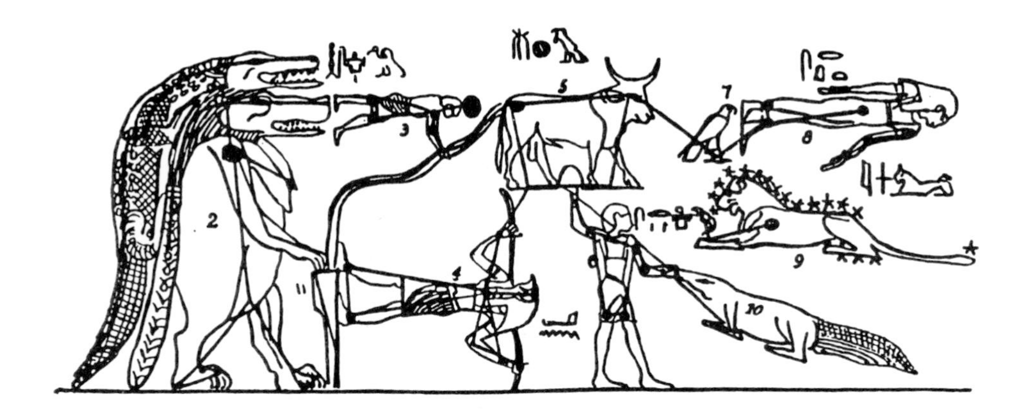

Fig. III.71a Mirror image of the northern constellations as given in the tomb of Seti I (cf. Fig. III.69) with the circles (representing stars) connected by heavy lines. Prepared by Biegel, *Zur Astrognosie der alten Ägypter*, Fig. 6a, after p. 36.

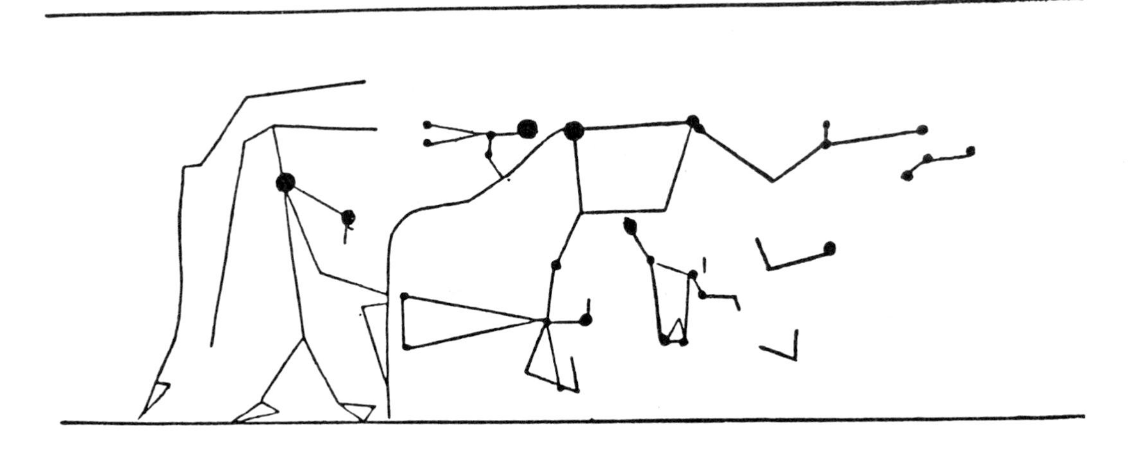

Fig. III.71b Biegel's diagram depicting her view of the origins of the ancient Egyptian northern constellations of Fig. III.71a by means of connecting lines between the circles representing the stars. Biegel, *ibid.*, Fig. 6b, after p. 36.

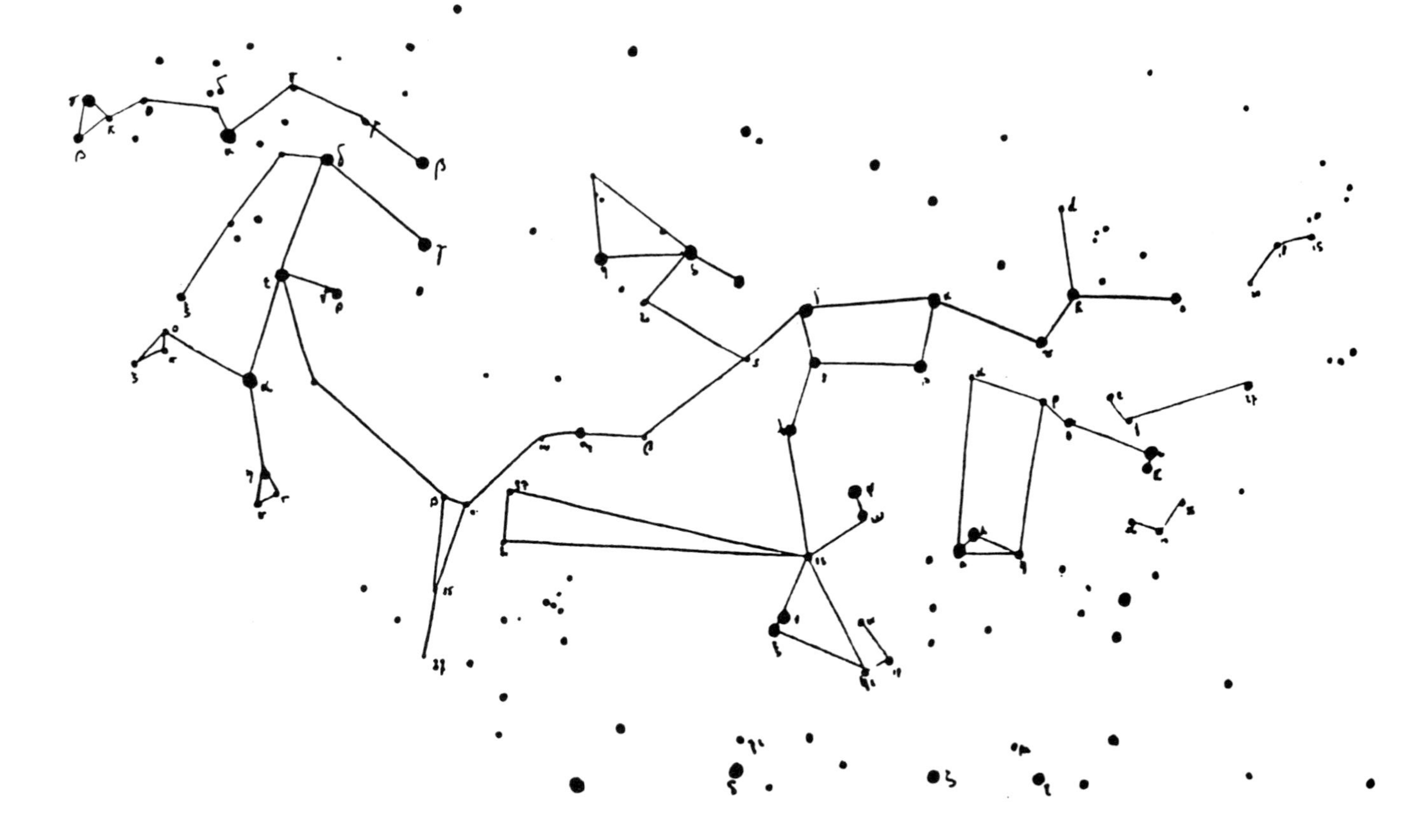

Fig. III.72 Details of the forming of the northern constellations according to Biegel, *ibid.*, Fig. 7, after p. 36.

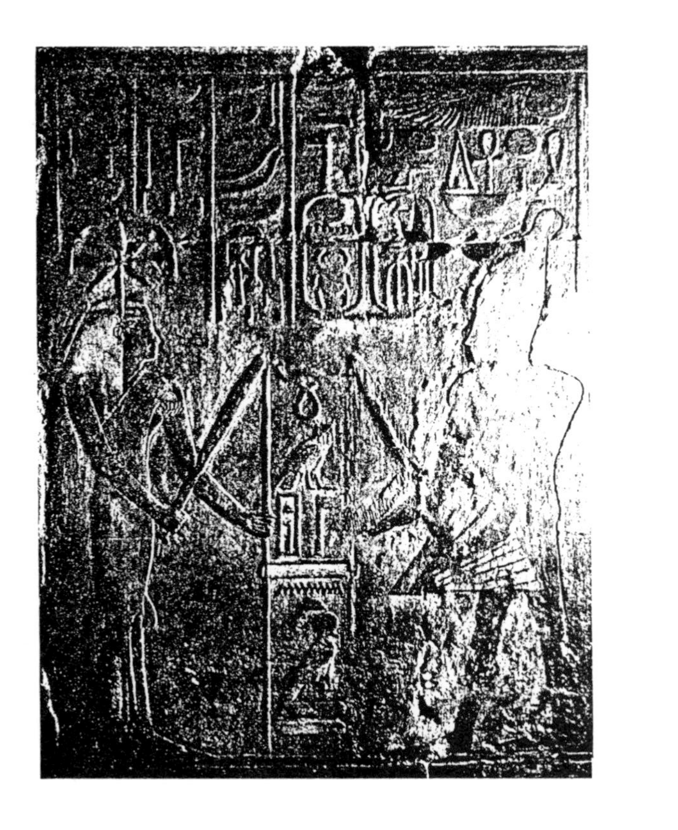

Fig. III.73a Depiction of the ceremony of "stretching the cord" by Seshat and Tuthmosis III, in the temple of Amada (ca. 125 miles south of Aswan). Taken from Borchardt, *Die altägyptische Zeitmessung*, Tafel 17 (2), opposite p. 54.

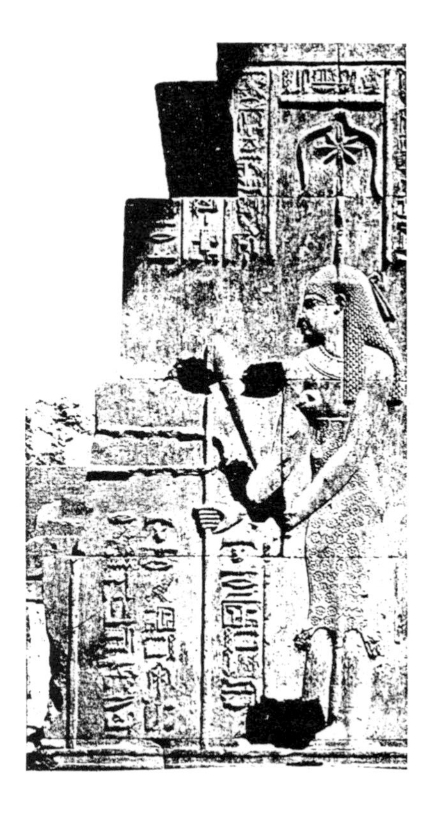

Fig. III.73b Another depiction of Seshat taking part in the ceremony of "stretching the cord," in the temple of Kom Ombo (ca. 150 B.C.). *Ibid.*, Tafel 17 (3).

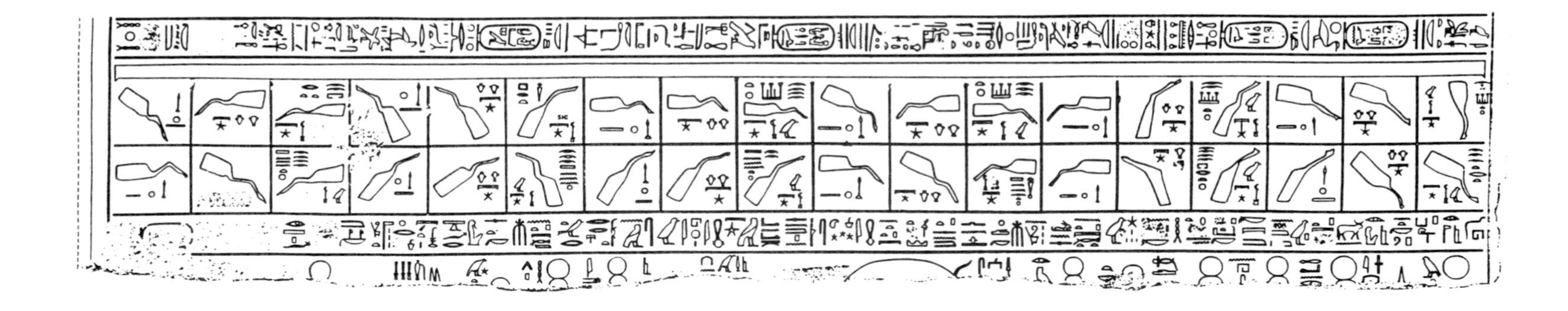

Fig. III.74 The successive positions, every ten days, of the Big Dipper. Depicted in a bull sarcophagus from Abu Yasin. Time of Nectanebo (360-43 B.C.). Astronomically useless. Taken from Neugebauer and Parker, *Egyptian Astronomical Texts*, Vol. 3 (Plates), the top part of Plate 24.

ENVIRONS D'ESNÉ (LATOPOLIS.)

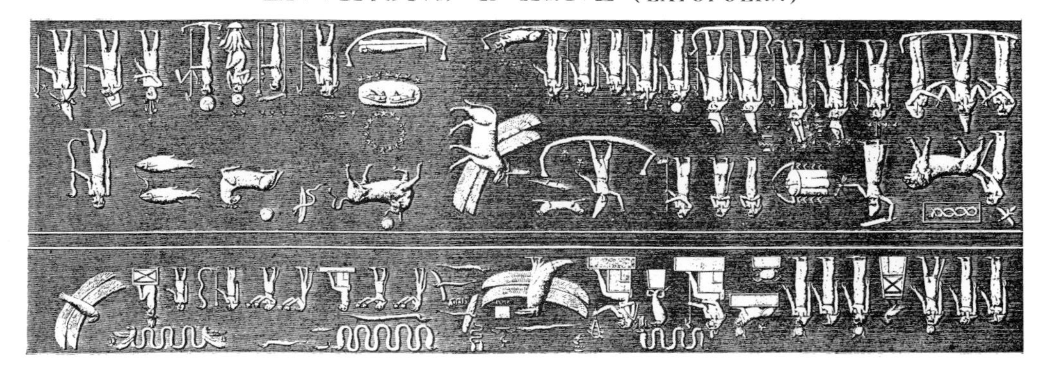

ZODIAQUE SCULPTÉ AU PLAFOND DU TEMPLE AU NORD D'ESNÉ.

Fig. III.75a The earliest Egyptian zodiac (from about 200 B.C.). It came from the Temple of Khnum near Esna that was dismantled in 1843 and is called "Esna A." Its blocks were used to build a canal. Though now lost, the zodiac is reproduced from *Description de l'Égypte, ou, Recueil des observations et des recherches qui ont été faites en Egypte pendant l'expédition de l'armée française*, Vol. I (Paris, 1809), Plate 87.

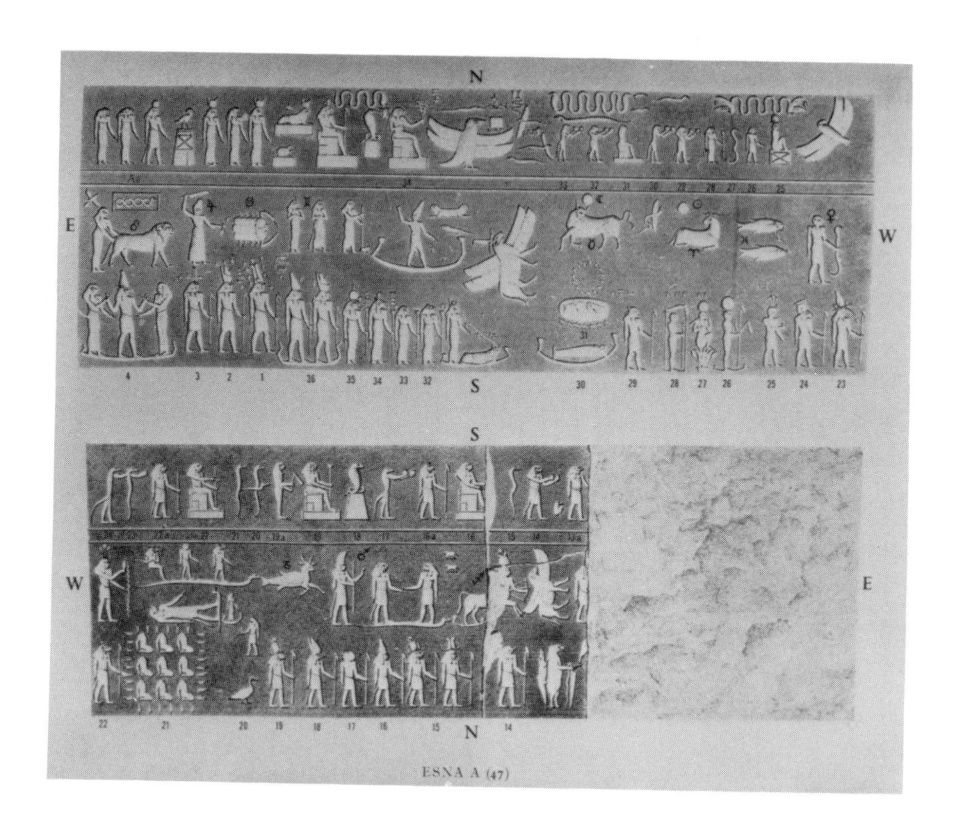

Fig. III.75b The same zodiac as III.75a, but with numbers and symbols added by Neugebauer and Parker, *Egyptian Astronomical Texts*, Vol. 3 (Plates), Plate 29, to identify the two sets of decans (represented here only by the figures of their associated deities), the zodiacal signs, and the deities representing the planets. Notice that about one-third of the lower figure (on the right side) has been almost completely destroyed. The upper figure has been inverted for easier comparison.

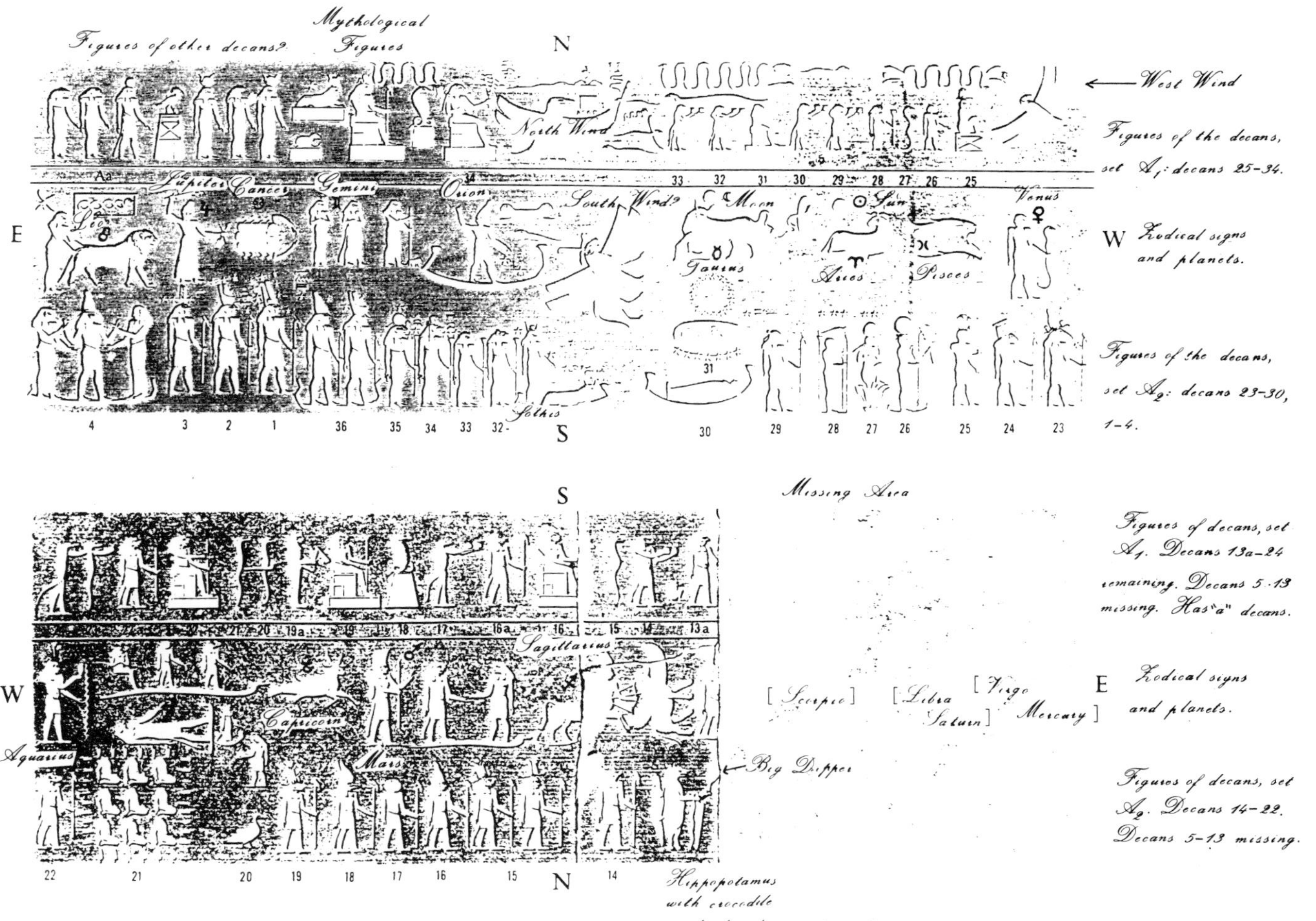

Fig. III.75c The same illustration as III.75b, with labels added by the author.

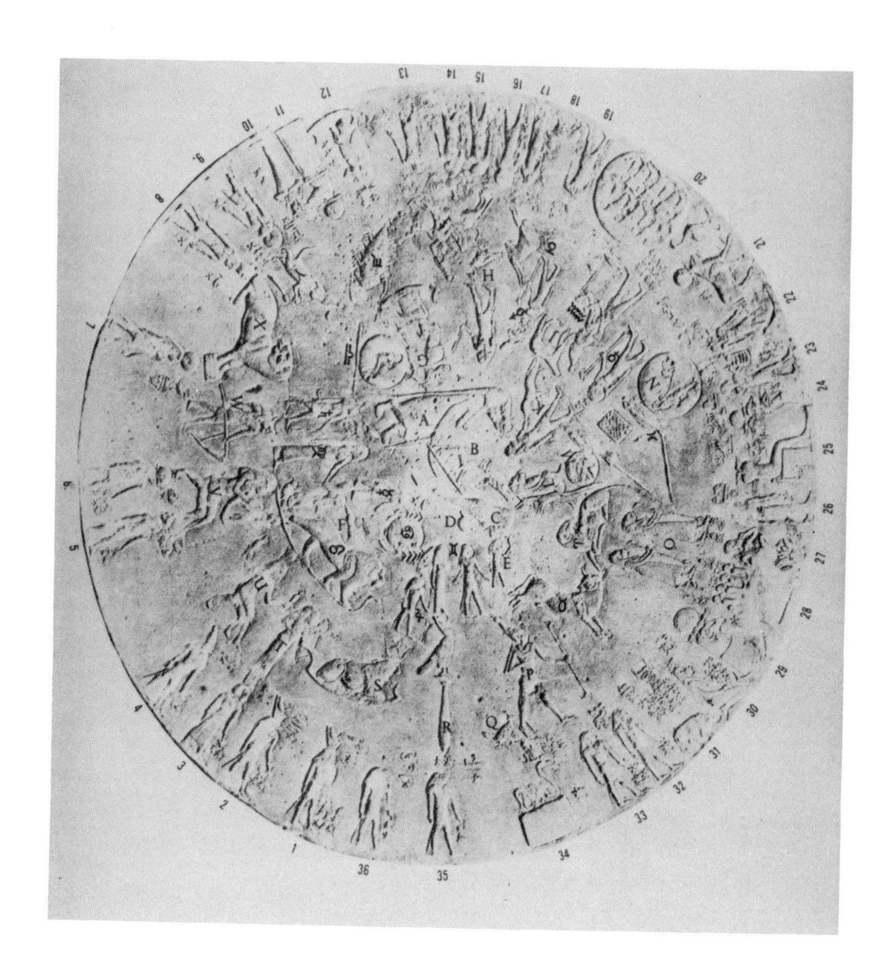

Fig. III.76a The round zodiac of Dendera ("Dendera B"), dated to late Ptolemaic times. The plate given here is from Neugebauer and Parker, *Egyptian Astronomical Texts*, Vol. 3 (Plates), Plate 35, and includes their additions of numbers, letters and symbols to identify decans, constellations, and zodiacal signs.

Fig. III.76b The zodiac of Fig. III.76a, as given in the *Description de l'Egypte*, Vol. 4, Plate 21. And see the drawing on the back of that plate.

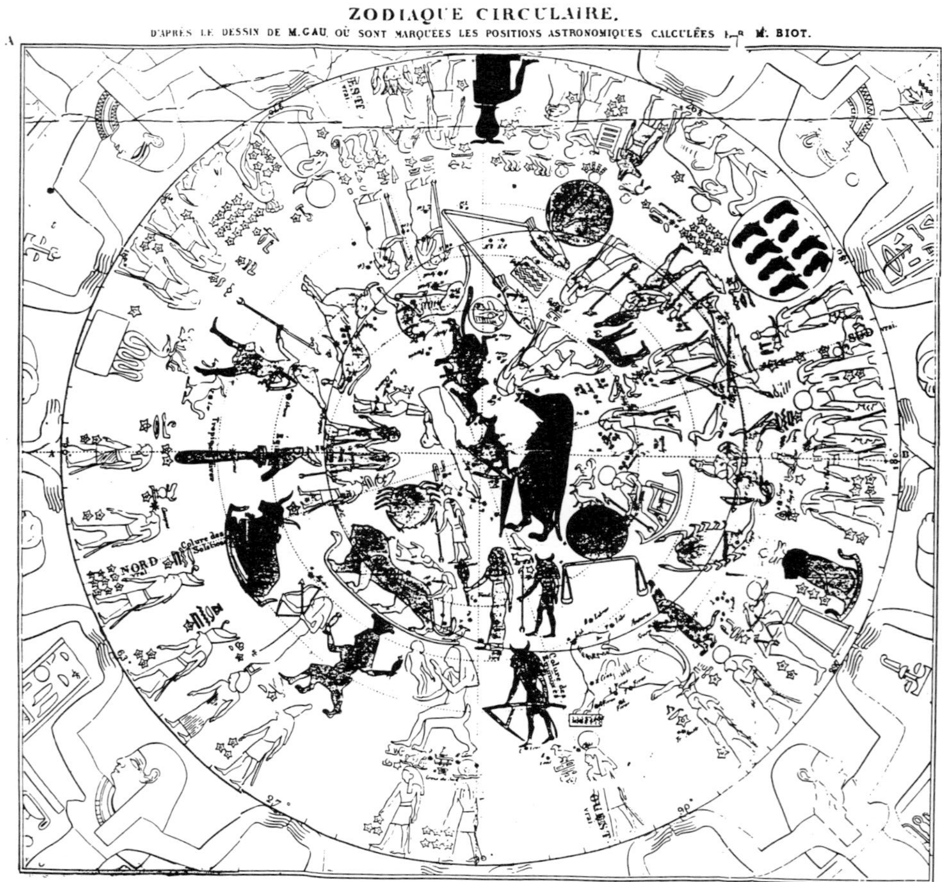

Fig. III.76c The Zodiac of Dendera B as redrawn by M. Gau where the astronomical positions are marked according to the calculations of M. Biot. Its orientation is erroneous by 90° and the ceiling image is reversed. Taken from the *Mémoires de l'Institut Royal de France, Académie des Inscriptions et Belles-Lettres*, Tome 16,2 (1846), Plate I.

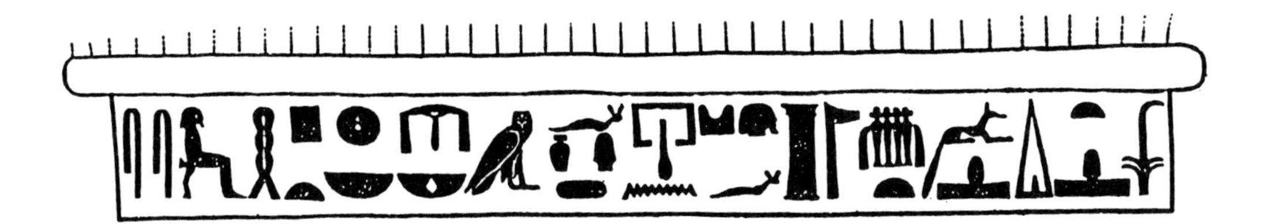

Fig. III.77 A list of festivals established for Ptahshepses (Dynasty VI) in his tomb at Saqqara. Text given by M.A. Murray, *Saqqara Mastabas*, Part 1, Plate XXVIII.

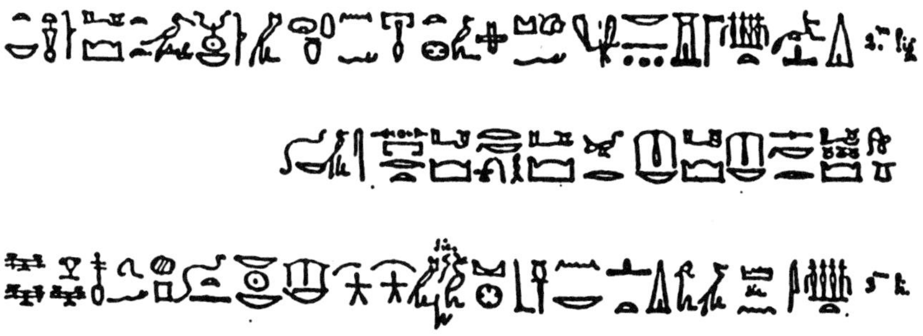

Fig. III.78 Festivals in the funerary offerings for Ptahshepses (Dynasty V) in his tomb at Saqqara. Text given by Mariette, *Les Mastabas de l'Ancien Empire*, p. 130.

Fig. III.79 A round offering table of Hetepherakhti. Dynasty 5. Text given by Murray, *Saqqara Mastabas*, Part 1, Plate 3, no. 4.

Fig. III.80 Circular offering table for Kaihap. Dynasty V (? Mariette dates to 2nd half of Dynasty IV). Text given by Mariette, *Les Mastabas de l'Ancien Empire*, p. 164.

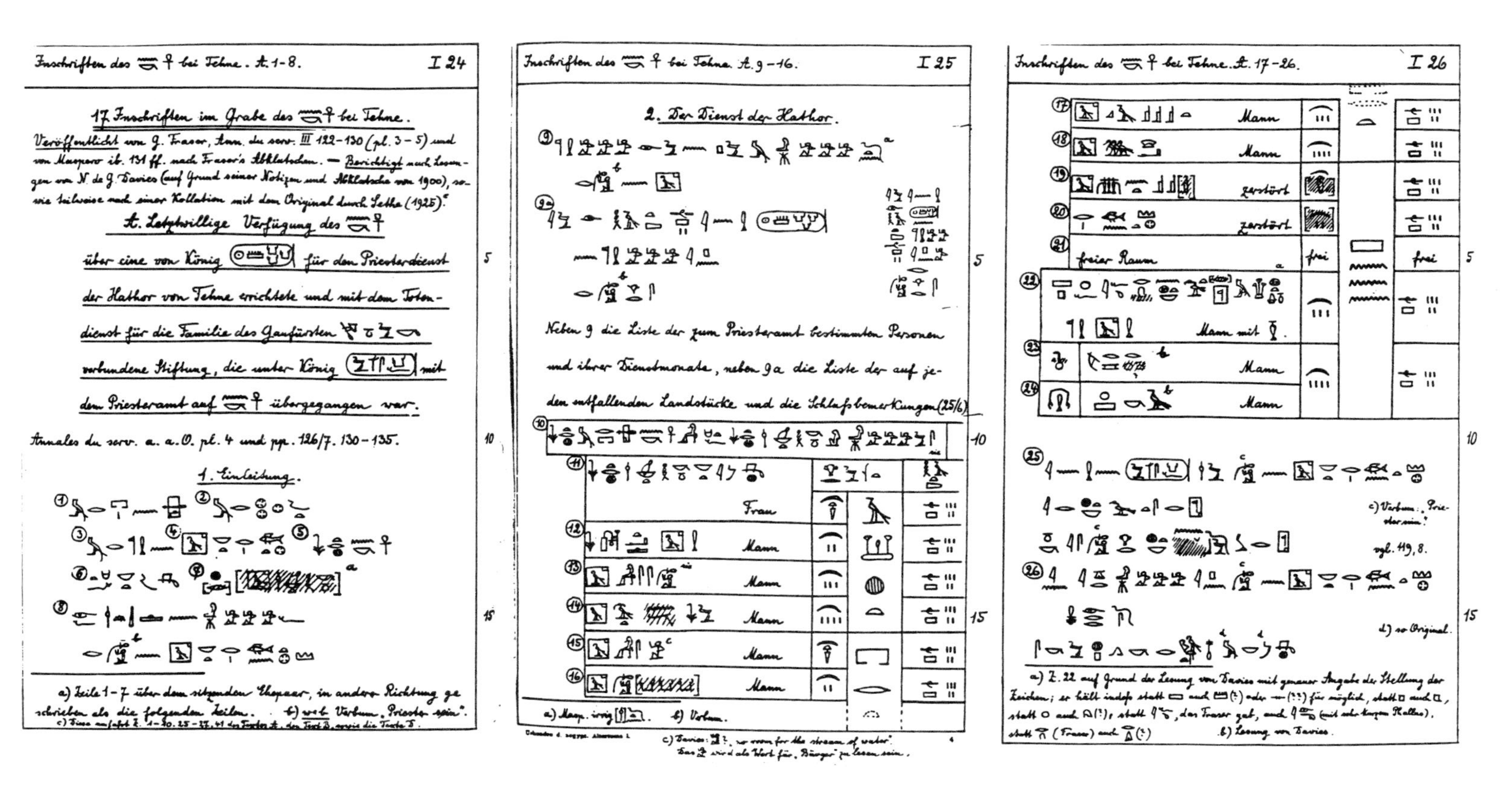
Inschriften des bei Tehne. A. 1-8. I 24

17. Inschriften im Grabe des bei Tehne.

Veröffentlicht von G. Fraser, Ann. du serv. III 122-130 (pl. 3-5) und von Maspero ib. 131 ff. nach Fraser's Abklatschen. — Berichtigt nach Lesungen von N. de G. Davies (auf Grund seiner Notizen und Abklatsche von 1900), sowie teilweise nach einer Kollation mit dem Original durch Sethe (1925).

A. Letztwillige Verfügung des

über eine von König für den Priesterdienst der Hathor von Tehne errichtete und mit dem Totendienst für die Familie des Gaufürsten verbundene Stiftung, die unter König mit dem Priesteramt auf übergegangen war.

Annales du serv. a. a. O. pl. 4 und pp. 126/7. 130-135.

1. Einleitung.

Inschriften des bei Tehne. A. 9-16. I 25

2. Der Dienst der Hathor.

Neben 9 die Liste der zum Priesteramt bestimmten Personen und ihrer Dienstmonate, neben 9a die Liste der auf jeden entfallenden Landstücke und die Schlafsbemerkungen (25/6).

Inschriften des bei Tehne. A. 17-26. I 26

Fig. III.81a Part of a long inscription from the tomb of Nekankh (Dynasty V) at Tehne. The text is taken from Sethe, *Urkunden des Alten Reichs* (=*Urkunden I*), Vol. 1, pp. 24-28.

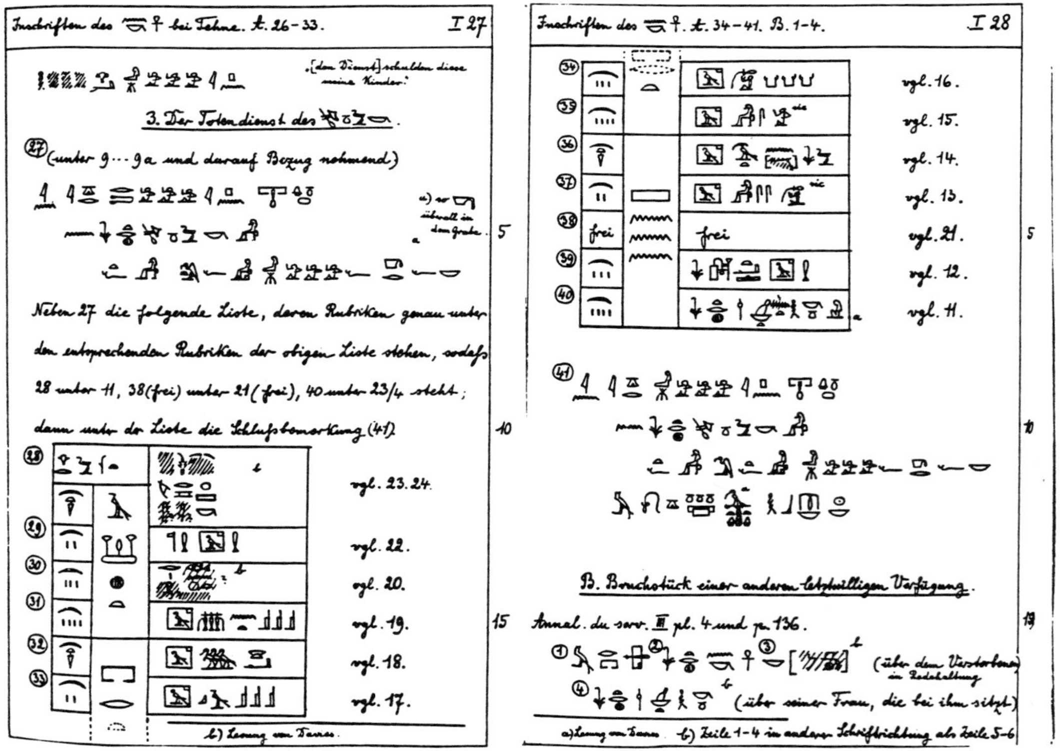

Inschriften des [hieroglyphs] bei Tehne. A. 26–33. I 27

a) [den Dienst] schulden diese meine Kinder.

3. Der Totendienst des [hieroglyphs].

(27) (unter 9 ... 9a und darauf Bezug nehmend)

a) so [hieroglyph] überall im dem Grabe.

Neben 27 die folgende Liste, deren Rubriken genau unter den entsprechenden Rubriken der obigen Liste stehen, sodaß 28 unter 11, 38 (frei) unter 21 (frei), 40 unter 23/4 steht; dann unter der Liste die Schlußbemerkung (41).

(28)	[hieroglyphs]	vgl. 23. 24.
(29)	[hieroglyphs]	vgl. 22.
(30)	[hieroglyphs]	vgl. 20.
(31)	[hieroglyphs]	vgl. 19.
(32)	[hieroglyphs]	vgl. 18.
(33)	[hieroglyphs]	vgl. 17.

b) Lesung von Tanas.

Inschriften des [hieroglyphs]. A. 34–41. B. 1–4. I 28

(34)	[hieroglyphs]	vgl. 16.
(35)	[hieroglyphs]	vgl. 15.
(36)	[hieroglyphs]	vgl. 14.
(37)	[hieroglyphs]	vgl. 13.
(38)	frei / frei	vgl. 21.
(39)	[hieroglyphs]	vgl. 12.
(40)	[hieroglyphs]	vgl. 11.

(41) [hieroglyphs]

B. Bruchstück einer anderen letztwilligen Verfügung.

Annal. du serv. III pl. 4 und p. 136.

(1) [hieroglyphs] (2) [hieroglyphs] (3) [hieroglyphs] (über dem Verstorbenen in Redehaltung)

(4) [hieroglyphs] (über seiner Frau, die bei ihm sitzt)

a) Lesung von Sanas. b) Zeile 1–4 in anderer Schriftrichtung als Zeile 5–6

Fig. III.81a (concluded).

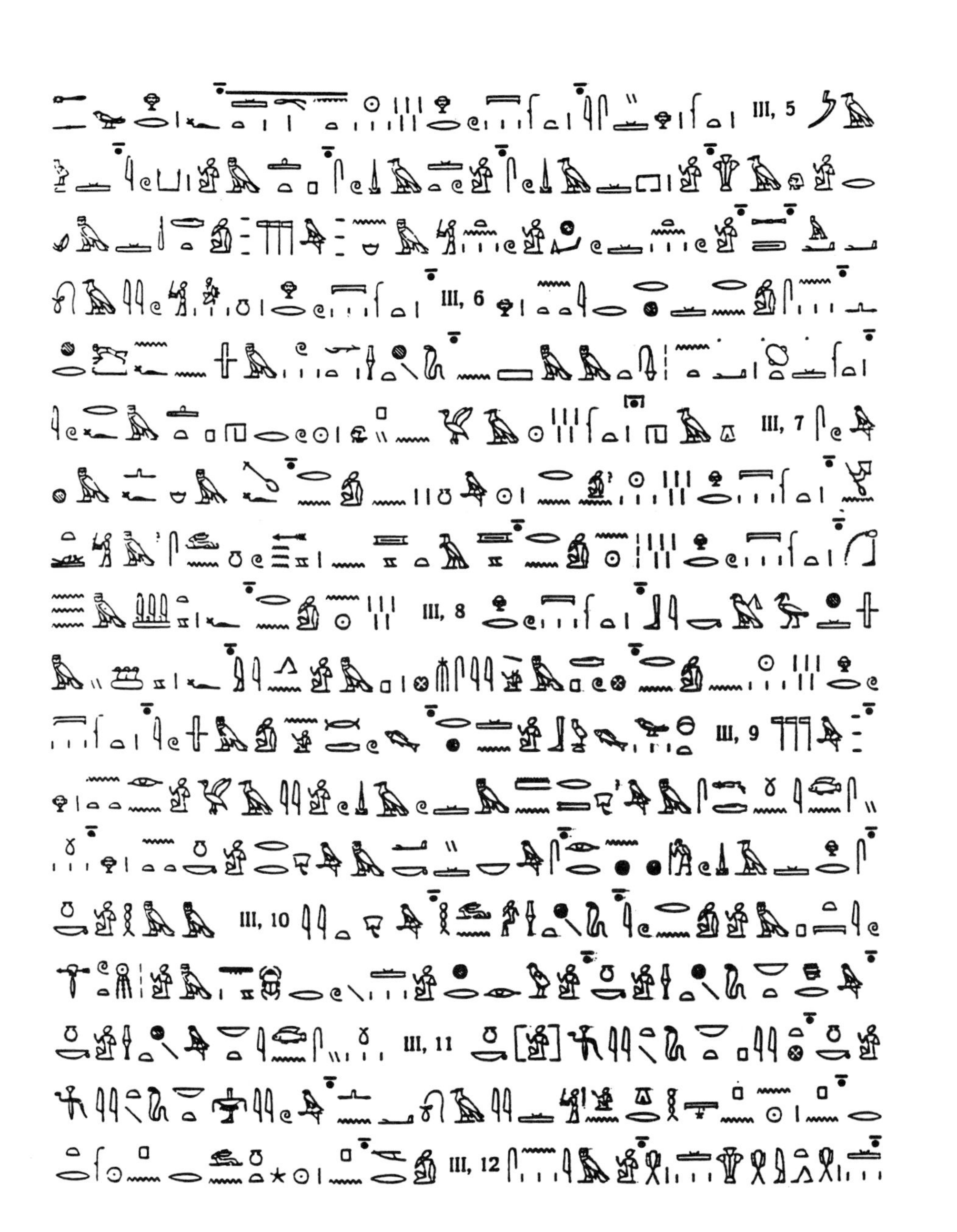

Fig. III.81b Hieroglyphic transcription of an extract on the epagomenal days from Papyrus Leiden I 346, III recto. Taken from Stricker, "Spreuken tot Beveiliging gedurende de Shrikkeldagen naar Pap. I 346," p. 65.

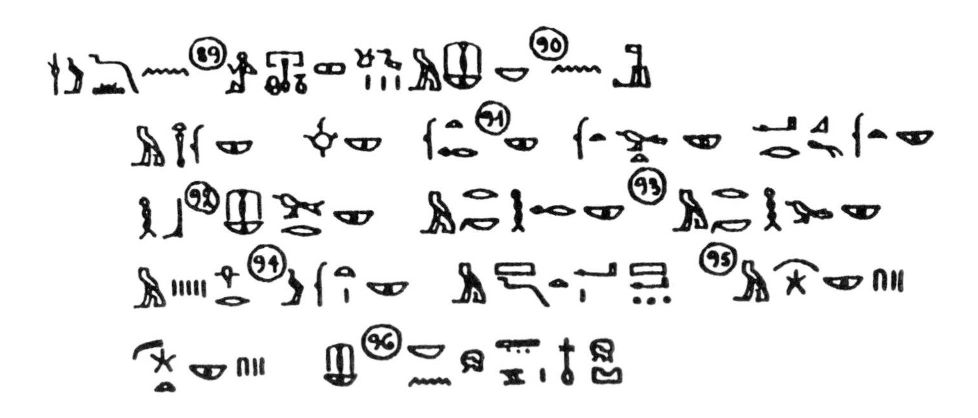

Fig. III.82 Festival list from the tomb of Khnumhotep II at Beni Hasan. Dynasty XII. Consult Sethe, *Historisch-biographische Urkunden des Mittleren Reiches (=Urkunden VII)*, Vol. 1, pp. 29-30.

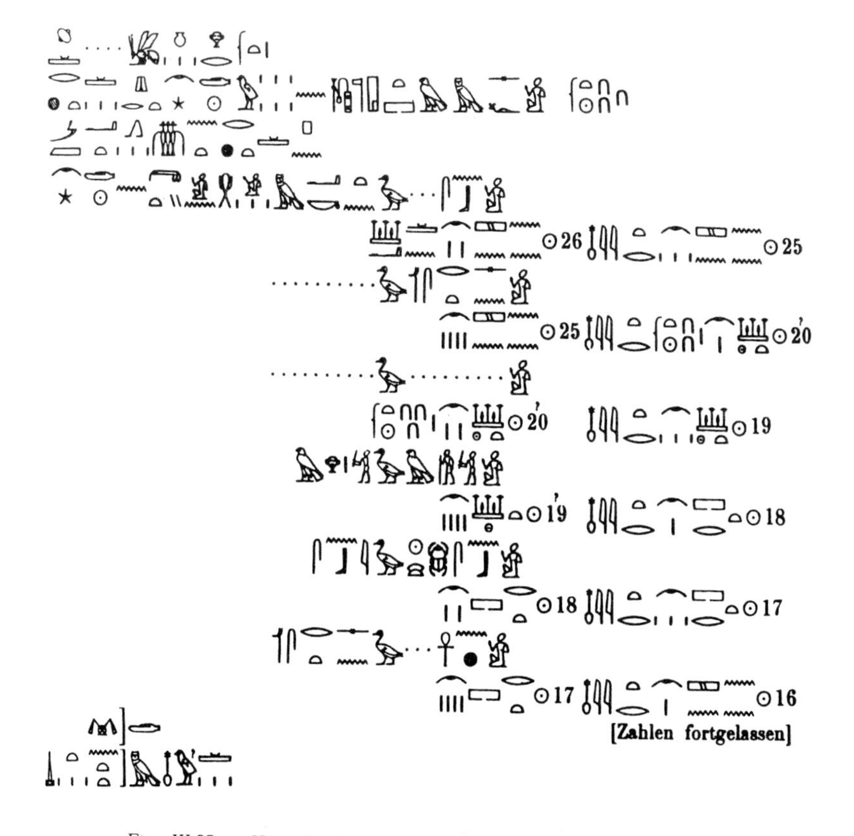

Fig. III.83a Hieroglyphic transcription of the hieratic text (P. Berlin 10056, Year 30/31) of the income-periods of Heremsaf, a scribe of the temple of the Pyramid of Illahun. Middle Kingdom. Transcription given by Borchardt, "Der zweite Papyrusfund von Kahun und die zeitliche Festlegung des mittleren Reiches der ägyptischen Geschichte," *ZÄS*, Vol. 37 (1899), pp. 92-93.

MONTH BEGINS ON			MONTH ENDS ON DAY BEFORE			DAYS
*II	Shemu	26	III	Shemu	25	30
#III	Shemu	26	IIII	Shemu	24	29
*IIII	Shemu	25	I	Akhet	19	30@
#I	Akhet	20	II	Akhet	19	30
*II	Akhet	20	III	Akhet	19	30
#III	Akhet	20	IIII	Akhet	18	29
*IIII	Akhet	19	I	Peret	18	30
#I	Peret	19	II	Peret	17	29
*II	Peret	18	III	Peret	17	30
#III	Peret	18	IIII	Peret	16	29
*IIII	Peret	17	I	Shemu	16	30
#[I	Shemu	17]	[II	Shemu	15]	[29]

*Pairs of recorded entries.

#Deduced, unrecorded entries based on the assumption that day 1 of each lunar month falls on the day after the second entry for the preceding month.

@Includes 5 epagomenal days between Year 30, IIII Shemu 30, and Year 31, I Akhet 1.

Fig. III.83b First possible reconstruction of the lunar year extending over civil years 30-31 of Amenemhet III (according to Parker, but Sesostris III according Krauss), with the assumption noted under the chart.

MONTH BEGINS ON			MONTH ENDS ON DAY BEFORE			DAYS
*II	Shemu	26	III	Shemu	25	29
#III	Shemu	25	IIII	Shemu	25	30
*IIII	Shemu	25	I	Akhet	19	29@
#I	Akhet	19	II	Akhet	20	31
*II	Akhet	20	III	Akhet	19	29
#III	Akhet	19	IIII	Akhet	19	30
*IIII	Akhet	19	I	Peret	18	29
#I	Peret	18	II	Peret	18	30
*II	Peret	18	III	Peret	17	29
#III	Peret	17	IIII	Peret	17	30
*IIII	Peret	17	I	Shemu	16	29
#[I	Shemu	16]	[II	Shemu	16]	[30]

*Pairs of recorded entries.

#Deduced, unrecorded entries based on the assumption that each lunar month ends on the day before the second entry for that month, and hence the entry for the first day of each month is the same as the second entry for the preceding month.

@Includes 5 epagomenal days between Year 30, IIII Shemu 30, and Year 31, I Akhet 1.

Fig. III.83c Second possible reconstruction of the lunar year extending over civil years 30-31 of Amenemhet III (according to Parker, but Sesostris III according to Krauss), with the assumption noted under the chart.

Calculated Date				Given Date			Duration
II	šmw	26	=	II	šmw	26 →	
							29
III	"	25	=	III	"	25	
							30
IIII	"	25	=	IIII	"	25	
							29
I	ꜣḫt	19	=	I	ꜣḫt	19	
							30
II	"	19	=	II	"	⟨19⟩	
							30
III	"	18		III	"	19	
							29
IIII	"	18	=	IIII	"	⌜18⌝	
							30
I	prt	18	=	I	prt	18	
							30
II	"	17		II	"	18	
							29
III	"	17	=	III	"	17	
							30
IIII	"	17	=	IIII	"	17	
							29
I	šmw	16	=	I	šmw	16	

Fig. III.83d Table comparing Parker's calculated lunar months for the year 30-31 with the given dates.

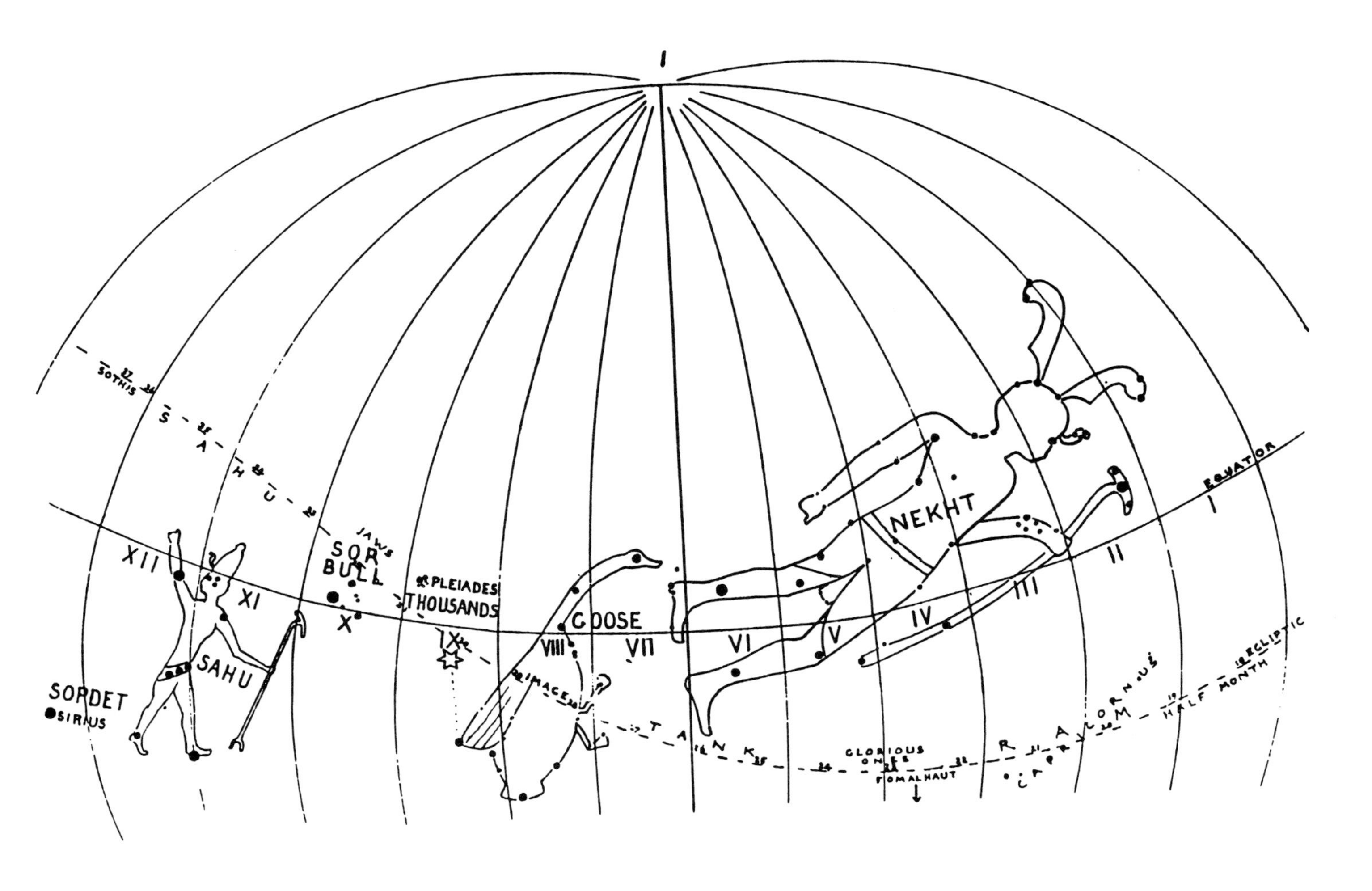

Fig. III.84a Petrie's reconstruction of the celestial area with decanal stars, constellations and hour-stars from the Ramesside Star Clock, and the so-called "Northern Constellations." First half. See Petrie, *Wisdom of the Egyptians*, Plate III.

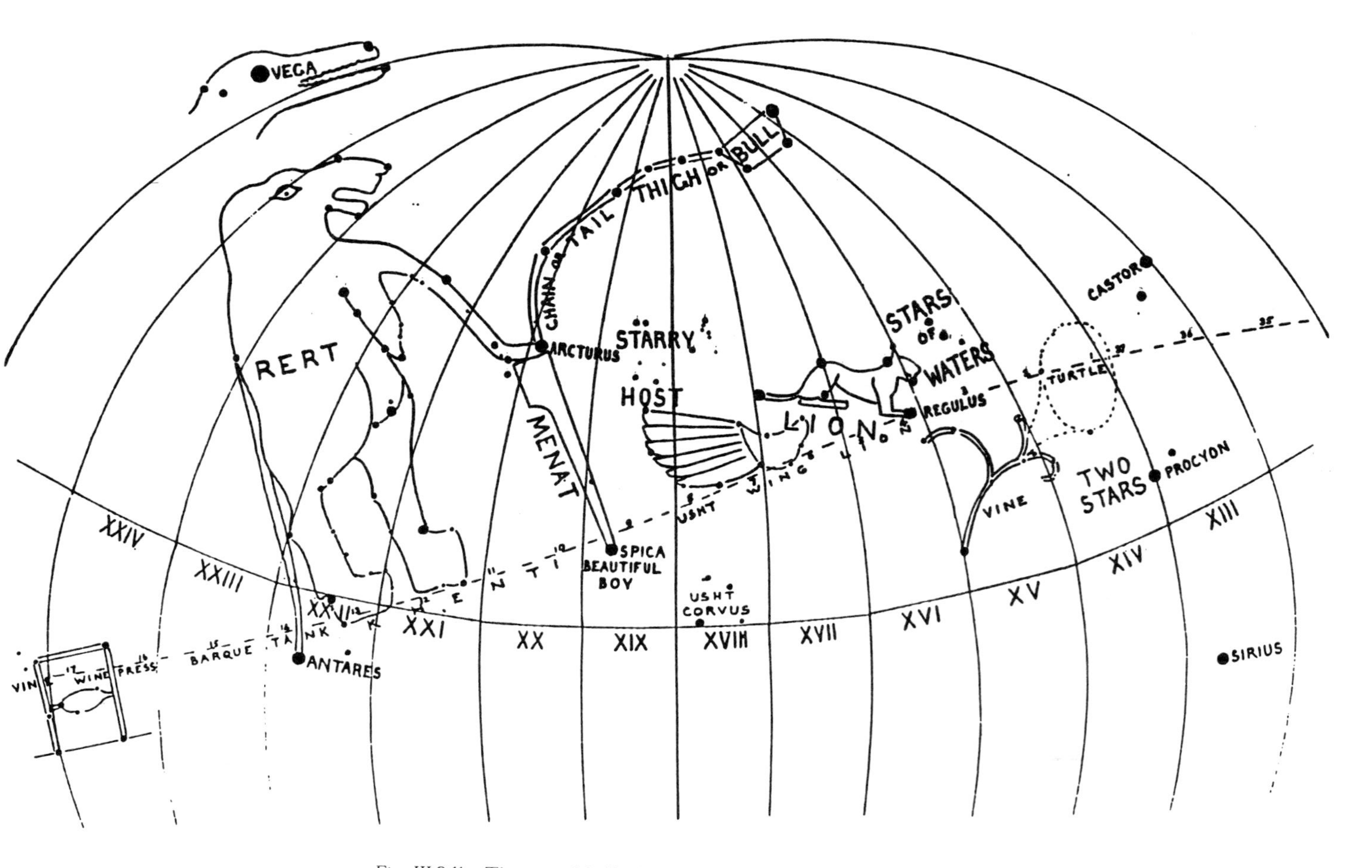

Fig. III.84b The second half of Petrie's reconstruction of the sky, including his enormous depiction of the Hippopotamus with her Mooring Post and the accompanying Big Dipper ("Thigh" or "Bull"). *Ibid.*, Plate IV.

	40	39	38	37	36	35	34	33	32	31	30	29	28	27	26	25	24	23	22	21	20	19	V	18	17	16	15	14	13	12	11	10	9	8	7	6	5	4	3	2	1	
T		27	14	1	36 ← 19 Decades																			18 ← 1 Decades																		T
1	A		15	2	↖	↖	↖	↖	↖	↖	↖	↖	↖	↖	↖	↖	↖	↖	↖	↖	↖	↖	W	↖	↖	↖	↖	↖	↖	↖	↖	↖	↖	↖	↖	↖	↖	↖	↖	↖	1	1
2	B	28	16	3	↖	↖	↖	↖	↖	↖	↖	↖	↖	↖	↖	↖	↖	↖	↖	↖	↖	↖		↖	↖	↖	↖	↖	↖	↖	↖	↖	↖	↖	↖	↖	↖	↖	↖	↖	2	2
3	C	29	17	4	↖	↖	↖	↖	↖	↖	↖	↖	↖	↖	↖	↖	↖	↖	↖	↖	↖	↖		↖	↖	↖	↖	↖	↖	↖	↖	↖	↖	↖	↖	↖	↖	↖	↖	↖	3	3
4	D	30	18	5	↖	↖	↖	↖	↖	↖	↖	↖	↖	↖	↖	↖	↖	↖	↖	↖	↖	↖	X	↖	↖	↖	↖	↖	↖	↖	↖	↖	↖	↖	↖	↖	↖	↖	↖	↖	4	4
5	[E]	31	19	6	↖	↖	↖	↖	↖	↖	↖	↖	↖	↖	↖	↖	↖	↖	↖	↖	↖	↖		↖	↖	↖	↖	↖	↖	↖	↖	↖	↖	↖	↖	↖	↖	↖	↖	↖	5	5
6	F	32	20	7	↖	↖	↖	↖	↖	↖	↖	↖	↖	↖	↖	↖	↖	↖	↖	↖	↖	↖		↖	↖	↖	↖	↖	↖	↖	↖	↖	↖	↖	↖	↖	↖	↖	↖	↖	6	6
R		7 + 6				36 + 35						20 + (19=H)				G + 31								26+C				B + A					Y	W	X			⊙				R
7	G	33	19	8	↖	↖	↖	↖	↖	↖	↖	↖	↖	↖	↖	↖	↖	↖	↖	↖	↖	↖	Y	↖	↖	↖	↖	↖	↖	↖	↖	↖	↖	↖	↖	↖	↖	↖	↖	↖	7	7
8	K	34	22	9	↖	↖	↖	↖	↖	↖	↖	↖	↖	↖	↖	↖	↖	↖	↖	↖	↖	↖		↖	↖	↖	↖	↖	↖	↖	↖	↖	↖	↖	↖	↖	↖	↖	↖	↖	8	8
9	J	35	23	10	↖	↖	↖	↖	↖	↖	↖	↖	↖	↖	↖	↖	↖	↖	↖	↖	↖	↖		↖	↖	↖	↖	↖	↖	↖	↖	↖	↖	↖	↖	↖	↖	↖	↖	↖	9	9
10	M	36	24	11	↖	↖	↖	↖	↖	↖	↖	↖	↖	↖	↖	↖	↖	↖	↖	↖	↖	↖	Z	↖	↖	↖	↖	↖	↖	↖	↖	↖	↖	↖	↖	↖	↖	↖	↖	↖	10	10
11	lo	1	25	12	↖	↖	↖	↖	↖	↖	↖	↖	↖	↖	↖	↖	↖	↖	↖	↖	↖	↖		↖	↖	↖	↖	↖	↖	↖	↖	↖	↖	↖	↖	↖	↖	↖	↖	↖	11	11
12	lal	4	26	13	M	K	J	H	G	F	E	D	C	B	A	2	1	36	35	34	33	32		31	30	29	28	26	25	24	23	22	19	20	18	17	16	15	14	13	12	12

Fig. III.85 A schematized chart of the inside of the lid from the inside coffin of Meshet (Dynasty IX or X). It shows the arrangement and diagonal displacement of the hour-decans, with numbers replacing the names of the decans and arrows showing the direction of their progressive displacement.

	1	2	3	4	5	6	7	8
T	[illegible]	[illegible]	[illegible]	[illegible]	[illegible]	[illegible]	[illegible]	[illegible]
1	[illegible]	[illegible]	[illegible]	[illegible]	[illegible]	[illegible]	[illegible]	[illegible]
2	[illegible]	[illegible]	[illegible]	[illegible]	[illegible]	[illegible]	[illegible]	[illegible]
3	[illegible]	[illegible]	[illegible]	[illegible]	[illegible]	[illegible]	[illegible]	[illegible]
4	[illegible]	[illegible]	[illegible]	[illegible]	[illegible]	[illegible]	[illegible]	[illegible]
5	[illegible]	[illegible]	[illegible]	[illegible]	[illegible]	[illegible]	[illegible]	[illegible]
6	[illegible]	[illegible]	[illegible]	[illegible]	[illegible]	[illegible]	[illegible]	[illegible]
R								
7	[illegible]	[illegible]	[illegible]	[illegible]	[illegible]	[illegible]	[illegible]	[illegible]
8	[illegible]	[illegible]	[illegible]	[illegible]	[illegible]	[illegible]	[illegible]	[illegible]
9	[illegible]	[illegible]	[illegible]	[illegible]	[illegible]	[illegible]	[illegible]	[illegible]
10	[illegible]	[illegible]	[illegible]	[illegible]	[illegible]	[illegible]	[illegible]	[illegible]
11	[illegible]	[illegible]	[illegible]	[illegible]	[illegible]	[illegible]	[illegible]	[illegible]
12	[illegible] (1)	[illegible]	[illegible]	[illegible]	[illegible]	[illegible]	[illegible]	[illegible]

(1) Sur l'original ces deux personnages sont accroupis et se tiennent par les deux mains. Je n'ai pas fait fondre un signe spécial.

	9	10	11	12	13	14	15	16
T	[illegible]	[illegible]	[illegible]	[illegible]	[illegible]	[illegible]	[illegible]	[illegible]
1	[illegible]	[illegible]	[illegible]	[illegible]	[illegible]	[illegible]	[illegible]	[illegible]
2	[illegible]	[illegible]	[illegible]	[illegible]	[illegible]	[illegible]	[illegible]	[illegible]
3	[illegible]	[illegible]	[illegible]	[illegible]	[illegible]	[illegible]	[illegible]	[illegible]
4	[illegible]	[illegible]	[illegible]	[illegible]	[illegible]	[illegible]	[illegible]	[illegible]
5	[illegible]	[illegible]	[illegible]	[illegible]	[illegible]	[illegible]	[illegible]	[illegible]
6	[illegible]	[illegible]	[illegible]	[illegible]	[illegible]	[illegible]	[illegible]	[illegible]
R								
7	[illegible]	[illegible]	[illegible]	[illegible]	[illegible]	[illegible]	[illegible]	[illegible]
8	[illegible]	[illegible]	[illegible]	[illegible]	[illegible]	[illegible]	[illegible]	[illegible]
9	[illegible]	[illegible]	[illegible]	[illegible]	[illegible]	[illegible]	[illegible]	[illegible]
10	[illegible]	[illegible]	[illegible]	[illegible]	[illegible]	[illegible]	[illegible]	[illegible]
11	[illegible]	[illegible]	[illegible]	[illegible]	[illegible]	[illegible]	[illegible]	[illegible]
12	[illegible]	[illegible]	[illegible]	[illegible]	[illegible]	[illegible]	[illegible]	[illegible]

	17	18	19	20	21	22	23	24
T	[illegible]	[illegible]	[illegible]	[illegible]	[illegible]	[illegible]	[illegible]	[illegible]
1	[illegible]	[illegible]	[illegible]	[illegible]	[illegible]	[illegible]	[illegible]	[illegible]
2	[illegible]	[illegible]	[illegible]	[illegible]	[illegible]	[illegible]	[illegible]	[illegible]
3	[illegible]	[illegible]	[illegible]	[illegible]	[illegible]	[illegible]	[illegible]	[illegible]
4	[illegible]	[illegible]	[illegible]	[illegible]	[illegible]	[illegible]	[illegible]	[illegible]
5	[illegible]	[illegible]	[illegible]	[illegible]	[illegible]	[illegible]	[illegible]	[illegible]
6	[illegible]	[illegible]	[illegible]	[illegible]	[illegible]	[illegible]	[illegible]	[illegible]
R								
7	[illegible]	[illegible]	[illegible]	[illegible]	[illegible]	[illegible]	[illegible]	[illegible]
8	[illegible]	[illegible]	[illegible]	[illegible]	[illegible]	[illegible]	[illegible]	[illegible]
9	[illegible]	[illegible]	[illegible]	[illegible]	[illegible]	[illegible]	[illegible]	[illegible]
10	[illegible]	[illegible]	[illegible]	[illegible]	[illegible]	[illegible]	[illegible]	[illegible]
11	[illegible]	[illegible]	[illegible]	[illegible]	[illegible]	[illegible]	[illegible]	[illegible]
12	[illegible]	[illegible]	[illegible]	[illegible]	[illegible]	[illegible]	[illegible]	[illegible]

	25	26	27	28	29	30	31	32
T								
1								
2								
3								
4								
5								
6								
R								
7								
8								
9								
10								
11								
12								

	33	34	35	36	37	38	39	40
T								VIDE
1							VIDE (?)	
2								
3								
4								
5								
6								
R								
7								
8								
9								
10								
11								
12								

Fig. III.86 (this and the facing page) The complete text of the hour-decans on the inside of the lid of the inside coffin of Meshet. Taken from Lacau, *Sarcophages antérieurs au novel empire*, Vol. 2, pp. 105-109.

Fig. III.87a. View toward the south wall of the Temple of Ramesses III at Medina Habu (Western Thebes). On this wall is inscribed the great calendar of the temple, with its lists of offerings to be furnished to the temple on the occasions of festivals. Taken from Nelson, *Work in Western Thebes, 1931-33*, Fig. 1, opp. p. 2.

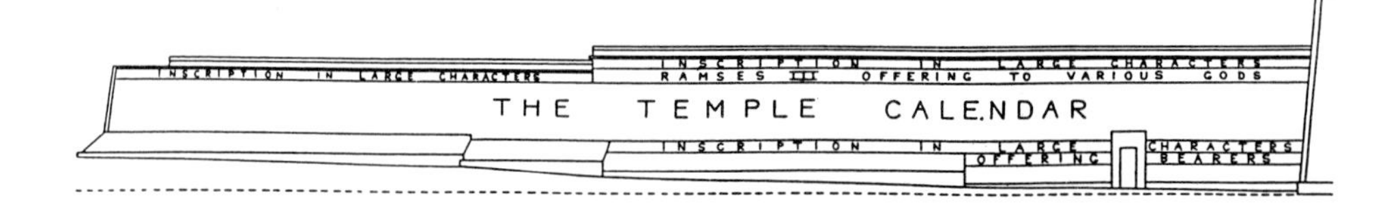

Fig. III.87b A diagram showing the arrangement of the calendar on the south wall. *Ibid.*, Fig. 2.

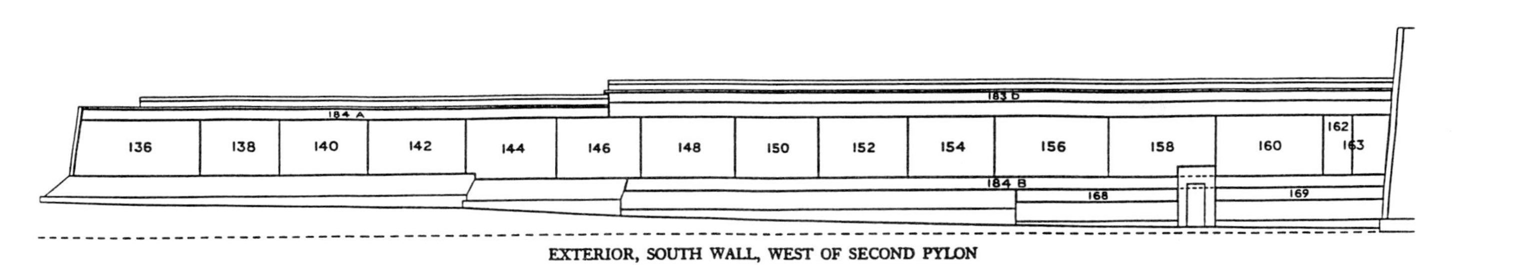

EXTERIOR, SOUTH WALL, WEST OF SECOND PYLON

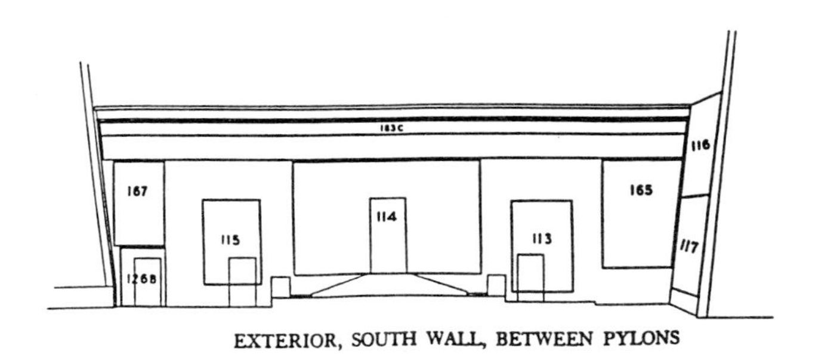

EXTERIOR, SOUTH WALL, BETWEEN PYLONS

Fig. III.87c The exterior south wall, with numbers referring to the plates in The Epigraphic Survey's publication of the temple calendar. Taken from Nelson et al., *Medinet Habu; Volume III, Plates 131-192*, Figs. 2 and 4.

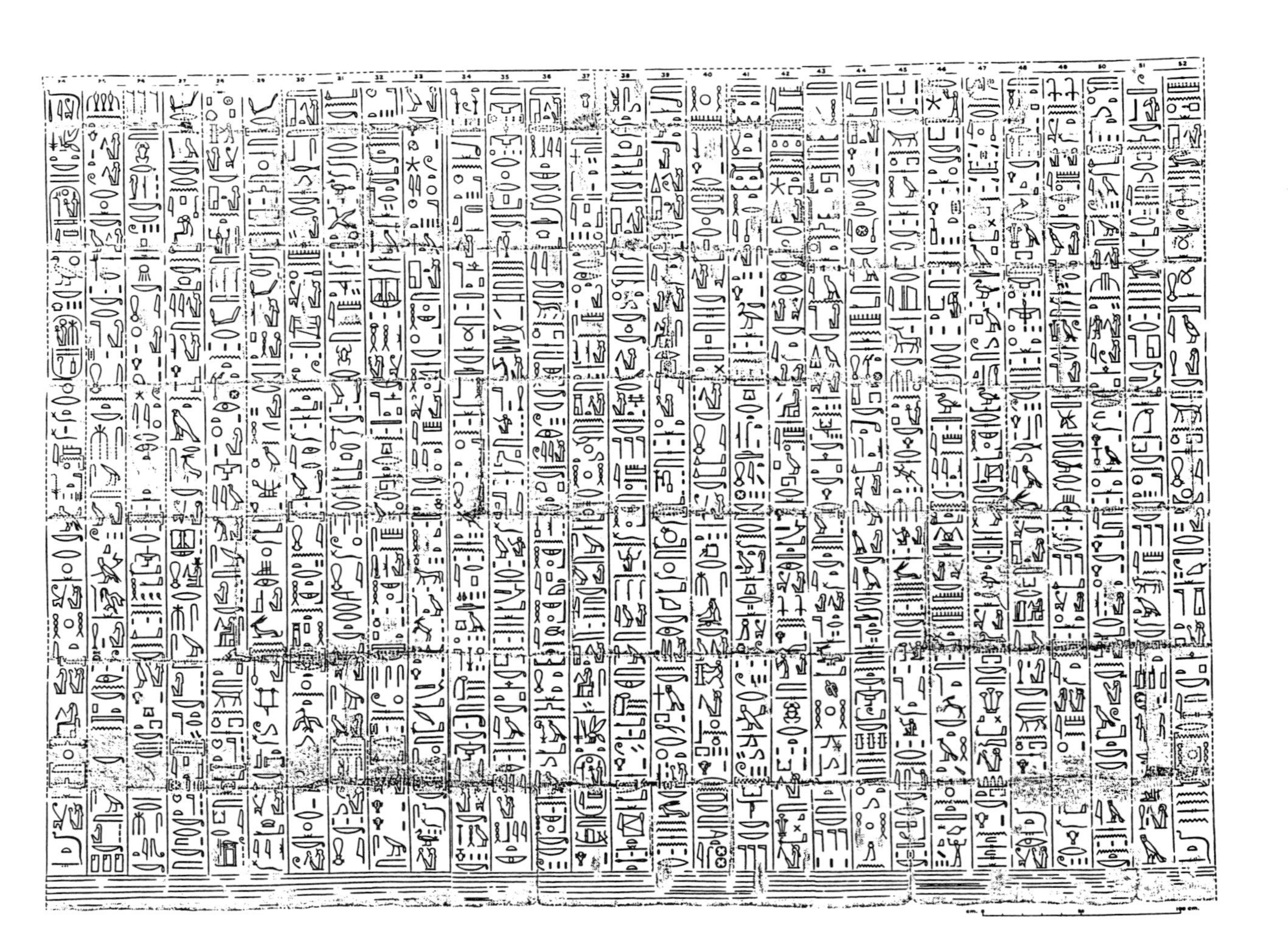

Fig. III.88 Ramesses III's Address to Amon-Re. Taken from *Medinet Habu; Volume III, Plates 131-192*, Plate 138: Calendar, lines 24-52.

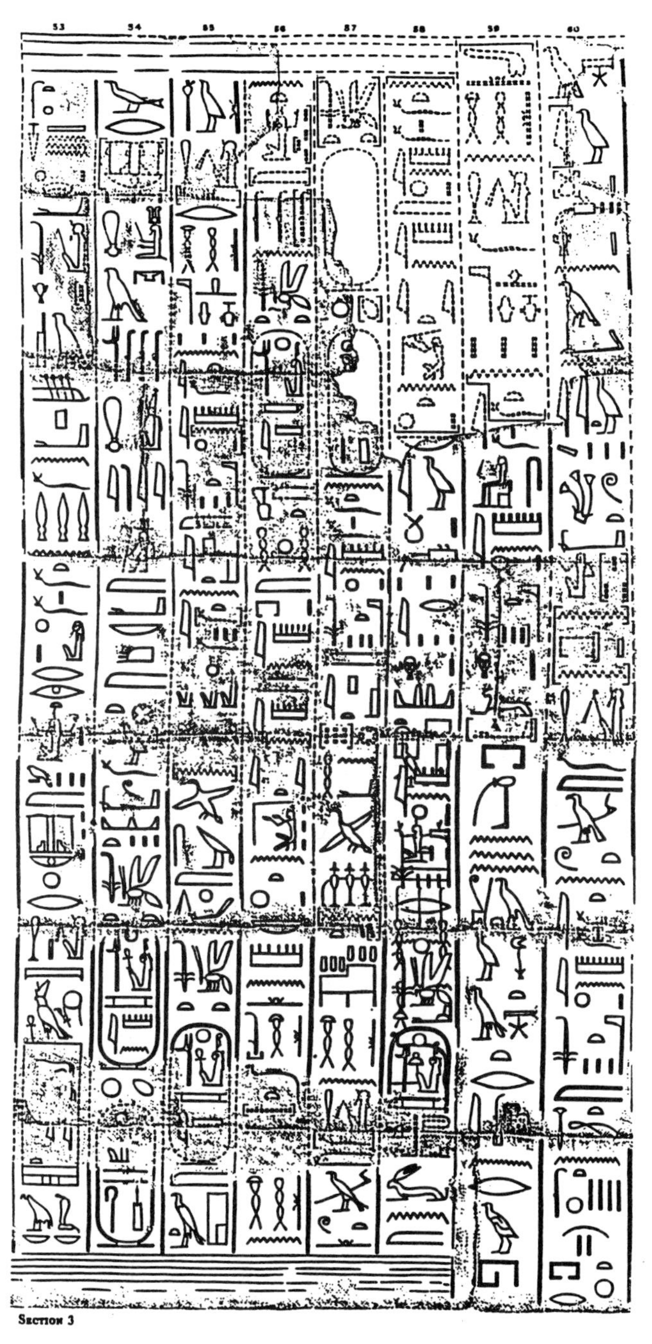

Fig. III.89 Decree by Ramesses III instituting the Calendar of Medina Habu for Amon-Re. Text from *ibid.*, left side of plate 140: Calendar, lines 53-60.

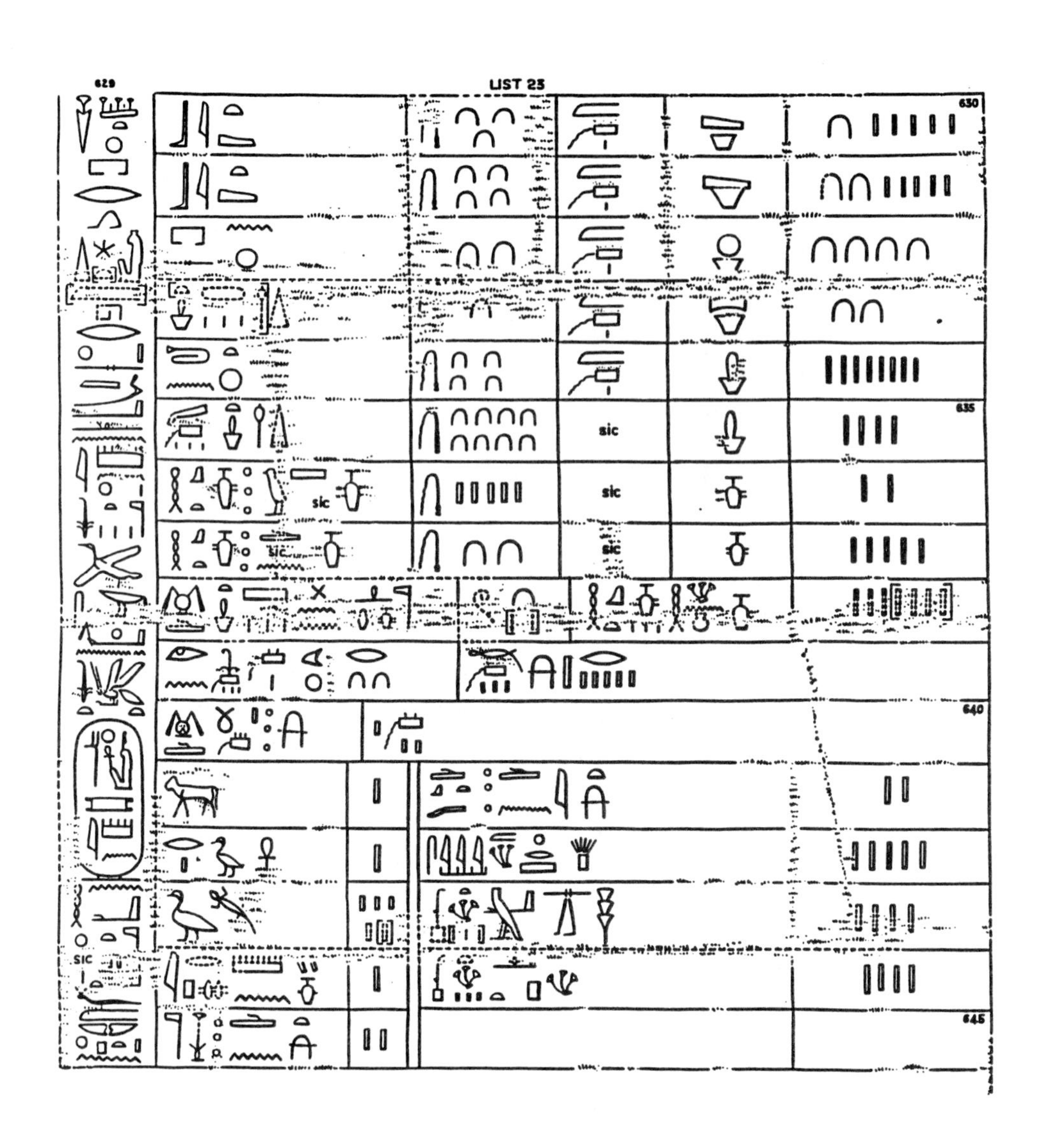

Fig. III.90 Feast of the Rising of Sothis. Text from *ibid.*, right side of plate 152: List 23, Calendar, lines 629-45.

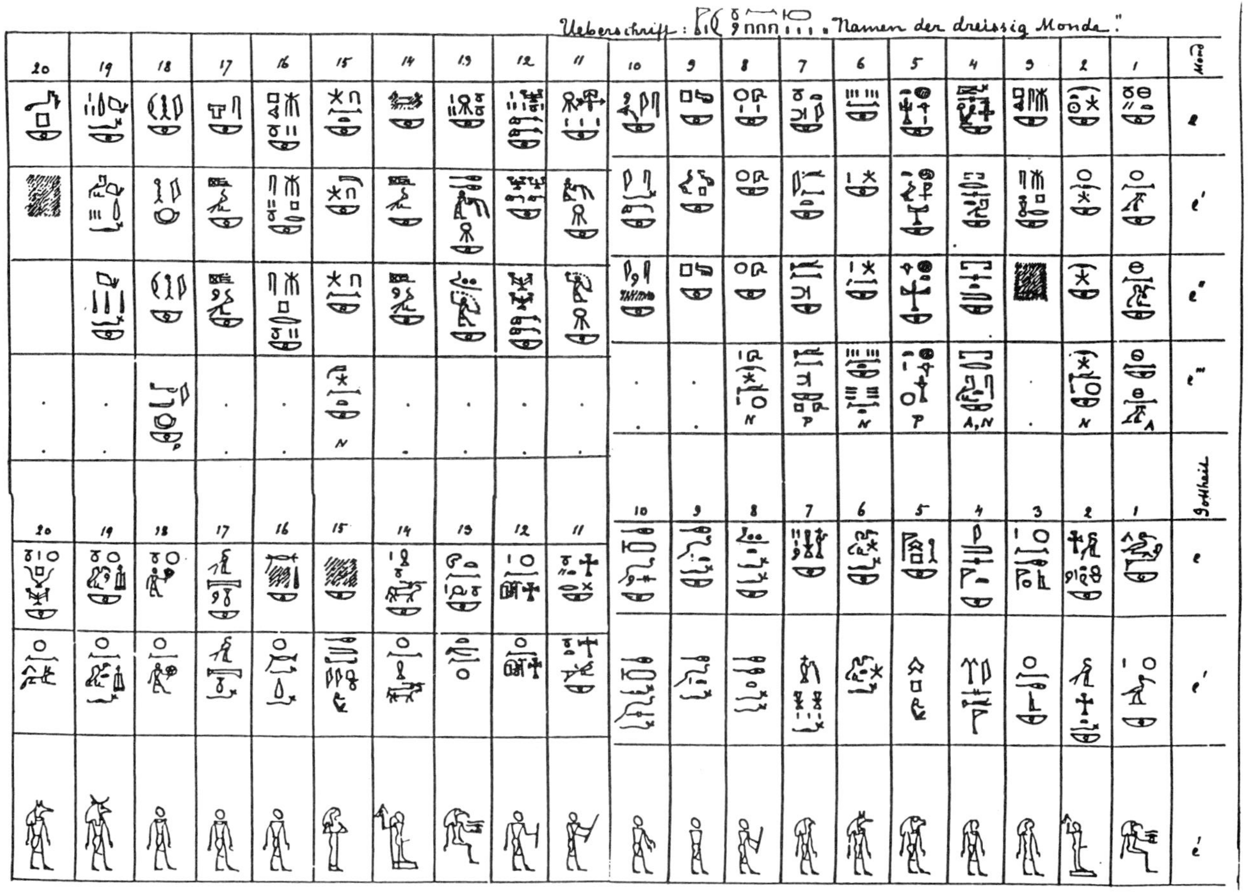

Fig. III.91a A table of the names of the thirty days of a lunar month prepared by Brugsch, *Thesaurus inscriptionum aegyptiacarum*, Abt. I, pp. 46-49. Sources of the table: e= ceiling of the pronaos of the temple of Dendera (time of Tiberius), e'= north wall of the pronaos of the temple of Edfu (Ptolemaic period), e''= second room of the north temple of Osiris on the roof of the temple of Dendera, e'''= variants from the Old Kingdom (A), the New Kingdom (N), and the Ptolemaic period (P).

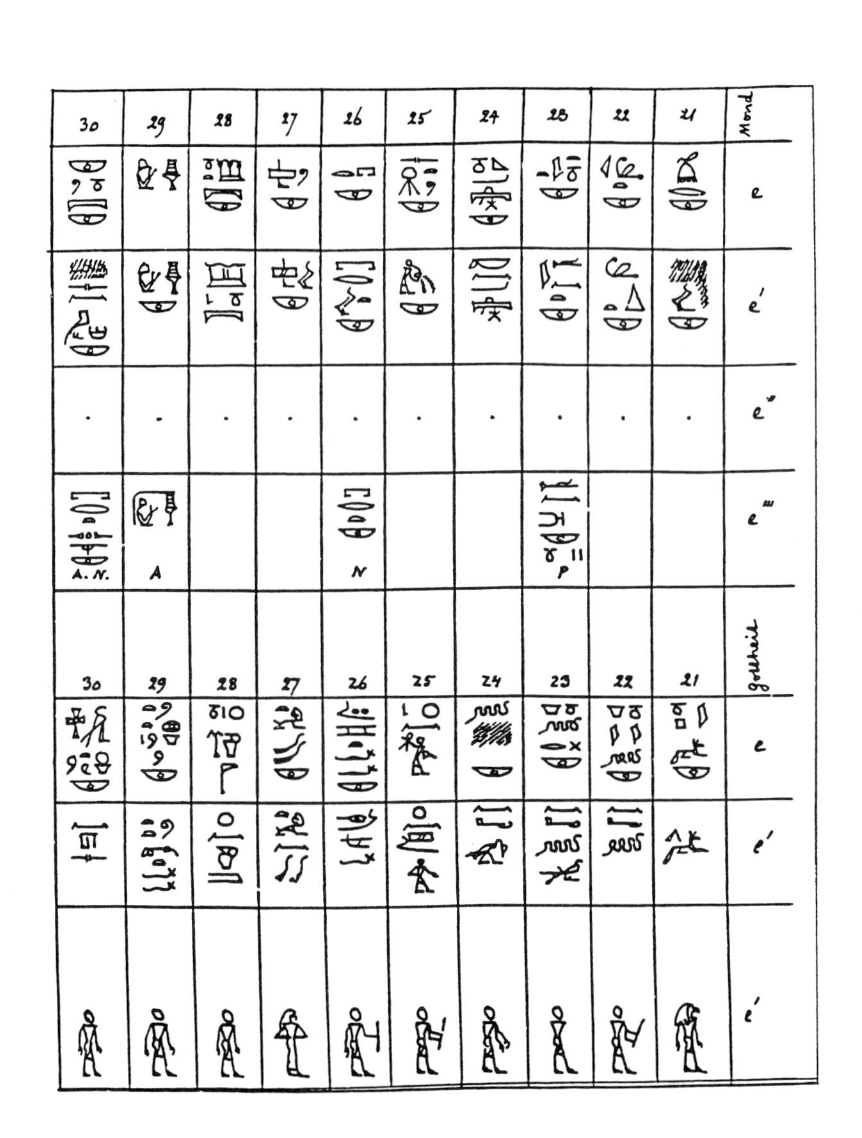

Fig. III.91a (concluded)

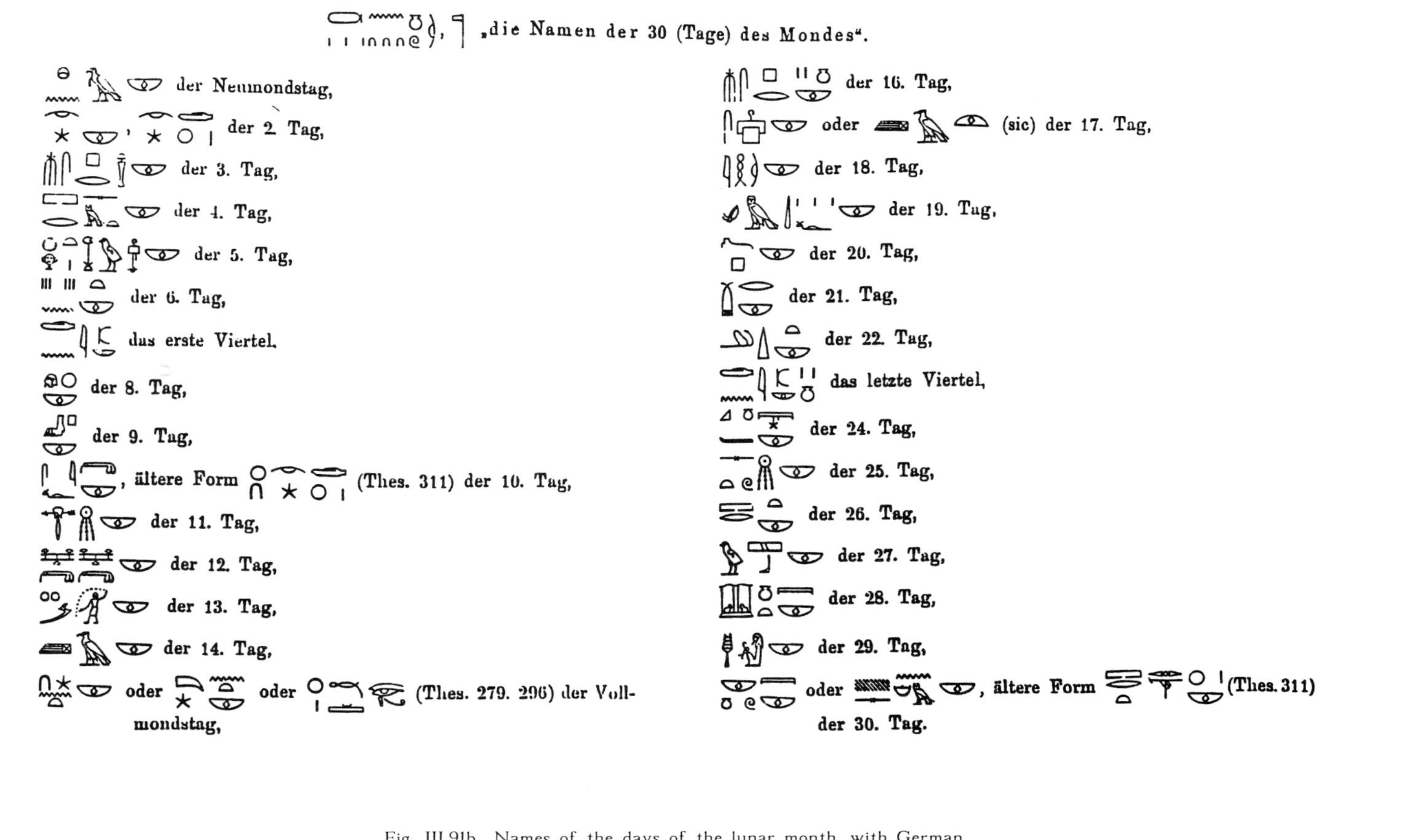

Fig. III.91b Names of the days of the lunar month, with German translations. Given by Brugsch, *Die Ägyptologie*, pp. 332-34.

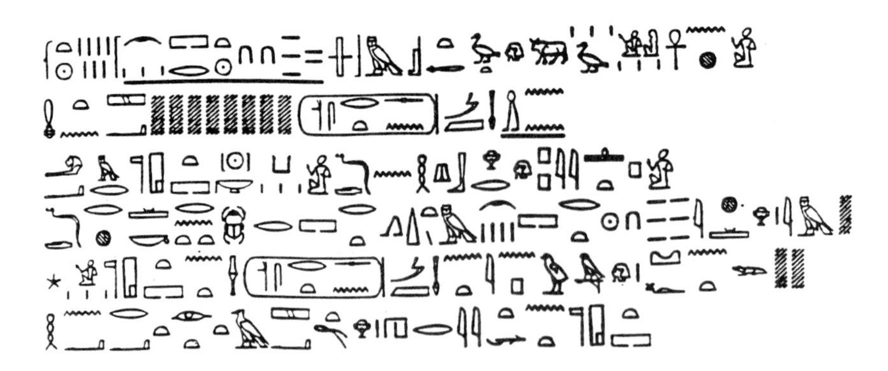

Fig. III.92a Reference to the rising of Sothis in year 7 of the reign of Sesostris III. The hieroglyphic transcription from the hieratic text of temple records from Illahun (P. Berlin 10012) is that of Borchardt, "Der zweite Papyrusfund von Kahun und die zeitliche Festlegung des mittleren Reiches der ägyptischen Geschichte," *ZÄS*, Vol. 37 (1899), p. 99.

(Berl. Pap. 10012, 18-21)

Fig. III.92b The hieratic text of P. Berlin 10012, lines 18-21, predicting the rising of Sothis in the 7th year of the reign of Sesostris III. The text is that of Möller, *Hieratische Lesestücke für den akademischen Gebrauch,* 1. Heft, p. 19.

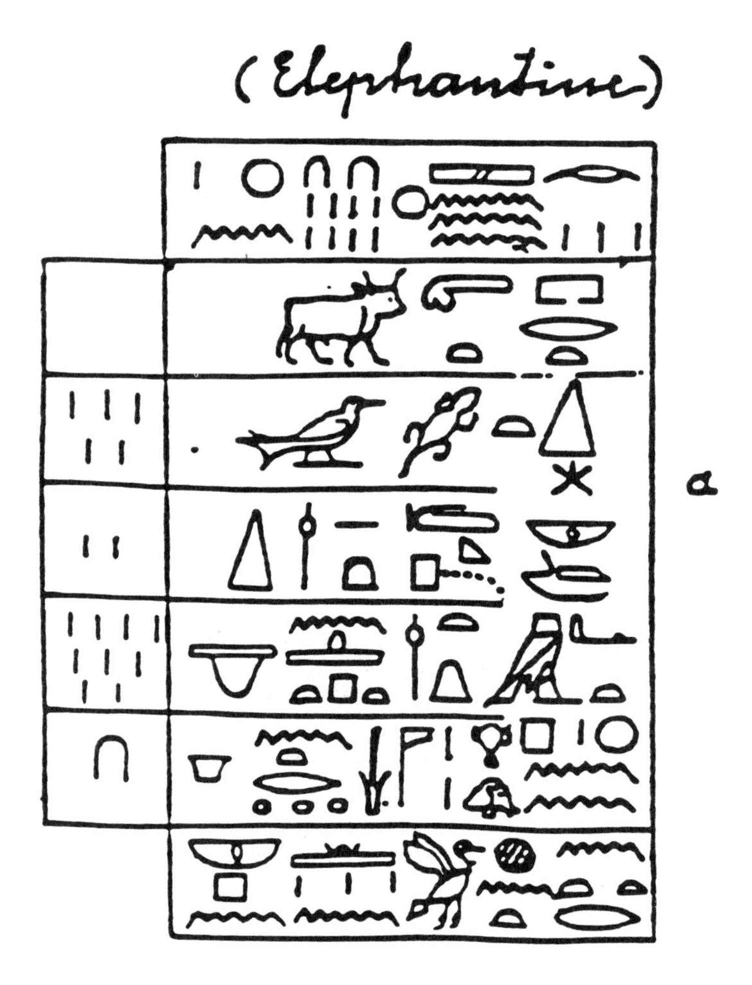

Fig. III.93 A notice of the rising of Sothis in an unknown year of Tuthmosis III. Given by Brugsch, *Thesaurus inscriptionum*, 2. Abtheilung, p. 363.

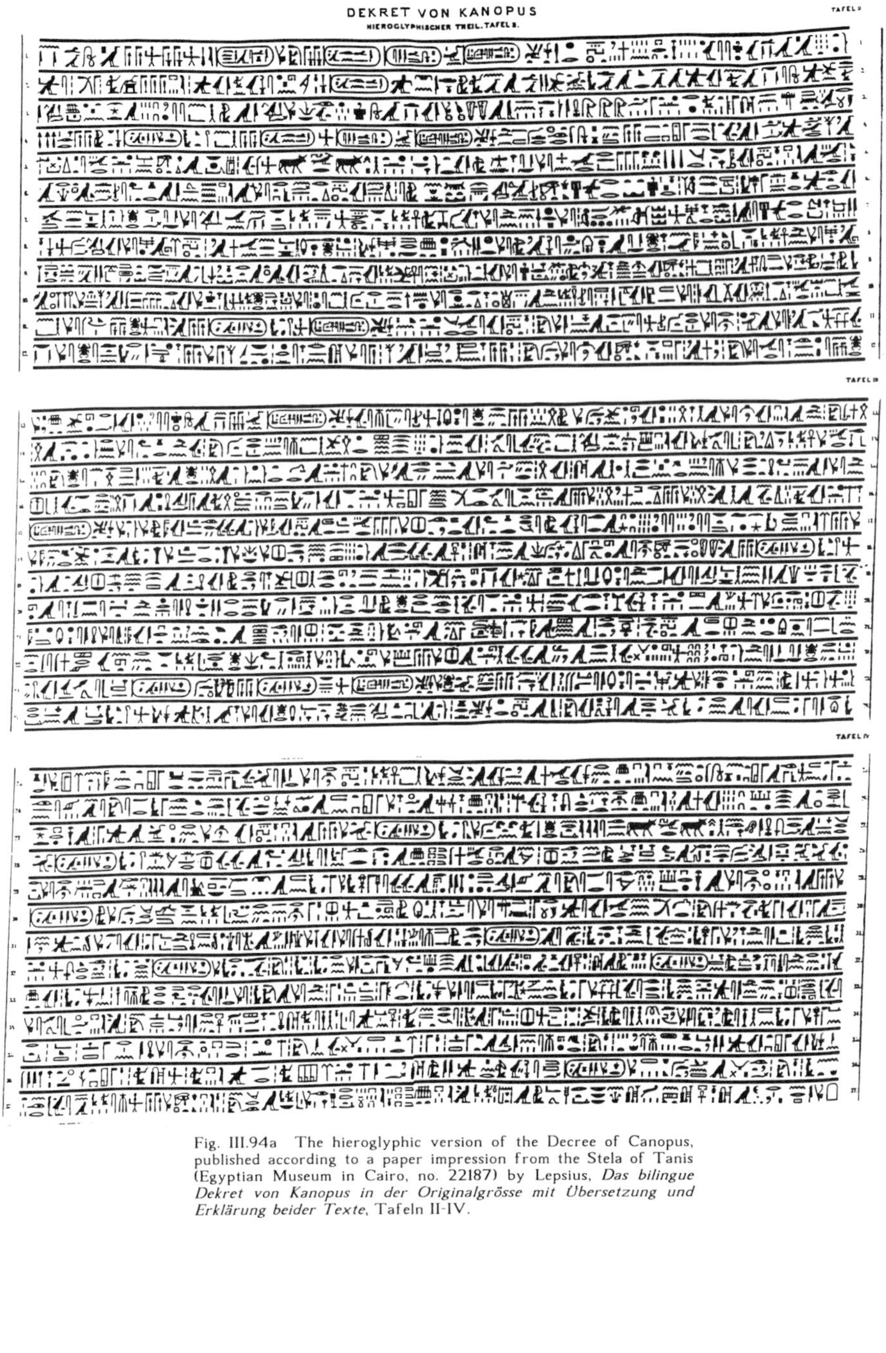

DEKRET VON KANOPUS

HIEROGLYPHISCHER THEIL. TAFEL II.

TAFEL II

TAFEL III

TAFEL IV

Fig. III.94a The hieroglyphic version of the Decree of Canopus, published according to a paper impression from the Stela of Tanis (Egyptian Museum in Cairo, no. 22187) by Lepsius, *Das bilingue Dekret von Kanopus in der Originalgrösse mit Übersetzung und Erklärung beider Texte*, Tafeln II-IV.

14. Die neue Priesterklasse soll die gleichen Rechte und die gleiche Ordnung haben wie die alten.

ἀντὶ δὲ τῶν εἴκοσι βουλευτῶν ἱερέων (108)

τῶν αἱρουμένων κατ' ἐνιαυτὸν ἐκ τῶν προυπαρχουσῶν τεσσάρων φυλῶν (109)

ἐξ ὧν πέντε ἀφ' ἑκάστης φυλῆς λαμβάνονται (110)

εἴκοσι καὶ πέντε τοὺς βουλευτὰς ἱερεῖς εἶναι (111)

προσλαμβανομένων ἐκ τῆς πέμπτης φυλῆς τῶν εὐεργετῶν θεῶν ἄλλων πέντε (112)

μετέχειν δὲ καὶ τοὺς ἐκ τῆς πέμπτης φυλῆς τῶν εὐεργετῶν θεῶν (113)

τῶν ἁγνειῶν (114)

καὶ τῶν ἄλλων ἁπάντων τῶν ἐν τοῖς ἱεροῖς (115)

καὶ φύλαρχον αὐτῆς εἶναι (116)

a) K. b) K. c–c) K. weggebrochen.
d) so T. für ; K. .

καθὰ καὶ ἐπὶ τῶν ἄλλων τεσσάρων φυλῶν ὑπάρχει (117)

15. Den „wohlthätigen Göttern" soll außer den früher eingerichteten monatlichen Festen auch alljährlich ein großes Fest gefeiert werden, gleich den großen aegyptischen Göttern, und zwar am Tage des Siriusfrühaufganges.

καὶ ἐπειδὴ (118) ἄγονται ἐν τοῖς ἱεροῖς ἑορταὶ τῶν εὐεργετῶν θεῶν (120)

καθ' ἕκαστον μῆνα (119) ἥ τε πέμπτη καὶ ἡ ἐνάτη καὶ ἡ πέμπτη ἐπ' εἰκάδι (122)

κατὰ τὸ πρότερον γραφὲν ψήφισμα (121)

τοῖς τε ἄλλοις μεγίστοις θεοῖς (123) συντελοῦνται ἑορταὶ (125)

καὶ πανηγύρεις δημοτελεῖς (126) κατ' ἐνιαυτόν (124)

ἄγεσθαι κατ' ἐνιαυτὸν πανήγυριν (127)

a) K. b) K. statt . c) K. . d) K. .
e) K. . f) K. . g–g) K. weggebrochen. h) K. .
i) … statt so aussieht: ●; vgl. ob. S. 128, i.

Fig. III.94b The text of the hieroglyphic and Greek versions of the calendric passages of the Decree of Canopus edited by Sethe, *Hieroglyphische Urkunden der Griechisch-römischen Zeit* (Leipzig, 1904), Vol. 2, pp. 136-43 and 152-54.

[hieroglyphs]
βασιλεῖ Πτολεμαίῳ καὶ βασιλίσσῃ Βερενίκῃ θεοῖς εὐεργέταις (129)

[hieroglyphs]
δημοτελῆ ἔν τε τοῖς ἱεροῖς καὶ καθ' ὅλην τὴν χώραν (128)

[hieroglyphs]
τῇ ἡμέρᾳ ἐν ᾗ ἐπιτέλλει τὸ ἄστρον τὸ τῆς Ἴσιος (130)

[hieroglyphs]
ἣ νομίζεται διὰ τῶν ἱερῶν γραμμάτων νέον ἔτος εἶναι (131)

16. Das Fest soll stets, wie im 9ten Jahre des Königs bei seiner Einrichtung, am 1sten Payni gefeiert werden, auch wenn der Siriusfrühaufgang auf einen anderen Kalendertag fallen sollte.

[hieroglyphs]
ἄγεται δὲ νῦν ἐν τῷ ἐνάτῳ ἔτει νουμηνίᾳ τοῦ Παῦνι μηνός (132)

[hieroglyphs]
ἐν ᾧ καὶ τὰ μικρὰ Βουβάστια καὶ τὰ μεγάλα Βουβάστια ἄγεται (133)

[hieroglyphs]
καὶ ἡ συναγωγὴ τῶν καρπῶν καὶ ἡ τοῦ ποταμοῦ ἀνάβασις γίνεται (134)

[hieroglyphs]
ἐὰν δὲ καὶ συμβαίνῃ (135)

a) K. ohne [glyph]. b) K. [glyph]. c) K. [glyph]. d) K. [glyph]. e) so K.; T. [glyph]. f) K. [glyph]. g) K. [glyph] ohne [glyph]. h) K. [glyph]. i) K. [glyph]. k) K. [glyph]. l) K. [glyph], vgl. S. 128. m) K. [glyph]. n–n) K. weggebrochen. o) K. [glyph].

[hieroglyphs]
τὴν ἐπιτολὴν τοῦ ἄστρου μεταβαίνειν εἰς ἑτέραν ἡμέραν (136)

[hieroglyphs]
διὰ τεσσάρων ἐτῶν (137)

[hieroglyphs]
μὴ μετατίθεσθαι τὴν πανήγυριν (138)

[hieroglyphs]
ἀλλ' ἄγεσθαι ὁμοίως τῇ νουμηνίᾳ τοῦ Παῦνι (139)

[hieroglyphs]
ἐν ᾗ καὶ ἐξ ἀρχῆς ἤχθη ἐν τῷ ἐνάτῳ ἔτει (140)

17. Wie das Fest gefeiert werden soll.

[hieroglyphs]
καὶ συντελεῖν αὐτὴν ἐπὶ ἡμέρας πέντε (141)

[hieroglyphs]
μετὰ στεφανηφορίας (142)

[hieroglyphs]
καὶ θυσιῶν (143)

[hieroglyphs]
καὶ σπονδῶν (144)

[hieroglyphs]
καὶ τῶν ἄλλων τῶν προσηκόντων (145)

a) K. [glyph]. b) so K.; T. [glyph]. c) K. [glyph]. d) K. [glyph]. e) K. [glyph]. f) K. [glyph]. g) K. [glyph]. h) K. [glyph]. i) K. [glyph]. k) in T. misraten. l) K. [glyph]. m) [glyph] ohne [glyph]. n–n) K. weggebrochen.

Fig. III.94b (continued)

Dekret von Kanopus. 20 – 21. II 140

18. Alle vier Jahre soll vor dem Neujahrstag (1. Thoth) außer den üblichen 5 Zusatztagen (Epagomenen) noch ein sechster Schalttag als „Fest der wohltätigen Götter" eingeschaltet werden, um die Verschiebung des Kalenderjahres zu beseitigen.

ὅπως δὲ καὶ (146)

αἱ ὧραι τὸ καθῆκον ποιῶσιν διὰ παντός (147)

κατὰ τὴν νῦν οὖσαν κατάστασιν τοῦ κόσμου (148)

καὶ μὴ συμβαίνῃ (149)

τινὰς τῶν δημοτελῶν ἑορτῶν των ἀγομένων ἐν τῷ χειμῶνι (150)

ἄγεσθαί ποτε ἐν τῷ θέρει (151)

τοῦ ἄστρου μεταβαίνοντος μίαν ἡμέραν διὰ τεσσάρων ἐτῶν (152)

Dekret von Kanopus. 21 – 23. II 141

ἑτέρας δὲ τῶν νῦν ἀγομένων ἐν τῷ θέρει (153)

ἄγεσθαι ἐν τῷ χειμῶνι ἐν τοῖς μετὰ ταῦτα καιροῖς (154)

καθάπερ πρότερόν τε συμβέβηκεν γενέσθαι (155)

καὶ νῦν ἂν ἐγίνετο (156)

τῆς συντάξεως τοῦ ἐνιαυτοῦ μενούσης ἐκ τῶν τριακοσίων καὶ ἑξήκοντα ἡμερῶν (157)

καὶ τῶν ὕστερον προσνομισθεισῶν ἐπάγεσθαι πέντε ἡμερῶν (158)

ἀπὸ τοῦ νῦν μίαν ἡμέραν ἑορτὴν τῶν εὐεργετῶν θεῶν ἐπάγεσθαι διὰ τεσσάρων ἐτῶν (159)

ἐπὶ ταῖς πέντε ταῖς ἐπαγομέναις πρὸ τοῦ νέου ἔτους (160)

ὅπως ἅπαντες εἰδῶσιν (161)

διότι τὸ ἐλλεῖπον πρότερον περὶ τὴν σύνταξιν τῶν ὡρῶν (162)

καὶ τοῦ ἐνιαυτοῦ (163)

Fig. III.94b (continued)

καὶ τῶν νομιζομένων περὶ τὴν ὅλην διακόσμησιν τοῦ πόλου (164)

δωρηθῶσιν (165) συμβέβηκεν (167)

καὶ ἀναπεπληρῶσθαι (166) διὰ τῶν εὐεργετῶν θεῶν (168)

19. Als die jugendliche Tochter des Königspaares Berenike noch während der Anwesenheit der Priester beim Hofe plötzlich starb, veranstalteten diese sogleich eine große Trauer.

καὶ ἐπειδὴ τὴν ἐγ βασιλέως Πτολεμαίου (169) γεγεννημένην θυγατέρα (171)

καὶ βασιλίσσης Βερενίκης θεῶν εὐεργετῶν (170)

καὶ ὀνομασθεῖσαν Βερενίκην (172)

ἣ καὶ βασίλισσα εὐθέως ἀπεδείχθη (173)

συνέβη ταύτην παρθένον οὖσαν ἐξαίφνης μετελθεῖν εἰς τὸν ἀέναον κόσμον (174)

a) K. b) K. c) K. d–d) K. weggebrochen. e) K. f) K. g) so K.; T. irrig h) so K.; T. i) K. k) so K.; T. l) T. irrig " statt o.

τῶν ἐκ τῆς χώρας παραγινομένων πρὸς αὐτὸν (τὸν βασιλέα) κατ' ἐνιαυτὸν ἱερέων (176) nach 175

ἔτι ἐνδημούντων παρὰ τῷ βασιλεῖ (175)

οὐ μέγα μὲν πένθος ἐπὶ τῷ συμβεβηκότι εὐθέως συνετέλεσαν (177)

20. Zugleich bewogen die Priester das Königspaar dazu, die verstorbene Prinzessin im Tempel von Kanopos als Göttin zusammen mit Osiris verehren zu lassen.

ἀξιώσαντες δὲ τὸν βασιλέα καὶ τὴν βασίλισσαν (178)

ἔπεισαν (179)

καθιδρῦσαι τὴν θεὰν μετὰ τοῦ Ὀσίριος ἐν τῷ ἐν Κανώπῳ ἱερῷ (180)

ὃ οὐ μόνον ἐν τοῖς πρώτοις ἱεροῖς ἐστιν (181)

ἀλλὰ καὶ ὑπὸ τοῦ βασιλέως καὶ τῶν κατὰ τὴν χώραν πάντων ἐν τοῖς μάλιστα τιμωμένοις ὑπάρχει (182)

a) K. b) K.; T. Brugsch liest, das wohl so, aber nicht gemacht wird. c) K.; T. (so). d) K. e) K. f) bei T. g) K. h–h) K. weggebrochen. i) K. ohne. k) K. ohne. l) K. m) K.

Fig. III.94b (continued)

Dekret von Kanopus 34–35. II 152

δῶσιν τῷ ᾠδοδιδασκάλῳ (265)

ὧν καὶ τὰ ἀντίγραφα καταχωρισθήσεται εἰς τὰς ἱερὰς βύβλους (266)

29. Wie die Priester selbst so sollen auch ihre Töchter ihren Unterhalt aus den Tempeleinkünften bekommen; das den Frauen (so) der Priester zu gebende Brot soll eine eigene Form haben und „Berenikebrot" genannt werden.

καὶ ἐπειδὴ τοῖς ἱερεῦσιν δίδονται αἱ τροφαὶ ἐκ τῶν ἱερῶν (267)

ἐπὰν ἐπαχθῶσιν εἰς τὸ πλῆθος (268)

δίδοσθαι ταῖς θυγατράσιν τῶν ἱερέων (269) (τὴν τροφὴν s. u. 272)

ἀφ' ἧς ἂν ἡμέρας γένωνται (271)

ἐκ τῶν ἱερῶν προσόδων (270)

a) so K.; T. [illegible]. b) so K.; T. [illegible] statt [illegible]. c–c) K. weggebrochen. d) K. [illegible].

Dekret von Kanopus 35–36. II 153

τὴν συνκριθησομένην τροφὴν ὑπὸ τῶν βουλευτῶν ἱερέων (272)

ἐν ἑκάστῳ τῶν ἱερῶν [var. ἑκάστῳ ἱερῷ] (273)

κατὰ λόγον τῶν ἱερῶν προσόδων (274)

καὶ τὸν διδόμενον ἄρτον ταῖς γυναιξὶν τῶν ἱερέων (275)

ἔχειν ἴδιον τύπον (276)

καὶ καλεῖσθαι Βερενίκης ἄρτον (277)

30. Der Beschluß soll in den drei Sprachen des Landes aufgezeichnet und in allen Tempeln allgemein sichtbar aufgestellt werden.

ἀναγραψάτωσαν (283) δὲ (279) τοῦτο τὸ ψήφισμα (284)

ὁ (278) ἐν ἑκάστῳ τῶν ἱερῶν καθεστηκὼς ἐπιστάτης (280)

καὶ ἀρχιερεύς (281)

a) K. [illegible]. b) K. [illegible]. c) K. [illegible] statt [illegible]. d–d) K. weggebrochen. e) so für [illegible] vor [illegible]. f) K. [illegible].

Fig. III.94b (continued)

καὶ οἱ τοῦ ἱεροῦ γραμματεῖς (282)

εἰς στήλην λιθίνην ἢ χαλκῆν (285)

ἱεροῖς γράμμασιν καὶ αἰγυπτίοις καὶ ἑλληνικοῖς (286)

καὶ ἀναθέτωσαν ἐν τῷ ἐπιφανεστάτῳ τόπῳ (287)

τῶν τε πρώτων ἱερῶν καὶ δευτέρων καὶ τρίτων (288) 5

ὅπως (289) φαίνωνται (291)

οἱ κατὰ τὴν χώραν ἱερεῖς (290) τιμῶντες τοὺς εὐεργέτας θεοὺς (292)

καὶ τὰ τέκνα αὐτῶν (293)

καθάπερ δίκαιόν ἐστιν (294). 9

---

a) K. … b) K. … ohne … c) so K.; T. … statt … d) K. ohne Suffix … e) K. … f) K. … g) K. … (10). h) so K.; T. … i) K. … k) K. … l) K. … ohne … m) K. … n) so K.; T. irrig … statt …

Fig. III.94b (concluded)

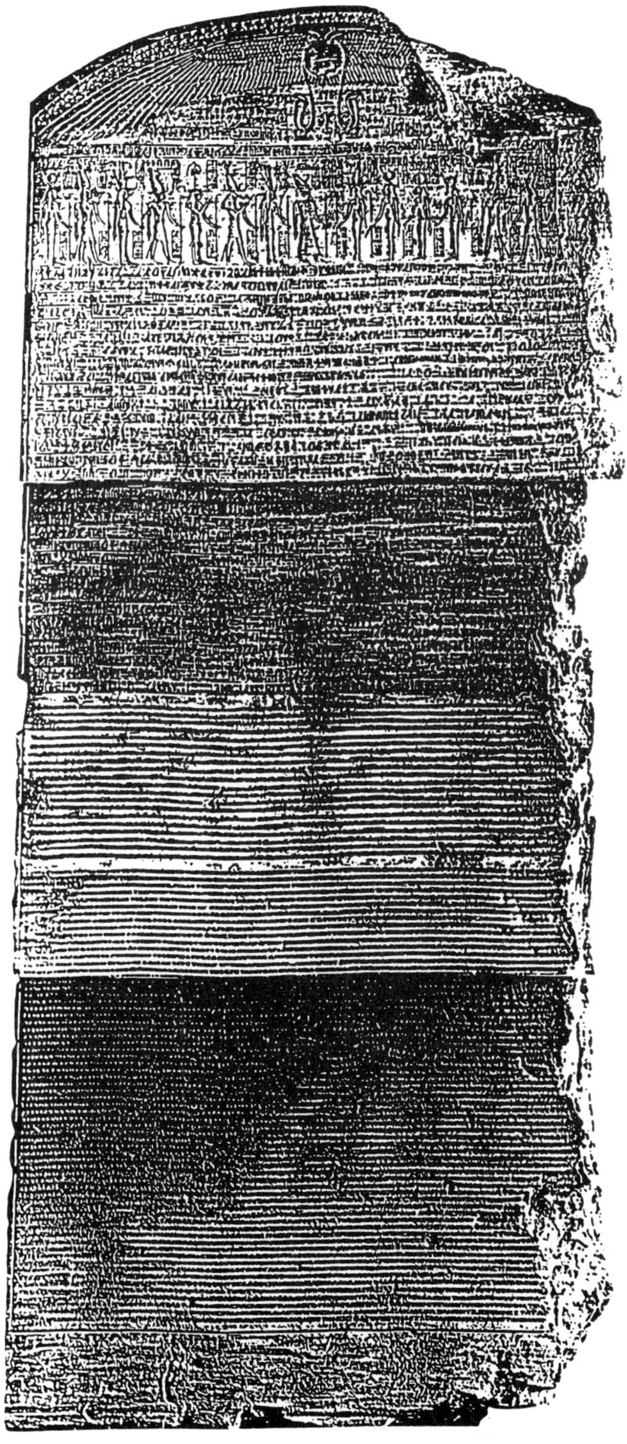

Fig. III.94c The Decree of Canopus inscribed on the Stela of Kom-el-Hisn (Egyptian Museum, Cairo, no. 22186). My figure was produced by joining Plates LIX, LX, and LXI from Ahmed Bey Kamal, *Stèles ptolémaiques et romaines*, Vol. 2.

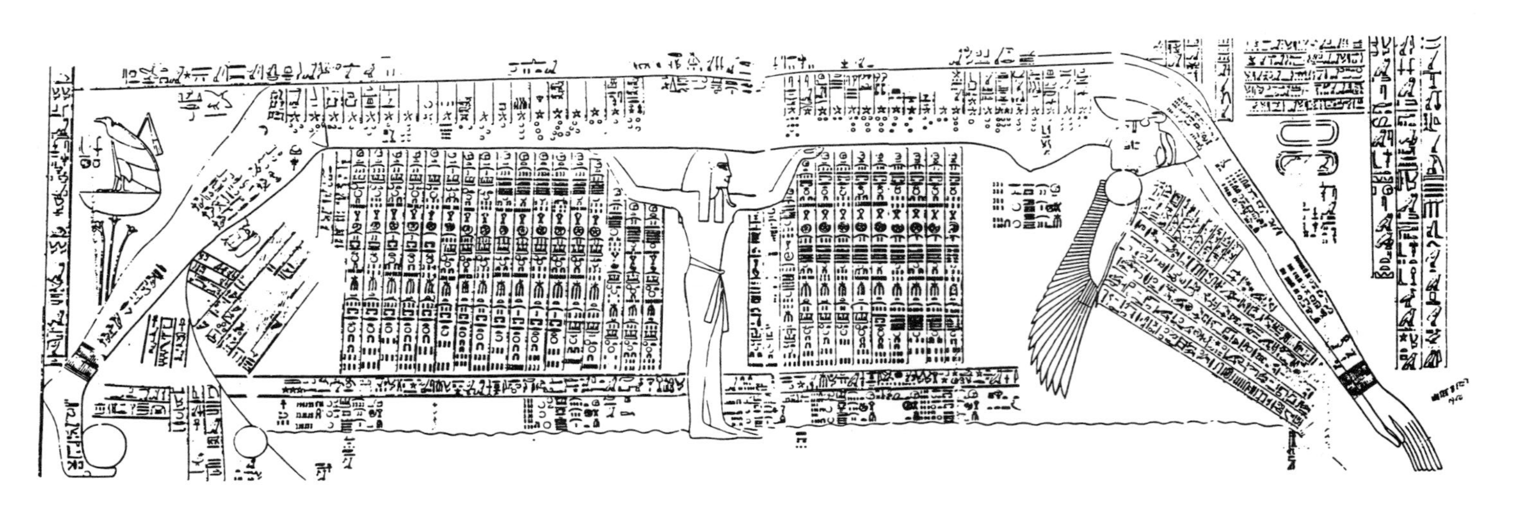

Fig. III.95a West side of the roof of the Sarcophagus Chamber in the Cenotaph of Seti I at Abydos. Group of the sky goddess Nut in vaulted position, the air god Shu supporting her, and stars with cosmological texts. Prepared by Walter Emery and reproduced from Frankfort, *The Cenotaph of Seti I at Abydos* (London, 1933), Vol. II: Plate LXXXI.

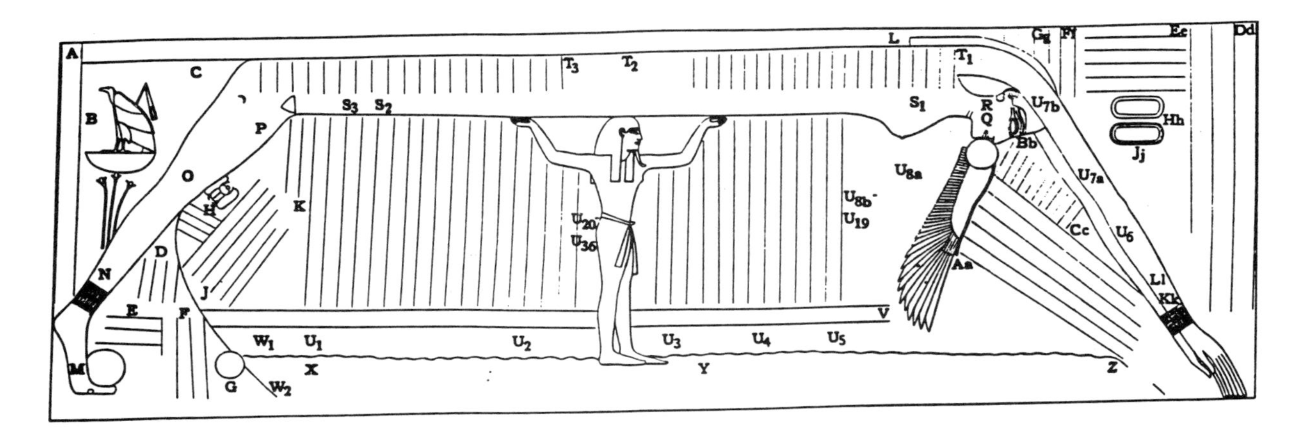

Fig. III.95b Chart locating the texts in the preceding illustration. Originally prepared by Lange and Neugebauer in their *Papyrus Carlsberg No. 1* (Copenhagen, 1940), p. 67, but given here as redrawn by Neugebauer and Parker, *Egyptian Astronomical Texts*, Vol. 1, p. 39.

A

S

omits omits R

P

D

S
R
P
Pa

C

S
omits R
P

B

S
R
omits P

G

S
R
P

E

S
R
P

F

S
R
P
Pa

H

S
R
P
Pa

[a] Surely restore
[b,c] Probably restore
[d] Emend for
[e] The beetle belonging here is just above the text in giant size

L

S
R
P

S
R
P

S
R
P

S
R
P

P

S
omits R
omits P

N

S
R
omits P

M

S
omits R
omits P

O

S
R
omits P

J

blank S
R
P

S
R
P

S
R
P

S
R
P
Pa

K

S
R
P
Pa

[a] Perhaps this was blank

T1,3

See Plates 30–31 S
See Plates 34–35 R
omits P

S

3 2 1 S
omits R
omits P

R

S
omits R
omits P

Q

S
omits R
omits P

T2

S
R
P

U

S
R
P
Pa

V

S
R
P

S
R
P

S
R
P

[a] These texts are to be dealt with in a later volume
[b] See the commentary to P and Pa and Text U
[c] In demotic only
[d] This ends at the leg of Shu

a Only the noteworthy variants are given b Text Aa continues on the next Plate

a R var. ends here b-c This column is Text Cc
d See the following note e-f
e-f These signs are dittography in S, and are out of place in R

Cc
Forms end of Text Aa S
Included in Text Aa R
Included in Text Aa P

Fig. III.95c Concordance of Texts of Seti I (**S**), Ramesses IV (**R**), P. Carlsberg I (**P**) and Ia (**Pa**). Taken from Neugebauer and Parker, *Egyptian Astronomical Texts*, Vol. 1, Plates 44-51.

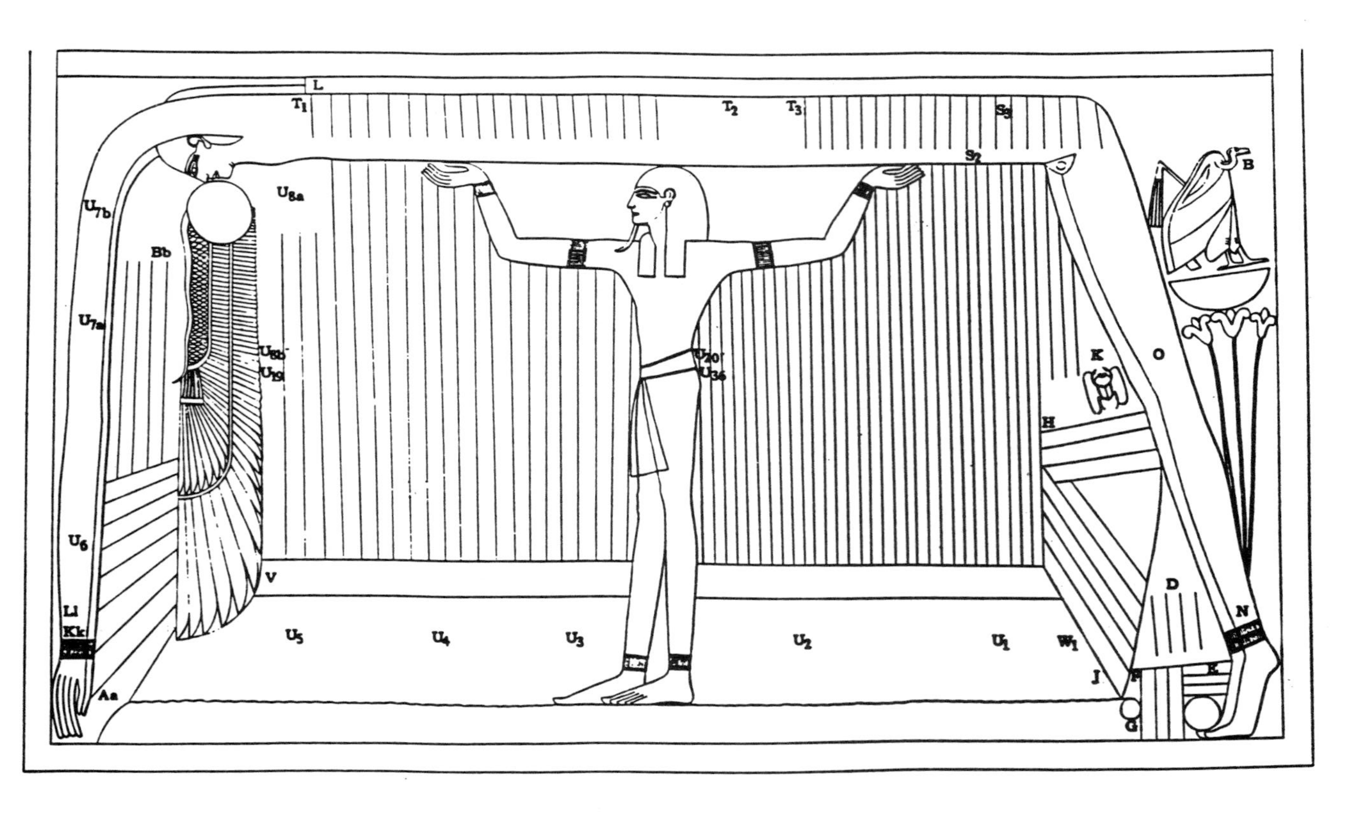

Fig. III.96a Chart locating the texts of the Cosmology of Seti I and Ramesses IV on the ceiling of the Sarcophagus Chamber in the Tomb of Ramesses IV in the Valley of the Kings (Western Thebes), as prepared by Neugebauer and Parker, *Egyptian Astronomical Texts*, Vol. I, p. 40.

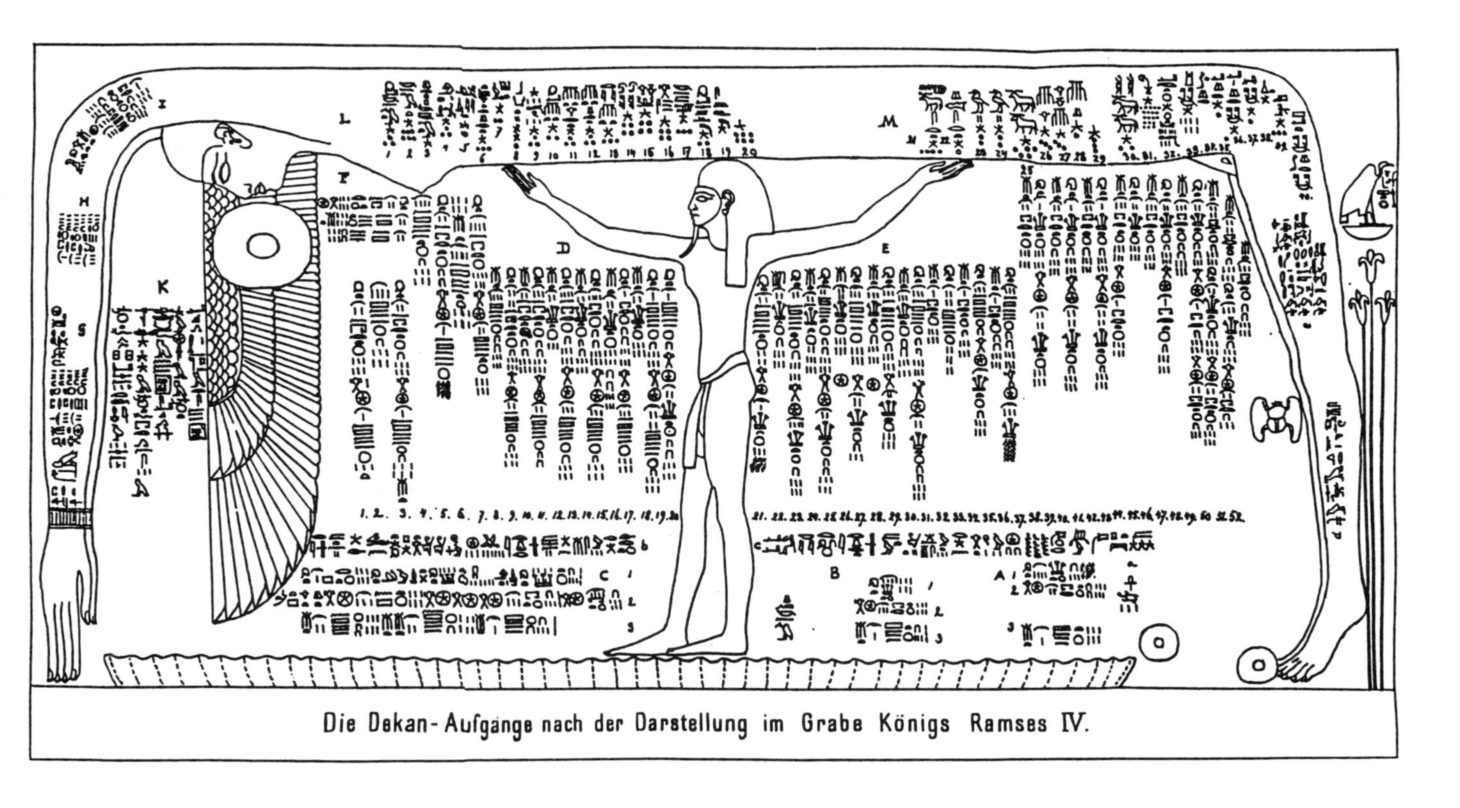

Die Dekan-Aufgänge nach der Darstellung im Grabe Königs Ramses IV.

Fig. III.96b Brugsch's drawing of the Figure of Nut with accompanying texts from the Sarcophagus Chamber in the Tomb of Ramesses IV. Taken from Brugsch, *Thesaurus inscriptionum aegyptiacarum*, following p. 174.

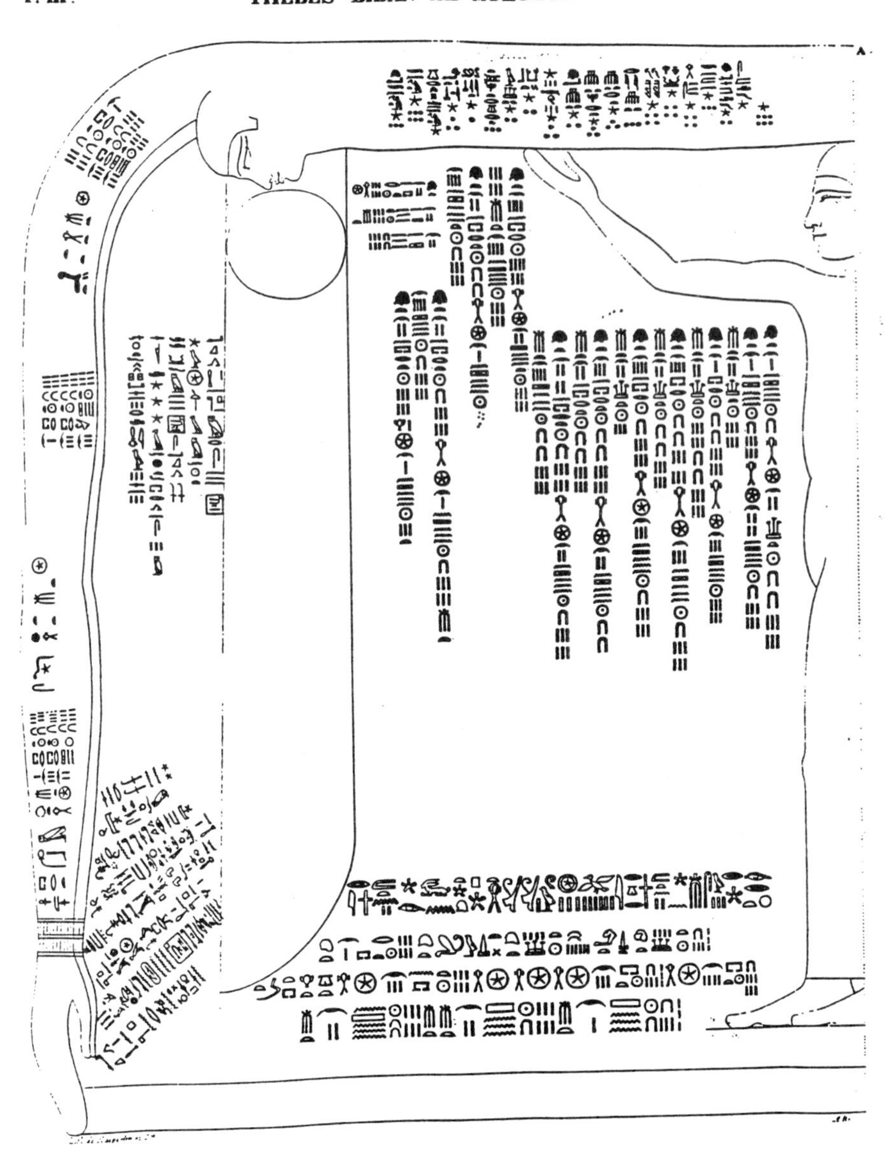

Fig. III.96c An earlier drawing of the same figure and text as in Fig. III.96b, taken from Champollion, *Monuments de l'Egypte et de la Nubie*, Vol. 3, Plates 75-76.

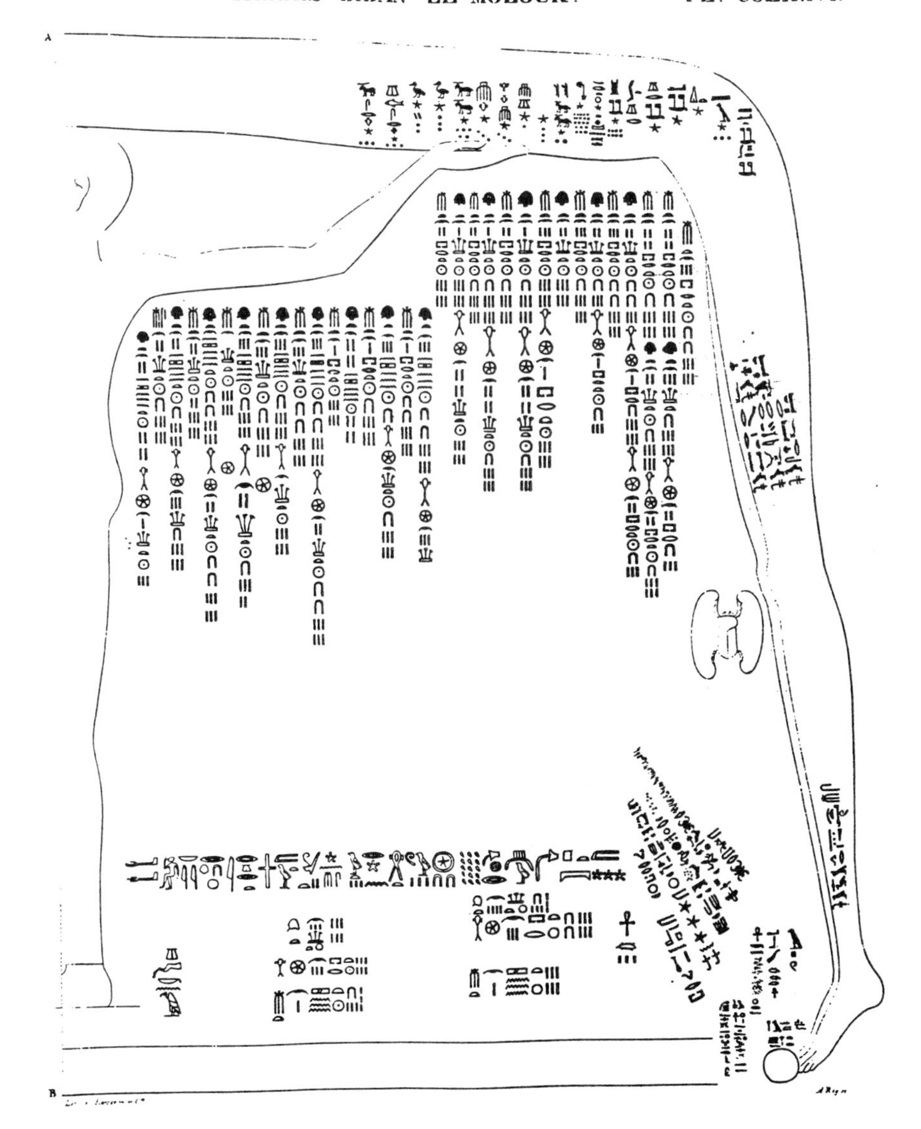

Fig. III.96c (continued)

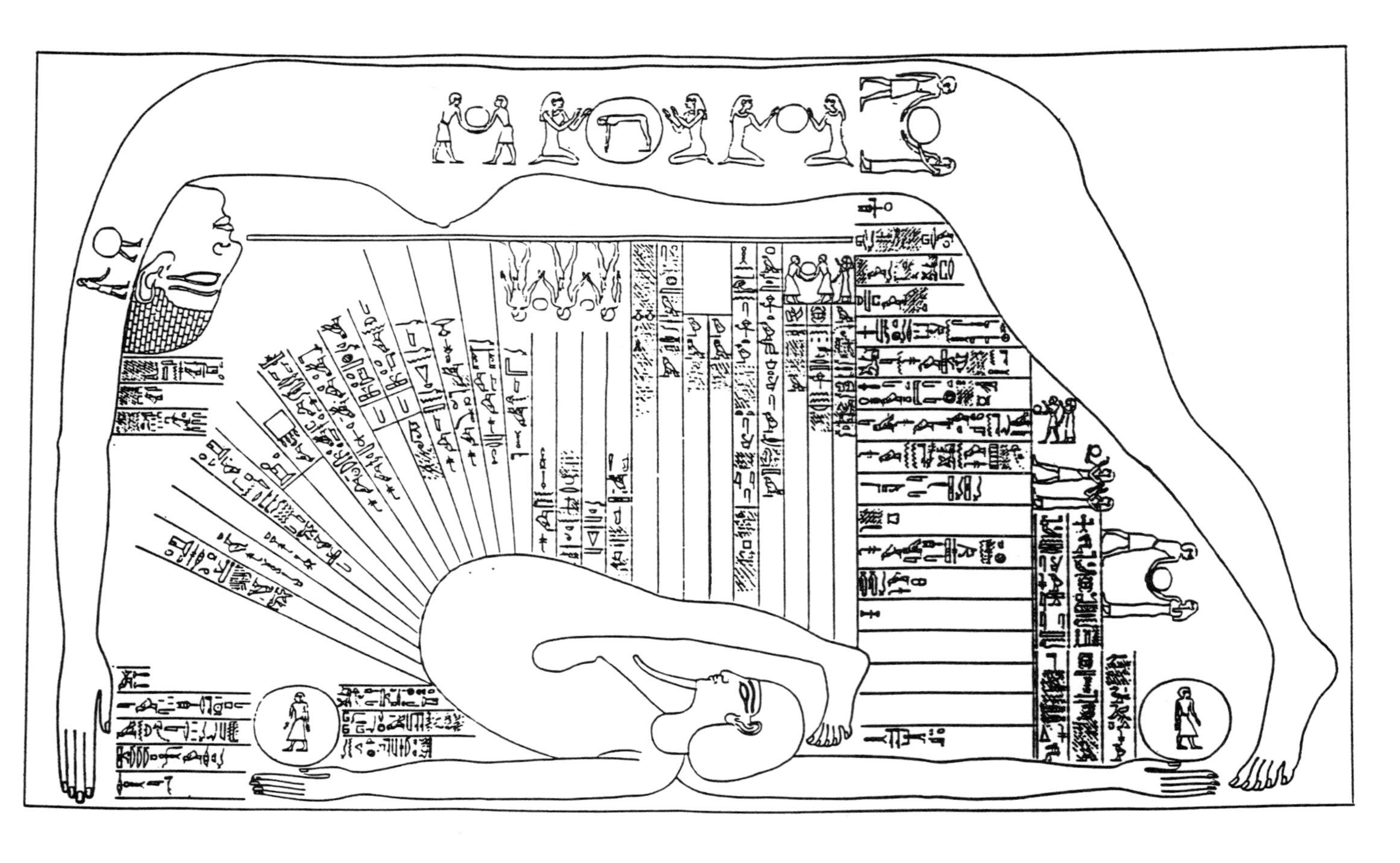

Fig. III.97 Figure of the sky godess Nut in the vaulted position, with the earth god Geb below, lying on his back in a contorted position. In the Western Chapel of Osiris (Chamber no. 2) on the roof of the Temple of Hathor at Dendera. Taken from Mariette, *Denderah. Description générale du grand temple de cette ville*, Vol. 4, Plate 76.

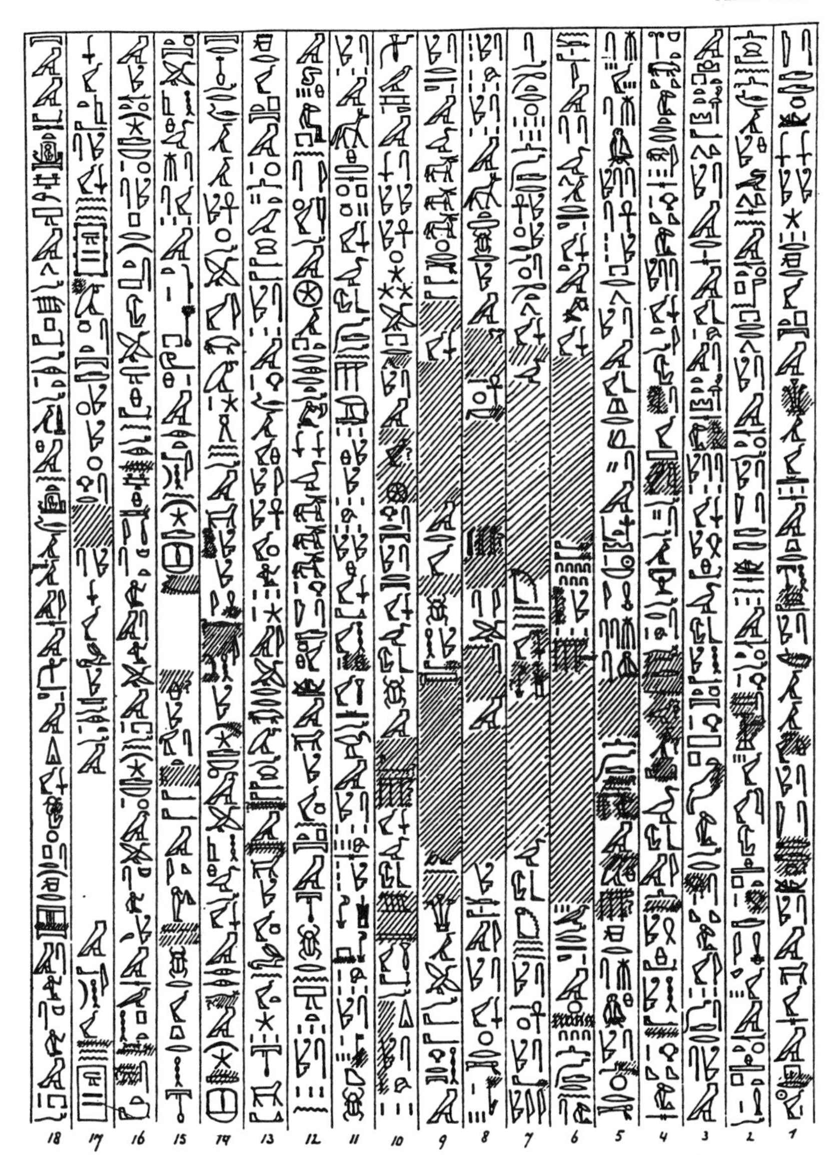

Fig. III.98a A drawing of the first 18 lines of the Dramatic Text near the figure of Nut in the west side of the roof of the Sarcophagus Chamber of the Cenotaph of Seti I. Taken from Frankfort, *The Cenotaph of Seti I at Abydos*, Vol. 2, Plate LXXXIV. The remainder of the text appears as Plate LXXXV, but is not given here since it contains material not used for the preparation of Document III.13.

PLATE 51

Jj

Hh

Ll

Kk

Dramatic Text

← indicates text faces left

PLATE 52

PLATE 53

a-a So De Buck, but very doubtful b Dittography c-c [illegible]

PLATE 54

a So De Buck b [illegible]

Fig. III.98b A hand copy of the first part of the Dramatic Text given in Neugebauer and Parker, *Egyptian Astronomical Texts*, Vol. 1, Plates 51-54.

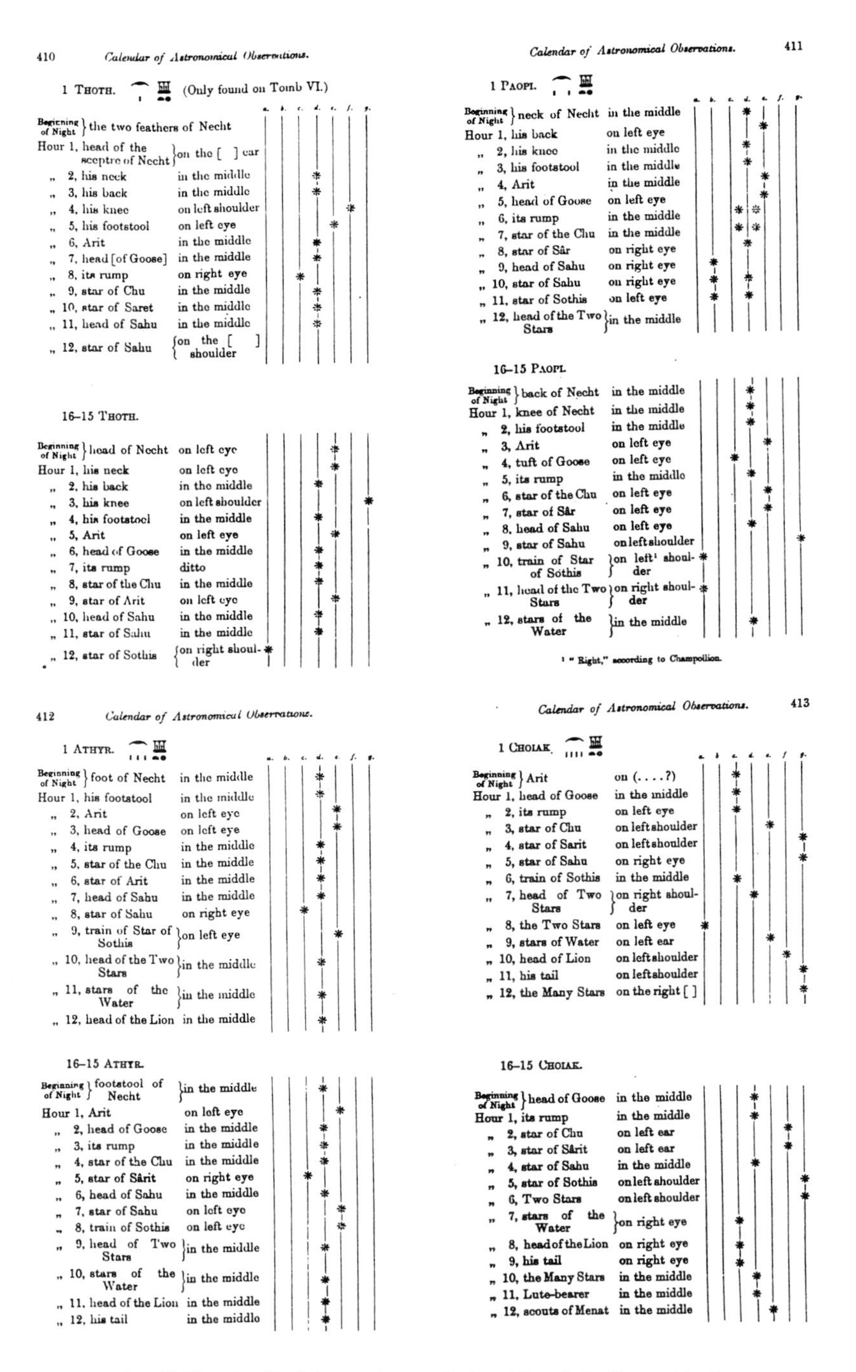

410 *Calendar of Astronomical Observations.*

1 THOTH. (Only found on Tomb VI.)

Beginning of Night	the two feathers of Necht	
Hour 1,	head of the sceptre of Necht	on the [ ] ear
„ 2,	his neck	in the middle
„ 3,	his back	in the middle
„ 4,	his knee	on left shoulder
„ 5,	his footstool	on left eye
„ 6,	Arit	in the middle
„ 7,	head [of Goose]	in the middle
„ 8,	its rump	on right eye
„ 9,	star of Chu	in the middle
„ 10,	star of Saret	in the middle
„ 11,	head of Sahu	in the middle
„ 12,	star of Sahu	on the [ ] shoulder

16–15 THOTH.

Beginning of Night	head of Necht	on left eye
Hour 1,	his neck	on left eye
„ 2,	his back	in the middle
„ 3,	his knee	on left shoulder
„ 4,	his footstool	in the middle
„ 5,	Arit	on left eye
„ 6,	head of Goose	in the middle
„ 7,	its rump	ditto
„ 8,	star of the Chu	in the middle
„ 9,	star of Arit	on left eye
„ 10,	head of Sahu	in the middle
„ 11,	star of Sahu	in the middle
„ 12,	star of Sothis	on right shoulder

Calendar of Astronomical Observations. 411

1 PAOPI.

Beginning of Night	neck of Necht	in the middle
Hour 1,	his back	on left eye
„ 2,	his knee	in the middle
„ 3,	his footstool	in the middle
„ 4,	Arit	in the middle
„ 5,	head of Goose	on left eye
„ 6,	its rump	in the middle
„ 7,	star of the Chu	in the middle
„ 8,	star of Sâr	on right eye
„ 9,	head of Sahu	on right eye
„ 10,	star of Sahu	on right eye
„ 11,	star of Sothis	on left eye
„ 12,	head of the Two Stars	in the middle

16–15 PAOPI.

Beginning of Night	back of Necht	in the middle
Hour 1,	knee of Necht	in the middle
„ 2,	his footstool	in the middle
„ 3,	Arit	on left eye
„ 4,	tuft of Goose	on left eye
„ 5,	its rump	in the middle
„ 6,	star of the Chu	on left eye
„ 7,	star of Sâr	on left eye
„ 8,	head of Sahu	on left eye
„ 9,	star of Sahu	on left shoulder
„ 10,	train of Star of Sothis	on left[1] shoulder
„ 11,	head of the Two Stars	on right shoulder
„ 12,	stars of the Water	in the middle

[1] "Right," according to Champollion.

412 *Calendar of Astronomical Observations.*

1 ATHYR.

Beginning of Night	foot of Necht	in the middle
Hour 1,	his footstool	in the middle
„ 2,	Arit	on left eye
„ 3,	head of Goose	on left eye
„ 4,	its rump	in the middle
„ 5,	star of the Chu	in the middle
„ 6,	star of Arit	in the middle
„ 7,	head of Sahu	in the middle
„ 8,	star of Sahu	on right eye
„ 9,	train of Star of Sothis	on left eye
„ 10,	head of the Two Stars	in the middle
„ 11,	stars of the Water	in the middle
„ 12,	head of the Lion	in the middle

16–15 ATHYR.

Beginning of Night	footstool of Necht	in the middle
Hour 1,	Arit	on left eye
„ 2,	head of Goose	in the middle
„ 3,	its rump	in the middle
„ 4,	star of the Chu	in the middle
„ 5,	star of Sârit	on right eye
„ 6,	head of Sahu	in the middle
„ 7,	star of Sahu	on left eye
„ 8,	train of Sothis	on left eye
„ 9,	head of Two Stars	in the middle
„ 10,	stars of the Water	in the middle
„ 11,	head of the Lion	in the middle
„ 12,	his tail	in the middle

Calendar of Astronomical Observations. 413

1 CHOIAK.

Beginning of Night	Arit	on (. . . . ?)
Hour 1,	head of Goose	in the middle
„ 2,	its rump	on left eye
„ 3,	star of Chu	on left shoulder
„ 4,	star of Sarit	on left shoulder
„ 5,	star of Sahu	on right eye
„ 6,	train of Sothis	in the middle
„ 7,	head of Two Stars	on right shoulder
„ 8,	the Two Stars	on left eye
„ 9,	stars of Water	on left ear
„ 10,	head of Lion	on left shoulder
„ 11,	his tail	on left shoulder
„ 12,	the Many Stars	on the right [ ]

16–15 CHOIAK.

Beginning of Night	head of Goose	in the middle
Hour 1,	its rump	in the middle
„ 2,	star of Chu	on left ear
„ 3,	star of Sârit	on left ear
„ 4,	star of Sahu	in the middle
„ 5,	star of Sothis	on left shoulder
„ 6,	Two Stars	on left shoulder
„ 7,	stars of the Water	on right eye
„ 8,	head of the Lion	on right eye
„ 9,	his tail	on right eye
„ 10,	the Many Stars	in the middle
„ 11,	Lute-bearer	in the middle
„ 12,	scouts of Menat	in the middle

Fig. III.99a An English translation of the tables of the Ramesside star clock by P. Le Page Renouf, "Calendar of Astronomical Observations Found in Royal Tombs of the XXth Dynasty," *Transactions of the Society of Biblical Archaeology*, Vol. 3 (1874), pp. 410-20.

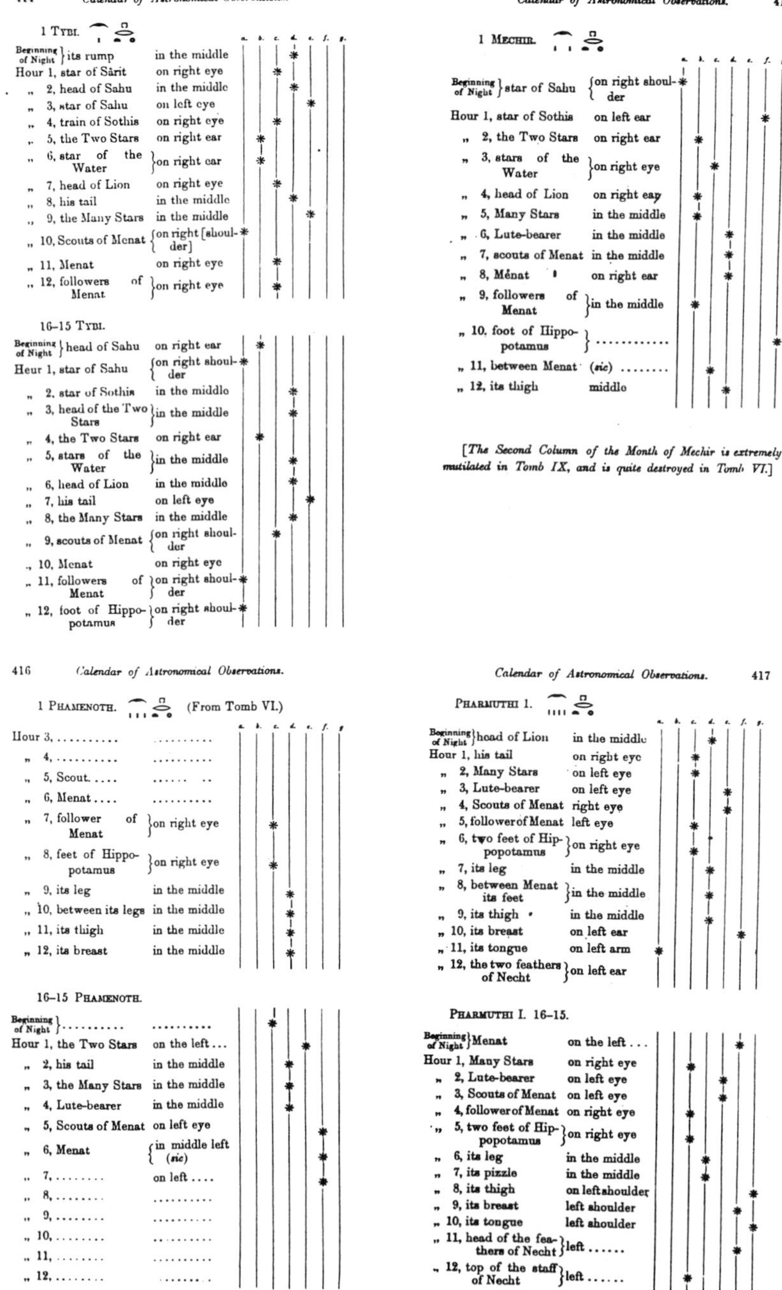

414 *Calendar of Astronomical Observations.*

1 Tybi.

			a.	b.	c.	d.	e.	f.	g.
Beginning of Night	its rump	in the middle				*			
Hour 1,	star of Sârit	on right eye			*				
„ 2,	head of Sahu	in the middle				*			
„ 3,	star of Sahu	on left eye					*		
„ 4,	train of Sothis	on right eye			*				
„ 5,	the Two Stars	on right ear		*					
„ 6,	star of the Water	on right ear		*					
„ 7,	head of Lion	on right eye			*				
„ 8,	his tail	in the middle				*			
„ 9,	the Many Stars	in the middle					*		
„ 10,	Scouts of Menat	on right [shoulder]	*						
„ 11,	Menat	on right eye			*				
„ 12,	followers of Menat	on right eye			*				

16–15 Tybi.

			a.	b.	c.	d.	e.	f.	g.
Beginning of Night	head of Sahu	on right ear		*					
Heur 1,	star of Sahu	on right shoulder	*						
„ 2,	star of Sothis	in the middlo				*			
„ 3,	head of the Two Stars	in the middle				*			
„ 4,	the Two Stars	on right ear		*					
„ 5,	stars of the Water	in the middle				*			
„ 6,	head of Lion	in the middlo				*			
„ 7,	his tail	on left eye					*		
„ 8,	the Many Stars	in the middle				*			
„ 9,	scouts of Menat	on right shoulder			*				
„ 10,	Menat	on right eye							
„ 11,	followers of Menat	on right shoulder	*						
„ 12,	foot of Hippopotamus	on right shoulder	*						

Calendar of Astronomical Observations. 415

1 Mechir.

			a.	b.	c.	d.	e.	f.	g.
Beginning of Night	star of Sahu	on right shoulder	*						
Hour 1,	star of Sothis	on left ear						*	
„ 2,	the Two Stars	on right ear		*					
„ 3,	stars of the Water	on right eye			*				
„ 4,	head of Lion	on right eay		*					
„ 5,	Many Stars	in the middle		*					
„ 6,	Lute-bearer	in the middle				*			
„ 7,	scouts of Menat	in the middle				*			
„ 8,	Mênat	on right ear				*			
„ 9,	followers of Menat	in the middle		*					
„ 10,	foot of Hippopotamus	....							*
„ 11,	between Menat	(*sic*)			*				
„ 12,	its thigh	middlo				*			

[*The Second Column of the Month of Mechir is extremely mutilated in Tomb IX, and is quite destroyed in Tomb VI.*]

416 *Calendar of Astronomical Observations.*

1 Phamenoth. (From Tomb VI.)

			a.	b.	c.	d.	e.	f.	g.
Hour 3,	....	....							
„ 4,	....	....							
„ 5,	Scout.....	....							
„ 6,	Menat....	....							
„ 7,	follower of Menat	on right eye			*				
„ 8,	feet of Hippopotamus	on right eye			*				
„ 9,	its leg	in the middle				*			
„ 10,	between its legs	in the middle				*			
„ 11,	its thigh	in the middle				*			
„ 12,	its breast	in the middle				*			

16–15 Phamenoth.

			a.	b.	c.	d.	e.	f.	g.
Beginning of Night	....	....			*				
Hour 1,	the Two Stars	on the left...					*		
„ 2,	his tail	in the middle				*			
„ 3,	the Many Stars	in the middle				*			
„ 4,	Lute-bearer	in the middle				*			
„ 5,	Scouts of Menat	on left eye						*	
„ 6,	Menat	in middle left (*sic*)						*	
„ 7,	....	on left....						*	
„ 8,	....	....							
„ 9,	....	....							
„ 10,	....	....							
„ 11,	....	....							
„ 12,	....	....							

Calendar of Astronomical Observations. 417

Pharmuthi 1.

			a.	b.	c.	d.	e.	f.	g.
Beginning of Night	head of Lion	in the middle				*			
Hour 1,	his tail	on right eye			*				
„ 2,	Many Stars	on left eye			*				
„ 3,	Lute-bearer	on left eye					*		
„ 4,	Scouts of Menat	right eye					*		
„ 5,	follower of Menat	left eye			*				
„ 6,	two feet of Hippopotamus	on right eye			*				
„ 7,	its leg	in the middle				*			
„ 8,	between Menat its feet	in the middle				*			
„ 9,	its thigh	in the middle				*			
„ 10,	its breast	on left ear						*	
„ 11,	its tongue	on left arm	*						
„ 12,	the two feathers of Necht	on left ear							

Pharmuthi I. 16–15.

			a.	b.	c.	d.	e.	f.	g.
Beginning of Night	Menat	on the left...						*	
Hour 1,	Many Stars	on right eye			*				
„ 2,	Lute-bearer	on left eye					*		
„ 3,	Scouts of Menat	on left eye					*		
„ 4,	follower of Menat	on right eye			*				
„ 5,	two feet of Hippopotamus	on right eye			*				
„ 6,	its leg	in the middle				*			
„ 7,	its pizzle	in the middle				*			
„ 8,	its thigh	on left shoulder							*
„ 9,	its breast	left shoulder						*	
„ 10,	its tongue	left shoulder							*
„ 11,	head of the feathers of Necht	left......						*	
„ 12,	top of the staff of Necht	left......			*				

Fig. III.99a (Continued)

1 Pachons.

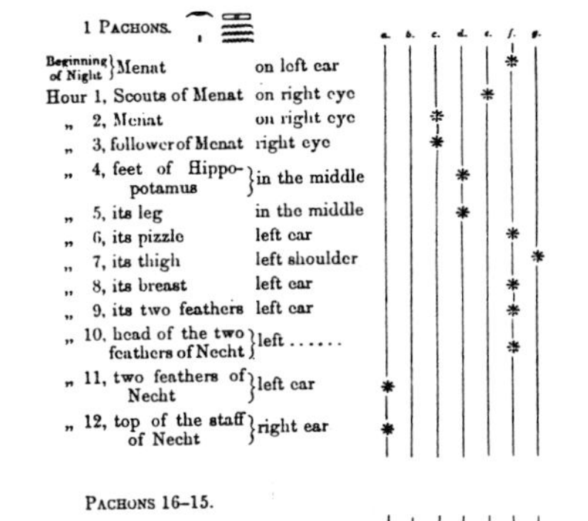

		a.	b.	c.	d.	e.	f.	g.
Beginning of Night — Menat	on left ear						*	
Hour 1, Scouts of Menat	on right eye					*		
„ 2, Menat	on right eye			*				
„ 3, follower of Menat	right eye			*				
„ 4, feet of Hippopotamus	in the middle				*			
„ 5, its leg	in the middle				*			
„ 6, its pizzle	left ear						*	
„ 7, its thigh	left shoulder							*
„ 8, its breast	left ear						*	
„ 9, its two feathers	left ear						*	
„ 10, head of the two feathers of Necht	left						*	
„ 11, two feathers of Necht	left ear	*						
„ 12, top of the staff of Necht	right ear	*						

Pachons 16–15.

		a.	b.	c.	d.	e.	f.	g.
Beginning of Night — Scout	on right eye			*				
Hour 1, Menat	on left eye					*		
„ 2, follower of Menat	on right eye			*				
„ 3, feet of Hippopotamus	in the middle				*			
„ 4, its leg	in the middle				*			
„ 5, its pizzle	on left shoulder							*
„ 6, its thigh	on ... shoulder							*
„ 7, its breast	.... ear		*					
„ 8, its two feathers	left ear						*	
„ 9, head of two feathers of Necht	right ear		*					
„ 10, two feathers of Necht	right ear		*					
„ 11, top of his staff	middle			*				
„ 12, his throat	.... shoulder	*						

1 Payni.

		a.	b.	c.	d.	e.	f.	g.
Beginning of Night — Menat	middle							*
Hour 1, feet of Hippopotamus	middle				*			
„ 2, its leg	right eye			*				
„ 3, its [pizzle]	middle				*			
„ 4, its thigh	middle				*			
„ 5, its breast	right eye			*				
„ 6, its tongue	right eye			*				
„ 7, its feathers	left ear							*
„ 8, head of feathers of Necht	left ear		*					
„ 9, two feathers of Necht	middle				*			
„ 10, his throat	middle				*			
„ 11, its breast	middle				*			
„ 12, his back	middle				*			

Payni 16–15.

		a.	b.	c.	d.	e.	f.	g.
Beginning of Night — feet of Hippopotamus	middle				*			
Hour 1, its leg	middle				*			
„ 2, its [pizzle]	middle				*			
„ 3, its thigh	middle				*			
„ 4, its breast	left eye					*		
„ 5, its tongue	left eye					*		
„ 6, feathers	left ear						*	
„ 7, two feathers of Necht	right ear		*					
„ 8, top	middle				*			
„ 9, his throat	middle				*			
„ 10, breast	left shoulder							*
„ 11, his back	left ear						*	
„ 12, his leg	middle				*			

1 Epiphi.

		a.	b.	c.	d.	e.	f.	g.
Beginning of Night — .... of Hippopotamus	right eye					*		
Hour 1, its thigh	middle				*			
„ 2, breast	right eye			*				
„ 3, tongue	middle				*			
„ 4, its feathers	middle				*			
„ 5, head of feathers of Necht	middle				*	*		
„ 6, two feathers of Necht	right eye					*		
„ 7, his neck	middle				*			
„ 8, breast	middle				*			
„ 9, his back	middle				*			
„ 10, his leg	right eye					*		
„ 11, his sebekes (?)	middle				*			
„ 12, his footstool	middle				*			

Epiphi 16–15.

		a.	b.	c.	d.	e.	f.	g.
Beginning of Night —	....							
Hour 1, its breast	middle				*			
„ 2, tongue	middle				*			
„ 3, its feathers	middle				*			
„ 4, head of Necht	middle				*			
„ 5, two feathers of Necht	left eye					*		
„ 6, his throat	middle				*			
„ 7, breast	middle				*			
„ 8, his back	middle				*			
„ 9, his leg	middle				*			
„ 10, ...	....			*				
„ 11, ...	....			*				
„ 12, ...	middle				*			

Fig. III.99a (Concluded)

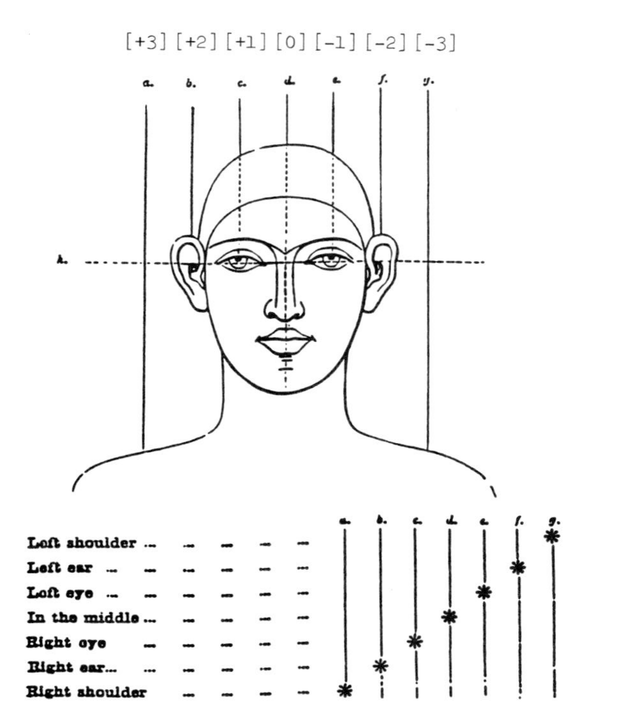

Fig. III.99b A diagram to illustrate the phrases used to indicate positions of stars transiting in the Ramesside star clock. Taken from Renouf, "Calendar of Astronomical Observations," p. 409. I have added the bracketed numbers and stripped from the original drawing a second set of letters a.-g., ambiguously used to list the body positions (see Introduction to Doc. III.14).

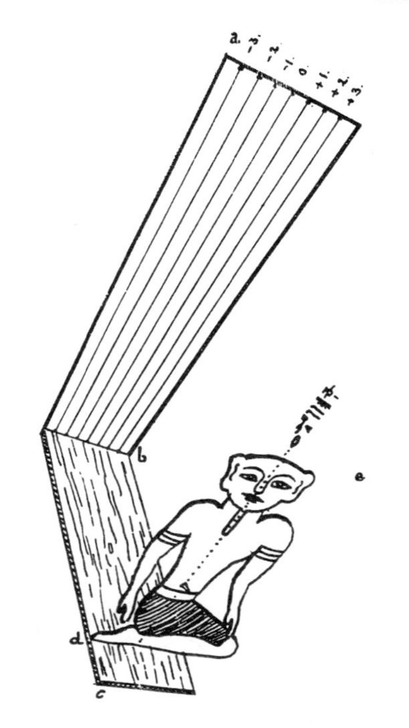

Fig. III.99c Stringed apparatus for sighting star transits, suggested for use with the Ramesside star clock by H. Schack-Schackenburg, *Ägyptologische Studien*, Vol. I (1902), 2. Heft, p. 63.

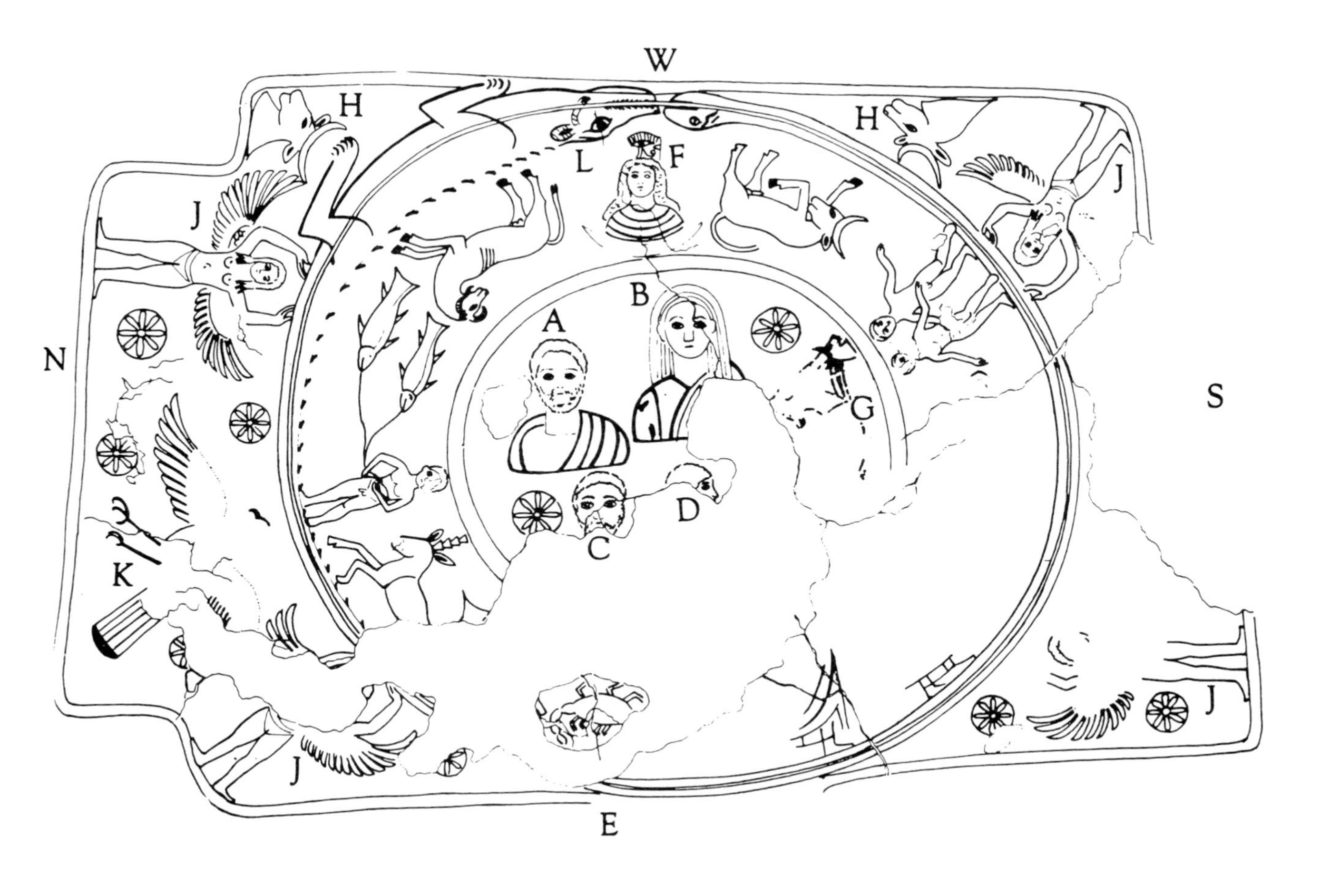

Fig. III.100a Zodiac from the ceiling of Chamber I of the Tomb of Petosiris. Date: 54-84 A.D. Taken from J. Osing et al., *Denkmäler der Oase Dachla. Aus dem Nachlass von Ahmed Fakhry* (Maiz am Rhein, 1982), Tafel 39 (cf. photograph, Tafel 38).

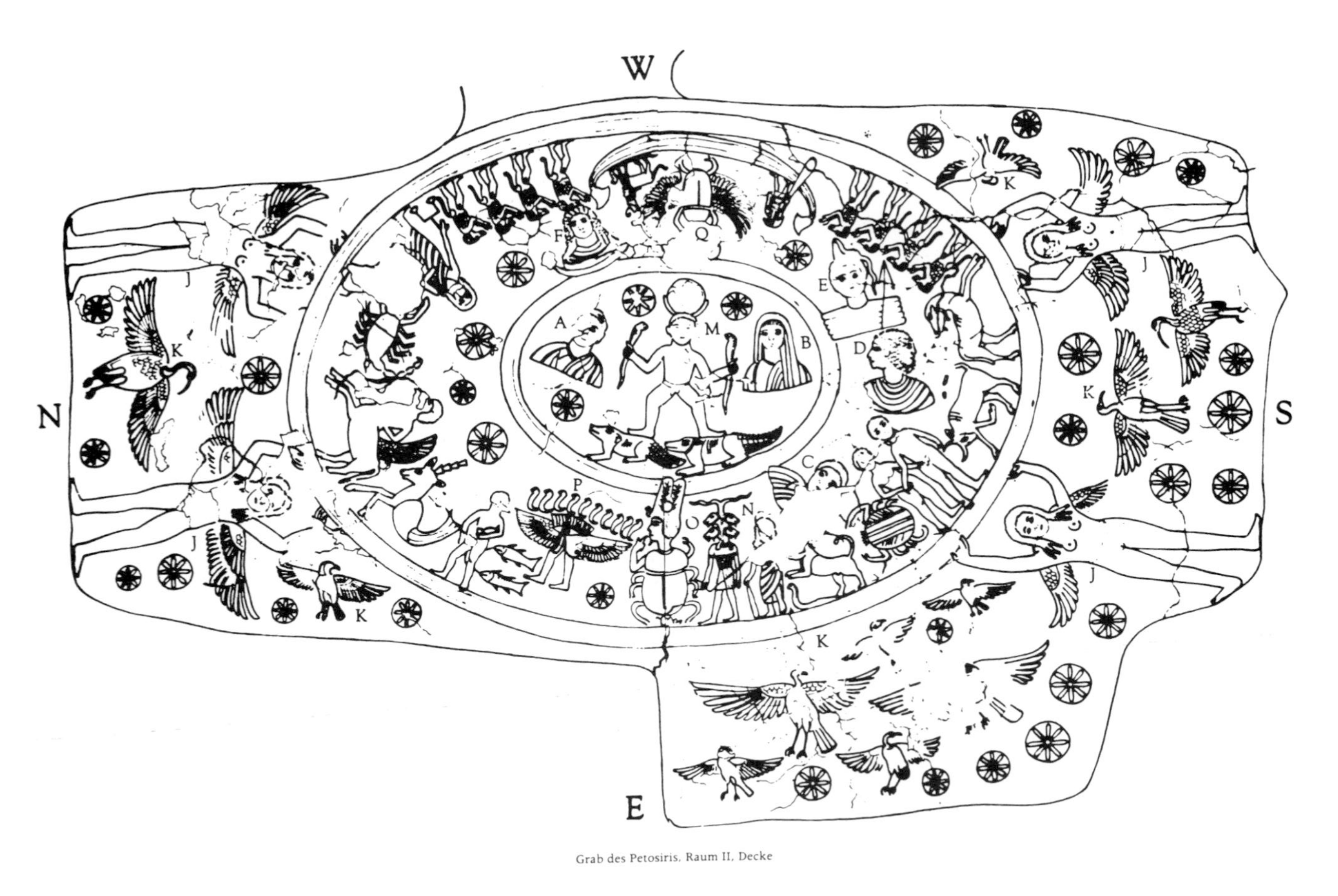

Fig. III.100b Zodiac from the ceiling of Chamber II of the Tomb of Petosiris. Taken from the work cited in Fig. III.100a, Tafel 41 (cf. Tafel 40).

Fig. III.101 Zodiac from the ceiling of the Tomb of Petubastis. Date: Roman period. Taken from the work cited in Fig. III.100a, Tafel 37 (cf. Tafel 36).

	Esna A_1 (Seti I B)	Esna A_2 (Tanis)
	tm(ꜣt)	*pḥwy ḏꜣt*
Virgo	*wšꜣt(ỉ) bkꜣt(ỉ)*	*tm(ꜣt)*
	ỉpsd	*wšꜣt(ỉ)*
	sbḫs	*bkꜣt(ỉ)*
Libra	*tpy-ꜥ ḫnt*	*ỉpsd*
	ḫnt ḥr(t)	*sbḫs*
	ḫnt ẖr(t)	*tpy-ꜥ ḫnt*
Scorpio	*tms (n) ḫnt*	*ḥry-ỉb wỉꜣ*
	spt(y) ḫnwy	*s(ꜣ)pt(ỉ) ḫnwy*
	ḥry-ỉb wỉꜣ	*sšm(w)*
Sagittarius	*sšmw*	*sꜣ sšm(w)*
	knm(w)	*knm(w)*
	tpy-ꜥ smd	*tpy-ꜥ smd*
Capricorn	*smd*	*pꜣ sbꜣ wꜥty*
	srt	*smd*
	sꜣ srt	*srt*
Aquarius	*ḫry ḫpd srt*	*sꜣ srt*
	tpy-ꜥ ꜣḫw(y)	*tpy-ꜥ ꜣḫw(y)*
	ꜣḫw(y)	*ꜣḫw(y)*
Pisces	*tpy-ꜥ bꜣw(y)*	*tpy-ꜥ bꜣw(y)*
	bꜣw(y)	*bꜣw(y)*
	ḫnt(w) ḥr(w)	*ḫnt(w) ḥr(w)*
Aries	*ḫnt(w) ẖr(w)*	*ḫnt(w) ẖr(w)*
	sꜣ ḳd	*ḳd*
	ḫꜣw	*sꜣ ḳd*
Taurus	*ꜥrt*	*ḫꜣw*
	rmn ḥry	*ꜥrt*
	ṯs ꜥrḳ	*rmn ḥry*
Gemini	*wꜥrt*	*ṯs ꜥrḳ*
	tpy-ꜥ spdt	*rmn ẖry*
	spdt	*wꜥr(t)*
Cancer	*št(w)*	*pḥwy ḥry*
	knm(t)	*knm(t)*
	ḫry ḫpd knm(t)	*ḫry (ḫpd) knm(t)*
Leo	*ḥꜣt ḏꜣt*	*ḥꜣt ḏꜣt*
	pḥwy ḏꜣt	*ḏꜣt*

Fig. III.102 A comparative listing of the two families of decans in the zodiac of Esna A. The "a" decans have been omitted from the A1 list. Taken from NP, Vol. 3, pp. 168-69.

	Seti I B	Hephaestion		Tanis
	spdt	σωθις		***knm(t)***
Cancer	*št(w)*	σιτ		*ḫry (ḫpd) knm(t)*
	knm(t)	χνουμις		*ḥꜣt ḏꜣt*
	ḫry ḫpd knm(t)	χαρχνουμις		*ḏꜣt*
Leo	*ḥꜣt ḏꜣt*	ηπη		*pḥwy ḏꜣt*
	pḥwy ḏꜣt	φουπη		*tm(ꜣt)*
	tm(ꜣt)	τωμ		*wšꜣt(ỉ)*
Virgo	*wšꜣt(ỉ) bkꜣt(ỉ)*	ουεστεβκωτ		*bkꜣt(ỉ)*
	ỉpsd	αφοσο	αφοσο	*ỉpsd*
	sbḫs	σουχωε	σουχωε	*sbḫs*
Libra	*tpy-ʿ ḫnt*	πτηχουτ	πτηχουτ	*tpy-ʿ ḫnt*
	ḫnt ḥr(t)	χονταρε		*ḥry-ib wỉꜣ*
	ḫnt ḥr(t)		στωχνηνε	*s(ꜣ)pt(ỉ) ḫnwy*
Scorpio	*tms (n) ḫnt*		σεσμε	*sšm(w)*
	spt(y) ḫnwy		σισιεμε	*sꜣ sšm(w)*
	ḥry-ib wỉꜣ	ρηουω		*knm(w)*
Sagittarius	*sšmw*	σεσμε		*tpy-ʿ smd*
	knm(w)	κομμε		*pꜣ sbꜣ wʿty*
	tpy-ʿ smd		σματ	*smd*
Capricorn	*smd*		σρω	*srt*
	srt		ισρω	*sꜣ srt*
	sꜣ srt		πτιαυ	*tpy-ʿ ꜣḫw(y)*
Aquarius	*ḫry ḫpd srt*		αευ	*ꜣḫw(y)*
	tpy-ʿ ꜣḫw(y)		πτηβυου	*tpy-ʿ bꜣw(y)*
	ꜣḫw(y)		βιου	*bꜣw(y)*
Pisces	*tpy-ʿ bꜣw(y)*		χονταρε	*ḫnt(w) ḥr(w)*
	bꜣw(y)	πτιβιου		*ḫnt(w) ḥr(w)*
	ḫnt(w) ḥr(w)	χονταρε		*ḳd*
Aries	*ḫnt(w) ḫrw*	χονταχρε		*sꜣ ḳd*
	sꜣ ḳd	σικετ		*ḫrw*
	ḫrw	χωου		*ʿrt*
Taurus	*ʿrt*	ερω		*rmn ḥry*
	rmn ḥry	ρομβρομαρε		*ṯs ʿrḳ*
	ṯs ʿrḳ	θοσολκ		*rmn ḥry*
Gemini	*wʿrt*	ουαρε		*wʿr(t)*
	tpy-ʿ spdt		φουορι	*pḥwy ḥry*

Fig. III.103 The decans of the zodiac of Hephaestion compared with the decans of the families designated Seti I B and Tanis. Taken from NP, Vol. 3, pp. 170-71.

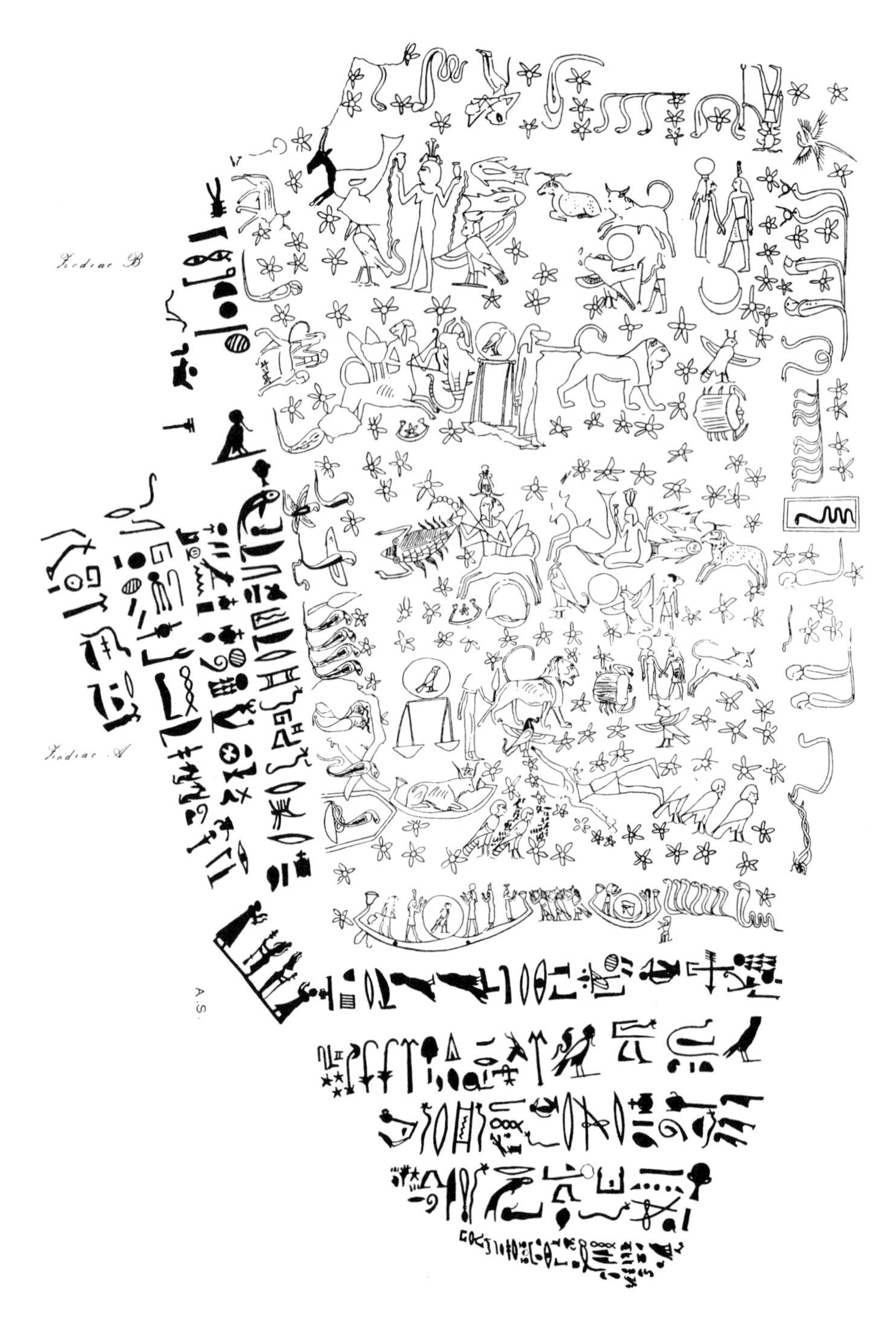

Fig. III.104 Two horoscopes and zodiacs (A and B) from the tomb of two brothers at Athribis. Roman, 2nd cent. A.D. (see NP, Vol. 3, pp. 96-98). Taken from Petrie, *Athribis*, Plate XXXVII.

A. Inscription on the Back Pillar

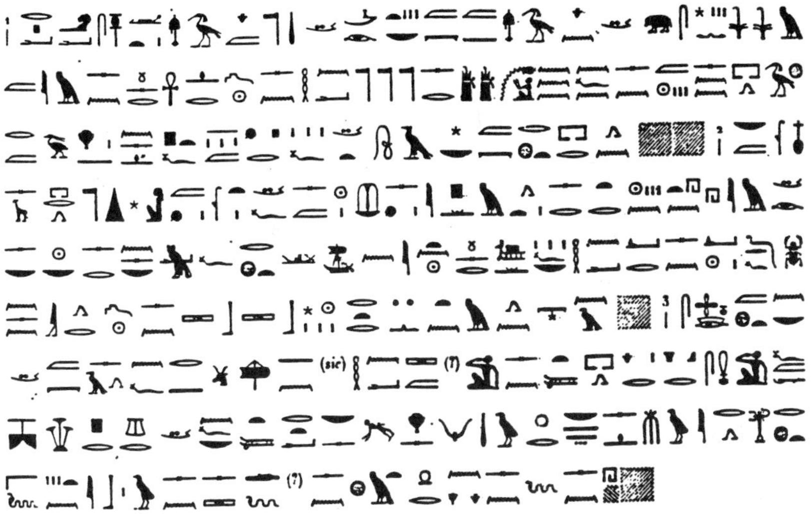

B. Inscription on the Left Side

Fig. III.105 Inscriptions from the back pillar (A) and the left side (B) of the statue of the astronomer Harkhebi (3rd cent. B.C., Cairo JE 38545). Taken from Daressy, "La statue d'un astronome," *ASAE*, Vol. 16 (1916), p. 2.

Fig. III.106a Location of petroglyphs at HK-61D in Hieraconpolis. Reproduced from Plate 33 of the unpublished paper of J.O. Mills, "Astronomy at Hierakonpolis" (excerpted in my Postscript).

Fig. III.106b A possible astronomical petroglyph at Hk-61D. Reproduced from Plate 34 of Mills' paper.